Artificial Neural Networks for Intelligent Manufacturing

Intelligent Manufacturing Series

Series Editor: Andrew Kusiak
Department of Industrial Engineering
The University of Iowa, USA

Manufacturing has been issued a great challenge – the challenge of Artificial Intelligence (AI). We are witnessing the proliferation of applications of AI in industry, ranging from finance and marketing to design and manufacturing processes. AI tools have been incorporated into computer-aided design and shop-floor operations software, as well as entering use in logistics systems.

The success of AI in manufacturing can be measured by its growing number of applications, releases of new software products and in the many conferences and new publications. This series on Intelligent Manufacturing has been established in response to these developments, and will include books on topics such as:

- design for manufacturing
- concurrent engineering
- process planning
- production planning and scheduling
- programming languages and environments
- design, operations and management of intelligent systems

Some of the titles are more theoretical in nature, while others emphasize an industrial perspective. Books dealing with the most recent developments will be edited by leaders in the particular fields. In areas that are more established, books written by recognized authors are planned.

We are confident that the titles in the series will be appreciated by students entering the field of intelligent manufacturing, academics, design and manufacturing managers, system engineers, analysts and programmers.

Titles available

Object-oriented Software for Manufacturing Systems
Edited by S. Adiga

Integrated Distributed Intelligence Systems in Manufacturing
M. Rao, Q. Wang and J. Cha

Artificial Neural Networks for Intelligent Manufacturing
Edited By C.H. Dagli

Artificial Neural Networks for Intelligent Manufacturing

Edited by

Cihan H. Dagli

Associate Professor
Department of Engineering Management
University of Missouri-Rolla
USA

CHAPMAN & HALL

London · Glasgow · New York · Tokyo · Melbourne · Madras

Published by Chapman & Hall, 2–6 Boundary Row, London SE1 8HN, UK

Chapman & Hall, 2–6 Boundary Row, London SE1 8HN, UK

Blackie Academic & Professional, Wester Cleddens Road, Bishopbriggs, Glasgow G64 2NZ, UK

Chapman & Hall Inc., One Peun Plaza, 41st Floor, New York NY10111, USA

Chapman & Hall Japan, Thomson Publishing Japan, Hirakawacho Nemoto Building, 6F, 1-7-11 Hirakawa-cho, Chiyoda-ku, Tokyo 102, Japan

Chapman & Hall Australia, Thomas Nelson Australia, 102 Dodds Street, South Melbourne, Victoria 3205, Australia

Chapman & Hall India, R. Seshadri, 32 Second Main Road, CIT East, Madras 600 035, India

First edition 1994

Typeset in 10/12 pts Times by Thomson Press (India) Ltd, New Delhi
Printed and Bound in Great Britain by Hartnolls Limited, Bodmin, Cornwall.

ISBN 0 412 48050 6

A catalogue record for this book is available from the British Library

Library of Congress Cataloging-in-Publication data
Artificial neural networks for intelligent manufacturing / edited by
Cihan H. Dagli. – 1st ed.
p. cm.
Includes index.
ISBN 0-412-48050-6 (acid-free paper)
1. Neural networks (Computer science) 2. Manufacturing processes.
I. Dagli, Cihan H., 1949– .
QA76.87.A7432 1994 93-35422
670′.285′63 – dc20 CIP

∞ Printed on permanent acid-free text paper, manufactured in accordance with the proposed ANSI/NISO Z 39.48–199X and ANSI Z 39.48–1984

To the memory of my father Kenan Dagli
To my mother Zuhre Dagli
and
My wife Refia and my sons Kenan Cagri and Mehmet Ediz

Contents

Contributors

Riadh Azouzi, Mechanical Engineering Department, Laval University, Quebec, Canada, G1K 7P4.

Ali Bahrami, Computer Information Systems, Department of Economics and Management, Rhode Island College, Providence, Rhode Island 02908.

John Y. Cheung, School of Electrical Engineering & Computer Science, The University of Oklahoma, Norman, OK 70319-0361.

Marie-Claude Cote, Mechanical Engineering Department, Laval University, Quebec, Canada, G1K 7P4.

Cihan H. Dagli, Engineering Management Department, University of Missouri-Rolla, Rolla, Missouri 65401

Benito Fernández R., Neuro-Engineering Research and Development (NERD) Laboratory, The University of Texas at Austin, Department of Mechanical Engineering, Austin, Texas, 78712-1063.

Joydeep Ghosh, Department of Electrical and Computer Engineering, The University of Texas at Austin, Austin, Texas 78712-1084.

Michel Guillot, Mechanical Engineering Department, Laval University, Quebec, Canada, G1K 7P4.

Mark R. Henderson, Associate Director, Computer-Integrated Manufacturing Systems Research Center, Associate Professor of Mechanical/Aerospace Engineering, Arizona State University.

A. Ikonomopoulos, Department of Nuclear Engineering and Center for Neural Engineering, The University of Tennessee and Oak Ridge National Laboratory, Knoxville, TN 37996-2300.

Soundar R.T. Kumara, Intelligent Design and Diagnostics Research Laboratory, Department of Industrial and Management Systems Engineering, The Pennsylvania State University, University Park, PA 16802

Gregory R. Madey, Kent State University, Graduate School of Management, Kent State University, Kent, Ohio 44242, USA.

César O. Malavé, Department of Industrial Engineering, Texas A & M University, College Station, TX 77843-3131.

Najwa S. Merchawi, Intelligent Design and Diagnostics Research Laboratory, Department of Industrial and Management Systems Engineering, The Pennsylvania State University, University Park, PA 16802

Madhusudhan Posani, University of Missouri, University Extension, Columbia, Missouri 65211.

Pipatpong Poshyanonda, Engineering Management Department, University of Missouri-Rolla, Rolla, Missouri 65401.

Satheesh Ramachandran, Department of Industrial Engineering, Texas A & M University, College Station, TX 77843-3131.

Vijay Shah, Kent State University, Graduate School of Management, Kent State University, Kent, Ohio 44242, USA.

Yung Shin, School of Mechanical Engineering, Purdue University, West Lafayette, IN 47907.

Deborah Stacey, Department of Computing and Information Science, University of Guelph, Guelph, Ontario Canada N1G 2W1.

L.H. Tsoukalas, Department of Nuclear Engineering and Center for Neural Engineering, The University of Tennessee and Oak Ridge National Laboratory, Knoxville, TN 37996-2300.

R.E. Uhrig, Department of Nuclear Engineering and Center for Neural Engineering, The University of Tennessee and Oak Ridge National Laboratory, Knoxville, TN 37996-2300.

Mahesh Kumar Vellanki, AT & T Bell Laboratories, Engineering Research Center, Hopewell, NJ 082525.

Jay Weinroth, Kent State University, Graduate School of Management, Kent State University, Kent, Ohio 44242. USA.

Preface

The quest for building systems that can function automatically has attracted a lot of attention over the centuries and created continuous research activities. As users of these systems we have never been satisfied, and demand more from the artifacts that are designed and manufactured. The current trend is to build autonomous systems that can adapt to changes in their environment. While there is a lot to be done before we reach this point, it is not possible to separate manufacturing systems from this trend. The desire to achieve fully automated manufacturing systems is here to stay.

Manufacturing systems of the twenty-first century will demand more flexibility in product design, process planning, scheduling and process control. This may well be achieved through integrated software and hardware architectures that generate current decisions based on information collected from manufacturing systems environment, and execute these decisions by converting them into signals transferred through communication network. Manufacturing technology has not yet reached this state. However, the urge for achieving this goal is transferred into the term 'Intelligent Systems' that we started to use more in late 1980s. Knowledge-based systems, our first efforts in this endeavor, were not sufficient to generate the 'Intelligence' required – our quest still continues. Artificial neural network technology is becoming an integral part of intelligent manufacturing systems and will have a profound impact on the design of autonomous engineering systems over the next few years.

This book introduces this newly-emerging technology and demonstrates its use, with examples, in intelligent manufacturing systems. After introducing the basic components of intelligent manufacturing systems and intelligent system architecture design techniques, the text provides sufficient coverage of basic artificial neural network architectures to be used successfully in manufacturing applications. The subsequent two parts of the book, discussed in more detail below, cover the use of neural networks in organization, coordination and execution levels of the manufacturing system hierarchy.

The second part of the book covers the development of artificial neural network architectures for solving fundamental problems of manufacturing; conceptual design, group technology, process planning, scheduling, automated assembly, inspection and manufacturing feature identification. It concludes with an excellent chapter that summarizes the performance analysis of artificial neural network methods. The third part concentrates on execution level applications that generally require extensive data collection. The use of artificial neural networks in process monitoring and control, adaptive control, fuzzy neurocontrol and continuous process diagnostics are examined.

As a whole, *Artificial Neural Networks for Intelligent Manufacturing* provides a major reference for all those interested in intelligent manufacturing systems. I believe that this is another step in our quest to build autonomous systems. I would like to thank all the contributors who have made this book possible.

Cihan H. Dagli
Rolla, Missouri, USA
October 1993

PART ONE

Intelligent manufacturing: Basic concepts and tools

CHAPTER 1

Intelligent manufacturing systems

Cihan H. Dagli

Engineering Management Department,

University of Missouri-Rolla, Rolla, Missouri 65401.

The quest for building systems that can function automatically has attracted a lot of attention over the centuries and created continuous research activity. As users of these systems we have never been satisfied; we demand more from the artifacts that are designed and manufactured. The current trend is to build autonomous systems that can adapt to changes in their environment. While there is much to be done before we reach this point it is not possible to separate manufacturing systems from this trend. The desire to achieve fully automated manufacturing systems is here to stay.

Manufacturing systems of the 21st century will demand more flexibility in product design, process planning, scheduling and process control. This may well be achieved through integrated software and hardware architectures that generate current decisions based on information collected from the manufacturing systems environment and then execute these decisions by converting them into signals transferred through a communication network. Manufacturing technology has not yet reached this state. However, the urge for achieving this goal is transferred into the term 'intelligent systems' that we started to use more in late 1980s (Carrico *et al.*, 1989; Chang, 1990; Coyne *et al.*, 1990; Jackson, 1990; Kusiak, 1988a; 1988b; 1990; Mockler, 1988; Schalkoff, 1990).

Knowledge-based systems, our first effort in this endeavor, were not sufficient to generate the 'intelligence' required, so our quest still continues (Turban and Liebowitz, 1992; Wright and Bourne, 1988). The trick is to merge the old with the new, for example integrating mathematical programming with artificial

Artificial Neural Networks for Intelligent Manufacturing. Edited by Cihan H. Dagli. Published in 1994 by Chapman & Hall, London. ISBN 0 412 48050 6

intelligence, and adding new approaches and technologies. Artificial neural networks is another technology that needs to be integrated to improve our quest for intelligent systems (Dagli, 1991). In this chapter, the characteristics of intelligent manufacturing systems and the impact of artificial neural networks in the design of autonomous manufacturing units to achieve our quest for building adaptive systems is discussed.

1.1 MANUFACTURING SYSTEMS AND STRATEGIES

Decisions complexity is an issue in manufacturing system design. This is due to the fact that excessive design and operation alternatives exist in the products to be produced and making the choice among appropriate combinations is not an easy task. Manufacturing strategies and systems that are evolved over the years produced solutions to this complex decision problem according to the competitive thrusts of the time and approaches adopted are converted into strategies which had paramount importance in shaping manufacturing environment. We can observe effectively these changes over the years and easily identify the evolution of various manufacturing systems and associated strategies.

Cost was the competitive thrust in the 1960s, creating manufacturing strategies such as cost minimization, stabilized production, product focus and high volume production. Hence, the manufacturing systems designed and operated during this decade used much specialized equipments, restricted product line, high product volume to reduce the setup time to benefit from economies of scale. They generally produced goods for stock due to excessive market demand. The concept of intelligent system design was not an issue at this time period, but high automation and high speed of production necessitated specialized equipments. The system used on the manufacturing floor was not adaptive to possible changes that can occur in product design and processes. On the other hand, they were extremely reliable and performed specialized tasks, which has been standardized for the particular product focus, over and over again. This type of strategy and manufacturing units were successful due to prevailing market structure and competitive thrust. There was no force neither from the market nor from the manufacturing environment to adapt to change. This structure created organized units within manufacturing entity which relied on labor intensive transactions reporting, operator dependent setup procedures, process control, and production monitoring. Due to the non-availability of computers at the shop floor and as most of the transactions were manual, only, accounting activities were being computerized during this time frame using mainframe computers with restricted memory. Unfortunately, due to this structure it was not possible to see an overall integration of various activities of manufacturing. It was difficult to visualize the whole plant as a single entity which has components that should work together harmoniously. This structure assumed fundamental manufacturing components as separate entities that can function independently without having concurrent activities. Marketing, accounting, production management, personnel management all evolved as separate entities and functions to be maintained in every manufacturing organization.

Easy access to the market and high demand for products were not prevailing factors in the 1970s. The ability to sell what you have produced while not taking the market into account led to a poor strategy during this time period. The market emerged as a competitive thrust during this decade and changed the manufacturing strategies that had depended on costs during the 1960s. Manufacturing companies did consider their markets and cost in determining their strategies for the 1970s. Functional integration became an important issue and was supported by the availability of mainframe computers. The concept of sharing computer resources and extensive availability of business-like computer languages (such as COBOL), and the development of custom data management programs provided the source for building closed-loop systems for manufacturing management. Manufacturing information system design became an important issue and was implemented using mainframe computers. There was a big trend toward upgrading manufacturing systems due to the availability of this type of computation power and difficulties met in selling products without considering the nature of the market effectively. These adopted strategies created the new forms of inventory management and production planning systems that we take for granted today. For example, during the 1960s excessive research work was concentrated on inventory control and identifying items to be selected for control through well-known techniques such as ABC analysis, to determine appropriate inventory items that would have a positive impact on total inventory costs. The literature is full of inventory control models developed for this purpose. As soon as the means became available for extensive data processing and manipulation, new techniques, such as material requirements planning, became popular to control dependent demand items in the factory. Ironically, these systems used none of the sophisticated inventory control models for determining order quantities; the simple economical order quantity model was sufficient for most of the cases. Availability of electronic data storage, even though it was on a shared basis through mainframcs, was a catalyst needed to create the integration that was forced by thc markct. This was the time where the concepts of master production planning and control, hierarchical approach to planning and scheduling, functional flow control and computer numerical control (CNC) were introduced. It became possible to talk about electronic company databases that can provide information as to the operation of the plant besides information that was always available in the form of costs obtained through accounting practices. Automation in accounting systems occurred during this decade. During the 1980s, it became more difficult to stay competitive in the market place because of exponential growth in communication systems around the world, and the emergence of the global market. Quality became the competitive thrust during this time frame, creating a new need to shape the manufacturing strategies. Process control, speed of material, new accounting systems, the ability to distribute overhead costs without depending on labor costs, all these became important issues and created new strategies and concepts that had not been a part of the manufacturing environment before. Material requirements planning systems evolved into manufacturing resource planning by integrating accounting information, capacity planning, scheduling, product routing and automating work and purchase order generation process based on forecasted demand for a given time bucket. The concept of just-in-time (JIT), optimized production technology, direct numerical control, statistical process control (SPC), total quality control (TQC), computer aided design (CAD) and computer aided manufacturing (CAM) became popular. The availability of personal computers speeded up this process, eliminating the need to share mainframes and providing extensive computational power at various locations of the plant. The next step was to join all these computational units through a communication network and create an electronic environment for communication and, in turn, paperless organization. All these changes, driven by competitive thrusts, created the basic foundation for the intelligent manufacturing environment.

Manufacturing strategies for the 1990s are based on flexibility as the competitive thrust. Responsiveness to market, new organization forms, introduction of new products, international thrusts, upgrade in manufacturing systems – all these are the strategies for this decade. Technical planning systems, decentralization, intelligent manufacturing systems, computer integrated manufacturing planning and simplification are the result of these new manufacturing strategies.

Flexibility requires integration among basic function of manufacturing, namely: product and part design, process planning, programming for machines, robots, automated guided vehicles, production planning, manufacturing receiving storage and shipping. It is possible to have organizations composed largely of specialists who direct their own performance through automated information and directions provided by peers, computers and main offices located miles away. Manufacturing flexibility demands customized high-

Table 1.1 Changes in manufacturing environment

Years	*Competitive thrust*	*Strategies*
1960s	Cost	High-volume product-focused manufacturing
1970s	Cost, market	Design of closed-loop manufacturing systems
1980s	Cost, market, quality	Extensive process monitoring and control in manufacturing
1990s	Cost, market, quality, flexibility	Integration of basic functions of manufacturing

quality inexpensive components. This requires autonomous machines, setup procedures and systems that have the ability to produce a good first part right at the first time. Flexibility necessitates integration, integration requires intelligence, and intelligence has to be backed up by real-time control. Manufacturing flexibility can be controlled by integrating basic functions of manufacturing through intelligent units which send and receive signals from various shop floor units and provide effective decisions through intelligent software architectures implemented. This is not an easy task to accomplish. However, it is not outside the scope of our quest of designing intelligent systems. Newly emerging technology, artificial neural networks, will be one of the contributors in the quest of achieving fully adaptive manufacturing systems. In the following sections, hierarchical levels in manufacturing and the impact of these levels to manufacturing flexibility and decision making, characteristics of intelligent systems and contributions of artificial neural networks to intelligent manufacturing systems are discussed.

1.2 HIERARCHICAL LEVELS IN MANUFACTURING

It is possible to identify three basic levels of control in manufacturing systems, namely; organization, coordination, and execution (Saridis, 1989). Although the number of levels in the hierarchy can change, the concepts used are common to many manufacturing systems (Fig. 1.1).

Organization level generally includes high-level decision-making activities such as planning (that requires extensive knowledge processing) and search (that depends heavily on database systems). The objective at this level is to identify basic courses of action without going into too much detail. The decisions made at this level provide basic guidelines to be followed by the coordination level. Basic manufacturing functions such as product design, process planning, master production scheduling, forecasting, and strategic business planning are the fundamental activities of this level. *Coordination level* serves as a mediator or interface between the execution and the organization level. This level involves making short-term decisions based on the inputs provided from the organization level. Manufacturing activities such as detailed scheduling, assembly operations, work and purchase order generation forms the typical activities of this level. *Execution level* provides appropriate control functions and detail feedback from actual operations of manufacturing units. It provides supply information to the coordination level for possible revisions of short-term decisions and also satisfactory database to create appropriate knowledge to modify long-term strategies provided by the organization level. Manufacturing functions such as process monitoring and control, adaptive control, manufacturing system diagnostic fall into this category. It is easy to see the details of this type of hierarchical structure by examining hierarchical production planning and control systems proposed in literature (Hax and Candea, 1984). For example, manufacturing resource

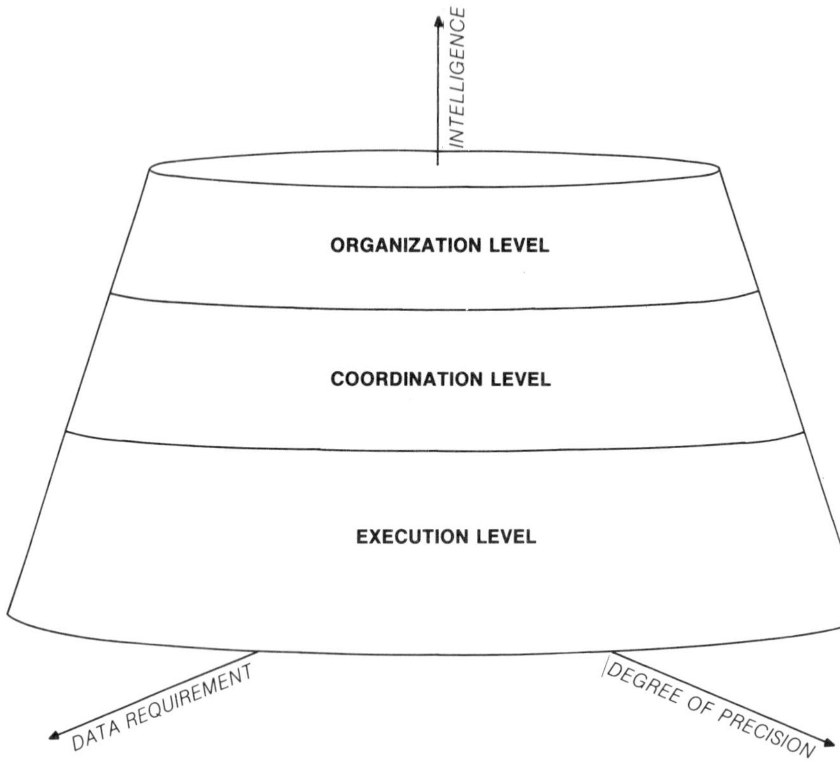

Fig. 1.1 Hierarchical levels in manufacturing.

planning and master production planning activities, based on demand forecast, constitute organization level decision-making functions. Capacity planning and detailed scheduling, based on the information provided by the master schedule, are actually coordination level activities of production planning. Machine loading, monitoring of production and maintaining detailed schedules forms the execution level activities. If we examine each level carefully, it is not difficult to observe the following characteristics regarding to the degree of intelligence, precision and data requirements (Fig. 1.2).

Organization level activities use abstract models and algorithms to perform extensive searches with little precision, and large time increments are used to represent the planning horizon. (Figure 1.3 depicts the change in time increments for each level.) Coordination level activities use localized models that depend on attribute measurements obtained from data integration to perform moderate searches with average precision; medium time increments are used for planning horizon representation. Execution level implements the decision made at the coordination level through precise operational procedures and maintains the accuracy of the actions by providing feedback to the system according to the inputs generated through sensory data and use actuators to perform specific operations required by the operational procedures. There is

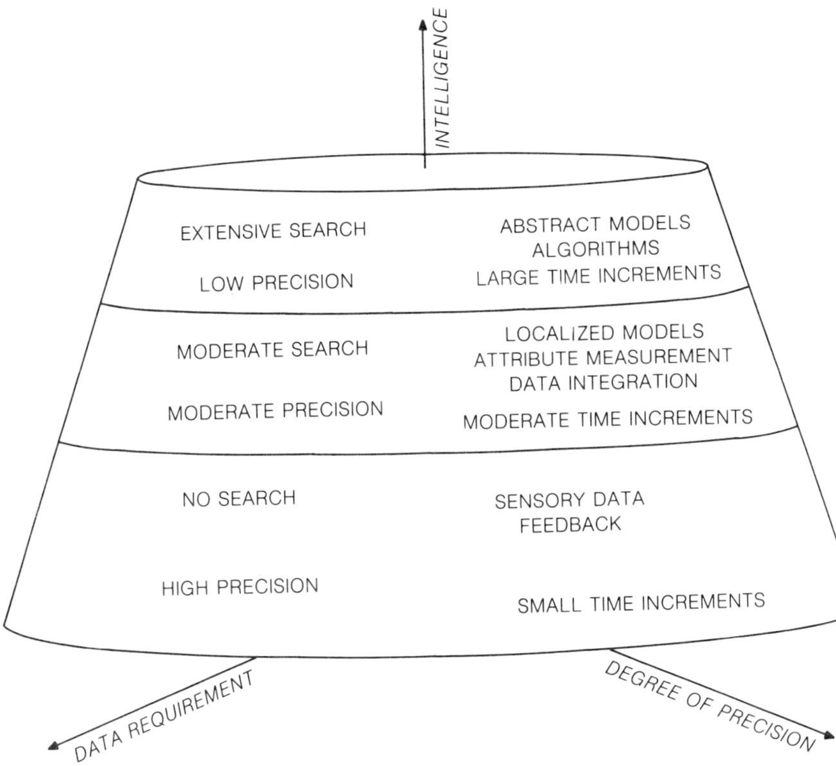

Fig. 1.2 Search and precision within hierarchical levels.

no search at this level of the hierarchy and small time increments are used to represent the planning horizon. This level provides all the raw data required to run a manufacturing system.

Figure 1.4 depicts interactions among hierarchical levels through the use of knowledge, data and information. It is evident that the nature of information flow changes as we move up along the levels. Lower levels of hierarchy are precise and require large quantities of data; higher levels of hierarchy are imprecise and demand an extensive search among large number of possible alternatives of action. This fact is studied by Saridis (1989) using the principle of entropy.

Saridis defines the **concept of knowledge flow** as an entity that connects hierarchical levels and represents data-handling and management, planning and decision performed by processing units, sensing and data acquisition obtained through sensors and actuators and formal languages that define the software. Hence, it is possible to express every action and decision of each level through subjective probabilistic models of fuzzy sets and compute their entropies to be able to provide an analytical measure of the total activity.

The classical optimal control approach utilizes a non-negative function to

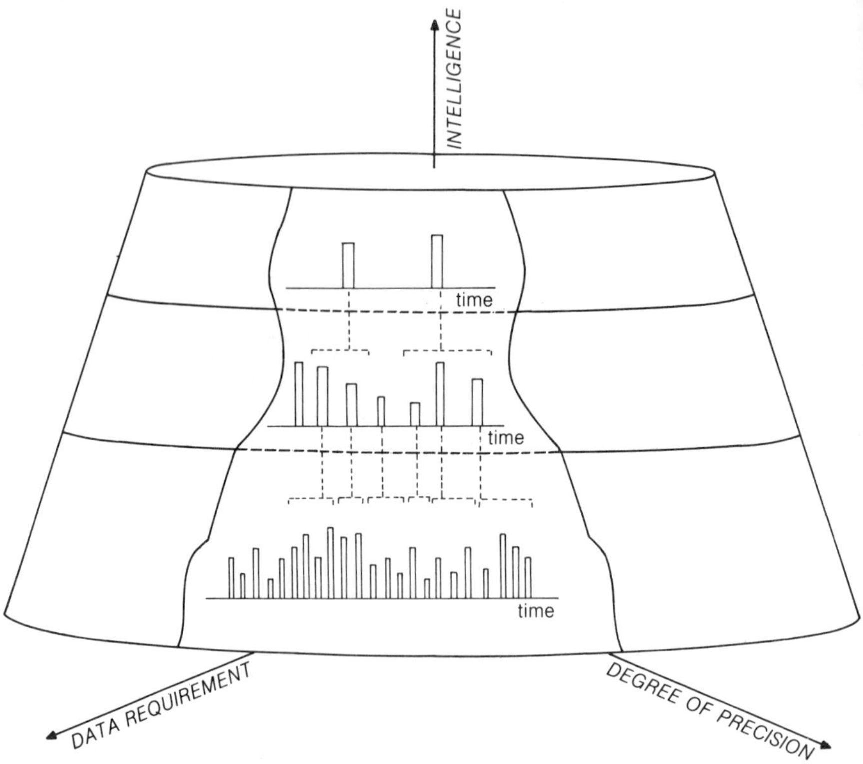

Fig. 1.3 Frequency of updates within hierarchical levels.

represent the states of the system – a specific control from the set of all feasible controls to define the performance measure for some initial conditions representing the state of the system $x(t)$ at time t. Minimization of this function yields the desired control law of the system.

It is shown by Saridis (1988), that the entropy for a particular control action $u(x, t)$, where x is the state of the system and t is the time, $H(u)$ is

$$H(u) = \int_{\text{system states}} p(x, u(x, t), t) \ln p(x, u(t), t) \, \mathrm{d}x \tag{1.1}$$

where $p(x, u(x, t), t)$ is an appropriate density function satisfying Jaynes' maximum entropy principle (Jaynes, 1957).

Entropy satisfies the additive property. Hence, it is possible to represent the entropy of a system by adding entropies of individual subsystems. Using this fact, Saridis (1989) summarizes the theory of hierarchically intelligent systems as follows.

> The theory of intelligent machines may be postulated as the mathematical problem of finding the right sequence of decisions and controls for a

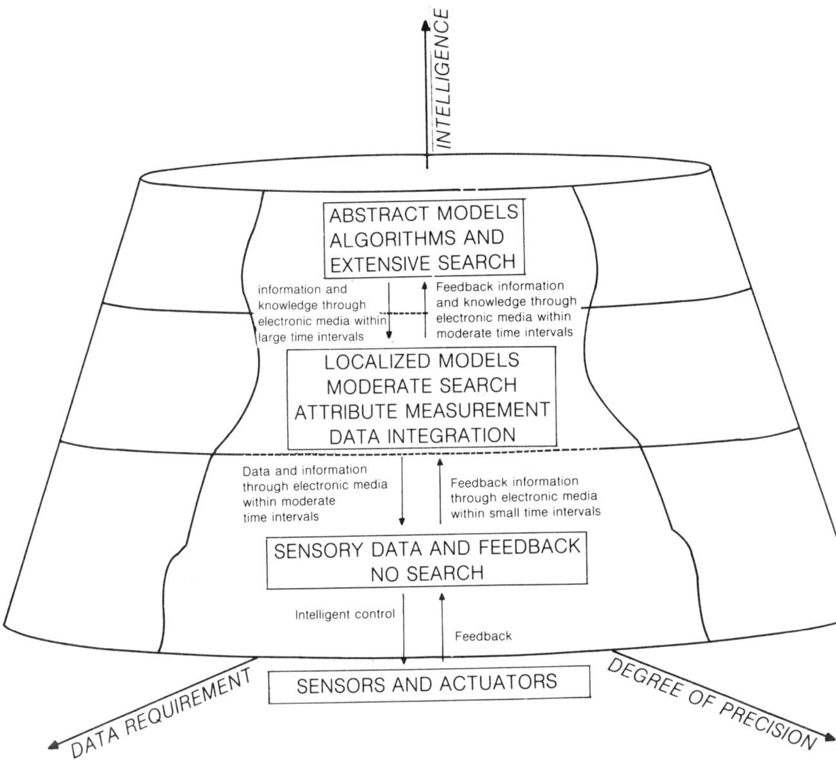

Fig. 1.4 Information, data and knowledge flow within hierarchical levels.

system structured according to the principle of increasing precision with decreasing intelligence such that it minimizes its total entropy.

Saridis (1989) proves this theory analytically as follows. Let K = knowledge representing information, and $P(K)$ = probability density of knowledge, then since Jaynes' principle is satisfied.

$$P(K) = \mathrm{e}^{-\alpha - K} \quad \alpha = \ln \int_x \mathrm{e}^{-K}\,\mathrm{d}x$$

Then knowledge (energy) can be defined as

$$K = -\alpha - \ln P(K)$$

The rate of knowledge (power) R represents the flow of knowledge throughout the system and defined as

$$R = \frac{K}{T}$$

where T is the fixed time interval.

Machine intelligence is defined as a setof rules which operates on a database of events to produce a flow of knowledge. Let knowledge state S_i, $i = 1, \ldots, n$ represent the state of events at the nodes of the hierarchical manufacturing system defining the task execution stages. Then, the knowledge between two knowledge states can be expressed as

$$K_{ij} = \tfrac{1}{2} W_{ij} S_i S_j$$

where W_{ij} are defined as state transition coefficients. Then the knowledge at state i can be written as

$$K_i = \frac{1}{2} \sum_j W_{ij} S_i S_j$$

If this knowledge is summed over all states, the total knowledge of the system can be computed as

$$K = \frac{1}{2} \sum_i \sum_j W_{ij} S_i S_j$$

yielding the following values for the rate of knowledge.

$$R_{ij} = \frac{K_{ij}}{T} \quad R_i = \frac{K_i}{T} \quad R = \frac{K}{T}$$

Knowledge can also be expressed as a probabilistic relation,

$$\ln P(K_i) = -\alpha_1 - K_i$$

where

$$P(K_i) = \mathrm{e}^{-\alpha_1 - K_i} \quad \text{and} \quad \mathrm{e}^{\alpha_1} = \sum_i \mathrm{e}^{-K_i}$$

satisfying Jaynes' principle.

Substituting $R_i T$ for K_i

$$P(R_i) = P(R_i T) = \mathrm{e}^{-\alpha_1 - R_i T} = \mathrm{e}^{-\alpha_1 - \mu_1 T}$$

Machine intelligence I can be defined as a set of rules operating on a database of events D, to produce a flow of knowledge R. Then, the principle of increasing precision with decreasing intelligence can be defined probabilistically as

$$P(I, D) = P(R)$$

Using Bayes' theorem

$$P(I/D)P(D) = P(R) \quad \text{or} \quad \ln P(I/D) + \ln P(D) = \ln P(R)$$

Taking the expected value of both sides,

$$\int_I P(I/D) \ln P(I/D) \, \mathrm{d}I + \int_D P(D) \ln P(D) \, \mathrm{d}D = \int_R P(R) \ln P(R) \, \mathrm{d}R$$

This expression can be written as

$$H(I/D) + H(D) = H(R)$$

where $H(I/D)$ is the entropy of intelligence for a given database of events, $H(D)$ is the entropy of database of events and $H(R)$ is the entropy of knowledge rate. If we keep the value of R constant, then an increase in the entropy of intelligence for a given database of events D requires a decrease in entropy of D, which proves the fact that increase in precision requires a decrease in intelligence.

This concept is used a lot in the design of intelligent manufacturing systems described in the following chapters. It is not difficult to see that the manufacturing level hierarchy determines the nature of the mapping problem to be solved by artificial neural networks. In the next section, characteristics of intelligence and intelligent manufacturing will be discussed in reference to an outline of the theory of intelligence proposed by Albus (1991).

1.3 CHARACTERISTICS OF INTELLIGENT MANUFACTURING SYSTEMS

Albus (1991) defines intelligence as follows and gives an outline for a theory of intelligence.

> The ability of a system to act appropriately in an uncertain environment, where appropriate action is that which increases the probability of success, and success is the achievement of behavior subgoals that support the system's ultimate goal.
>
> Both the criteria of success and the system's ultimate goal are defined external to the intelligent system. For an intelligent machine system, the goals and success criteria are typically defined by designers, programmers and operators. For intelligent biological creatures, the ultimate goal is gene propagation and success criteria are defined by the processes of natural selection.

Albus identifies three degrees of intelligence which is determined by

- the computational power of the system
- the complexity of the algorithms used by the system for processing sensory input and sophistication of system elements
- the information and values the system has stored in the memory.

He states that intelligence can be observed to grow and evolve, through the growth in computational power and accumulation of knowledge of how to behave in uncertain environments. According to Albus, there are four system elements of intelligence:

- sensory processing
- world model

- value judgment and
- behavior generation.

These elements also characterize elements of intelligent manufacturing systems, as they are valid for any intelligent system.

Actuators are used to produce output from an intelligent system; they excite transducers, generate forces to sensors, and move tools and manipulators. Sensors are used to receive inputs to the system; they measure attributes such as temperature, pressure and velocity to monitor the internal and external states of the system. An intelligent system may have thousands or more actuators and sensors that need to be coordinated in order to perform specific tasks and evaluate the sensory information.

According to Albus, the system's perception of the state of the world is achieved through sensory processing. This basic element of the intelligent systems compares observations obtained through the sensors and evaluates them using various processing algorithms for integrating similarities and expectations both over time and space to detect events, recognize features and identify valid relationships with the world model. Sensory processing algorithms combine sensory inputs obtained both in time and space to define consistent and unified perception of the world.

The best estimate of the state of the world is captured within the **world model** of the intelligent system. This model contains knowledge bases and a management information system for information and data manipulation. It also has the simulation capability to generate and predict future states of the system. It provides information to the behavior generation system (so that it can make intelligent plans) and to the value judgment system (so it can assess feasible courses of action based on cost, benefit, risk, priority, and uncertainty). The world model is constantly updated and defined through the sensory processing system in time as the intelligent system evolves.

The **value judgment system** evaluates the observed state of the world and plans possible courses of action generated by the world model. It identifies good and bad, important and unimportant events, certain and uncertain, and provides the basis of decision-making for selecting one action as opposed to another.

According to Albus, the **behavior generation system** selects goals and plans. Plans are executed through converting tasks into sub-tasks and sequencing them to achieve specific goals. The system selects the plans with the best evaluations for execution and monitors the execution of the plans and modifies existing plans if required.

Intelligent systems necessitate an architecture that interconnects these basic elements to interact and communicate with each other – providing a media for communication between modules to generate, evaluate, execute various tasks and collect data and activate actuators. Various system architectures are proposed for intelligent systems and some of them are implemented. In all of them, a **sensory processing system** maintains an internal model of the external world, and performs the situation assessment function while the behavior generating system controls actuators to achieve goals in the context of this perception of the external world. Both of these systems interact with the value judgment and world model system to reason about time and space in the former and to assign values to events, attributes for the latter. Most intelligent system architectures proposed replicate and distribute this structure over a hierarchical computing structure.

Detail of these architectures and the various interactions among basic intelligent system elements are given in Albus (1991). Intelligent manufacturing systems that can provide flexibility and have an ability to adapt to changes in their environment need to utilize all of these basic components of intelligence with minor modifications for specific applications; Artificial neural network models will be an important part of these systems. Albus (1991) states this fact with the following theorem.

> All of the processors described previously for the behavior generation, world model, sensory processing, and value judgment modules, whether implicit or explicit, can be implemented in neural net or connectionist architectures, and hence could be implemented in a biological neuronal substrate.

This theorem, in its conceptual state forms a basis for the use of artificial neural networks in intelligent systems. This concept is demonstrated and examined in detail in later chapters.

1.4 SUMMARY

In this chapter, the need for intelligent manufacturing systems that can respond to flexibility in product design, and process planning imposed by global markets is discussed. Following a brief overview of evaluation of manufacturing systems and strategies during the last three decades, the need for manufacturing system flexibility is stressed. Three basic levels of manufacturing hierarchy (organization, coordination and execution) are identified. Interactions among levels are studied using entropy concepts and the principle of increasing precision with decreasing intelligence is proved.

Characteristics of intelligent manufacturing systems are explained in reference to the outline for a theory of intelligence proposed by Albus. Four basic elements of intelligence are defined and the need for an intelligent systems architecture that integrates these elements is stressed. Implicit and explicit use of artificial neural networks for modeling all of the processes of intelligent systems elements are described and stated as a theorem.

In the chapters that follow, explicit examples that demonstrate the use of artificial neural networks in organization, coordination and execution levels of manufacturing hierarchy are discussed and possible benefits of this new approach are demonstrated.

REFERENCES

Albus, J.S. (1991) Outline for a theory of intelligence. *IEEE Transactions on Systems, Man, and Cybernetics*, **21**(3), 473–509

Carrico, M.A., Girard, J.E. and Jones, J.P. (1989) *Building Knowledge Systems*, McGraw-Hill, New York.

Chang, T. (1990) *Expert Process Planning for Manufacturing*, Addison-Wesley, New York.

Coyne, R.D., Roserman, M.A., Radford, A.D., Balachandran, M. and Gero, J.S. (1990) *Knowledge-based Design Systems*, Addition-Wesley, Reading, MA.

Dagli, C.H. (1991) Editorial. *Journal of Intelligent Manufacturing*, **2**, i.

Hax, A.C. and Candea, D. (1984) *Production and Inventory Management*, Prentice Hall, Englewood Cliffs, New Jersey.

Jackson, P. (1990) *Introduction to Expert Systems*, Addison-Wesley, Wokingham, England.

Jaynes, E.T. (1957) Information theory and statistical mechanics. *Physics Review*, **106**, 4.

Kusiak, A. (1988a) *Artificial Intelligence: Implications for CIM*, IFS (Publications) Ltd., Bedford, England.

Kusiak, A. (1988b) *Expert Systems*, Strategies and Solutions in Manufacturing Design and Planning, SME Publication, Dearborn, Michigan.

Kusiak, A. (1990) *Intelligent Manufacturing Systems*, Prentice Hall, Englewood Cliffs, New Jersey.

Mockler, R.J. (1988) *Knowledge-based Systems for Management Decisions*, Prentice Hall, Englewood Cliffs, New Jersey.

Saridis, G.N. (1988) Entropy formulation of optimal and adaptive control. *IEEE Transactions of Automatic Control*, **AC-33**(8), 713–21.

Saridis, G.N. (1989) Analytic formulation of the principle of increasing precision with decreasing intelligence for intelligent machines. *Automatica*, **25**(3), 461–7.

Schalkoff, R.J. (1990) *Artificial Intelligence: An Engineering Approach*, McGraw-Hill, NY.

Turban, E. and Liebowitz, J. (1992) *Managing Expert Systems*, Idea Group Publishing, Harrisburg, PA.

Wright, P.K. and Bourne, D.A. (1988) *Manufacturing Intelligence*, Addison-Wesley Publishing Company Inc., Reading, MA.

CHAPTER 2

Intelligent systems architecture: Design techniques

Deborah Stacey
Department of Computing and Information Science,
University of Guelph, Guelph,
Ontario Canada N1G 2W1.

2.1 INTRODUCTION

The purpose in modern intelligent systems design is to specify, design and implement systems that have a high degree of machine intelligence. **Machine intelligence** can be defined as the ability to emulate or duplicate the sensory processing and decision making capabilities of human beings in computing machines (Barr and Feigenbaum, 1981). Intelligent systems need the ability to learn autonomously and to adapt in uncertain or partially-known environments if they are to progress past the academic domain and into a full engineering implementation. Different approaches have been utilized that either take advantage of one particular artificial intelligence methodology or exploit the complementary properties of several techniques to achieve a common goal.

Classical artificial intelligence paradigms have sought to achieve competence in domains such as natural language understanding, problem solving in limited domains, abstract mathematical reasoning and common-sense reasoning. These are all tasks associated with high-level, abstract human thought. By contrast, industrial applications such as visual inspection systems, autonomous vehicles and flexible manufacturing systems demand intelligent systems that must work with dynamic, non-stationary sensory processing and reasoning situations in real-time. These domains have various characteristics and requirements that make them difficult problems and often take them outside

Artificial Neural Networks for Intelligent Manufacturing. Edited by Cihan H. Dagli.
Published in 1994 by Chapman & Hall, London. ISBN 0 412 48050 6

the capability of classical artificial intelligence methodologies. The tasks are complex with many components interacting with one another requiring large amounts of diverse and difficult to obtain information and data. Explicit control decisions or appropriate actions are often required in real-time. A system is usually expected to provide a sufficient level of explanation of its decisions and actions and often must provide an interactive environment for a human partner or operator who will provide the ultimate final decision or action to be taken.

Complexity is not the only hurdle to be overcome; most often the problem's complexity cannot be easily judged as it is often ill-structured. Goals or goal states are poorly defined and there may not be an obvious path from an initial to a final state. All of the system's input information must be put together into an integrated environment capable of accommodating the volume and diversity of this information and of producing some form of result or assessment. This may have to be done in an environment where the available information or domain data is incomplete, uncertain or conflicting in nature (Serra and Zararini, 1990)

A modern industrial or manufacturing system requires many different cooperating knowledge agents and forms of expertise to be fused together to accomplish a common purpose or goal. This data fusion may involve the combining and rectifying of sensor information from very diverse sources that can differ in format and level of detail. The reasoning and decision-making components of the system may also be composed of numerous modules that employ various and different methodologies and computational paradigms.

2.2 KNOWLEDGE-BASED SYSTEMS

A basic tenet of classical artificial intelligence (AI) is that problem solving can be formalized as a state-space search; the task is one of finding a path from an initial to an end state. The theory of production systems has been one of the results of this concept. It is useful to structure problem-solving systems in a way that facilitates a description of the search process. Production systems were proposed for modelling human problem-solving behaviour by Newell and Simon (1972). They are often referred to as rule-based systems and consist of a set of rules, a memory and a control strategy.

- The set of rules or problem-solving operators is usually in the form of IF ⟨some condition is met⟩ THEN ⟨execute some action⟩. The conditional (or IF part) of the rule is generally referred to as the left-hand side (LHS) and the resultant part is called the right-hand side (RHS).
- The working memory or database contains the current state description and other information.
- The control strategy chooses between rules when more than one is applicable. This strategy essentially controls the order in which the production rules are fired and resolves conflicts if more than one rule is

applicable. The control strategy and rules combine to form an inference engine which performs limited domain reasoning.

Artificial intelligence theoreticians have developed a powerful paradigm in the classical production system. Its contributions to the field of intelligent systems design include its uses as

- a powerful knowledge representation scheme
- a bridge connecting classical AI research to the main work in expert systems
- a heuristic model for human behaviour
- its capacity for justification or explanation.

Production system characteristics incorporate

- expressiveness and intuitiveness
- simplicity
- modularity and modifiability
- knowledge intensiveness.

The knowledge base is very effectively separated from the inference engine, thus allowing the inference engine to be general purpose.

But, the production system paradigm is not without its problems. While each rule in the system may be very clear, the combined operation and effect of the control program may be relatively opaque. This opacity is rooted in the lack of hierarchy within the set of production rules and a dynamic working memory which must be queried to follow the operation of the analysis. The rule base can also lead to inefficiency. Often the addition of more and more rules causes the system to run slower and slower. And while rule-based systems have been a very successful strategy in many domains, this basic system formulation can lead to a combinatorial explosion of possible choices and actions as the size of the rule base grows. But perhaps the most critical failure of most production systems are their inability to learn. Unless there is some special provision for adding and modifying rules based on experience, the system will not be able to learn.

Domains that have been successfully modelled by production systems are ones in which the structure of the knowledge resembles the structure of production rules and in which the actions that are required are relatively independent and where the domain knowledge itself is distinct from the application to which it will be put. The success that was achieved with production systems led to a further exploration of problems and domains normally thought to require human specialists for their solution. This led to the development of the **expert system**. Not only must the system model a human expert in its decision-making capability but it must also have the capability of explaining or justifying its conclusions. The task of extracting knowledge from human experts and transmitting it to expert systems has become known as **knowledge engineering** (Adeli, 1990).

No matter what AI methodology is used to achieve an expert system, there are certain shared characteristics.

- The system performs at a level generally recognized as equivalent to that of a human expert in the field.
- The system is highly domain specific; it knows a great deal about a narrow range of knowledge.
- The system can explain its reasoning since to provide a useful service in most domains, it must have the ability to justify its own advice, analysis or conclusion.
- If the information with which it is working is probabilistic or fuzzy, the system can correctly propagate uncertainties and provide a range of alternate solutions with associated likelihoods.

Expert systems mark a paradigm shift in classical AI – a shift from general purpose, knowledge-sparse, weak methods to domain-specific, knowledge-rich techniques in which knowledge is explicitly represented. Unfortunately, this methodology develops systems that have a great reliance on human experts for knowledge acquisition. The knowledge acquired from human experts is often of unknown reliability and usability. Some expert tasks involve approximate and common-sense reasoning whose mechanisms are poorly understood at the present time. The difficulty involved in deciphering rules for cognitive tasks is related both to the complexity of the tasks involved and the lack of theory and methodologies in the field of cognitive science and applied psychology.

Because of the general lack of learning mechanisms in expert systems, they show difficulties with the integration of new, partial or conflicting information into their already existing knowledge bases. Related to this characteristic is a general inability to handle dynamic environments and non-stationary data. For the most part, the techniques for handling imprecise, noisy or conflicting data during the knowledge acquisition phase and the inferencing phase of an expert system's operations are still immature.

Successful intelligent systems designs have been achieved with production systems and expert systems methodologies. To summarize the limitations of these techniques, an examination of the basic philosophy behind knowledge acquisition and learning must be done. The exploration of the elemental questions of the basis for machine intelligence will naturally end in the exploration of the field of artificial neural networks and the development of the hybrid intelligent system.

In nature, experience changes the state of an organism in such a way that the new state enables it to do better in subsequent situations. The classical name for this process is learning. An important parameter in the classification of learning strategies is the degree of inference required on the part of the learner. A widely accepted classification scheme spanning the spectrum from no inference required to a great deal is

- rote learning
- learning by instruction
- learning by deduction
- learning by analogy
- learning by induction.

Rote learning is the lowest level of both human and machine learning. In machine learning it corresponds to knowledge being represented directly in the system by programming or simple database systems. No further processing or transformations on the data are required in order to use it. The performance of systems employing rote learning may be enhanced by selectively forgetting knowledge which is rarely used or proves to be incorrect.

In **learning by instruction**, knowledge is acquired from a teacher or textbook and is transformed by the learning system into an internal representation applicable to the problem at hand. The primary responsibility of structuring and representing knowledge remains with the teacher, but the learner must perform some inference in order to transform the knowledge into a readily usable representation. The learner's role in this mode of learning may be considered to be that of performing a syntactic reformulation of the knowledge provided by the primary source.

Learning by deduction shifts even more responsibility for transforming knowledge into a usable form from the teacher to the learner. The constraints on knowledge representation from the source are relaxed. The learner draws deductive inferences from the knowledge and reformulates it in the form of useful conclusions which preserve the information content of the original data. Deductive learning includes knowledge reformulation, compilation, and organizational procedures such as chunking which preserve the truth of the original formulation.

The mode of learning known as **learning by analogy** combines deductive and inductive learning. The first step is the inductive inference required to find the common substructure between the problem domain and one of the analogous domains stored in the learner's existing knowledge base. The next step is to map the solution from the selected analogous domain to the problem domain using deductive logic.

Inductive learning (Holland *et al.*, 1987) is the classification of a set of experiences into categories or concepts. The concepts to be learned are given by the teacher or emerge as similarities in experience are noticed. From one or more instances in which a particular action was the appropriate action in response to a situation, it is inferred that a generalized form of this action is appropriate in response to this general situation type. This type of learning can be broken down into many sub-categories. **Learning by example** involves the process of concept acquisition in which general concept descriptions are inferred from a set of examples provided by a teacher, the environment, or the knowledge base of the learner itself. In **learning by experimentation**, the acquired concept should be general enough to explain all of the positive examples provided but restricted enough to exclude any counter-examples. If the source of examples is the environment, the learner may have the option of performing experiments from which it receives feedback. This form of learning also goes under the name of **stimulus-response learning.** **Learning by observation and discovery** takes place without a teacher. It is a form of descriptive generalization in which the learner infers general rules and regularities which explain the observations. Inference techniques employed may include curve fitting of data, classification of observations and objects, and conceptual clustering in which simple concepts are classified. The observations may take place in a passive mode in which the learner does not interact with the environment or in an active mode in which the learner may experiment on the environment to perform stimulus-response learning. The objective is the discovery of economical theories to account for the observations.

A philosophical problem presents itself in inductive learning: do general conclusions ever actually follow from specific observations in this way? Machine learning theory makes the assumption that some inductive inferences are correct and some are not and the problem is to recognize the useful ones and quickly eliminate or change the useless ones. The problem for the designers of intelligent systems is to find a pattern, or concept definition that applies to one set of situations (the positive instances of a concept) and does not apply to another (the negative instances). Once started, a learning program can be thought of as wandering around the space of possible generalizations. **Generalization and specialization** provide a set of methods for revising rules, theories or observations, but merely knowing in which direction to make a revision does not tell us much about how to do the revision. To know the right move to make toward the correct rule or theory, one would have to know the correct rule or theory already. It is a search problem where the correctness of a step toward increasing or decreasing generality cannot be judged until the final goal is reached and so false steps are unavoidable. The search space of possible generalizations can be huge.

There is no hope of getting through the search space of possible generalizations in a particular domain unless the following problems are solved.

- The situation-identification problem
- The description problem
- The credit-assignment problem.

The **situation-identification problem** involves the figuring out of what constitutes the current situation since we cannot take this to be the set of everything the machine believes. In other words, the system must figure out what features of the situation to ignore in revising rules. In any real situation there are thousands of true but irrelevant facts. Any algorithm will bog down unless given just a few facts to work on. Whatever filters those facts automatically constrains what can be learned.

The **description problem** involves the choice of the most appropriate knowledge representation for the information in a particular domain. Knowledge representation is a set of syntactic and semantic conventions that make it possible to describe knowledge. Syntax can be defined as the symbols to be used in a language and the way those symbols may be arranged. Semantics is the meaning which is embodied in the language's symbols and therefore those symbol arrangements which are allowed. A knowledge representation should be adequately expressive and a reasonable efficient medium in which to reason. It should be understandable and maintainable. Classical AI knowledge representations can be classified as having declarative and/or procedural characteristics. In **declarative knowledge representations** most of the knowledge is represented as a static collection of facts. General procedures are applied to domain specific knowledge to make deductions or inferences. The **procedural approach** treats the bulk of knowledge as procedures. This view makes it much easier to encode heuristic knowledge. A piece of knowledge in this type of system specifies how, and under what conditions, it is to be used. A good representation make inference smoother. The notation used for expressing concept definitions must have the property that similar situations should have similar descriptions in the notation. Therefore, a short generalization will capture all the positive instances.

One of the most difficult problems to solve in AI is the **credit-assignment problem**. Consider a chess program that has just lost a game. This one bit of information tells the machine that it has made one or more mistakes along the way, but it does not indicate where the first mistake was made. This type of credit-assignment information can come from multiple sources: careful experiments, a teacher who already knows or a technique that assigns proportional blame (or credit) according to an algorithm that is trying to minimize or maximize an objective function based on the problem space.

Solving these problems requires imposing *a priori* constraints on the learning problem and constraints on what can be learned. There is no way around this requirement. When the situation-identification problem, description problem and the credit assignment problem have natural solutions, then an expert system or knowledge-based system works well.

Other factors that have to be considered when designing an intelligent system are the presence of noisy data and the element of time. Suppose that the teacher or environment occasionally lies about a concept or that the learner occasionally misidentifies the rule that gets the blame for a failure. Most inductive learning algorithms will fall apart under these circumstances because they take every positive and negative instance of a concept equally seriously. The algorithm will generalize or specialize the pattern to account for this anomaly and the resulting alteration to the pattern will last forever.

A related problem is the notion that a concept may change over time. A strategy that retains all positive and negative instances, and tries to find an eternal pattern that accounts for all of them, is probably on the wrong track. The goal of most knowledge-based systems is to find necessary and sufficient conditions for something to fall under a concept. The goal of a problem solver should be to maximize its chances of success. Unfortunately, this goal is hard to formalize.

2.3 ARTIFICIAL NEURAL NETWORKS

The artificial neural network (ANN) approach to machine intelligence (Serra and Zanarini, 1990) is centred upon the study of the brain and its emergent properties while AI's symbolic approach is centred upon the study of the mind, independent of the structure and functioning of its physical support. AI uses dedicated symbols (local representation) to represent each concept and these symbols are manipulated by chains of inference or rules, usually under a centralized control. Artificial neural network systems, using distributed representations, reach their conclusions by applying evolutive rules to numerical values. They lack centralized control in the classical sense; all local units are simultaneously working. While AI systems possess a well-defined termination condition or goal, for ANN systems this is unnecessary. While every rule in an AI system has a precise, pre-assigned meaning, it is common that several neurons in an ANN model have no pre-defined meaning. They evolve and/or specialize during learning in a manner which is often difficult for a human observer to predict or understand.

Classical AI systems tend to be characterized as having

- a limited domain,
- trouble handling contradictions and time varying knowledge, and
- little or no capacity for learning.

The overall quality of system performance rapidly declines as soon as one leaves the specific domain for which the system was designed. This is often referred to as **brittleness.** Artificial neural network systems, on the other hand, tend toward system robustness, resistance to noise, contradictions and incompleteness, and have the possibility of automated learning.

One of the early theoreticians of artificial intelligence, Hebert Simon, maintained that the intelligent processes of the human brain could be simulated by a physical symbol system. This philosophy naturally contributed to a focusing on the problem of how strategies and constraints interact to produce effective behavior in intelligent tasks. It also led Simon to maintain that the human cognitive system and the machine intelligences that emulated it were basically serial in nature.

Another more recent philosophy has challened some of these premises of classical AI. In a 1985 article, *The Challenge of Open Systems*, Carl Hewitt postulated that the physical systems that we wish to model are actually open systems. Open systems are always subject to communications and constraints from outside. They are characterized by

- continuous change and evolution
- decentralized decision making

- perpetual inconsistency among knowledge bases
- the need for negotiation among system components and
- the inadequacy of the closed-world assumption.

The science of complexity centres upon the development of the concept of self-organization, which expresses precisely the possibility of highly organized behavior even in the absence of a pre-ordained design. This self-organization can arise from the interaction between microscopic or local dynamics. Natural systems such as lasers and convection currents in fluids are examples of complex systems that display this type of behavior. At the global or macroscopic level, system behavior appears purposeful or meaningful whereas at the microscopic level, not only are the transition functions and the interconnections amongst elements important, but so is the detailed information about initial conditions. In many ways this is typical of what some researchers theorize about the organization of the brain which gives rise to high-level cognitive functions.

Artificial neural networks are systems constructed to make use of some of the organizational principles felt to be used by the brain. In the **generic neural network** or **connectionist model** there are three main components: the *neuron*, the network topology and the learning algorithm or strategy.

The **neurons** or **processing elements (PEs)** (Simpson, 1990) are the components where most, if not all, of the computation is done in most ANN systems. The PEs receive inputs from other PEs or from the environment via synapses or interconnections which can be excitatory or inhibitory. If the input to the PE is excitatory then it is more likely that it will pass this message or action potential to the other PEs to which it is connected. Obviously, inhibitory inputs act in the opposite manner. These inputs can be formulated as an input vector $\mathbf{A} = (a_1, \ldots, a_i, \ldots, a_n)$ where a_i is the activity level of the ith PE or input. Associated with each connected pair of PEs is an adjustable value or weight. The weights connected to the jth PE can be expressed as a vector of the form, $\mathbf{W}_j = (w_{1j}, \ldots, w_{ij}, \ldots, w_{nj})$ where w_{ij} represents the connection strength from the PE a_i to the PE a_j.

A neuron contains an internal potential or limit and when this exceeds a threshold, it can propagate an action potential down its connections to other neurons. Many neurons can be active simultaneously and thus neural networks can be considered to be parallel computational systems. The internal threshold value θ_j is modulated by the weights that are associated with the PE's inputs. The output value of a PE can thus be defined as

$$f\left(\sum_{i=1}^{n} a_i w_{ij} - w_{0j}\theta_j\right)$$

Because a PE's input domain is potentially infinite, threshold functions are used to confine the PE's output to a pre-specified range or value. There are four common functions: linear, non-linear, step and sigmoid

1. **Linear function**: $f(x) = \alpha x$ where α is a real-valued constant that regulates the magnification of the inputs.

2. **Nonlinear ramp function**: the linear function is bounded to the range $[-\gamma, +\gamma]$ such that γ $(-\gamma)$ is the PE's maximum (minimum) output value or saturation level.
3. **Step threshold function**: the threshold function only responds to the sign of the input; $f(x) = +\gamma$ if $x > 0$ and $f(x) = -\delta$ otherwise, with γ and δ being positive scalars, often 0 and 1.
4. **Sigmoid threshold function**: a bounded, monotonic, non-decreasing S-shaped function that provides a graded, non-linear response. Examples include the logistic sigmoid function

$$S(x) = \frac{1}{1 + e^{-x}}$$

with saturation levels of 0 to 1 and the hyperbolic tangent function $S(x) = \tanh(x)$ with saturation levels of -1 and 1.

The processing elements of an artificial neural network are connected together and overall system behavior is determined by the structure and strengths of these connections. This network structure consists of the processing elements arranged in groups or layers. A single layer of interconnected neurons produces an autoassociative system and is often used as a **content-addressable memory**. Connections between PEs in the same layer are referred to as **lateral connections** and those that loop back and connect to the same neuron are called **recurrent connections**. Multi-layer systems contain input and output PE layers that receive or emit signals to the environment and neurons which are neither, called **hidden units.** This hidden layer provides networks with the ability to perform non-linear mappings as well as contributing to the complexity of reliable training of the system. Inter-field connections or connections between PEs in different layers can propagate signals in one of two ways:

- feedforward signals only allow information to flow along connections in one direction, while
- feedback signals allow information to flow in either direction and/or recursively.

Artificial neural networks can provide content addressable memory (CAM) which stores data at stable states in the weight matrix and associative memory (AM) which provides output responses from input stimuli. In other words, AM stores pattern pairs (Input_1, Output_1),...,(Input_m, Output_m) in the weight matrix. In this heterasssociative memory system, the ANN recall mechanism is a function $g(\quad)$ that takes the weight matrix **W** and an input stimulus Input_k and produces the output response Output_k. The two primary recall mechanisms are nearest-neighbor recall and interpolative recall. **Nearest-neighbor recall** finds the stored input that most closely matches the stimulus and responds with the corresponding output using a distance measure such as Hamming or Euclidian distance. **Interpolative recall** takes the stimulus and interpolates from the entire set of stored inputs to produce the corresponding output.

Learning in an artificial neural network is defined to be any change in the memory (weight matrix) and can be categorized as supervised or unsupervised. **Unsupervised learning** or self-organization is a process that does not incorporate an external teacher and relies only upon local information and internal control strategies. Examples include

- adaptive resonance theory (ART1 and ART2) (Carpenter and Grossberg, 1988)
- Hopfield networks (Hopfield, 1982)
- bidirectional associative memory (BAM) (Kosko, 1987)
- learning vector quantization (LVQ) (Kohonen, 1988)
- counter-propagation (Hecht-Nielsen, 1987).

Supervised learning, which includes

- the perceptron (Minsky and Papert, 1969)
- adaline,
- madaline (Widrow and Lehr, 1990)
- back-propagation,
- the Boltzmann machine (Ackley *et al*., 1985)

incorporates an external teacher and/or global information and includes such techniques as error-correction learning, reinforcement learning and stochastic learning. **Error-correction learning** adjusts the connection weight matrix in proportion to the difference between the desired and computed values of each PE in the output layer. **Reinforcement learning** is a technique by which the weights are reinforced for properly performed actions and punished for inappropriate ones where the performance of the output layer is captured in a single scalar error value. **Stochastic learning** works by making a random change in the weight matrix and then determining a property of the network called the resultant energy. If the change has made this energy value lower than it was previously then the change is accepted, otherwise the change is accepted according to a pre-chosen probability distribution. This random acceptance of changes that temporarily degrade the performance of the system allows it to escape from local energy minima in its search for the optimal system state.

In a multilayer net using supervised learning, the input stimuli can be recoded into an internal representation and the outputs generated are then representative of this internal representation instead of just the original pattern pair. The network is provided with a set of example input–output pairs (a training set) and the weight matrix modified so as to approximate the function from which the training set has been drawn. In the ideal case, the net would then be able after training to generalize or produce reasonable output for input stimulation that it has never been exposed to. Currently the most popular technique for acomplishing this type of learning in an ANN system is the multilayer perceptron employing back-propagation (BP) learning. It evolved from Rosenblatt's perceptron; a two-layer supervised ANN which provides nearest-neighbor pattern matching via the perceptron error-correction procedure. This procedure effectively works to place a hyperplane between two classes of data in an n-dimensional pattern space. It has been shown that this algorithm will find a solution for any linearly separable problem in a finite amount of time.

Error back-propagation through non-linear systems has existed in variational calculus for many years but the first application of gradient descent to the training of multilayered nets was by Amari in 1967 who used a single hidden-layer system to perform non-linear classification. Werbos (1974) discovered **dynamic feedback** and Parker (1982) talked about **learning logic**, but the greatest impact on the neural network community came when Rumelhart, Hinton and Williams (Rumelhart *et al.*, 1985) published their version of the back-propagation algorithm.

One of the major reasons for the development of the back-propagation algorithm was the need to escape one of the constraints on two-layer ANNs which is that similar inputs lead to similar output. But while ANNs like the perceptron may have trouble with non-linear mappings, there is a guaranteed learning rule for all problems that can be solved without hidden units. Unfortunately, it is known that there is no equally powerful rule for multilayer perceptrons.

Back-propagation consists of a three-layer perceptron with feedforward connections from the input layer to the hidden layer and from the hidden layer to the output layer. This function-estimating ANN stores pattern pairs using a multilayer gradient descent error-correction learning algorithm. It achieves its internal representation of the training set by minimizing a cost function. The most commonly used cost function is the squared error or difference between the computed and desired output values for each output PE across all patterns in the training set. Other cost functions include the entropic cost function, linear error and the Minknowski-r or rth power of the absolute value of the error. In all cases, the changes made to the weight matrix are derived by computing the change in the cost function with respect to the change in each weight. The most basic version of the algorithm minimizes the squared error and is also known as the generalized delta rule (Simpson, 1990).

Step 1. Assign small random values to all weights on connections between all layers of the network as well as to all PE thresholds.

Step 2. For each pattern pair in the training set:

(a) read the environmental stimuli into the PEs of the input layer and proceed to calculate the new activations for the PEs in the hidden layer using

$$\text{hidden}_i = f\left(\sum_{h=1}^{n} \text{input}_h w_{hi} + \theta_i\right)$$

where $f(\ )$ is the logistic sigmoid threshold function, there are n input PEs and θ_i is the threshold for the ith hidden PE;

(b) use these new hidden-layer activations and the weights on the connections between the hidden layer and the output to calculate the new output activations using

$$\text{output}_j = f\left(\sum_{i=1}^{n} \text{hidden}_i w_{ij} + \Gamma_j\right)$$

where $f(\ )$ is the logistic sigmoid threshold function, there are n hidden PEs and Γ_j is the threshold for the jth output PE;

(c) determine the difference between the computed and desired values of the output layer activations using

$$\text{diff}_j = \text{output}_j(1 - \text{output}_j)(\text{desired}_j - \text{output}_j)$$

and calculate the error between each PE in the hidden layer relative to this diff_j using

$$\text{err}_i = \text{hidden}_i(1 - \text{hidden}_i) \sum_{j=1}^{n} w_{ij}\text{diff}_j$$

(d) modify each connection between the hidden and output layers with $\Delta w_{ij} = \alpha\, \text{hidden}_i \text{diff}_j$ which is the amount of change to be made to the weight on the connection from the ith PE in the hidden layer to the jth PE in the output layer. α is a positive constant that controls the rate of modification or learning;

(e) perform a similar modification to the weights on the input to hidden layer connections with $\Delta w_{hi} = \beta\, \text{input}_h \text{err}_i$ where β is a positive learning rate constant; and

(f) adjust the thresholds on all hidden and output PEs by using $\Delta\theta_i = \beta\, \text{err}_i$ for the hidden units and $\Delta\Gamma_j = \alpha\, \text{diff}_j$ for the output units.

Step 3. Repeat step 2 until all the diff_js are either zero or sufficiently low.

Step 4. After the BP-ANN has been trained, recall consists of two feedforward operations which create hidden PE values

$$\text{hidden}_i = f\left(\sum_{h=1}^{n} \text{input}_h w_{hi} + \theta_i\right)$$

and then use them to create new output PE values

$$\text{output}_j = f\left(\sum_{i=1}^{n} \text{hidden}_i w_{ij} + \Gamma_j\right)$$

Back-propagation is guaranteed only to find the local, not the global error minimum. And while this technique has proven extremely successful for may practical applications, it is based on gradient descent which can proceed very slowly because it is working only with local information. Practical implementation factors that must be considered include

- the number of units in the hidden layer
- the size of the learning rate constants
- the amount of data that is necessary to create the proper mapping.

Once these issues have been addressed, the power of back-propagation is realized in a system that has the ability to store many more patterns than the number of dimensions inherent in the size of its input layer. It also has the ability to acquire arbitrarily complex non-linear mappings. This is all possible if the application allows for a reasonably long training time in an off-line mode.

Current research in the area of back-propagation improvements is looking at:

- optimizing the number of units in the hidden layer and the effect of the inclusion of more than one layer of hidden units;
- improving the rate of learning by dynamic manipulation of the learning rates and by the use of techniques such as momentum;
- the effects of dynamically changing and modular connection topologies;
- analyzing the scaling and generalization properties of this particular ANN model; and
- employing higher-order correlations and arbitrary threshold functions.

2.4 HYBRID INTELLIGENT SYSTEMS

The next generation of intelligent systems design will integrate the computational paradigms of expert or knowledge-based systems, artificial neural systems, fuzzy logic (Kosko, 1992), uncertainty management and genetic algorithms (Holland, 1975), among other techniques (Kandel and Langholz, 1992). This approach is a reflection of the ineffectiveness of purely symbolic, numeric or distributed representations and of the methodologies for dealing with complex and imprecise data and goals that are part of most real world problems.

Two design methodologies, derived from the field of knowledge-based systems, which allow for the incorporation of multiple intelligence techniques and data fusion are the hierarchical approach and the blackboard approach.

The **hierarchical approach** for intelligent systems design involves the use of different levels or layers. High-level goals are decomposed or subdivided into more elementary or simpler subgoals or subproblems. This division keeps occurring until the subgoals are relatively easily obtainable. Information from the various lower level modules is made available to higher levels usually by means of a communications mechanism that also allows the system to set up a feedback loop; information at the lower levels can affect the higher levels which in turn modify the information being fed to the lower levels. This approach has a readily accessible and describable control structure; control

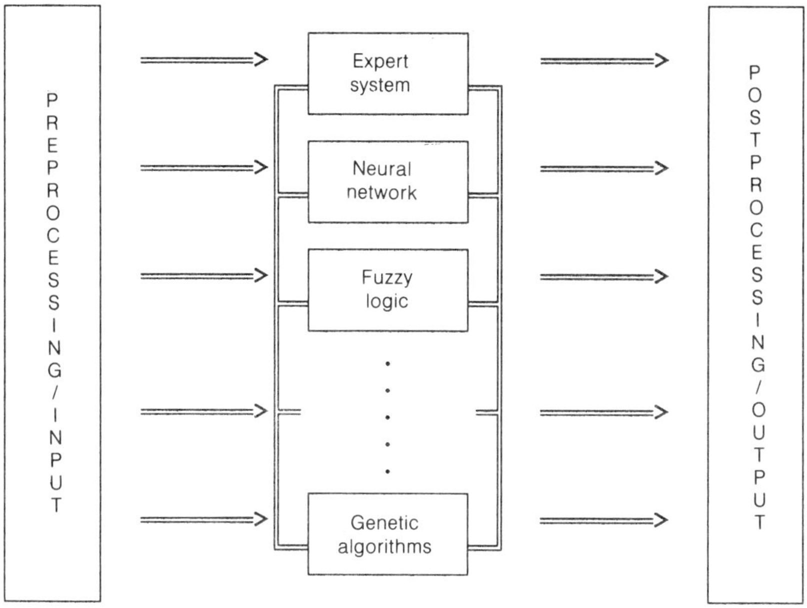

Fig. 2.1 Hybrid intelligent systems.

is generally top-down in nature. This makes systems design relatively straightforward. Unfortunately, the feedback which is possible in this paradigm is harder to describe and control. If the system is designed to have many levels then feedback is difficult to implement; just how much feedback is feasible and between which layers. Often this forces designers to constrain the feedback to a restricted number of layers and usually only adjacent layers. This may result in an under-utilization of information and hamper the problem solving ability of the system as an entity.

A **blackboard system** is a highly structured, opportunistic problem-solving system that prescribes the organization of knowledge in domain and solution spaces (Nii, 1986; 1986a). This model grew out of the work in the early 1970s (1971–1976) at Carnegie-Mellon University on HEARSAY-II (Erman *et al.*, 1980). HEARSAY-II was a speech understanding system developed for the DARPA Speech Understanding Project. The reasoning part of a blackboard system is called opportunistic since it applies pieces of knowledge in either a forward or backward manner at the most opportune time. In backward reasoning the inference steps are applied from the goal states backwards to some initial condition, whereas in forward reasoning, the inference steps are applied toward a goal state from some previous state.

Three components are said to make up the typical blackboard system: the knowledge sources, the blackboard data structure, and a control mechanism. The **knowledge sources** or agents consist of the domain knowledge that is needed to solve the problem. It is partitioned into separate and independent

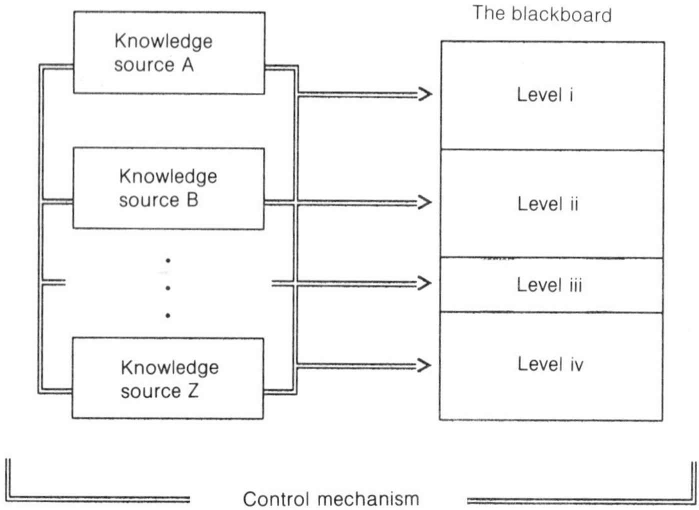

Fig. 2.2 Blackboard structure in intelligent systems.

modules utilizing declarative and procedural knowledge representations. These modules perform transformations on the knowledge in the solution space (in the blackboard) using deterministic algorithms and/or heuristics. This reasoning take place when it is appropriate to apply a particular piece of knowledge or procedure to produce an incremental solution or partial solution instance. To accomplish all of this, the knowledge agent must know under what conditions it can modify the information in the blackboard.

The **blackboard data structure** stores the system's state information in a global database. As such, it contains all full or partial possible solutions that the system, through its knowledge agents has generated to date. This solution space is organized as one or more application-dependent analysis levels, usually hierarchies, where information at each level represents partial solutions. All communication between knowledge sources takes place through the blackboard's database and the knowledge agents also use this resource to monitor their possible participation in the solution space. In this type of system, the goal state is the solution (or a set of solutions) to a particular problem.

A conceptually implicit **control mechanism** exists in all blackboard models. It organizes opportunistic responses to the blackboard by the knowledge sources and determines the focus of attention of the system. This mechanism also determines when to terminate the problem-solving process – either with a solution or the knowledge that the system cannot solve the problem with the available knowledge. Blackboard implementations on serial computers constrain the knowledge agent's solution building to just one step at a time in some order controlled by the control mechanism or monitor. Parallel processors and techniques should allow truly concurrent blackboards in line with the original formulation of this conceptual framework.

And the blackboard is just that, a conceptual, and not a computational, framework. It allows many different extensions and variations to this prescriptive model. The application itself and the various knowledge agents available to build the system will determine the final implemented form of the blackboard. The concept embraces the notion of problem decomposition with different agents attacking areas of the problem to which they are most suited. This allows for very flexible knowledge application and the utilization of multiple inferencing techniques and approaches. It is particularly true in the case of a concurrent version of a blackboard system. But, the system is still dependent on how the original problem was stated and how it was broken down into subproblems. This decomposition will determine the speed, resource allocation and probability or possibility of even finding a solution.

A more recent view of hybrid intelligent systems consists of three distinct phases (Kandel and Langholze, 1992).

- The first phase is procedural preprocessing of the input data in which the techniques of expert and fuzzy systems predominate.
- The next stage is comprised of many different modules: machine learning, neural networks, fuzzy logic systems, genetic learning and expert systems. These various modules perform the main operations of an intelligent system: inferencing, learning, adapting and planning.
- The third and final phase is one of procedural postprocessing which produces the major output of the system as a whole.

Hybrid systems utilizing ANNs and other AI or pattern-recognition techniques are becoming extremely attractive alternatives to homogeneous systems. Two basic approaches to the design of such systems are:

- the controller (master–slave) paradigm, and
- the integrated module approach.

Typically, the **controller (master–slave) paradigm** has been used to design systems where a rule-based expert system selects and prepares data for presentation to one or more ANNs. An example is LAM, developed at DuPont Polymers (Sherald, 1991). LAM is a window (glass laminates) design system that has a rule-based expert subsystem for logical inferencing and ANN subsystems for plausible reasoning. The expert system handles the relatively straightforward task of calculating glass strength and thickness while the ANNs derive estimates for sound transmission and solar control. Another variation of the paradigm is the case of an expert system deciding which neural network, from a collection of nets, is the most appropriate for a particular problem instance. Both versions have the property that each part of the system is decoupled from the others in the training and/or development phase. Litle integration is achieved between parts of the system – a pipeline architecture is affected which attempts to maximize the benefits of each technique while minimizing the possible coupling or correlation between components of each subsystem.

The **integrated module approach** starts with separate modules (some of which are ANNs) but attempts to force more collaboration between subsystems and their components. This can be accomplished by such design techniques as decentralized or distributed control, integrated training schedules and feedback. Few systems have been developed because of their complexity but those which come closest to this description are ones which have subsystems solving local- or sub-problems. These local solutions are combined at some later stage to produce a global solution. An example is the hybrid coder for hidden Markov models (HMMs) using a recurrent ANN developed at the Centre de Recherche Informatique de Montreal (Bengio *et al.*, 1990). A hybrid coder was developed for obtaining descriptions of speech patterns. The pattern space of speech signals is very large and as such is not suitable for effective modeling by HMMs. The hybrid approach consists of multiple subsystems to quantize the parameters of the speech signal. Subsystems or coders provide unsupervised learning by means of a *k*-means clustering algorithm on vectors of Mel-scaled Cepstral coefficients and their time derivatives as well as supervised learning by means of a recurrent ANN to generate hypotheses about gross phonetic categories. The quantized parameters are then fed to the hypothesizer unit which is driven by HMMs.

The reasoning behind the hybrid approach is that subsystems should be constructed using the most appropriate methodology. In the hybrid coder system it was recognized that recurrent ANNs perform discriminative learning and make assumptions about the form of the recognition function, while HMMs make assumptions about the form of the data and possess a very fast and powerful learning algorithm. These differences determine the methodology that is most suitable for a particular phase of the analysis.

An industrial vision system developed at the Unversity of Waterloo (McAndless *et al.*, 1991) is an example of a hybrid. ANNs are integrated into the system in areas where traditional machine vision techniques are inadequate or not available. Calculations that are easy to perform by traditional means are not replaced by ANNs. As in most hybrid systems, this vision system starts out by loosely integrating separately developed ANN modules into an already existing traditional machine vision system. The problem of 3D object representation was done manually in the original system. The ANN subsystem constructs a representation after exposure to multiple views of an object and then coordinates the eventual recognition of the object. The integration of the ANN into the recognition part of the system requires a level of integration comparable to and even exceeding that obtained in the hybrid speech coder.

2.5 MANUFACTURING SYSTEMS IMPLEMENTATIONS

The human activity of manufacturing can be monitored and controlled at various levels. The physical plant itself can be broken down into work centers which can be further subdivided into manufacturing cells that are composed of individual machines, hardware implementation. ANNs have aided in the achievement of intelligent control systems that can be applied to complex non-linear dynamical systems in the presence of structural failures. Multilayer nets have been used to identify and control the plant, and as pattern recognizers for the detection of failure in the system (Narendra and Mukhopadhyay, 1992).

Trajectory planning includes such activities as mobile robot path planning, obstacle avoidance by multi-jointed robotic arms, and the guidance and organization of robotic hand movements. Many researchers have used the ANNs that employ some notion of an energy function which is to be minimized, such as the Boltzmann machine (Jorgenson, 1987) and Hopfield nets (Tsutsumi and Matsumoty 1987; Tsutsumi, *et al.*, 1988). Seshadri (1988) employed Hopfield nets to formulate the problem of mobile robot path planning as a travelling salesman problem where the Hopfield net has had some previous success (Hopfield and Tank, 1985; Wilson and Pawley, 1988; Van den Bout and Miller, 1988). Multilayer perceptrons with back-propagation learning can be designed to function as a classification network to classify robotic hand grips based on the characteristics of the object to be grasped (Liu *et al.*, 1988).

The utility of artificial neural networks in the area of control and planning for computer integrated manufacture (CIM) is a function of their ability as classifiers. Their achievement of successful classification in complex non-linear situations has compensated for their need, in most cases, to learn from a teacher in an off-line mode. The teacher is not that great a handicap since most other alternatives to ANNs employ some version of an expert or global information source, but it will be necessary to achieve on-line learning for those CIM components that most need to adapt to the dynamic environment of the manufacturing plant.

2.6 SUMMARY

The next generation of intelligent systems design will integrate the computational paradigms of expert or knowledge-based systems and artificial neural networks, as well as other methodologies such as fuzzy logic and genetic algorithms. This is a reflection of the ineffectiveness of the physical symbol system approach of classical AI where concepts are given local representations and these symbols are manipulated by chains of inference or rules, usually under a centralized control. Real world problems such as those in CIM and flexible manufacturing have complex and imprecise data and goals and classical AI systems tend to have a limited domain, trouble handling contradictions and time varying knowledge, and little or no capacity for learning.

The purpose in modern intelligent systems design is to specify, design and implement systems that have a high degree of machine intelligence so that these systems can learn autonomously and adapt in uncertain or partially known environments. The artificial neural network approach to machine intelligence has shown some promise in this regard. Using distributed representations, ANNs reach their conclusions by applying evolutive robots and tools. This organization can usually be treated conceptually as a hierarchy. The most common activity in the system is that of planning. Actions are selected, resources are scheduled and the goals to be achieved are ones involving production, quality and economics. All of this needs to be done in a dynamic environment and ideally the intelligent manufacturing system should be able to respond to changes in real-time.

Computer-integrated manufacturing (CIM) has made use of several AI and ANN methodologies. These techniques have been used to capture and store engineering expertise, add to this expertise with new methods, and design intelligent systems that react in real-time to the changing factory environment. Not unlike AI itself, factory automation has been divided into two areas of emphasis: the imitation of human actions by machines and the accomplishing of goals with *non-human* strategies. These different approaches are combined in the engineering-based expert system approach which both explicitly models the physics of a factory situation to provide the proper solution and also captures the capabilities of human operators.

An example of a traditional expert system approach to an aspect of CIM is an expert system for the planning of mechanical assembly using robots (Chang, 1988). The purpose of the system is the planning of robot actions using heuristics to order, prioritize and manage the uncertainties in the sequencing of subgoals and constraints for mechanical assembly. The system is composed of a knowledge source, a control structure and a blackboard. The knowledge source is a knowledge base of workpiece structures, assembly operators and assembly principles that are all domain dependent. Control is divided into two activities: structure analysis which employs analysis principles to analyze workpieces and generate assembly sequences and constraints; and plan generation that chooses operations for the assembly sequences that satisfy the constraints. A typical task for the assembly of a power supply produced a final assembly plan of 169 robot operations and was generated in 4 seconds (1.2 for analysis and 2.8 for plan generation) on an AMDAHL 470 V/7A using Standard LISP.

A major domain for the use of artificial neural networks in manufacturing has been in the area of robotics (Horne *et al.*, 1990). Here they have been used for control, trajectory planning, sensing, and task planning. They have had a great impact on the area of control. Control systems have been designed with neural networks to make use of prior control system information such as input/output data points and the specifications between the data points. This data can be systematically stored in an ANN and used in the design of control subsystems. The technique has been used in dynamical system (plant and controller) modeling and fault detection and identification (Sartori and Antsakus, 1992). The ANNs describe the relationship between the input/output pairs so as to obtain an initial approximation to the system's behavior. The advantage that ANNs have over traditional mathematical methods is their inherent parallelism that could eventually be translated into rules to numerical values in a highly parallel manner. Unlike most AI techniques, ANNs tend toward system robustness, resistance to noise, contradictions and incompleteness, and have the possibility of automated learning.

But alone, even artificial neural network systems cannot solve the problem of completely and effectively modelling open systems such as those found in CIM. Open systems are always subject to communications and constraints from outside. They involve continuous change, decentralized decision making, and perpetual inconsistency among knowledge sources. One technique or paradigm on its own is not sufficient to model this complex phenomenon. Different approaches must be utilized to take advantage of several intelligent methodologies and exploit the complementary properties of classical AI and artificial neural networks (among other techniques) to achieve the common goal of an intelligent, dynamic system.

REFERENCES

Ackley, D.H., Hinton, G.E. and Sejnowski, T.J. (1985) A learning algorithm for Boltzmann machines. *Cognitive Science*, **9**, 147–69

Adeli, H. (ed) (1990) *Knowledge Engineering: Volume II Applications*, McGraw-Hill, New York.

Amari, S. (1967) A theory of adaptive pattern classifiers. *IEEE Transactions Electronic Computers*, EC–16, pp. 279–307.

Barr, A. and Feigenbaum, E.A. (1981) *The Handbook of Artificial Intelligence*, Morgan Kaufmann, Los Altos, CA.

Bengio, Y., Cardin, R., De Mori, R. and Normandin, Y. (1990) A hybrid coder for hidden Markov models using a recurrent neural network. *Proceedings IEEE ICASSP*, 537–40

Carpenter, G.A. and Grossberg, S. (1988) The ART of adaptive pattern recognition by a self-organizing neural network. *Computer*, March, 77–88

Chang, T.C., Anderson, D.C. and Mitchell, O.R. (1988) *QTC – An Integrated Design/Manufacturing Vision Inspection System for Prismatic Port.* Proceedings of the ASME 1988 Computers in Engineering Conference, Vol. 1, July 31–August 3, 417–26

Erman, D.L., Hayes-Roth, F., Lesser, V.R. and Reddy, D.R. (1980) The HEARSAY-II speech understanding system: Integrating knowledge to resolve uncertainty. *ACM Computing Survey*, **12**, 213–53

Hecht-Nielsen, R.(1987) Counter-Propagation Networks. *IEEE First International Conference on Neural Networks*, **II**, 19–32

Hewitt, C. (1985) The challenge of open systems. *Byte*, **10**(4), 223–42

Holland, J.H. (1975) *Adaptation in natural and artificial systems*, Basic Books, New York.

Holland, J.H., Holyoak, K.J., Nisbett, R.E. and Thagard, P.R. (1987) *Induction: Processes of Interence, Learning, and Discovery*, MIT Press, Cambridge, MA.

Hopfield, J.J. (1982) Neural networks and physical systems with emergent collective computational abilities. *Proceedings of the National Academy of Sciences*, USA, **79**, 2554–8.

Hopfield, J.J. and Tank, D.W. (1985) Neural computation of decisions in optimization problems. *Biological Cybernetics*, **52**, 141–52.

Horne, B., Jamshidi, M. and Vadiee, N. (1990) Neural networks in robotics: a survey. *Journal of Intelligent and Robotic Systems*, **3**, 51–66.

Jorgenson, C.C. (1987) Neural network representation of sensor graphs in autonomous robot path planning. *IEEE Conf. on Neural Networks*, **4**, 507–16.

Kandel, A. and Langholz, G. (eds) (1992) *Hybrid Architectures for Intelligent Systems*, CRC Press, Boca Raton.

Kohonen, T. (1988) Statistical pattern recognition with neural networks: Benchmark studies. *Proceedings of the Second Annual IEEE International Conference on Neural Networks*, **1**.

Kosko, B. (1987) Bidirectional associative memories. *IEEE Trans. on Systems, Man, and Cybernetics*, SMC–17.

Kosko, B. (1992) *Neural Networks and Fuzzy Systems: A Dynamical Systems Approach to Machine Intelligence*, Prentice Hall, Englewood Cliffs.

Liu, H. and Bekey, G.A. (1988) Building a generic architecture for robot hand control. *IEEE Conf. on Neural Networks*, **II**, 567–74.

McAndless, E., Stacey, D., Rueb, K. and Wong, A. (1991) A hybrid neural network/rule-based approach to a real-time machine vision system, in *Intelligent Engineering Systems Through Artificial Neural Networks* (eds Dagli, Kumara and Shin), pp. 909–14

Minsky, M.L. and Papert, S.S. (1969) *Perceptrons*, MIT Press, Cambridge, MA.

Narendra, K.S. and Mukhopadhyay, S. (1992) Intelligent control using neural networks. *IEEE Control Systems*, April, 11–18

Newell, A. and Simon, H. (1972) *Human Problem Solving*, Prentice Hall, Englewood Cliffs.

Nii, Penny H. (1986) Blackboard systems: The blackboard model of problem solving and the evolution of blackboard architectures. *The AI Magazine*, Summer, 38–53

Nii, Penny H. (1986a) Blackboard systems: blackboard application systems, blackboard systems from a knowledge engineering perspective. *The AI Magazine*, August, 82–106.

Parker, D.B. (1982) Learning-Logic, Invention report S81-64, File 1, Office of Technology Licensing, Stanford University, October.

Rumelhart D.E., Hinton G.E., and Williams R.J. (1985) *Learning internal representations by error propagation*, ICS Report 8506, Institute for Cognitive Science, University of California at San Diego, September.

Sartori, M.A. and Antsaklis, P.J. (1992) Implementations of learning control systems using neural networks. *IEEE Control Systems*. April, 49–57.

Serra, R. and Zanarini, G. (1990) *Complex Systems and Cognitive Processes*, Springer-Verlag, Berlin.

Seshadri, V. (1988) A neural network architecture for robot path planning in *Proc. Second International Symp. on Robotics and Manufacturing: Research, Foundation, and Applications*, ASME Press, pp. 249–56.

Sherald, M. (1991) Mission possible if you combine neural networks and expert systems. *PC AI*, May/June, 56–57

Simpson, P.K. (1990) *Artificial Neural Systems: Foundations, Paradigms, Applications, and Implementations*, Pergamon Press, New York.

Tsutsumi, K. and Matsumoto, H. (1987) Neural computation and learning strategy for manipulator position control. *IEEE Conf. on Neural Networks*, **4**, 525–34.

Tsutsumi, K., Katayama, K. and Matsumoto, H. (1988) Neural computation for controlling the configuration of 2-dimensional truss structure. *IEEE Conf. on Neural Networks*, **II**, 575–86.

Van den Bout, D.E. and Miller, T.K. (1988) A travelling salesman objective function that works. *Proceedings of the IEEE First International Conference on Neural Networks*, San Diego, CA, **2**, 299–303.

Werbos, P. (1974) Beyond Regression: New Tools for Prediction and Analysis in the Behavioral Sciences, Ph.D. thesis, Harvard University.

Widrow, B. and Lehr, M.A. (1990) Thirty years of adaptive neural networks: Perceptron, madaline, and backpropagation. *Proceedings of the IEEE*, **78**(9), September, 1415–42.

Wilson, G.V. and Pawley, G.S. (1988) On the stability of the travelling salesman problem algorithm of Hopfield and Tank. *Biological Cybernetics*, **58**, 63–70.

CHAPTER 3

Basic artificial neural network architectures

Cihan H. Dagli and Pipatpong Poshyanonda
Engineering Management Department, University of Missouri-Rolla, Rolla, Missouri 65401.

Different artificial neural network paradigms and architectures are proposed for the solution of manufacturing problems in the literature. A quick review of this literature highlights the fact that two paradigms (backpropagation and adaptive resonance theory (ART)) are used extensively to create intelligent manufacturing systems that can adapt to changing conditions. However, generally, these paradigms are not discussed in detail in papers if they are used in specific manufacturing application, so this chapter aims to fill this gap. After an introduction of basic artificial neural networks, three classical artificial neural network architectures – perceptron, backpropagation and adaptive resonance theory (ART1) – are examined.

3.1 BASIC CONCEPTS

The **neuron** is the fundamental unit of the biological nervous system. It is a simple processing unit which receives and processes the signal from other neurons through its input paths called **dendrites**. An activity of a neuron is an 'all-or-none' process. If the combined signal is strong enough, it generates the output signal to its output path (called **axon**) which splits up and connects to other neurons' input paths through a junction referred to as a **synapse**. The amount of signals transferred depends on the synaptic strength of the junction which is chemical in nature. This synaptic strength is found to be modified during the learning process of the brain, therefore, it can be considered as a

Artificial Neural Networks for Intelligent Manufacturing. Edited by Cihan H. Dagli.
Published in 1994 by Chapman & Hall, London. ISBN 0 412 48050 6

memory unit of each interconnection. Artificial neural networks are mathematical representation of this biological process as far as it is understood today. In the mathematical model, neurons are represented as processing elements (PEs), input paths are defined as interconnections, the process combining signals and generating the output of neurons is modeled through a transfer function, synaptic strength of each connection is represented as weight and the change in synaptic strength is defined as learning.

Artificial neural network architecture defines a given configuration of these basic elements, while the artificial neural network paradigm is characterized through learning algorithms that determines the changes in weights.

3.1.1 Processing elements

A processing element (PE) is a simple device that approximates the function of a biological neuron. Although biological details could not be represented by the computer model, PEs retain enough structure to work like a biological neuron. In artificial neural network models, PEs handle many basic functions such as input signals evaluation, summation of signals and comparison to a threshold value to determine the output value (Fig. 3.1). Implementation of a neuron could be a one-bit processor, a floating point processor, or an operation amplifier depending on the network model selected. Furthermore, several PEs can carry out their computations simultaneously due to the parallel structure of the system.

Inputs and outputs

Each PE can receive many input signals simultaneously, but there is only one output signal (either positive or negative) which depends on the input signals, weights and threshold for that PE. Some network models have an extra input

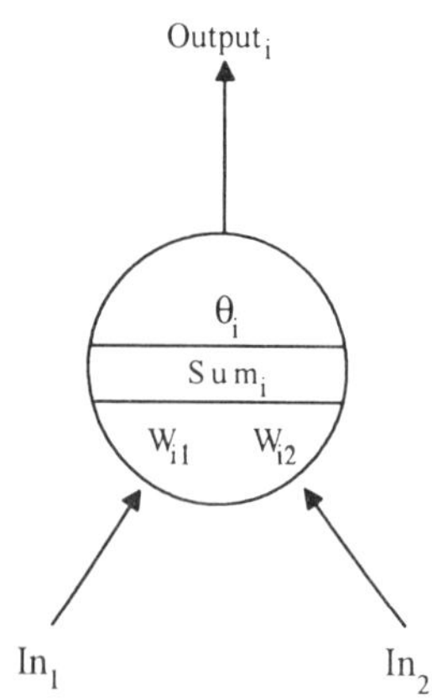

Fig. 3.1 A processing element.

called a **bias term**, or **forcing term** for other influences from outside the network. Three types of PEs can be classified based on their inputs and outputs: input, output and hidden PEs. **Input PEs** are those that receive input from external sources to the system, while **output PEs** are those that send the signals out of the system. Hidden processing elements have their inputs and outputs within the system. They are shown in Fig. 3.11 in reference to a three-layer perceptron.

Transfer function

This is the mechanism of translating input signals to an output signal for each PE. The first step is to create the net input value for a PE. Some inputs may be more important than the others, so there is a corresponding weight with each input presented to a PE. These weights are strength of connection between PEs, and normally, they are represented in terms of vectors such as $\mathbf{w}_i = (w_{i1}, \ldots, w_{in})$. After a PE receives all of its input signals, it computes the total input being received from its input paths according to these weights. The most commonly used method is to use the summation function i.e.

$$\mathrm{sum}_i = \sum_{i=1}^{n} w_{ij} \mathrm{In}_j$$

where sum_i is the net weighted input received by a PE_i, w_{ij} is the weight value on the interconnection from PE_j to PE_i and In_j is the input signal from PE_j.

The second step of the translation is to create the activation value for each PE through activation function. The activation function can be a simple linear function or more complex e.g. a sigmoid function. Other functions that can be used include threshold logic, step function, binary step function, or hyperbolic tangent function. The activation function is continuous and non-linear, and the most commonly used is the sigmoid function, an S-shaped curve as shown in Fig. 3.2, which is bounded within a specific range, such as (0, 1) and has a continuous first derivative.

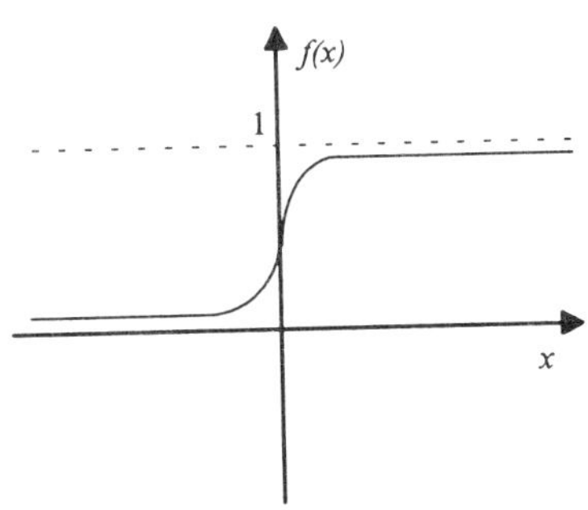

Fig. 3.2 The sigmoid function: $f(x) = \dfrac{1}{1 + e^{-x}}$.

We note that, if as in Fig. 3.2

$$f(x) = \frac{1}{1 + e^{-x}}$$

then

$$\frac{df(x)}{dx} = \frac{1}{(1 + e^{-x})}\left(1 - \frac{1}{(1 + e^{-x})}\right)$$

i.e.

$$f'(x) = f(x)(1 - f(x))$$

The transfer function is used to convert the activation value for each PE into its output value. Most networks use the threshold value in determining the output of the PE.

$$\text{Output}_i = \begin{Bmatrix} f(\text{sum}_i) & \text{if} \quad \text{sum}_i > \theta_i \\ 0 & \text{otherwise} \end{Bmatrix}$$

where θ_i is threshold value corresponding to each PE. In other words, the PE will generate the output signal if the net weighted input is greater than its threshold value, otherwise PE will not generate any signal, or even an inhibitory signal.

3.1.2 Interconnections

The interconnection scheme is a part of the network architecture definition, which propagates a signal from one PE to another, or even to PE itself. It is unidirectional and there is a weight value assigned to each connection, these weight values forming the memory of the network. Excitory connections, usually positive signals, enhance the PE's input value, while inhibitory connections (typically a negative signal) decrease the PE's input value.

There are three different connection schemes that connect PEs in a network.

- Intrafield connections connects PEs in the same layer.
  Interfield connections connect between PEs in different layers.
- Recurrent connections loop and connect back to the PE itself.

If the information flows in one direction, the interconnections are called feedforward. Feedback lets the information flow among PEs in either direction and/or recursively. In Fig. 3.11 feedforward interfield connections are given for a back propagation neural network architecture.

3.1.3 Learning rules

The learning rule is the last attribute used to define an artificial neural network paradigm. Learning is considered as a rate of change in the weights, or memory of the network. Two learning categories are defined; supervised and unsupervised. **Supervised learning** depends on training data set that contains input vectors and corresponding output vectors. These vector pairs are used to determine the non-linear mappings that exist on the data set. The weights of the network (which represents the memory) is updated based on the degree of network error made in successive presentation of the training vector pairs to the network until the rate of change in weights is close to zero indicating the fact that total error is also approaching to zero. This type of learning is used in artificial neural network architectures that need to be trained off-line.

Unsupervised learning, incorporates no external teacher and relies upon local information and internal control. The network self-organizes using internal rules. This is also known as **competitive learning.**

These basic elements are sufficient to define various neural network paradigms. It is not difficult to observe a sharp increase in artificial neural network paradigms proposed in literature during the last decade. This is primarily due to excellent generalization capability of backpropagation paradigm which provided successful real-life applications of the technology. It is not feasible to cover all these paradigms in this chapter. Interested readers should refer to Aleksander (1989), Aleksander and Morton (1991), Blum (1992), Caudill and Butler (1992), Dayhoff (1990), Eberhart and Dobbins (1990), Freeman and Skapura (1991), Khanna (1990), Kosko (1992a, 1992b), McClelland and Rumelhart (1989), Nelson and Illingworth (1991), Rumelhart *et al.* (1986), Simpson (1990), Wesserman (1989) and Zurada (1992) for detailed information on these paradigms. In this chapter, three paradigms namely perceptron, which is a linear separator, backpropagation that uses supervised learning and adaptive resonance theory (ART) that uses unsupervised learning is covered.

3.2 PERCEPTRON

Perceptron (as proposed by McCullock and Pitts in 1943) is the earliest artificial neural network paradigm. The architecture of the perceptron is simple: it uses the feedforward scheme, usually two layers, with supervised learning as shown in Fig. 3.3.

The PE computes the net weighted sum of input signals and then compares to the threshold value θ. If the computed value is greater than the threshold value the output is 1; otherwise is -1. The transfer function of the perceptron is

$$\text{sum}_i = \sum_{i=1}^{n} w_{ij}\,\text{In}_j \qquad \text{Output}_i = \begin{Bmatrix} +1 & \text{if} & \text{sum}_i \geqslant \theta \\ -1 & \text{if} & \text{sum}_i < \theta \end{Bmatrix}$$

A major characteristic of perceptron is the linear separation due to its threshold function. Figure 3.4 shows regions that can be identified by perceptron easily.

To be able to divide the area into two regions for the problem in Fig. 3.4(a), only one PE of perceptron is required, since the whole area is divided into two

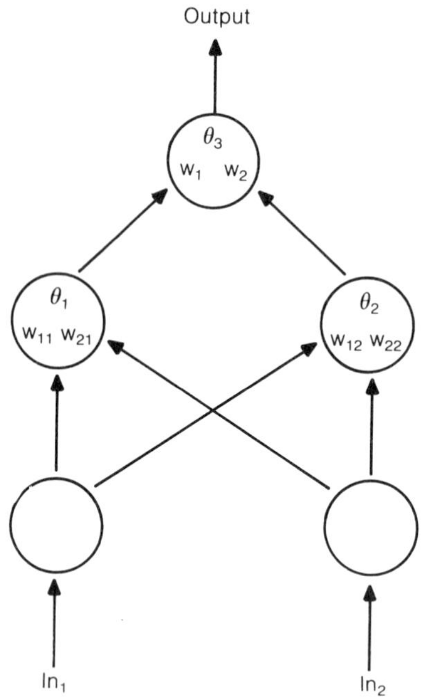

Fig. 3.3 An elementary perceptron network.

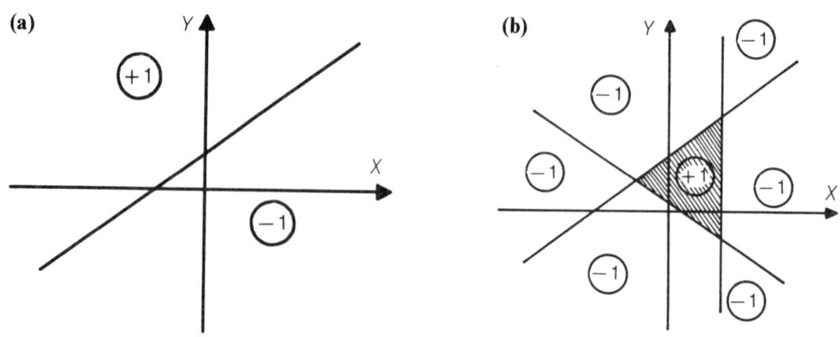

Fig. 3.4 (a) Two separated regions defined by a perceptron; (b) a selected region defined by the intersection of areas created by perceptrons.

separate regions by a single line. However, the threshold area Fig. 3.4(b) is formed by many lines, thus, we need more than one perceptron to generate them – a solution to this problem is shown in Fig. 3.5. In Fig. 3.5, θ_1, θ_2 and θ_3 are the threshold values and w_{1j}, w_{2j} and w_{3j} are the interconnection weights for processing elements 1, 2 and 3 respectively. Each of the processing elements are connected to the third layer processing element through equal weights of one, and a threshold value of 2.5 to be able to perform and operation. X and Y represents x–y coordinates of the point selected from the

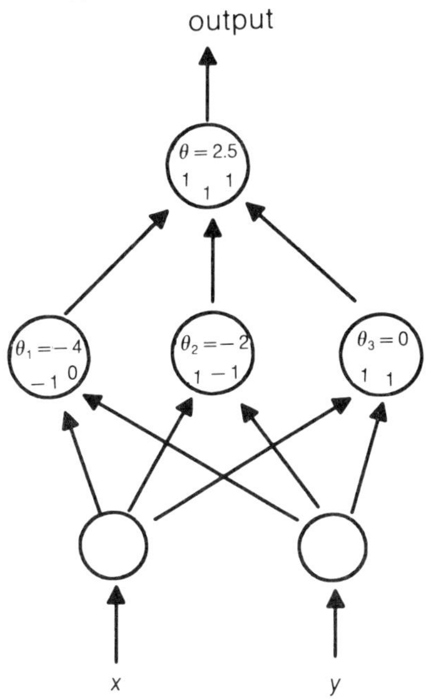

Fig. 3.5 A solution perceptron network for the problem given by Fig. 4(b).

regions specified in Fig. 3.4(b). Each perceptron separates the area into two regions by a line, but the solution is the intersection of these areas. Therefore, one more perceptron is needed to combine the outputs of these PEs to identify the hatched area. This special perceptron combines the outputs from other perceptrons with the corresponding weight of 1, and the threshold value of $n - 0.5$.

The perceptron is able to perform an exclusive OR operation, e.g. in identifying the truth table (Table 3.1) for the regions given in Fig. 3.6.

Let us identify the region A using a three-layer perceptron. Hence we need to compute the weights (w_{ij}s) associated with the interconnections between the input units which represents X and Y coordinates of the points given in Fig. 3.4(a) and a corresponding threshold value (θ_i) to be able to make the distinction between two regions identified as '+1' and '−1' in the figure. These

Table 3.1 Truth table

Point 1	*Point 2*	*Outcome*
In A	In B	0
In A	Out B	1
Out A	In B	1
Out A	Out B	0

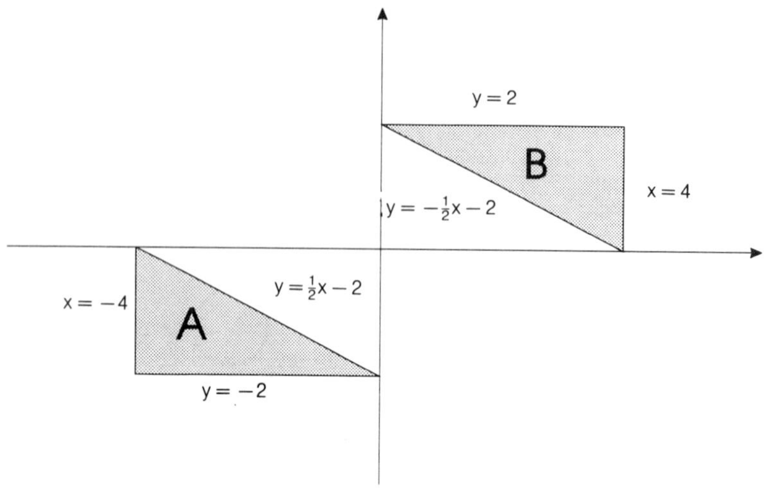

Fig. 3.6 Perceptron example.

values also correspond to output values of the perceptron.

$$\begin{array}{llll} y & < -\frac{1}{2}x - 2 & x > -4 & y > -2 \\ \frac{1}{2}x + y + 2 & < 0 & x + 4 > 0 & y + 2 > 0 \\ x + 2y + 4 & < 0 & & \\ -x - 2y - 4 & > 0 & & \end{array}$$

$$w^1_{11} = -1, w^1_{21} = -2, \theta^1_1 = 4, \quad w^1_{12} = 1, w^1_{22} = 0, \theta^1_2 = -4$$
$$w^1_{13} = 0, w^1_{23} = 1, \theta^1_3 = -2, \quad w^2_{11} = w^2_{21} = w^2_{31} = 1, \theta^2_1 = 2.5$$

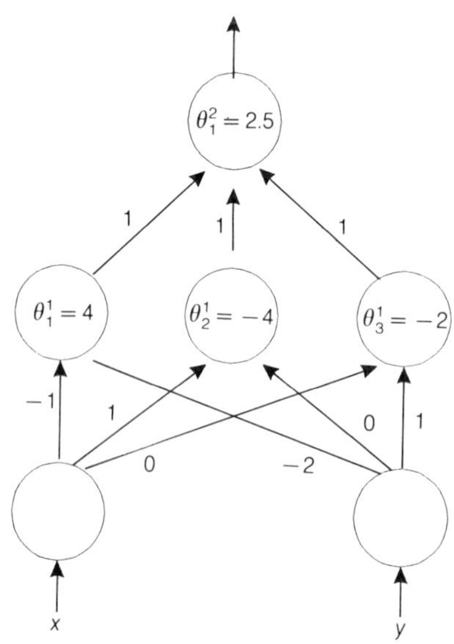

Fig. 3.7 Perceptron for region A.

Similarly for region B:

$$y > -\tfrac{1}{2}x + 2 \qquad x < 4 \qquad y < 2$$
$$y + \tfrac{1}{2}x - 2 > 0 \qquad -x + 4 > 0 \qquad -y + 2 > 0$$
$$2y + x - 4 > 0$$
$$x + 2y - 4 > 0$$

$$w^1_{11} = 1, w^1_{21} = 2, \theta^1_1 = 4, \quad w^1_{12} = -1, w^1_{22} = 0, \theta^1_2 = -4$$
$$w^1_{13} = 0, w^1_{23} = -1, \theta^1_3 = -2, \quad w^2_{11} = w^2_{21} = w^2_{31} = 1, \theta^2_1 = 2.5$$

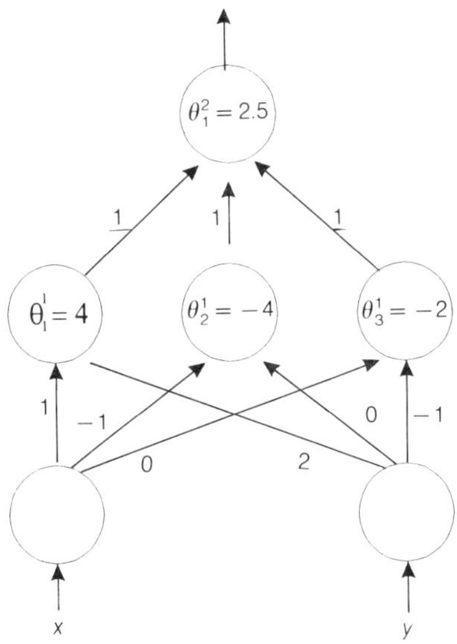

Fig. 3.8 Perceptron for region B.

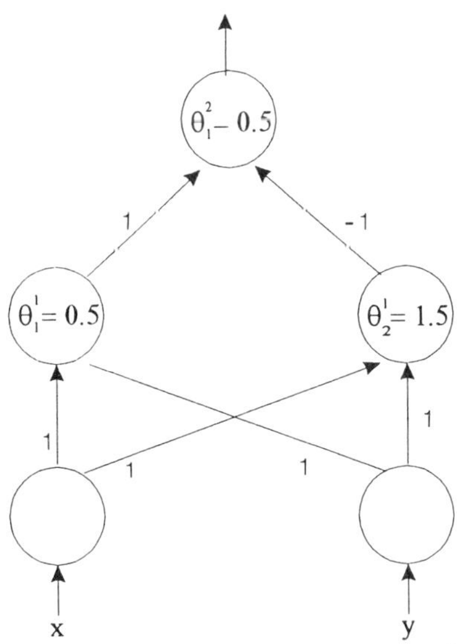

Fig. 3.9 Perceptron for Exclusive-OR.

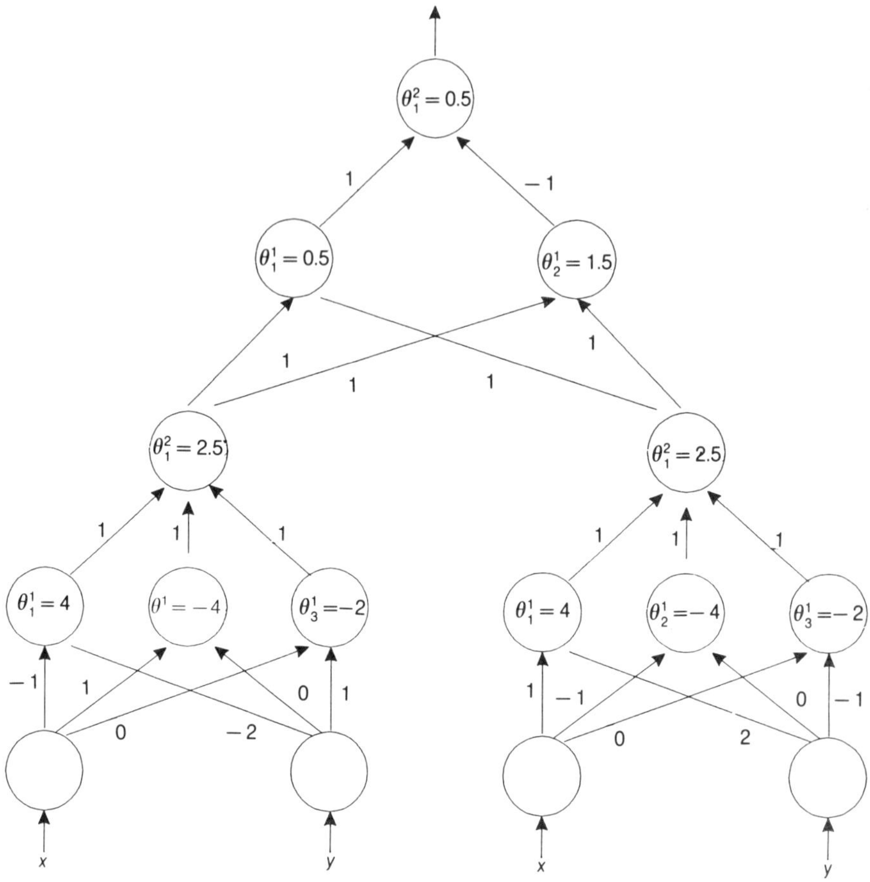

Fig. 3.10 Combined perceptrons for identifying outcomes.

To be able to produce the result as in Table 3.1 which is an exclusive-OR operation, we need another three-layer perceptron. The following weights and threshold values can perform this function which is illustrated in Fig. 3.9.

$$w_{11}^1 = 1, w_{12}^1 = 1, \theta_1^1 = 0.5$$
$$w_{12}^1 = 1, w_{22}^1 = 1, \theta_2^1 = 1.5$$
$$w_{11}^2 = 1, w_{21}^2 = -1, \theta_1^2 = 0.5$$

Combination of these three perceptrons can identify the outcomes given in the truth table above.

3.3 BACKPROPAGATION

A problem arises with the perceptron under non-linear separation of classes, since it uses a step function as a threshold, which brings a discontinuity to the network. The linear separation characteristic of perceptron thus restricts its use for non-linear mapping. The backpropagation paradigm is proposed to overcome this difficulty; it is also a feedforward network, and uses a supervised learning scheme with a different threshold function and learning rule. The elementary architecture of the back propagation network has three layers, which are fully connected with feedforward processing (Fig. 3.11). Although this architecture is common, it is possible to have more than three layers (more than one hidden layer), and restricted connections between two subsequent layers.

In supervised learning, the knowledge of the desired output for each PE is sufficient to calculate the error produced by the network. However, in the case of more than two layers it is not possible to identify the desired output of the PEs of the hidden layer(s). The errors produced in the output layer need to be propagated back to these hidden layers. However, each unit in the hidden layers receive only a portion of the error produced in the output layer based on the contribution of the PE to the output layer. This propagation process provides a means to update the weights and train the network and is known as the **generalized delta rule**. Addition of a non-linear transfer function to a multi-layer perceptron, and adaptation of the generalized delta rule provide generalization capability to the network architecture and the ability of non-linear separation. Interested readers should refer to Aleksander and Morton (1991), Blum (1992), Caudill and Butler (1992), Dayhoff (1990), Freeman and Skapura (1991), Zurada (1992) for a detailed description of the backpropagation paradigm. In the following sections, the generalized delta rule is described and backpropagation training is demonstrated on an example.

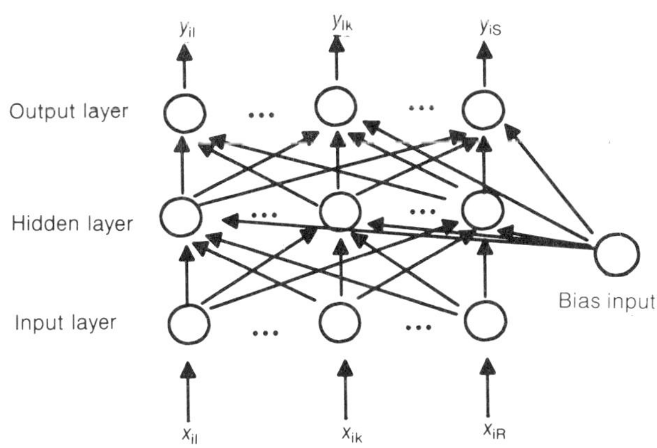

Fig. 3.11 The three-layer backpropagation neural network architecture.

3.3.1 Generalized delta rule

Let M be the number of pattern pairs presented to the network, R be the number of input layer nodes and S be the number of output layer nodes. Pattern pairs $(\mathbf{x}_1, \mathbf{y}_1)\cdots(\mathbf{x}_M, \mathbf{y}_M)$ are presented as inputs to the network to determine the non-linear mapping between x_i and y_i vectors where $\mathbf{x}_i = (x_{i1}, \ldots x_{iR})$, $\mathbf{y}_i = (y_{i1}, \ldots, y_{iS})$ for $i = 1, \ldots, M$.

Let L be the number of layers in the network, and each layer be denoted by l.

$$\begin{aligned} l &= 0 && : \text{ input layer} \\ l &= 1, \ldots, L-1 && : \text{ hidden layer(s)} \\ l &= L && : \text{ output layer} \end{aligned}$$

There are K_l nodes where $l = 1, \ldots, L$ in each layer. Inputs to the kth processing element in layer lth are In_{ij}^l for $j = 1, \ldots, K_{l-1}$. These inputs are also the output values from the previous layer PEs which are called Out_{ij}^{l-1}. Together with the threshold value, θ_k^l, the net input value, sum_{ik}^l, for the kth processing element in layer l is

$$\text{sum}_{ik}^l = \sum_{j=1}^{K_{l-1}} w_{kj}^l \, \text{In}_{ij}^l + \theta_k^l$$

where w_{kj}^l are the weights of connections between processing element j and k. The output produced by the kth processing element in layer lth is obtained by passing the value of sum through a transfer function, $f_k^l(x)$.

$$\text{Out}_{ik}^l = f_{ik}^l(\text{sum}_{ik}^l)$$

The error at the output layer, L, of the network is the difference between the computed and the desired outputs, which can be represented as

$$\text{Err}_i = \frac{1}{2} \sum_{k=1}^{K_L} (y_{ik} - \text{Out}_{ik}^L)^2$$

and this error is minimized in determining the weight updates. To demonstrate how the weights are updated, let us assume a network of two hidden layers ($L = 3$). The factor $\frac{1}{2}$ is used to help in differentiating the equation.

$$\begin{aligned} \text{Err}_i &= \frac{1}{2} \sum_{k=1}^{K_3} (y_{ik} - \text{Out}_{ik}^3)^2 \\ &= \frac{1}{2} \sum_{k=1}^{K_3} (y_{ik} - f_k^3(\text{sum}_{ik}^3))^2 \\ &= \frac{1}{2} \sum_{k=1}^{K_3} \left(y_{ik} - f_k^3 \left(\sum_{j=1}^{K_2} w_{kj}^3 \text{Out}_{ij}^2 + \theta_k^3 \right) \right)^2 \\ &= \frac{1}{2} \sum_{k=1}^{K_3} \left(y_{ik} - f_k^3 \left(\sum_{j=1}^{K_2} w_{kj}^3 f_j^2(\text{sum}_{ij}^2) + \theta_k^3 \right) \right)^2 \end{aligned}$$

$$=\frac{1}{2}\sum_{k=1}^{K_3}\left(y_{ik}-f_k^3\left(\sum_{j=1}^{K_2}w_{kj}^3f_j^2\left(\sum_{u=1}^{K}w_{ju}^2\text{Out}_{iu}^1+\theta_j^2\right)+\theta_k^3\right)\right)^2$$

$$=\tfrac{1}{2}\sum_{k=1}^{K_3}\left(y_{ik}-f_k^3\left(\sum_{j=1}^{K_2}w_{kj}^3f_j^2\left(\sum_{u=1}^{K}w_{ju}^2f_u^1(\text{sum}_{iu}^1)+\theta_j^2\right)+\theta_k^3\right)\right)^2$$

$$=\frac{1}{2}\sum_{k=1}^{K_3}\left(y_{ik}-f_k^3\left(\sum_{j=1}^{K_2}w_{kj}^3f_j^2\left(\sum_{u=1}^{K}w_{ju}^2f_u^1\left(\sum_{v=1}^{K_0}w_{jv}^1\text{Out}_{iv}^0+\theta_u^1\right)+\theta_j^2\right)+\theta_k^3\right)\right)^2$$

where $\text{Out}_{iv}^0=\text{Int}_{iv}^1=x_{iv}$, then we can use this equation to calculate the gradient of the error Err_i with respect to the weights, w_{kj}^l.
For the output layer (3):

$$\frac{\partial \text{Err}_i}{\partial w_{kj}^3}=-(y_{ik}-\text{Out}_{ik}^3)\frac{\partial \text{Out}_{ik}^3}{\partial \text{sum}_{ik}^3}\frac{\partial \text{sum}_{ik}^3}{\partial w_{kj}^3}$$
$$=-(y_{ik}-\text{Out}_{ik}^3)f_k^{3'}(\text{sum}_{ik}^3)\,\text{In}_{ij}^3$$

Then the rate of change in weights for a given value of learning rate α is

$$\Delta_i w_{kj}^3=\alpha(y_{ik}-\text{Out}_{ik}^3)f_k^{3'}(\text{sum}_{ik}^3)\,\text{In}_{ij}^3$$
$$=\alpha\delta_{ik}^3\,\text{In}_{ij}^3 \quad \text{where } \delta_{ik}^3=(y_{ik}-\text{Out}_{ik}^3)f_k^{3'}(\text{sum}_{ik}^3)$$

The negative sign cancels out since the direction of the change is negative gradient.

For the second hidden layer (2):

$$\frac{\partial \text{Err}_i}{\partial w_{ju}^2}=-\sum_{k=1}^{K_3}(y_{ik}-\text{Out}_{ik}^3)\frac{\partial \text{Out}_{ik}^3}{\partial \text{sum}_{ik}^3}\frac{\partial \text{sum}_{ik}^3}{\partial \text{Out}_{ij}^2}\frac{\partial \text{Out}_{ij}^2}{\partial \text{sum}_{ij}^2}\frac{\partial \text{sum}_{ij}^2}{\partial w_{ju}^2}$$
$$=-\sum_{k=1}^{K_3}(y_{ik}-\text{Out}_{ik}^3)f_k^{3'}(\text{sum}_{ik}^3)w_{kj}^3f_j^{2'}(\text{sum}_{ij}^2)\,\text{In}_{iu}^2$$

$$\Delta_i w_{ju}^2=\alpha f_j^{2'}(\text{sum}_{ij}^2)\,\text{In}_{iu}^2\sum_{k=1}^{K_3}(y_{ik}-\text{Out}_{ik}^3)f_k^{3'}(\text{sum}_{ik}^3)w_{kj}^3$$

$$=\alpha f_j^{2'}(\text{sum}_{ij}^2)\,\text{In}_{iu}^2\sum_{k=1}^{K_3}\delta_{ik}^3w_{kj}^3$$

$$=\alpha\delta_{ij}^2\,\text{In}_{iu}^2 \quad \text{where } \delta_{ij}^2=f_j^{2'}(\text{sum}_{ij}^2)\sum_{k=1}^{K_3}\delta_{ik}^3w_{kj}^3$$

For the first hidden layer (1):

$$\frac{\partial \text{Err}_i}{\partial w_{uv}}=-\sum_{k=1}^{K_3}(y_{ik}-\text{Out}_{ik}^3)\frac{\partial \text{Out}_{ik}^3}{\partial \text{sum}_{ik}^3}\frac{\partial \text{sum}_{ik}^3}{\partial \text{Out}_{ij}^2}\frac{\partial \text{Out}_{ij}^2}{\partial \text{sum}_{ij}^2}\frac{\partial \text{sum}_{ij}^2}{\partial \text{Out}_{iu}}\frac{\partial \text{Out}_{iu}}{\partial \text{sum}_{iu}}\frac{\partial \text{sum}_{iu}}{\partial w_{uv}}$$
$$=-\sum_{k=1}^{K_3}(y_{ik}-\text{Out}_{ik}^3)f_k^{3'}(\text{sum}_{ik}^3)w_{kj}^3\cdot\sum_{j=1}^{K_2}f_j^{2'}(\text{sum}_{ij}^2)w_{ju}^2\cdot f_u^{1'}(\text{sum}_{iu}^1)\cdot x_{iv}$$

$$\Delta_i w_{uv}^1 = \alpha f_u^{1'}(\text{sum}_{iu}^1) x_{iv} \sum_{k=1}^{K_3} \delta_{ik}^3 w_{kj}^3 \cdot \sum_{j=1}^{K_2} f_j^{2'}(\text{sum}_{ij}^2) w_{ju}^2$$

$$= \alpha f_u^{1'}(\text{sum}_{iu}^1) x_{iv} \sum_{j=1}^{K_2} \delta_{ij}^2 w_{ju}^2$$

$$= \alpha \delta_{iu}^1 x_{iv} \quad \text{where } \delta_{iu}^1 = f_u^{1'}(\text{sum}_{iu}^1) \sum_{j=1}^{K_2} \delta_{ij}^2 w_{ju}^2$$

$\Delta_i w_{kj}^3$, $\Delta_i w_{ju}^2$ and $\Delta_i w_{uv}^1$ are used to update the weights during the training process. Training is terminated when the values of these variables are close to zero. The process of training can be summarized as follows:

1. Filter the input pattern to the network.
2. Calculate output values for each processing elements in the network by using

$$\text{Out}_{ik}^l = f_{ik}^l(\text{sum}_{ik}^l) \quad \text{where sum}_{ik}^l = \sum_{j=1}^{K_l} w_{kj}^l \, \text{In}_{ij}^l + \theta_k^l$$

3. Calculate the error produced by the network

$$\text{Err}_i = \sum_{k=1}^{K_L} (y_{ik} - \text{Out}_{ik}^L)^2$$

4. For the output layer, $l = L$, calculate the values of weight changes using

$$\Delta_i w_{kj}^L = \alpha \delta_{ik}^L \, \text{In}_{ij}^L \quad \text{where } \delta_{ik}^L = (y_{ik} - \text{Out}_{ik}^L) f_k^{L'}(\text{sum}_{ik}^L)$$

5. For the hidden layers, $l = 1, \ldots, L-1$, calculate the values of weight changes using

$$\Delta_i w_{ju}^l = \alpha \delta_{ij}^l \, \text{In}_{iu}^l \quad \text{where } \delta_{ij}^l = f_j^{l'}(\text{sum}_{ij}^l) \sum_{k=1}^{K_{l+1}} \delta_{ik}^{l+1} w_{kj}^{l+1}$$

6. Update all weights according to their values of changes.

$$w_{ju}^l = w_{ju}^l + \Delta_i w_{ju}^l \quad \text{for } l = 1, \ldots, L$$

7. Repeat these steps until the error produced by the network is sufficiently low.

Let us demonstrate this process with an example. Consider the four-layer perceptron given in Fig. 3.12.

A single binary training pair, (1010, 1) is used to demonstrate the mechanism of the weights update with learning rate of 0.5. Different transfer functions are used for different nodes. These functions are summarized in Table 3.2.

Before we can start with calculations, the binary pattern pair needs to be changed to an analog number that can reflect these binary numbers instead of 0.0 and 1.0. We use 0.1 to represent '0', and 0.9 for '1'. Furthermore, we need to find the derivatives of the transfer functions, as given in Table 3.3.

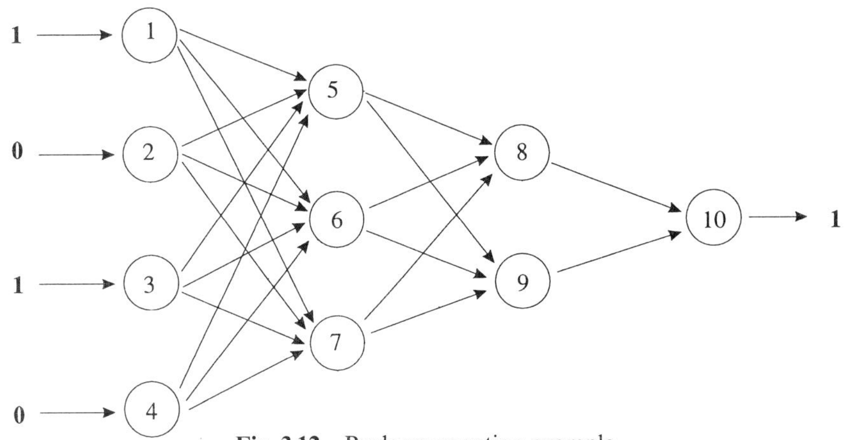

Fig. 3.12 Back propagation example.

Table 3.2 Transfer functions

Node	*Transfer function*
5	$(1+e^{-3x})^{-1}$
6	$(1+e^{-4x})^{-1}$
7	$(1+e^{-5x})^{-1}$
8	$(1+e^{-6x})^{-1}$
9	$(1+e^{-7x})^{-1}$
10	$(1+e^{-8x})^{-1}$

Table 3.3 Derivatives of the transfer functions

Node	*Transfer function*	*Derivation of the function*
5	$(1+e^{-3x})^{-1}$	$\dfrac{3e^{-3x}}{(1+e^{-3x})^2}$
6	$(1+e^{-4x})^{-1}$	$\dfrac{4e^{-4x}}{(1+e^{-4x})^2}$
7	$(1+e^{-5x})^{-1}$	$\dfrac{5e^{-5x}}{(1+e^{-5x})^2}$
8	$(1+e^{-6x})^{-1}$	$\dfrac{6e^{-6x}}{(1+e^{-6x})^2}$
9	$(1+e^{-7x})^{-1}$	$\dfrac{7e^{-7x}}{(1+e^{-7x})^2}$
10	$(1+e^{-8x})^{-1}$	$\dfrac{8e^{-8x}}{(1+e^{-8x})^2}$

1. Randomize the values of weights, $\mathbf{w}^l$ and threshold, θ^l, between the value of -0.5 and 0.5

$$\mathbf{w}^1 = \begin{bmatrix} 0.182 & -0.100 & -0.088 \\ 0.376 & -0.391 & -0.321 \\ 0.206 & 0.416 & 0.055 \\ -0.233 & 0.442 & -0.192 \end{bmatrix} \qquad \boldsymbol{\theta}^1 = [0.282 \quad 0.107 \quad -0.155]$$

$$\mathbf{w}^2 = \begin{bmatrix} 0.058 & -0.249 \\ -0.281 & -0.332 \\ -0.352 & 0.471 \end{bmatrix} \qquad \boldsymbol{\theta}^2 = [-0.187 \quad 0.478]$$

$$\mathbf{w}^3 = \begin{bmatrix} 0.231 \\ -0.399 \end{bmatrix} \qquad \boldsymbol{\theta}^3 = [0.399]$$

2. Apply an input pattern to the network.

$$\mathbf{I} = [0.9 \quad 0.1 \quad 0.9 \quad 0.1]$$

3. Calculate the output of the first hidden layer ($l = 1$),

$$\text{Output}^1 = [0.871 \quad 0.831 \quad 0.234]$$

4. Calculate the output of the second hidden layer ($l = 2$),

$$\text{Output}^2 = [0.062 \quad 0.661]$$

5. Calculate the output of the output layer ($l = 3$),

$$\text{Output}^3 = [0.768]$$

6. Calculate δ^3 using the error value, the difference between the desired output and the network output produced, and the derivatives of the transfer functions in $l = 3$,

$$\delta^3 = [0.188]$$

7. Calculate $\Delta\mathbf{w}^3$, weight changes in $l = 3$,

$$\Delta\mathbf{w}^3 = \begin{bmatrix} 0.058 \\ 0.0623 \end{bmatrix}$$

8. Calculate δ^2 using $\mathbf{w}^3$ and the derivative of the transfer functions in $l = 2$,

$$\delta^2 = [0.0152 \quad -0.1179]$$

9. Calculate $\Delta\mathbf{w}^2$, weight changes in $l = 2$,

$$\Delta\mathbf{w}^2 = \begin{bmatrix} 0.0066 & -0.0515 \\ 0.0063 & -0.0489 \\ 0.0018 & -0.0139 \end{bmatrix}$$

10. Calculate δ^1 using $\mathbf{w}^2$ and the derivative of the transfer functions in $l = 1$,

$$\delta^1 = [0.0100 \quad 0.0197 \quad -0.0547]$$

11. Calculate $\Delta\mathbf{w}^1$, weight change in $l = 1$,

$$\Delta\mathbf{w}^1 = \begin{bmatrix} 0.0045 & 0.0089 & -0.0246 \\ 0.0005 & 0.0010 & -0.0027 \\ 0.0045 & 0.0089 & -0.0246 \\ 0.0005 & 0.0010 & -0.0027 \end{bmatrix}$$

12. Update the weight matrices.

$$\mathbf{w}^1 = \begin{bmatrix} 0.1865 & -0.0911 & -0.1126 \\ 0.3765 & -0.3900 & -0.3237 \\ 0.2105 & 0.4249 & 0.0796 \\ -0.2325 & 0.4430 & -0.1947 \end{bmatrix}$$

$$\mathbf{w}^2 = \begin{bmatrix} 0.0646 & -0.3005 \\ -0.2747 & -0.3809 \\ -0.3502 & 0.4571 \end{bmatrix}$$

$$\mathbf{w}^3 = \begin{bmatrix} 0.2368 \\ -0.3367 \end{bmatrix}$$

13. Apply an input pattern to the network and calculate the new output value.

This process continues until sufficiently low error value is obtained for a given training set.

3.3.2 Implementation consideration

Many points need to be stressed in backpropagation implementation. The major consideration is the architecture of the network. The number of input and output layer nodes depend on representation code selected for a specific application. There are no constraints about the number of hidden layer(s); a network can have only one, or as many hidden layers as selected. However, there is no evidence that the network with more hidden layers can perform better. In addition, each of the hidden layer(s) can have as many nodes as required. If the performance of backpropagation networks is compared based on hidden layers criterion, the networks with larger hidden layer(s) distribute the weights effectively over the layers and provide better performance than the smaller ones at the expense of increased training time. The training time is directly dependent on the size of the network, the larger the network size, the longer the training time.

The choice of learning rate value is another important implementation consideration. All PEs in the network may use the same learning rate, or the rate can vary for each layer. The rate can remain the same over the whole training period, or it may change after a certain period. At the beginning of the training period, the learning rate can take large values which increases the convergence rate. After a certain period of time, the learning rate should be reduced to create smaller weight changes, otherwise, it may cause a 'jumpy' behavior as networks converge to the optimum error value (Parlos *et al.*, 1992).

The training process can be terminated when the maximum error produced by the network approaches the value which is smaller than the acceptable error of each problem. However, it is possible to be trapped by local minimum during the training process preventing the network from converging to a desirable error value. This important problem is dealt with extensively in the literature (Dagli *et al.*, 1991; 1992; Cybenko, 1989) and algorithms that reduce training time and prevent trapping by local minimum prematurely are proposed.

The Quickprop learning algorithm (Fahlman, 1988) modifies the standard backpropagation learning algorithm by using the previous weight change to generate the weight change in the next training cycle. However, the new weight change is not allowed to be greater in magnitude than the 'maximum growth factor' times the previous weight change for that particular weight.

The adaptive learning algorithm (Parlos *et al.*, 1992) also updates the weights according to the direction of gradient but the difference between this algorithm and the normal backpropagation learning algorithm is the learning rate. To accelerate the convergence of the training, its learning rate is modified to be a function of error gradient. This dependence of the learning rate produces an advantage to the algorithm because it allows the error to dictate the dynamics of the functional resulting in 'faster' convergence. A disadvantage of the algorithm is when the optimal value is approached, it is possible to have 'jumpy' behavior of the weights. This algorithm, therefore, is more suitable with analog problems rather than binary problems.

There is another algorithm called the fast backpropagation algorithm which is a modification of the normal backpropagation algorithm (Karayiannis, 1992). After the initial adaptation cycles, this algorithm uses the minimization of an alternative criterion. An advantageous characteristic of the algorithm is it can be used for the training of the networks whose output is either binary or analog with a fast convergence. These problems are also discussed at length in Chapter 12.

3.4 ADAPTIVE RESONANCE THEORY

The ability of a network to learn new patterns without forgetting the old ones is an important attribute. If a back propagation paradigm is used to recognize new patterns without forgetting the previously trained ones, there is a need to retrain the network with the new pattern including all the patterns that are previously used to train the network. This process may require a significant amount of time. However, it is possible to perform this function easily using adaptive resonance theory (ART) paradigm proposed by Carpenter and Grossberg (1987) which is an extension of the competitive learning. A feedback mechanism between the competitive layer and the input layer is added to be able to learn new information without destroying the old ones. In an ART network, the information is sent back and forth between layers until the resonance state is achieved, and no learning occurs before the resonance state is accomplished. The time required for the network to achieve the resonance state is much less than the time used to change the weight values of the PEs. There are two possible situations: if the input information was previously learned by the network, the resonance state can be obtained quickly and the stored pattern is updated; if the input information is not immediately recognized, the rest of the stored patterns are rapidly searched for a match – if no match is found, the network goes into the resonance state to store the input information as a new pattern. Freeman and Skapura (1991) gives an excellent description of ART, but in the following sections, their description is summarized. Interested readers should refer to their books for details.

3.4.1 Components of an ART network

In general, an ART network is a neural network structure that has two layers of PEs connected to each other through top-down and bottom-up connections. All activities produced over the nodes in the network are called the short-term memory (STM) since they are created to associate with each input pattern which exist over a short period. Long-term memory (LRM) is the weight associated with connection between layers, and these weights are the encoded information that remains as a part of the network.

The architecture of the ART system is illustrated in Fig. 3.13 and is considered to have two major subsystems; attentional subsystem and orienting subsystem. The **attentional subsystem** responds to those familiar patterns, and creates the learned top-down LTM to stabilize the learned bottom-up LTM of the familiar patterns. However, the isolated attentional subsystem is incapable of creating new classes or unfamiliar patterns; the **orienting subsystem** is needed for determining whether a pattern is familiar and well represented in the LTM or if a new class needs to be created for an unfamiliar pattern.

3.4.2 Pattern matching

An ART network tries to classify a new input pattern into a previous stored class or a new class. First, an input pattern I_1 is presented to an ART network through F_1 layer PEs, the orienting subsystem, A and the gain control, G. In Fig. 3.13, it can be seen that there are three possible input signals for F_1 nodes:

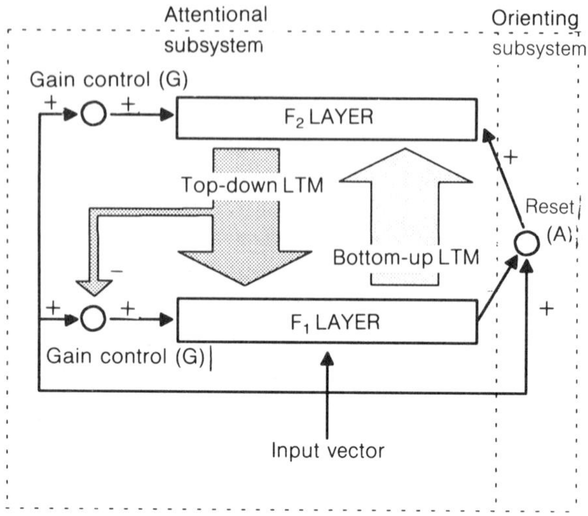

Fig. 3.13 The architecture of an ART system.

input pattern, or vector gain control signal and the pattern created from F_2 layer. In ART system, a PE is active when at least two out of three of its inputs are active; this rule is called the 2/3 rule. Since I is now active and the gain control receives a signal from the input pattern, I, hence, G is also active. Because the use of 2/3 rule, active signals from G and I make F_1 PEs active. Pattern X, therefore, is generated across F_1 PEs. Pattern S is created as the output of F_1 PEs that is sent to A as an inhibitory signal to cancel the excitory signal created from the input pattern I. The cancellation of these signals causes the orienting subsystem, A, to remain inactive.

Pattern S is also sent through the bottom-up LTM, Z_{ij}, to F_2, F_2 PEs calculate their input value by the sum of the product of vector S and the weights W. Because of an active gain control signal and the pattern S from F_1 layer, pattern Y is developed across F_2 as an activity stored in STM. F_2 PEs compete with each other to find a winning PE, because F_2 is a competitive layer. The ouput pattern U is generated as a result of the competition across F_2 PEs which creates an inhibitory signal sent to the gain control system. Inhibitory signal generated from gain control halts the activity on F_1 layer. However, there is one possible input signal created from pattern U through the top-down LTM connections, Z_{ji}, which is called pattern V. Pattern V and input pattern I together make F_1 PEs stay active. The pattern that remains on F_1 PEs is the intersection of I and V. Therefore, the pattern X^*, which is $I \cap V$, is developed. If there is a mismatch between previous pattern X, and the new developed pattern X^*, a new pattern S^* is generated as a result of F_1 PEs. S^*, which is different from the previous pattern S, is sent as an inhibitory signal to A. A has become active since the two input signals cannot cancel each other. A sends a reset signal to all the PEs in F_2 layer. This signal results in the active PEs becoming inactive and staying inactive for an extended period of time. Other PEs which are not active do not respond to this signal. This mechanism prevents the same PE winning the competition in the next matching cycle. After the active PE in F_2 layer is reset, there is no longer an inhibitory signal coming to the gain control system. The next matching cycle, therefore, continues. The search will continue until the matched pattern is found, or all F_2 PEs with the previously stored pattern are visited. If there is no match found, the network will assign a new F_2 PE to the new pattern and will begin to learn. In an ART system, the learning process, which is the LTM modification, does not start or stop. When signals are sent over the connections, the LTMs are subject to change. However, the time required to update the LTM is much longer than the searching process. But when a match is established, the network will go into the resonance state. The connections will stay active long enough to modify the LTMs.

3.4.3 Adaptive resonance theory networks

Many networks have been created from the adaptive resonance theory. ART1 is the earliest ART network, operating on binary inputs, which limits their applications. Freeman and Skapura (1991) provide a calculation of the bounds and the parameters and development of these training steps and derivation of equations, and the ART1 processing mechanism using a fast learning algorithm is summarized below.

1. Let N_1 and N_2 be the number of nodes in the F_1 and F_2 layers respectively.
2. A, B, C, D, and L are constant parameters. ρ is the vigilance value used in the matching process. These parameters must be chosen according to the following constraints:

$$A, C, D \geqslant 0$$
$$\max\{D, 1\} < B < D + 1$$
$$L > 1$$
$$0 < \rho \leqslant 1$$

3. Initialize the long-term memory trace according the following equations, for $i = 1, \ldots, N_1$ and $j = 1, \ldots, N_2$.

$$\text{Top-down weights: } Z_{ji} > \frac{B-1}{D}$$

$$\text{Bottom-up weights: } 0 < Z_{ij} < \frac{L}{L - 1 + N_1}$$

4. Apply input vector (I) to the F_1 layer and calculate the activation (X) across the F_1 layer, for $i = 1, \ldots, N_1$

$$X_i = \frac{I_i}{1 + A(I_i + B) + C}$$

5. Calculate the output vector (S) of the F_1 layer, for $i = 1, \ldots, N_1$

$$S_i = \begin{cases} 1 & \text{if } X_i > 0 \\ 0 & \text{if } X_i \leqslant 0 \end{cases}$$

6. Propagate pattern S through bottom-up LTM to F_2 layer and produce the activation pattern, Y, across F_2, for $j = 1, \ldots, N_2$

$$Y_j = \sum_{i=1}^{N_1} S_i Z_{ij}$$

7. Since F_2 PEs use competitive learning, they compete with each other until there is only one winning neuron. The pattern U is created as the output of the competition, for $j = 1, \ldots, N_2$

$$U_j = \begin{cases} 1 & \text{If } Y_j = \max\{Y\} \\ 0 & \text{otherwise} \end{cases}$$

8. Propagate pattern U through top-down LTM back to F_1 layer, pattern V is produced according to pattern U from F_2 layer, for $i = 1, \ldots, N_1$

$$V_i = \sum_{i=1}^{N} U_j Z_{ji}$$

9. Calculate the new activation (X^*) from all inputs of the F_1 layer, for $i = 1, \ldots, N_1$

$$X_i^* = \frac{I_i + DV_i - B}{1 + A(I_i + DV_i) + C}$$

10. Recalculate the output pattern of the F_1 layer (S) by using the same equation as in Step 5, but with pattern X^* instead of X.
11. Where $|P|$ is the number of non-zero elements in the pattern P, the degree of matching, M, between input pattern I and the produced pattern S^* is calculated using:

$$M = \frac{|S^*|}{|I|}$$

12. If the degree of matching, M, is less than the value of the vigilance parameter, ρ, then the active node, J, in the F_2 layer is set to inactive and we resume at Step 4. Otherwise, long-term memory traces, bottom-up and top-down, corresponding to the node J in the F_2 layer are updated by the following steps.
13. Update the bottom-up weights, for $i = 1, \ldots, N_1$

$$Z_{iJ} = I_i\left(\frac{L}{L - 1 + |S|}\right)$$

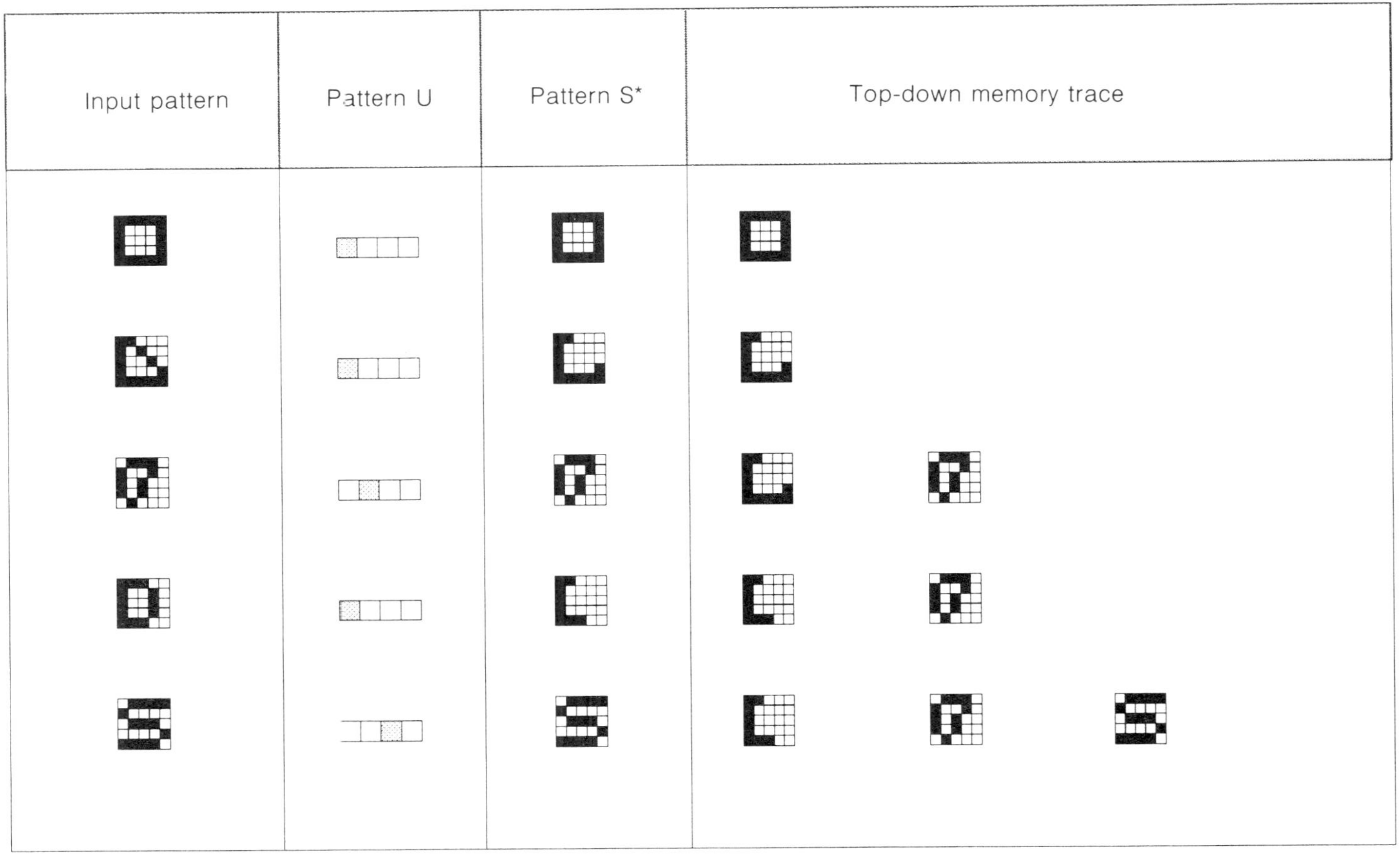

Fig. 3.14 Example of pattern matching of ART-1 network, using vigilance value of 0.6.

14. Update the top-down weights, for $i = 1, \ldots, N_1$

$$Z_{Ji} = I_i$$

15. Reset all nodes in the F_2 layer. Apply new input to the network and then go to Step 4.

Let us demonstrate fast learning in ART1 with an example. Figure 3.14 demonstrates the pattern matching process of ART1. It graphically summarizes the learning process of an ART1 network for vigilance parameter of 0.6 with respect to five different input patterns. The first input pattern in this figure is used in the example to demonstrate the calculations of an ART1 network.

1. Parameter N_1 is selected to be 25, N_2 to be 4.
2. Initialize values of the parameters; A, B, C, D and L, are selected as follows,

$$A = 1, \quad B = 2, \quad C = 1, \quad D = 1.5 \quad \text{and} \quad L = 2$$

3. Initialize the long-term memory trace as

Top-down weights: $$Z_{ji} = \frac{2-1}{1} = 1$$

Bottom-up weights: $$Z_{ij} < \frac{2}{2-1+25} = 0.077$$

for $i = 1, \ldots, N_1$, and $j = 1, \ldots, N_2$

4. Apply the input pattern and calculate X vector across F_1 lays,

$I =$ 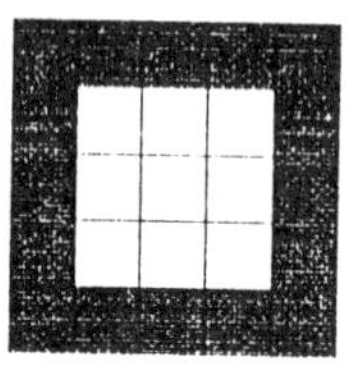 $= [1\ 1\ 1\ 1\ 1\ 1\ 0\ 0\ 0\ 1\ 1\ 0\ 0\ 0\ 1\ 1\ 0\ 0\ 0\ 1\ 1\ 1\ 1\ 1\ 1]$

$$X = [0.2\ 0.2\ 0.2\ 0.2\ 0.2\ 0.2\ 0\ 0\ 0\ 0.2\ 0.2\ 0\ 0\ 0\ 0.2\ 0.2\ 0\ 0\ 0\ 0.2\ 0.2\ 0.2\ 0.2\ 0.2\ 0.2]$$

5. Calculate output vector S of F_1 layer

$$S = [1\ 1\ 1\ 1\ 1\ 1\ 0\ 0\ 0\ 1\ 1\ 0\ 0\ 0\ 1\ 1\ 0\ 0\ 0\ 1\ 1\ 1\ 1\ 1\ 1]$$

6. Calculate vector Y across F_2 layer

$$Y = [1.232 \quad 1.232 \quad 1.232 \quad 1.232]$$

7. Since there is only one winning neuron, let pick the first neuron as a winning neuron. Pattern U is created as

$$U = [1 \quad 0 \quad 0 \quad 0]$$

8. Calculate pattern V

$V = [1\ 1\ 1\ 1\ 1\ 1\ 1\ 1\ 1\ 1\ 1\ 1\ 1\ 1\ 1\ 1\ 1\ 1\ 1\ 1\ 1\ 1\ 1\ 1\ 1] =$

9. Recalculate pattern X^*

$$\begin{aligned} X^* = [&0.111\ \ 0.111\ \ 0.111\ \ 0.111\ \ 0.111 \\ &0.111\ \ -0.14\ \ -0.14\ \ -0.14\ \ 0.111 \\ &0.111\ \ -0.14\ \ -0.14\ \ -0.14\ \ 0.111 \\ &0.111\ \ -0.14\ \ -0.14\ \ -0.14\ \ 0.111 \\ &0.111\ \ 0.111\ \ 0.111\ \ 0.111\ \ 0.111] \end{aligned}$$

10. Recalculate pattern S^*

$S^* = [1\ 1\ 1\ 1\ 1\ 1\ 0\ 0\ 0\ 1\ 1\ 0\ 0\ 0\ 1\ 1\ 0\ 0\ 0\ 1\ 1\ 1\ 1\ 1\ 1\ ] =$

11. Calculate degree of matching, $M = \frac{16}{16} = 1$.
12. Compare the degree of matching, $M = 1$, with the vigilance parameter, $\rho = 0.6$. $M > \rho$, therefore, we continue with Step 13–14.
13. Update the bottom-up weights,

$$\begin{aligned} Z_{i1} = [&0.059\ \ 0.059\ \ 0.059\ \ 0.059\ \ 0.059\ \ 0.059\ \ 0\ \ 0\ \ 0\ \ 0.059 \\ &0.059\ \ 0\ \ 0\ \ 0\ \ 0.059\ \ 0.059\ \ 0\ \ 0\ \ 0\ \ 0.059 \\ &0.059\ \ 0.059\ \ 0.059\ \ 0.059\ \ 0.059] \end{aligned}$$

14. Update the top-down weights,

$Z_{1i} = [1\ 1\ 1\ 1\ 1\ 1\ 0\ 0\ 0\ 1\ 1\ 0\ 0\ 0\ 1\ 1\ 0\ 0\ 0\ 1\ 1\ 1\ 1\ 1\ 1] -$

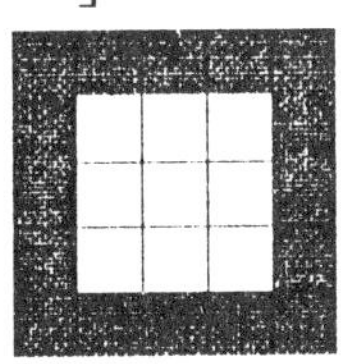

Later, the ART1 paradigm was modified to handle analog inputs, as ART2, and recently, a family of ART paradigms, such as ART3, Fuzzy ART and ARTMAP have been proposed. Carpenter and Grossberg (1991) summarize these paradigms extensively.

3.5 SUMMARY

In this chapter basic elements of artificial neural networks have been introduced and their role in designing various neural network paradigms stressed. Three basic paradigms, perceptron, back propagation, and ART1, were introduced and learning procedures used by these networks demonstrated with examples. The basic understanding given in this chapter is essential to follow the various manufacturing applications of artificial neural networks discussed in subsequent chapters.

REFERENCES

Aleksander, I. (ed.) (1989) *Neural Computing Architectures*, MIT Press, Cambridge, Mass.

Aleksander, I. and Morton, H. (1991) *An Introduction to Neural Computing*, Chapman & Hall, London.

Blum, A. (1992) *Neural Networks in C++: An Object-Oriented Framework for Building Connectionist Systems*, Wiley, NY.

Carpenter, G.A. and Grossberg, S. (1987) A massively parallel architecture of a self organising neural pattern recognition machine. *Comput. Graphics Image Process.*, **37**, 54–115.

Carpenter, G.A. and Grossberg, S. (1991) *Pattern Recognition by Self-Organizing Neural Networks*, MIT Press, Cambridge, MA.

Caudill, M. and Butler, C. (1992) *Understanding Neural Networks: Computer Explorations*, Vol. I & II, MIT Press, MA.

Cybenko, G. (1989) Approximations by superpositions of a sigmoidal function, *Math. of Control, Signals & Systems*, Center for Supercomputing Research and Development, Report no. 856, University of Illinois, Urbana, IL, 61801–2932.

Dagli, C.H., Kumara, S.R.T. and Shin, Y.C. (eds) (1991) *Intelligent Engineering Systems Through Artificial Neural Networks*, ASME Press, NY.

Dagli, C.H. Burke, L.I. and Shin, Y.C. (eds) (1992) *Intelligent Engineering Systems Through Artificial Neural Networks*, Vol. II, ASME Press, NY.

Dayhoff, J. (1990) *Neural Network Architectures: An Introduction*, Van Nostrand Reinhold, NY.

Eberhart, R.C. and Dobbins, R.W. (ed.) (1990) *Neural Network PC Tools: A Practical Guide*. Academic Press, CA.

Fahlman, S.E. (1988) Faster learning variations on back-propagation: an empirical study, in *Proceedings of the 1988 Connectionist Model Summer School*. (eds D. Touretzky, G. Hinton and T. Sejnowski), Morgan Kaufmann, San Mateo, California, pp. 38–51.

Freeman, J.A. and Skapura, D.M. (1991) *Neural Networks, Algorithms, Applications and Programming Techniques*, Addison-Wesley, MA.

Hertz, J., Krough, A. and Palmer, R. (1991) *Introduction to the Theory of Neural Network Computation*, Addison-Wesley, CA.

Karayiannis, N.B. (1992) The fast back propagation algorithm, in *Intelligent Engineering Systems Through Artificial Neural Networks*, Vol. II (eds C.H. Dagli, S.R.T. Kumara and Y.C. Shin), ASME Press, NY.

Khanna, T. (1990) *Foundations of Neural Networks*, Addison-Wesley, MA.

Kosko, B. (1992a) *Neural Networks and Fuzzy Systems: A Dynamical Systems Approach to Machine Intelligence*, Prentice Hall, NY.

Kosko, B. (1992b) *Neural Networks for Signal Processing*, Prentice Hall, NY.

McClelland, J.L. and Rumelhart, D.E. (1989) *Exploration in Parallel Distributed Processing: A Handbook of Models, Programs, and Exercises*, MIT Press, MA.

Nelson, M.M. and Illingworth, W.T. (1991) *A Practical Guide to Neural Nets*, Addison-Wesley, MA.

Parlos, A.G., Fernandez, B., Atiya, A.F., Muthusami, J. and Tsai, W.K. (1992) An adaptive backpropagation learning algorithm for multilayer neural networks, *Intelligent Engineering Systems Through Artificial Neural Networks*, Vol. II (eds C.H. Dagli, *et al.*), ASME Press, NY.

Rumelhart, D.E., McClelland, J.L. and the PDP Research Group (1986) *Parallel Distributed Processing*, Vols I & II, MIT Press, MA.

Simpson, P.K. (1990) *Artificial Neural Systems, Foundations, Paradigms, Applications and Implementations*, Pergamon Press, New York.

Wesserman, P.D. (1989) *Neural Computing*, Van Nostrand Reinhold, NY.

Zurada, J.M. (1992) *Introduction to Artificial Neural Systems*, West Publishing Company, MN.

CHAPTER 4

Hybrid intelligent systems: Tools for decision making in intelligent manufacturing

Gregory R. Madey, Jay Weinroth and Vijay Shah
Kent State University, Graduate School of Management, Kent State University, Kent, Ohio 44242, USA.

4.1 OVERVIEW

Many intelligent computing techniques have been developed over the last decade. Some of these include neural networks, fuzzy systems, genetic algorithms, rule-based expert systems, inductive expert systems, and the wide array techniques lumped under the heading of artificial intelligence. The application of these intelligent computing techniques to support planning and control in manufacturing can be done at three levels: organization, coordination, and execution. Because of the different nature of the problems at each level, different types of solutions may be required. Until recently, problem solvers have typically used **single-technique-based tools** to build these solutions, e.g., an expert system based solution, a neural network based solution, or a linear programming solution. Simple problems that fit the assumptions of a single-technique-based solution may be easily solved by such an approach. However, most real world manufacturing problems are not simple. They may neither fit the assumptions of a single technique nor be effectively solved by the strengths and capabilities of a single technique. One approach to deal with these complex real world problems is to integrate the use of two or more techniques in order to combine their different strengths and overcome each other's weaknesses to generate **hybrid solutions**.

Artificial Neural Networks for Intelligent Manufacturing. Edited by Cihan H. Dagli.
Published in 1994 by Chapman & Hall, London. ISBN 0 412 48050 6

Hybrid solutions can be developed in a variety of ways. We analyze these approaches by presenting an extension of a classification model for intelligent hybrid systems that was developed by Goonatilake and Khebbal (1992). We expand their model into two parts: one on the underlying technologies used in manufacturing-problem solving, and a second on how hybrid intelligent systems might be developed. The first, the problem-solving taxonomy, specifies four types of problem-solving processing:

- information processing
- numeric processing
- symbolic processing
- subsymbolic processing.

The second part specifies that hybrids solutions can be developed by either, or both, of two broad approaches: function-replacement and coupling-intercommunication. Each of these approaches for hybrid development has a continuum of possible implementations defined by two extremes. The **function-replacement continuum** is defined by

- embedded function-replacement at one extreme and
- polymorphic function-replacement at the other extreme.

The **coupling-intercommunication continuum** is defined by

- weak or loose coupling at one extreme and
- strong or tight coupling at the other extreme.

We survey hybrid intelligent systems that have been reported in the literature and present a hybrid intelligent system that was developed by embedding neural network technology into a discrete event simulation of a manufacturing organization's human resource policies and their effect on continuous improvement practices.

4.2 MANUFACTURING DECISION-MAKING PROBLEMS: ORGANIZATION, COORDINATION AND EXECUTION LEVELS

Manufacturing decision-making is involved in both planning and control and can be analyzed at the organization, coordination and execution levels, as shown in Fig. 4.1, adapted from Georgantzas and Acar (1992). Examples of typical decision activities of manufacturing management at the organization level include

- market positioning of the firm's products
- design of those products and their manufacturing processes
- capacity planning, location and distribution decisions
- long-range demand forecasting
- development of human-resources policies.

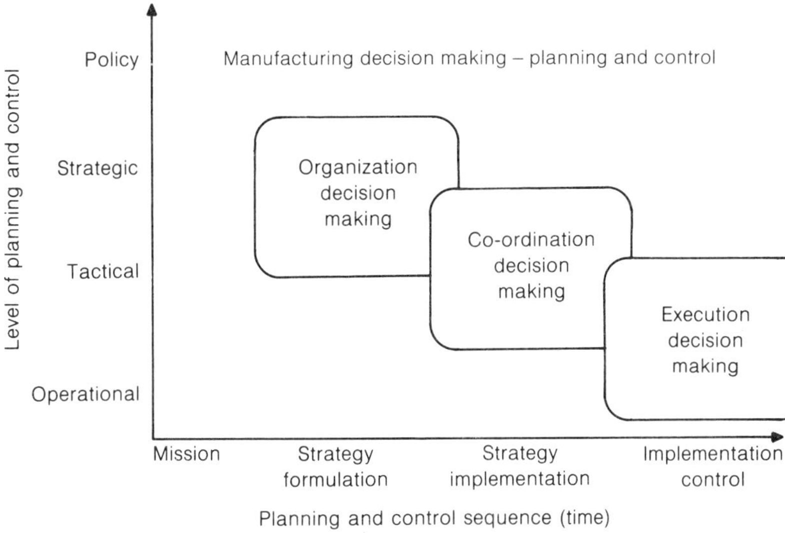

Fig. 4.1 Organization, coordination and execution level decision-making in manufacturing involve both planning and control. (Adapted from Georgantzas and Acar 1992.)

Decision activities at the coordination level include

- machine-part formulation
- process planning
- design of inventory replenishment policies
- manufacturing resource planning
- scheduling.

Decisions activities that occur at the execution level involve

- inspection and testing
- process monitoring and adaptive control
- fault analysis and diagnosis.

4.2.1 Summary of current manufacturing problem-solving

Figure 4.2 displays some general manufacturing problem areas and some techniques currently used to support decision-making in those problem areas. These techniques were developed in part by the industrial engineering, operations research, and engineering management disciplines and have been the standard techniques taught in most manufacturing management courses. Using the hybrid problem-solving taxonomy described later in this chapter (e.g. Fig. 4.5) these currently used techniques are based primarily on numeric and information processing technologies and tend to be single-technology solutions. Two important developments in manufacturing management have

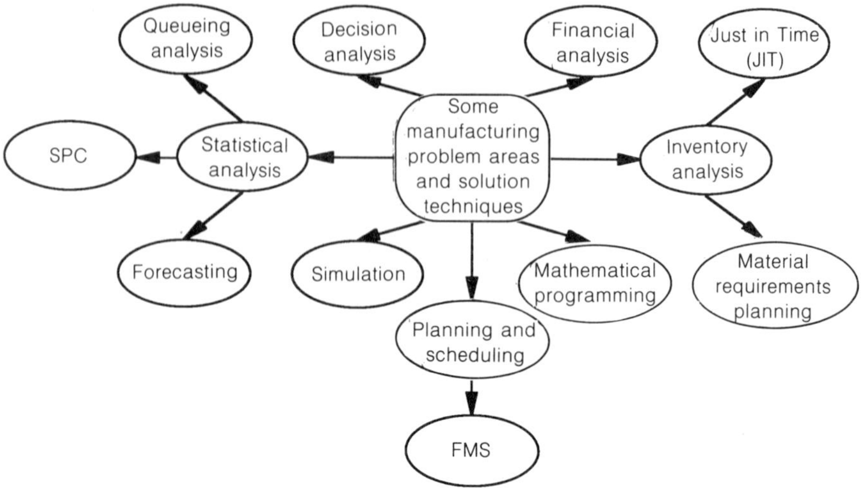

Fig. 4.2 Some typical manufacturing problems and solutions used at the organization, coordination and execution decision-making levels.

evolved recently. First, new problem-solving technologies, called **intelligent technologies**, have been developed and are being applied to manufacturing problem-solving. Second, single-technology solutions are being enhanced by integrating two or more problem solving technologies together into hybrid solutions for manufacturing problem-solving. The following sections discuss hybrid intelligent problem-solving systems with an emphasis on those systems based on neural networks and applied to manufacturing problem-solving.

4.2.2 Potential contributions of artificial neural networks

Although there are many reports of artificial neural networks being investigated for use on manufacturing problems, most of these reports are by investigators in academia, government, and R&D laboratories. It is not clear from these early reports whether neural networks are actually being used in real-world manufacturing situations or whether these are the early, hopeful reports of researchers advocating the use of this new technology. We are at that point in the life cycle of the investigation of this new technology where there are few reported real-world implementations, perhaps because either

- the technology is being used by manufacturing firms but those firms wish to keep that usage a secret for competitive reasons, or
- it is not yet being used in real manufacturing applications.

Looking at these early reports, still gives an indication of what some of the potential contributions of artificial neural networks could be at the organization, coordination, and execution levels of manufacturing decision-making. At the organization level, Murtaza and Fisher (1991) describe an integrated system for managerial support, Madey *et al.* (1992) describe a hybrid neural system applied to modeling continuous improvement systems in manufacturing, and Harston (1990) reviews a variety of business and operations management applications. At the coordination level numerous researchers have reported on the use of neural networks for forecasting (Kang, 1991; Khotanzad and Fowler, 1991), load balancing (Mehra and Wah, 1991), and scheduling (Rabelo *et al.*, 1990; Yih *et al.*, 1991). Most of the reported investigations of neural networks for intelligent manufacturing are at the execution level. Examples of these include applications in adaptive control (Guha, 1991; Steck *et al.*, 1991; Rabelo and Avula, 1991), vision and inspection (Cios *et al.*, 1991; Cho and Conners, 1991), process monitoring and quality control (Guillot and El Quafi, 1991; Smith, 1991), and fault detection, fault analysis and fault diagnosis (Osyk, 1991; Keyvan and Rabelo, 1991).

4.3 HYBRID INTELLIGENT SYSTEMS DEVELOPED OUT OF NEURAL NETWORKS

There are several different levels at which hybrid neural network systems are developed, and this can cause confusion. For example, at the hardware level, neural network chips consisting of analog and digital designs are often called **hybrid synapse chips** (Moopenn, *et al.*, 1990). We exclude this type of hybrid system from our discussion below. At the simple neural network level, a new artificial neural paradigm that is built out of two or more simpler neural paradigms is often called a **hybrid neural network**. For example, the counter-propagation neural model is a hybrid of the Kohonen adaptive vector quantization neural network and the Grossberg outstar (Hecht-Nielsen, 1987). The Hamming network is based on the perceptron at one layer and a Hopfield-like network at another layer (Lippmann *et al.*, 1987; Maren, 1990). In these examples, two simple neural network models are tightly coupled together to produce a new neural paradigm.

If several neural nets are loosely coupled and formed into a hierarchy or a 'network of networks' to solve a particular problem, the resulting system is often called a hybrid neural net or a **hybrid intelligent system**. Gersho and Reiter (1990) describe such a system for information retrieval in the telecommunications industry. Their hybrid system used a multi-layer network comprised of a Kohonen feature map coupled with either a backpropagation network or a Hamming network. Nygard (Nygard *et al.*, 1990) developed what he called, 'a modular system of neural networks' that was used to select solution techniques for vehicle routing problems. It consisted of seven backpropagation neural networks. He used a hidden layer of one network for feature extraction and a second network to recommend which of the remaining five networks would be used to make the solution technique recommendation.

Intelligent technologies for manufacturing problem-solving include the

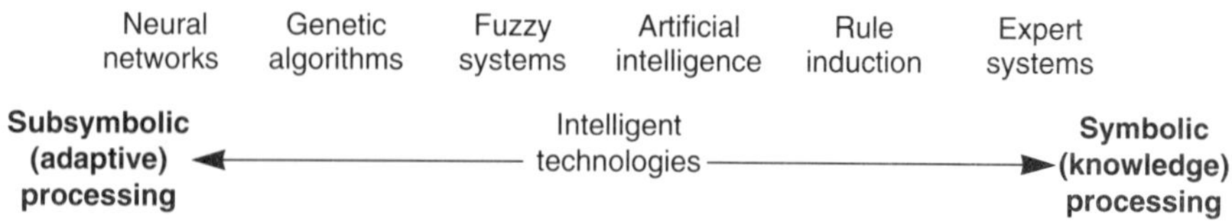

Fig. 4.3 Intelligent technologies for the manufacturing fall on a continuum between subsymbolic processing at one end and symbolic processing at the other.

various techniques of artificial intelligence, expert systems, neural networks, fuzzy systems, rule induction, and genetic algorithms. We observe in Fig. 4.3 that these technologies fall on a continuum formed by subsymbolic processing, such as neural networks, at one extreme and symbolic processing, such as expert systems, at the other extreme. Some even newer intelligent technologies are appearing under the names of **artificial life** and **nanotechnology**, but are beyond the scope of this chapter. Hybrid intelligent systems are problem-solving systems that are based on the integration of several technologies where at least one is an intelligent technology.

An example of a hybrid intelligent system would be an expert system using a neural network to perform a pattern-matching function. In this example, both technologies – expert systems and neural networks – are intelligent technologies. Dagli and Lammers (1989) describe a conceptual hybrid intelligent system for job scheduling that integrates a database system, an expert system with knowledge about conventional scheduling algorithms and heuristics, and two types of neural networks, one for problem recognition and another for schedule optimization. It is also possible to build a system that uses an intelligent technology integrated with a conventional problem-solving technique, e.g., neural networks and discrete-event simulation. This hybrid intelligent system might be used to model a manufacturing system with learning and adaptative behavior more realistically represented (Madey *et al.*, 1992). In these examples, we have at least one intelligent technology integrated with other approaches to produce a hybrid intelligent system. Several examples of how such systems might be integrated are displayed in Fig. 4.4.

Until recently, advocates for one technology have tried to demonstrate the superiority of that favored technology by applying it as a solitary problem-solving approach. They would identify problem situations where their pure approach excelled over competing approaches and advocated its sole use as a solution technique. But most manufacturing problems are sufficiently complex and fail to meet perfectly the assumed conditions for a single approach. Thus, rarely will one technology suffice for the broad spectrum of problems encountered in manufacturing.

With this in mind, we see that the motivation for hybrid systems, and intelligent hybrid systems in particular, is to merge the strengths of the various technologies to produce systems that effectively solve manufacturing problems

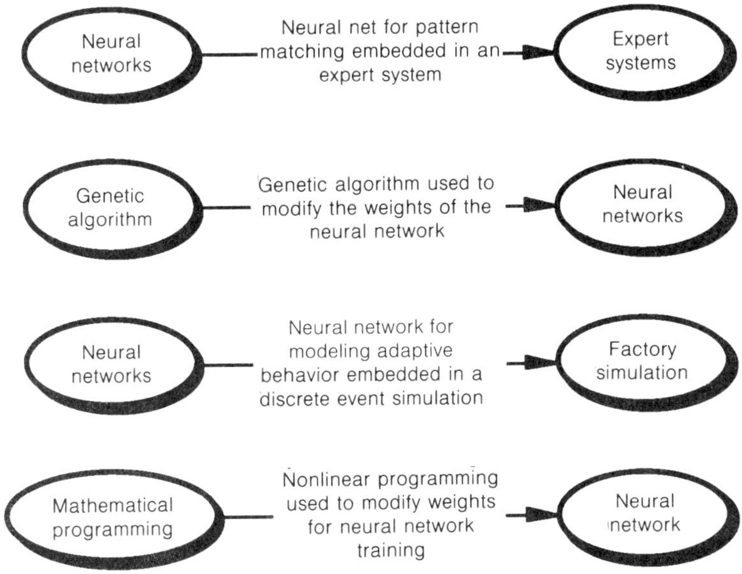

Fig. 4.4 Examples of how intelligent technologies could be integrated into hybrid intelligent systems.

that may not perfectly fit the required assumption of single approaches. Hopefully, with a large enough toolbox of problem-solving technologies (including the more recently developed intelligent technologies), more effective manufacturing problem-solving systems can be developed. These technologies can be mixed and matched so that the strengths of each compensate for the weaknesses of the others.

4.3.1 A classification model for hybrid intelligent systems

For purposes of this chapter, we define hybrid intelligent technologies to be any combination of problem-solving technologies that include at least one intelligent technology that could be positioned along the continuum displayed in Fig. 4.3. In Fig. 4.5 we display a hierarchical taxonomy of the components that could make up a hybrid system. The taxonomy consists of two parts: one for the underlying technologies used in problem-solving, and a second identifying how hybrid intelligent systems might be developed. The first part, the problem-solving technologies, specifies four types of problem-solving processing: information processing, numeric processing, symbolic processing, and subsymbolic processing. Information and numeric processing are grouped together as conventional problem-solving technologies while symbolic and subsymbolic processing are grouped together as intelligent problem-solving technologies. The second part of the taxonomy in Fig. 4.5 specifies that hybrid solutions can be developed by either, or both, of two broad approaches:

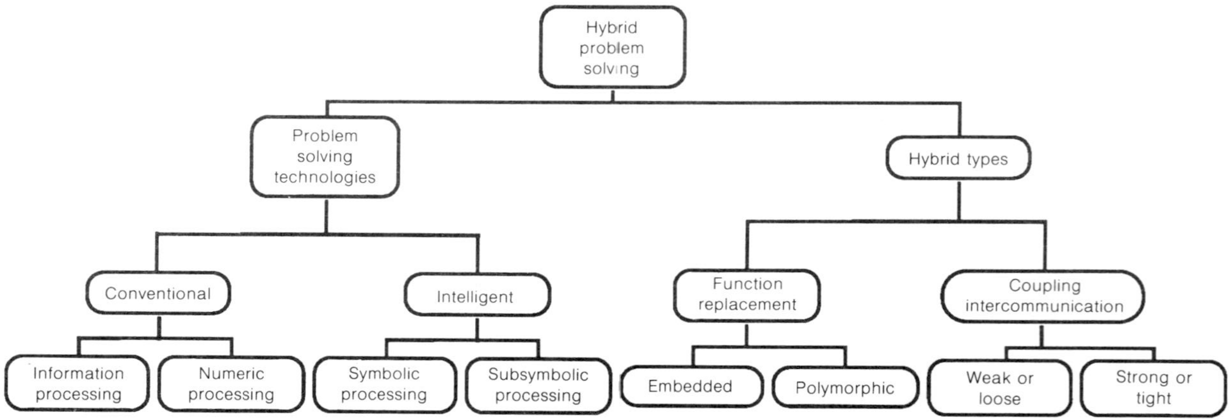

Fig. 4.5 Classification taxonomy for hybrid systems for problem solving.

function-replacement and coupling-intercommunication. Each of these approaches for hybrid development has a continuum of possible implementations defined by two extremes. The function-replacement continuum is defined by embedded function-replacement at one extreme and polymorphic function-replacement at the other extreme. The coupling-intercommunication continuum is defined by weak or loose coupling at one extreme and strong or tight coupling at the other extreme. In summary, the left-hand side of Fig. 4.5 addresses what hybrid systems are made out of, while the right-hand side identifies how hybrid systems are developed out of the technologies on the left-hand side.

The four types of problem-solving technologies displayed on the left-hand side of the hierarchy in Fig. 4.5 could be viewed as dimensions in a universe of problem-solving techniques. These include the two classes of intelligent technologies, subsymbolic and symbolic processing, and the two classes of conventional problem-solving technologies, numeric and information processing. This universe of problem-solving technologies is depicted in Fig. 4.6 with each of the four technologies connected to the other three, suggesting that they all define the extremes of problem-solving continuums. Similar to the subsymbolic/symbolic continuum that defined the class of intelligent technologies and represented along the base of the triangle of Fig. 4.6, a numeric/symbolic continuum contains the solution techniques that consist of a hybrid combinations of numeric and symbolic problem-solving techniques. Similar continuums containing the other combinations of numeric, information,

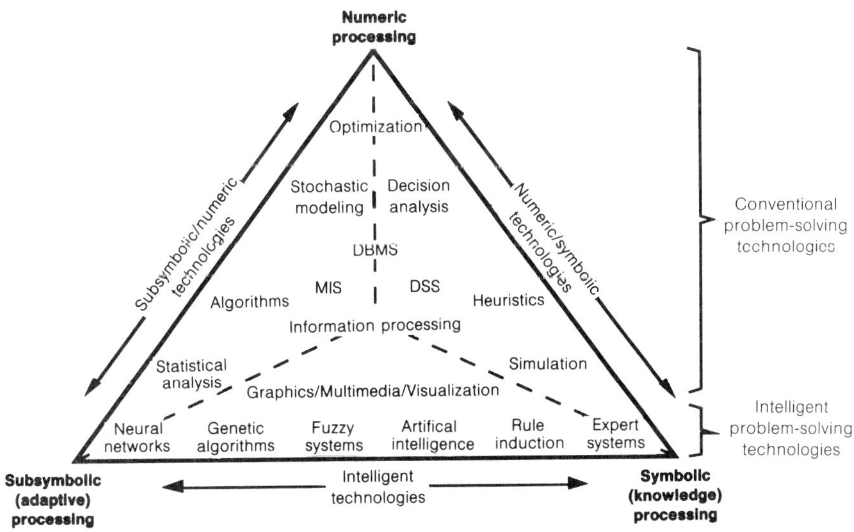

Fig. 4.6 Problem-solving methods for manufacturing include techniques for numeric processing, symbolic processing, subsymbolic processing and information processing.

and subsymbolic processing techniques are also defined in Fig. 4.6. The other problem-solving techniques listed in Fig. 4.6 (e.g. optimization, schochastic modeling, decision analysis, etc.) are positioned in the figure to represent how they might relate to the four main problem-solving technologies. The current manufacturing problems and their solutions displayed in Fig. 4.2 are primarily formed out of the conventional problem-solving technologies that are positioned in the upper part of the triangle in Fig. 4.6. We define a hybrid intelligent system to be any problem-solving system formed by one of the intelligent technologies that could be positioned along the base of the triangle of Fig. 4.6 and any other problem-solving technology.

Next, we define the ways that the integration of these problem-solving technologies can be implemented by adapting and extending frameworks developed by Goonatilake and Khebbal (1992) and Reisman (1988). Two dimensions characterize hybrid systems: the degree of function-replacement by one technology in the other; and the level of coupling or intercommunication between the two technologies. Although hybrids are developed by integrating more than two technologies, we simplify the discussion here by considering the integration of only two technologies at a time.

Function-replacement, the first dimension, describes how one technology is integrated with another. Several examples include:

- the weight-adjustment during the training of a neural network is achieved by replacing the conventional error minimization techniques by genetic algorithms (Montana and Davis, 1989; Dominic *et al.*, 1991);
- a neural network is used to induce symbolic rules for an expert system, either by replacing manual knowledge engineering or by replacing an induction technique such as ID3 (McMillan *et al.*, 1991; Yang and Bhargava, 1990; Noordewier, *et al.*, 1991)
- symbolic rules are replaced by a neural network for faster performance (Hruska *et al.*, 1991), or a neural network model is constructed so that it is functionally equivalent to a rule-based expert system (Ajjanagadde and Shasi, 1991).

The level or degree of function replacement can be defined by two endpoints of a continuum. At one extreme, the degree of replacement is minor, and is an embedding of one technology into another. At the other extreme the replacement is so complete that the first technology essentially replaces the second technology. This method is called a **polymorphic replacement** by Goonatilake and Khebbal (1992). Figure 4.7 displays these two extremes and the function-replacing continuum they define.

The **coupling or intercommunication** between the two technologies, the second dimension that characterizes hybrid systems, also can be defined in terms of two extremes on a continuum. At one extreme, the two technologies are loosely coupled, maintain their identity, and share data with one another, either directly or through an intermediary mechanism, such as a blackboard system. At the other extreme, the two technologies are tightly coupled and

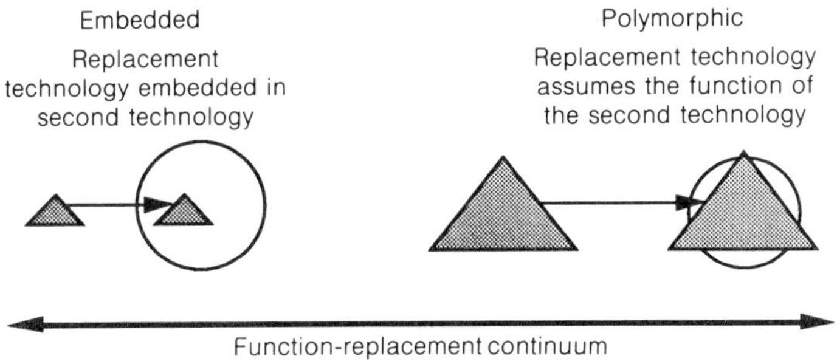

Fig. 4.7 The function-replacement approach to developing hybrid intelligent manufacturing.

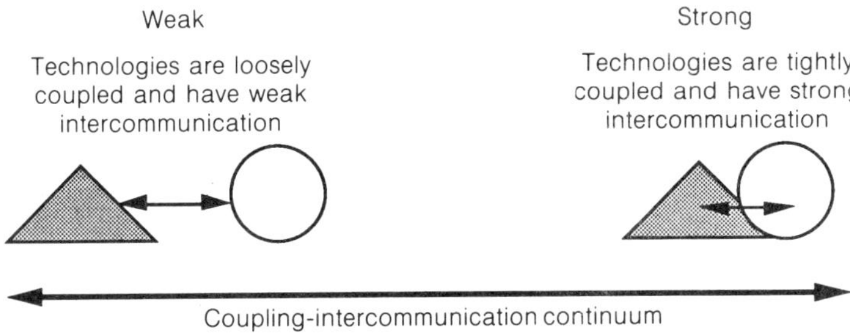

Fig. 4.8 The coupling-intercommunication approach to developing hybrid intelligent systems for intelligent manufacturing.

often assume an identity together as a new paradigm. Figure 4.8 displays these two extremes and the coupling-intercommunication continuum they define.

4.4 SURVEY OF NEURAL NETWORK HYBRID INTELLIGENT SYSTEMS

We summarize some of the reported investigations into hybrid neural networks in this section. Many other hybrids have been reported, but we only identify a few representative examples from a few of the types of intelligent hybrid systems that are possible. These representative intelligent hybrids are listed in Table 4.1, and grouped by the shared technology that the hybrid is developed from, cited, summarized under the comments column, and classified on one or both of the hybrid continuums displayed in Figs 4.7 and 4.8. Although the hybrids could fall anywhere on the continua displayed in Figs 4.7 and 4.8, we classified each hybrid identified in Table 4.1 with whichever end of the continuum it was most strongly associated with. Note that many of the hybrids are developed by both function replacement and coupling-intercommunication.

Table 4.1 Representative intelligent hybrid systems reported in the literature, arranged by shared technology and classified according to the methods used to develop the hybrid

Shared technology	*Citation*	*Comments*	*Hybrid Classification*
Expert systems	Kohonen (1987)	Neural network used for rule induction.	Function replacement – polymorphic
	Bochereau and Bourgine (1990)	Rules extracted out of a trained multilayer neural network. Used for explanation of neural network or as first step in knowledge engineering.	Function replacement – polymorphic; weakly coupled – intercommunicating
	Rabello (1990) Rabello *et al.* (1990)	Neural networks used to classify a scheduling problem which in turn is used by expert system schedulers.	Weakly coupled – intercommunicating
	Bradshaw *et al.* (1989)	Multilayered neural network that functions as an expert system.	Function replacement – polymorphic
	Hruska *et al.* (1991)	Rules of an expert system are converted into a neural network. The weights learned in the neural network are interpreted as certainty factors in the expert system.	Function replacement – embedding; weakly coupled – intercommunicating
	Gallant (1988)	Neural network functions as an expert system.	Function replacement – polymorphic
	Ajjanagadde and Shastri (1991)	Neural network functions as an expert system.	Function replacement – polymorphic
	McAndless *et al.* (1991)	Neural network and rule-based approach applied to real time machine vision.	Weakly coupled – intercommunicating
Rule induction	Krovidy and Wee (1990)	Used rule induction (ID3) to extract rules from a database and a Hopfield network to optimize a sequence of actions.	Weakly coupled – intercommunicating
	McMillan *et al.* (1991)	A neural network is used to induce explicit, symbolic, condition-action rules from examples.	Function replacement – polymorphic
	Yang and Bhargava (1990)	A modified perceptron network is used to induce rules for an expert system.	Function replacement – embedding

Multiple neural network	Gersho and Reiter (1990)	Hierarchy of multiple neural networks.	Weakly coupled – intercommunicating
	Nygard *et al.* (1990)	Network of neural networks.	Weakly coupled – intercommunicating
	Hecht-Nielsen (1987)	Counterpropagation neural model: a hybrid of the Kohonen adaptive vector quantization neural network and the Grossberg outstar.	Strongly coupled – intercommunicating
	Lippmann *et al.* (1987)	The Hamming network is based on the perceptron at one layer and a Hopfield-like network at another layer.	Strongly coupled – intercommunicating
Fuzzy systems	Kosko (1987, 1992)	Fuzzy cognitive maps used to store knowledge that is represented as signed digraphs.	Function replacement – polymorphic
	Amano (1989)	Neural networks used for feature detection and fuzzy logic for the decision procedure.	Weakly coupled – intercommunicating
	Hayashi (1991)	Neural network with automated extraction of fuzzy IF-THEN rules.	Function replacement – embedding
Genetic algorithms	Maricic and Nikolov (1990)	Genetic algorithms used to determine neural network parameters.	Function replacement – embedding
	Spears and DeJong (1990)	Neural networks and genetic algorithms are used as heuristics in NP-complete problems.	Weakly coupled – intercommunicating
	Stacey and Kremer (1991)	Neural network learning implemented using genetic algorithms.	Function replacement – embedding
	Montana and Davis (1989)	Weight-adjustment during the training of a neural network by replacing the conventional error minimization techniques by genetic algorithms.	Function replacement – embedding
	Dominic *et al.* (1991)	Weight-adjustment during the training of a neural network by replacing the conventional error minimization techniques by genetic algorithms.	Function replacement – embedding

Table 4.1 (*Contd.*)

Shared technology	*Citation*	*Comments*	*Hybrid Classification*
Mathematical optimization	Hsiung *et al.* (1991)	Successive quadratic programming is used to replace the gradient descent learning algorithm of back-propagation.	Function replacement – embedding
	Subramanian (1990)	Generalized reduced gradient (GRG2) is used to replace the gradient descent learning algorithm of back-propagation.	Function replacement – embedding
	Denton (1991)	Limited-memory variable metric methods used to replace the gradient descent learning algorithm of back-propagation.	Function replacement – embedding
Discrete event simulation	Madey and Weinroth (1990), Weinorth and Madey (1990)	Neural network used to model human behavior in a factory simulation.	Function replacement – embedding; weakly coupled – intercommunicating
Multiple integrated technologies	Dagli and Lammers (1989)	Integration of DBMS, expert system, back-propagation neural network for classification of scheduling problem and Hopfield network for schedule optimization.	Weakly coupled – intercommunicating
	Kasabov (1992)	Integration of neural networks with a rule-based system developed in CLIPS to build a fuzzy expert system.	Weakly coupled – intercommunicating; function replacement – polymorphic
	Kasabov and Petrov (1992)	Integration of neural networks with a logic programming system developed in Prolog to build a fuzzy expert system.	Weakly coupled – intercommunicating; function replacement – polymorphic

4.5 CASE STUDY OF THE DEVELOPMENT OF A HYBRID INTELLIGENT SYSTEM FOR DECISION MAKING IN MANUFACTURING

In this section we describe a neural network that is embedded into a factory simulation for modeling continuous improvement policies to assist planning and decision-making at the organization level. In a component of the simulation, we consider the problem-solving context in which an experienced human operator continually improves the production process at his machine. Specifically, he does this by recalibrating the settings on the machinery to achieve higher production with fewer defectives resulting in better gross profit for the operation. This problem-solving environment is receiving attention in research involving artificial sensors and emulation of responses of human operators and research on the use of neural networks for modeling human performance (Noaker, 1988; Fix, 1990; Kraiss and Kütelwesch, 1990).

The motivation for embedding a neural network in a factory simulation was to provide the model builder with a more accurate and efficient method of modeling skilled human performance in the continuous improvement policy simulation, in particular that of an operator at a manufacturing workstation. A component of the simulation model, as shown in Fig. 4.9, was developed by which optimal or near-optimal solutions can be found for a production control problem, in the usual way of trial-and-error search by the human simulationist. In this way, a set of expert answers is obtained for a specific problem. In this case, by using the model, the researcher was able to determine the best values at which the machine should be recalibrated, assuming that the situation is specifically characterized by certain input values. The best values are determined by simple trial-and-error, as is done so often in use of simulation models. Adjustment of the value of an input parameter in a direction which produces improvement in the performance criteria or criterion is continued in the direction of initial change until no further improvement is apparent or even until performance begins again to deteriorate. Other input values at various intervals from the apparently best

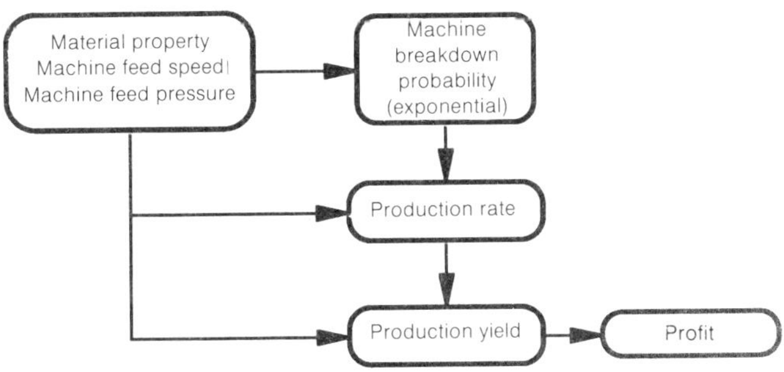

Fig. 4.9 Simulation of a manufacturing workstation.

value are also tested, and degraded performance of consistent degree and direction confirms the location of the optimal or near-optimal result.

Such a time-consuming trial-and-error process must be repeated each time one wants to use the simulation model in the traditional way for a new problem scenario. However, by submitting the scenario input values and the target answers discovered through model experimentation to a neural net as a training set, the objective was to demonstrate the elimination of the need for further manual experimentation with the simulation model. Instead, the neural net would provide an answer sufficiently close to the best one that could be found by further manual experimentation with the model, for all new input values.

In our model, gross profit for a produced unit is 25% of the cost of the unit, and units which emerge from the machinery in a defective condition are simply lost as scrap. Therefore, the penalty cost of a defective unit is four times the positive revenue for a good unit produced. The number of units produced for a given time period is a function of the speed at which the machine is set to operate, less the amount of downtime experienced. Speed of the machine may range from the fastest rate of one-half minute per unit to the slowest rate of one and one-half minutes per unit. This time value is constant for all units produced at a given setting. The time between machine breakdowns is a negative exponential function of a multiple of 25 times the amount of time required per unit, and repair times are uniformly distributed between one and two minutes in length.

The probability that the machine will produce a defective unit is a function of three factors: the speed at which the machine is operating, the pressure which the surfaces of the machinery apply to the material being processed, and a density property of the raw material. Where the machinery is operating to some degree out of control, as concerns settings for optimal gross profit, the economic losses could range from minor to quite serious ones. In such a case, the experienced human operator must correct the machine settings. This is done by observing four input values: number of good units produced, number of defective units produced, current setting on the machine for operating speed, and current setting on the machine for pressure. This relationship is represented in Fig. 4.10.

The corrections made, constituting output values in the recalibration process, are the degree of positive or negative change in the settings for machine speed of operation and for machine pressure exerted on the material. The critical factor of the density of the particular batch of raw material being used for the production run is not directly ascertainable. However, the experienced operator has learned to recognize are appropriate recalibrations from the four input values of good and bad units and current pressure and machine speed.

It may be observed from this description that our intention was to present the neural net with a somewhat ambiguous situation requiring discovery of relationship patterns that were not evident from the data without some form

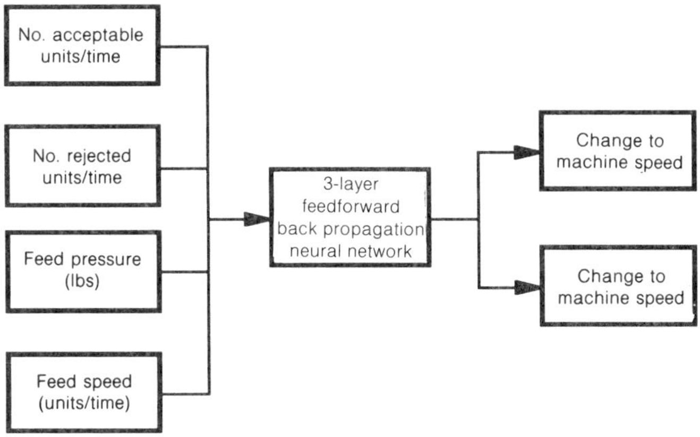

Fig. 4.10 The input/output mapping learned by the neural network.

of analysis. That is, we wanted to give the neural net a problem somewhat like a real world puzzle with some genuine unknowns, and, quite importantly, to need the assistance of a simulation model to imitate the expertise of an experienced operator by finding best solutions for given states.

At the same time, it was necessary to generate a set of results which we knew would have an underlying pattern to discover. The larger part of the pattern in the data was produced by using a regression equation to generate the probability of a defect as a function of machine speed and pressure, with an indirect way of representing the effect of material density. The effect of material density was represented by allowing a coefficient in the equation to vary randomly within limits that prevent illegitimate outcome values from the equation. This indirect approach was taken in order to generate a pattern whose shape we could not entirely predict beforehand. A second element of the pattern of results also unknown beforehand, was the correct strategy for producing the best gross profit in any random situation generated by the simulation model.

Twenty case situations were randomly created using the simulation model. By trial-and-error experimentation using the same random number streams and systematically varying the settings for simulated pressure and speed, the best results achievable for each given situation were discovered. This resulted in 20 sets of input/output data to be taught to a neural net. Using a leave-k-out procedure (with $k = 5$), 5 sets of data were set aside to be used afterward for a test of the network's degree of success in imitating the choices found by human trial-and-error. One data set was discarded as a statistical outlier. While the use of exponential distributions in the model caused this result as one consistently within the possible extremes of the range of model outcomes, the degree of change to machine speed and pressure were so large as to constitute a possibly disruptive influence on the rest of the training set, consisting as it did of only 14 additional sets. The net was trained, therefore, on just the

Table 4.2 Examples of input and output values used for training the neural network

Case	*1*	*2*	*3*	*4*	*5*
Good units	812	1062	961	1107	1246
Bad units	123	55	55	184	64
Pressure (lbs.)	10.0	10.45	10.65	10.67	10.35
Speed (minutes per unit)	1.47	1.23	1.35	1.06	1.04
Profit	$3200	$8240	$7410	$3710	$9900
Change to pressure	0%	– 4.306%	– 6.103%	– 6.803%	– 3.382%
Change to speed	2.041%	– 55.285%	– 59.259%	41.509%	– 51.923
New profit	$3660	$11 160	$9590	$3940	$12 470

Inputs are good units, bad units, pressure, and speed. Outputs are change to pressure and change to speed. Profit is the criterion for target values.

remaining 14 sets of data. The optimal corrections found by trial-and-error for 5 of the 20 simulated cases are shown in Table 4.2. It may be noted from these data that only the unknown density of the material being processed can explain the required changes for machine calibration, yet the neural network does discover this relationship.

In addition to the usual confirmation that the trained network's answers were within a small percentage of error with respect to the submitted target values, the capability of the net to reproduce individual target values was also compared with that of a good regression analysis of the 14 training cases. The least squares estimate derived here has a multiple R^2 of 0.98 ($p < 0.001$). The regression equation performed more effectively with respect to recommendations for recalibration of machine pressure. However, this target value turned out to be virtually deterministic in our particular simulation model, and the net's accuracy was reasonably good.

Table 4.3 Comparison of performance of neural network and regression equation in achieving target values

		Average error	*Maximum error*
Neural network:			
Machine speed	Training set[1]	6.435%	16.702%
	Test set[2]	11.052%	17.731%
Pressure	Training set[1]	0.65%	1.294%
	Test set[2]	3.317%	13.33%
Regression equation:			
Machine speed	Training set[1]	13.413%	102.374%
	Test set[2]	19.82%	52.582%
Pressure	Training set[1]	0.189%	0.357%
	Test set[2]	0.138%	0.268%

[1](14 exemplars)
[2](5 exemplars)

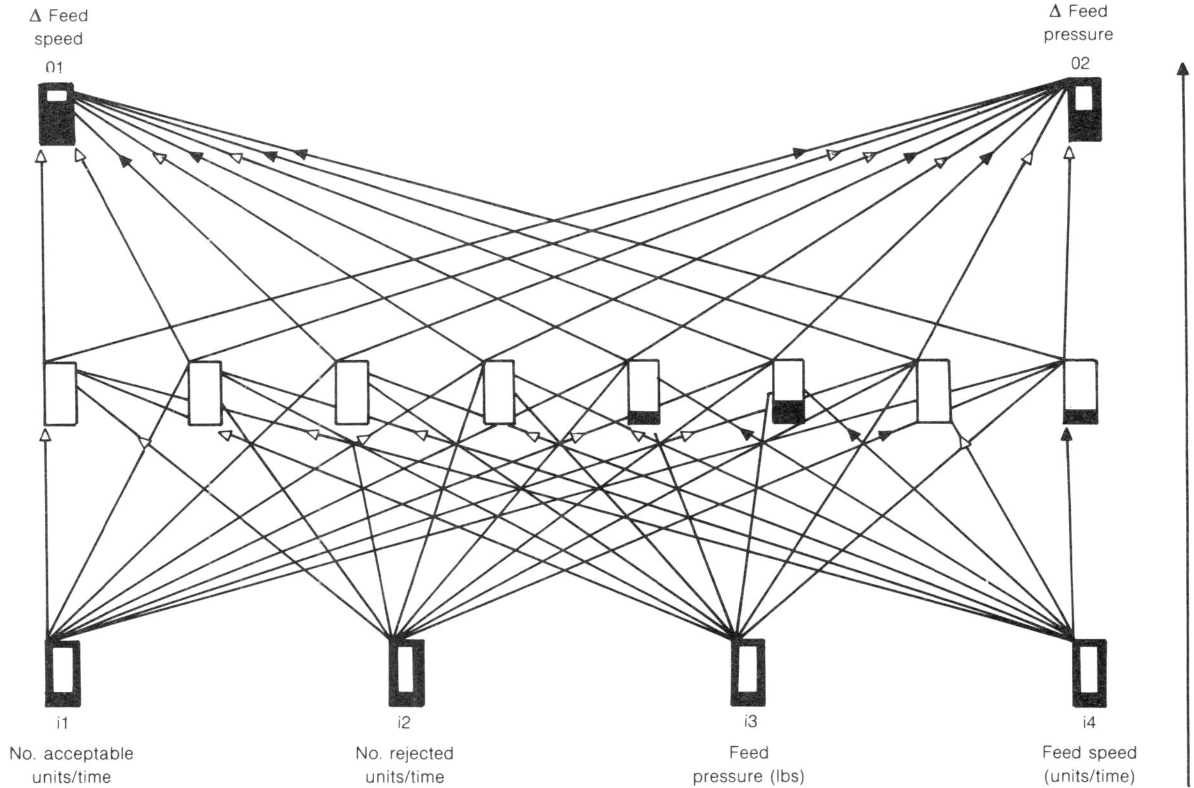

Fig. 4.11 The neural network trained to simulate the intelligent control of a machine tool.

With respect to recommendations for recalibration of machine speed, the regression equation averaged an error rate of 13.4% for the 14 known cases and 19.8% for the 5 unknown cases, with maximum errors of 102.4% and 52.6% respectively. The neural net's avarage rate of error was 6.4% for the 14 known cases and 11.1% for the 5 unknown cases, with maximum errors of 16.7% and 17.7%. Table 4.3 compares the results of the network's predictions, both for the 14 cases which provided the training set and for the 5 cases which had not been presented to the network, against the predictions of a regression analysis of the 14 training cases.

In an additional test of the effectiveness of the net's recommendations, 20 new cases were simulated at random. The model ran two parallel simulations for a simulated run of five days, for each of the new cases. One of these simulations uses the net's recommendations, while the other simulation continues with initial, uncontrolled settings. Identical random number streams are maintained by using duplicate generators with identical seeds.

Overall, the model when running under the net's recommendations produced simulated results over 12% better than when running without them. With subtraction of the cases with results in which the net's recommendations were worse than the accidental settings, gross profit was improved by over 18%. This latter situation corresponds to a situation where a real-time simulation model is installed in an actual production process, and particular recommendations which promise to have negative results are overridden.

The overall simulation was developed using GPSS/H from Wolverine Software running on either an IBM mainframe or PC (Banks *et al.*, 1989; Schriber, 1974). The neural network was trained off-line using the neural network simulator MacBrain (Jensen and Chait, 1988). A standard back-propagation network with a three layer (4–8–2) architecture was employed (see Fig. 4.11) The network required approximately 5000 epochs of the entire training set (14 examples) to reach acceptable performance (approximately ± 5% of the optimal control output). Additional experiments using higher-order neural networks (HONNs) were conducted. They produced comparable results but with significant improvement in training time.

4.6 SUMMARY

The complexity of real world manufacturing problems caused in part by different needs at the different problem-solving levels of organization, coordination, and execution, has created the need for hybrid systems. Hybrid systems combine the strength of two or more technologies and provide the next generation of solution methodologies for manufacturing planners and problem solvers. To design these hybrid systems, properly, it is necessary for those problem-solvers to be cognizant of the broad spectrum of solution technologies, especially the intelligent technologies developed over the last decade, and to understand the various approaches by which these can be integrated into hybrid systems. We have provided a framework that emphasizes that solution technologies need not come from one problem-solving paradigm, but that hybrid solutions can be built out of numeric processing, information processing, subsymbolic (neural networks) processing, and symbolic (knowledge-based) processing technologies.

These hybrid solutions can be developed in a variety of ways. We have presented an extension of a classification model for intelligent hybrid systems that was developed by Goonatilake and Khebbal (1992). Two general approaches can be used to design hybrid systems: function-replacing integration (where a technology is embedded into another) and intercommunicating integration (whereby the respective technologies maintain their respective functional forms, but share data through different degrees of coupling).

As an example of an embedded hybrid, we summarize a manufacturing organization level problem solving approach, using a neural network embedded into a discrete event simulation of human resource policies that would support continuous improvement practices in the factory. This intelligent hybrid system enabled more efficient development and more realistic modeling of the problem under evaluation.

REFERENCES

Ajjanagadde, V. and Shastri, L. (1991) Rules and variable in neural networks. *Neural Computation*, **3**(1), 121–34.

Amano, A., Aritsuka, T., Hatoka, N. and Ichikawa, A. (1989) On the use of neural networks and fuzzy logic in speech recognition. *IJCNN*, Washington, 301–6.

Banks, J., Carson II, J.S. and Sy, J.N. (1989) *Getting Started with GPSS/H*, Wolverine Software Corporation, Annandale, VA.

Bochereau, L. and Bourgine, P. (1990) Rule extraction and validity domain on a multilayer neural network, in *Proceedings of the International Conference on Neural Networks – Vol. 1*, IEEE, New York, 97–105.

Bradshaw, G., Fozzard, R. and Ceci, L. (1989) A connectionist expert system that actually works, in *Advances in Neural Information Processing Systems 1* (ed. D.S. Touretzky), Morgan Kaufman Publishers, San Mateo, CA. pp. 248–55.

Cho, T.-H. and Conners, R.W. (1991) A neural network approach to machine vision systems for automated industrial inspection, in *Proceedings of the International Conference on Neural Networks – Vol. I*, IEEE, New York, pp. 205–10.

Cois, K.J., Tijia, R.E., Liu, N. and Langenderfer, R.A. (1991) Study of continuous ID3 and radial basis function algorithms for the recognition of glass defects, in *Proceedings of the International Conference on Neural Networks – Vol. I*, IEEE, New York, pp. 49–54.

Dagli, C. and Lammers, S. (1989) Possible applications of neural networks in manufacturing, in *Proceedings of the International Joint Conference on Neural Networks – Vol. 2*, IEEE, New York, pp. 605.

Denton, J.W. (1991) A Robust and Efficient Training Algorithm for Feedforward Neural Networks, unpublished dissertation, Graduate School of Management, Kent State University, Kent, Ohio.

Dominic, S., Das, R., Whitley, D. and Anderson, C. (1991) Genetic reinforcement learning for neural networks, in *Proceedings of the International Conference on Neural Networks – Vol. 2*, IEEE, New York, pp. 71–6.

Fix, E. (1990) Modeling human performance with neural networks, in *Proceedings of the International Conference on Neural Networks – Vol. 1*, IEEE, New York, pp. 247–52.

Gallant, S.I. (1988) Connectionist Expert System. *Communications of the ACM*, **31**(2), 152–69.

Georgantzas, N.C. and Acar, W. (1993) *Scenario-driven planning: managing strategic uncertainty*, Quorum Books, Westport, Ct.

Gersho, M. and Reiter, R. (1990) Information retrieval using hybrid multi-layer neural networks, in *Proceedings of the International Conference on Neural Networks – San Diego – Vol. 2*, IEEE, New York, pp. 111–17.

Goonatilake, S. and Khebbal, S. (1992) Intelligent hybrid systems, in *Proceedings of the First Singapore International Conference on Intelligent Systems*, Singapore, September, pp. 207–12.

Guha, A. (1991) A hybrid neural network approach to adaptive control and optimization, in *Intelligent Engineering Systems through Artificial Neural Networks*, (eds C.H. Dagli, S.R.T. Kumara and Y.C. Shin), ASME Press, New York, pp. 505–10.

Guillot, M. and El Quafi, A. (1991) On-line identification of tool breakage in metal cutting process by use of neural networks, in *Intelligent Engineering Systems through Artificial Neural Networks* (eds C.H. Dagli, S.R.T. Kumara and Y.C. Shin), ASME Press, New York, pp. 701–10.

Harston, C.T. (1990) Business with neural networks, in *Handbook of Neural Computing Applications* (ed. A.J. Maren), Academic Press, San Diego, CA, pp. 391–400.

Hayashi, Y. (1991) A neural expert system with automated extraction of fuzzy if-then rules and its application to medical diagnosis, in *Advances in Neural Information Processing Systems* 3 (eds R.P. Lippman, J.E. Moody and D.S. Touretzky), Morgan Kaufman Publishers, San Mateo, CA, pp. 578–84.

Hecht-Nielsen, R. (1987) Counter-propagation networks. *Applied Optics*, **26**, 4979–84.

Hruska, S.I., Kuncicky, D.C. and Lacher, R.C. (1991) Hybrid learning in expert networks, in *Proceedings of the International Conference on Neural Networks – Vol. II*, IEEE, New York, pp. 117–20.

Hsiung, J.T., Suewatanakul, W. and Himmelblau, D.M. (1991) Should back propagation be replaced by more effective optimization algorithms?, in *Proceedings of the International Conference on Neural Networks – Vol. I*, IEEE, New York, pp. 353–6.

Jensen, M. and Chait, D. (1988) *MacBrain 2.0 – Adaptive Simulation of Complex Systems*, Neuronics, Inc., Cambridge, MA.

Kang, S. (1991) An Investigation of the Use of Feedforward Neural Networks for Forecasting, unpublished dissertation, Graduate School of Management, Kent State University, Kent, Ohio.

Kasabov, N. (1991) Incorporating neural networks into production systems and a practical approach towards the realisation of fuzzy expert systems. *Computer Science of India Journal*, **21**(2), 26–34.

Kasabov, N. and Petrov, S. (1992) Approximate reasoning with hybrid connectionist logic programming systems, in *Artificial Networks II* (eds I. Alexander and J. Taylor), Elsevier Science Publishers, pp. 749–52.

Khotanzad, A. and Fowler, M. (1991) Neural network based time series forecasting, in *Intelligent Engineering Systems through Artificial Neural Networks* (eds C.H. Dagli, S.R.T. Kumara and Y.C. Shin), ASME Press, New York, pp. 813–18.

Keyvan, S. and Rabelo, L.C. (1991) Nuclear reactor pump diagnosis via noise analysis/artificial neural networks, in *Intelligent Engineering Systems through Artificial Neural Networks* (eds C.H. Dagli, S.R.T. Kumara and Y.C. Shin), ASME Press, New York, pp. 651–6.

Kohonen, T. (1987) Self-learning inference rules by dynamically expanding context, in *Proceedings of the IEEE First International Conference on Neural Networks – Vol. 2* (eds M. Caudill and C. Butler), IEEE, New York, pp. 3–9.

Kosko, B. (1987) Adaptive inference in fuzzy knowledge networks, in *Processing of the IEEE First International Conference on Neural Networks – Vol. 2* (eds M. Caudill and C. Butler), IEEE, New York, pp. 261–8.

Kosko, B. (1992) *Neural Networks and Fuzzy Systems*, Prentice Hall, Englewood Cliffs, NJ.

Kraiss, K.F. and Küttelwesch, H. (1990) Teaching neural networks to guide a vehicle through an obstacle course by emulating a human teacher, in *Proceedings of the International Conference on Neural Networks – Vol. 1*, IEEE, New York, pp. 333–7.

Krovvidy, S. and Wee, W.G. (1990) A knowledge based neural network approach for waste water treatment system, in *Proceedings of the International Conference on Neural Networks – Vol. 1*, IEEE, New York, pp. 327–32.

Lippmann, R.P., Gold, B. and Malpass, M.L. (1987) A comparison of Hamming and Hopfield neural nets for pattern classification. *MIT Lincoln Lab. Tech. Report 769*, Lexington, MA.

Madey, G.R. and Weinroth, J. (1990) Neural networks and general purpose simulation theory, in *Proceedings of the International Conference on Neural Networks – Washington D.C. – Vol. 2*, IEEE, New York, pp. 647–50.

Madey, G.R., Weinroth, J. and Shah, V. (1992) Integration of neurocomputing and system simulation for modeling continuous improvement systems in manufacturing, *Journal of Intelligent Manufacturing*, **3**, 193–204.

Maren, A.J. (1990) Hybrid and complex networks, in *Handbook of Neural Computing Applications* (ed. A.J. Maren), Academic Press, San Diego, CA, pp. 203–17.

Maricic, B. and Nikolov, Z. (1990) GENNET – System for computer aided neural network design using genetic algorithms, in *Proceedings of the International Conference on Neural Networks – Washington D.C. – Vol. 1*, IEEE, New York, pp. 103–5.

McAndless, E., Stacey, D., Rueb, K. and Wong, A.K.C. (1991) A hybrid artificial neural network/rule-based approach to a real-time machine vision system, in *Intelligent Engineering Systems through Artificial Neural Networks* (eds C.H. Dagli, S.R.T. Kumara and Y.C. Shin), ASME Press, New York, pp. 909–14.

McMillan, C., Mozer, M. and Smolensky, P. (1991) Learning explicit rules in a neural network, in *Proceedings of the International Conference on Neural Networks – Vol. 2*, IEEE, New York, pp. 83–8.

Mehra, P. and Wah, B.W. (1991) Learning load-balancing strategies using artificial neural networks, in *Intelligent Engineering Systems through Artificial Neural Networks* (eds C.H. Dagli, S.R.T. Kumara and Y.C. Shin), ASME Press, New York, pp. 855–60.

Montana, D. and Davis, L. (1989) Training feedforward neural networks using genetic algorithms, in *Proceedings of 11th International Joint Conference on Artificial Intelligence*, pp. 762–7.

Moopenn, A., Duong, T. and Thakoor, A.P. (1990) Digital-analog hybrid synapse chips for electronic neural networks, in *Advances in Neural Information Processing Systems 2* (ed D.S. Touretzky), Morgan Kaufman Publishers, San Mateo, CA, pp. 769–76.

Murtaza, M.B. and Fischer, D.J. (1991) An integrated system for managerial decision support, in *Intelligent Engineering Systems through Artificial Neural Networks*, (eds C.H. Dagli, S.R.T. Kumara and Y.C. Shin), ASME Press, New York, pp. 933–8.

Noaker, P.M. (1988) Smart sensing on the factory floor. *Production*, **100**, 42–8.

Noordewier, M.O., Towell, G.G. and Shavlik, J.W. (1991) Training knowledge-based neural networks to recognize genes in a DNA sequence, in *Advances in Neural Information Processing Systems 3* (eds R.P. Lippman, J.E. Moody and D.S. Touretzky), Morgan Kaufman Publishers, San Mateo, CA, pp. 530–6.

Nygard, K., Juell, P. and Kadaba, N. (1990) Neural networks for selecting vehicle routing heuristics. *ORSA Journal on Computing*, **2**(4), 353–64.

Osyk, B.A. (1991) The Use of Neural Networks for Fault Analysis in a Computer Integrated Manufacturing System, unpublished dissertation, Graduate School of Management, Kent State University, Kent, Ohio.

Rabelo, L.C. (1990) A hybrid artificial neural networks and knowledge-based expert systems approach to flexible manufacturing systems scheduling, unpublished dissertation, University of Missouri-Rolla.

Rabelo, L.C., Alptekin, S. and Kiran, A.S. (1990) Synergy of artificial neural networks and knowledge-based expert systems for intelligent FMS scheduling, in *Proceedings of the International Conference on Neural Networks – Vol. 1*, IEEE, New York, pp. 359–66.

Rabelo, L.C. and Avula, X.J.R. (1991) Intelligent control of a robotic arm using hierarchical neural network systems, in *Proceedings of the International Conference on Neural Networks – Vol. 2*, IEEE, New York, pp. 747–51.

Reisman, A. (1988) On alternative strategies for doing research in the management and social sciences. *IEEE Transactions on Engineering Management*, **35**(4), 215–20.

Stacey, D.A. and Kremer, S. (1991) The Guelph Darwin Project: The evolution of neural networks by genetic algorithms, in *Proceedings of the International Conference on Neural Networks – Vol. 2*, IEEE, New York, p. A-957.

Schriber, T.J. (1974) *Simulation using GPSS*, J. Wiley & Sons, New York.

Smith, A.E. (1991) Quality control using backpropagation: an injection molding application, in *Intelligent Engineering Systems through Artificial Neural Networks* (eds C.H. Dagli, S.R.T. Kumara and Y.C. Shin), ASME Press, New York, pp. 729–34.

Spears, W.M. and De Jong, K.A. (1990) Using neural networks and genetic algorithms as heuristics for NP-complete problems, in *Proceedings of the International Conference on Neural Networks – Washington D.C. – Vol. 1*, IEEE, New York, pp. 118–21.

Steck, J., Krishnamurthy, K., McMillin, B. and Leineinger, G. (1991) Neural modelling and control of a distillation column, in *Proceedings of the International Conference on Neural Networks – Vol. 2*, IEEE, New York, pp. 771–7.

Subramanian, V. (1990) Sampling Issues for Classification using Neural Networks, unpublished dissertation, Graduate School of Management, Kent State University, Kent, Ohio.

Weinroth, J. and Madey, G.R. (1990) An evaluation of embedded neural networks within simulations, in *Proceedings of the Conference on AI and Simulation, Society for Computer Simulation*, San Diego, C.A, January.

Yang, Q. and Bhargava, V.K. (1990) Building expert systems by a modified perceptron network with rule-transfer algorithms, in *Proceedings of the International Conference on Neural Networks – San Diego – Vol. 2*, IEEE, New York, pp. 111–17.

Yih, Y., Liang, T-P. and Moskowitz, H. (1991) A hybrid approach for crane scheduling, in *Intelligent Engineering Systems through Artificial Neural Networks*, (eds C.H. Dagli, S.R.T. Kumara and Y.C. Shin), ASME Press, New York, pp. 867–72.

PART TWO

Neurocomputing for intelligent manufacturing: Organization and coordination level applications

CHAPTER 5

Conceptual design problem

Ali Bahrami[1], *Cihan H. Dagli*[2]

[1] *Computer Information Systems, Department of Economics and Management, Rhode Island College, Providence, Rhode Island 02908.*

[2] *Department of Engineering Management, University of Missouri-Rolla, Rolla, Missouri, 65401.*

5.1 CHARACTERISTICS OF DESIGN PROBLEM

Design or problem solving is a natural human activity. We have been designing and acting as designers (sometimes unconsciously) throughout our lives. Design begins with the acknowledgment of needs and dissatisfaction with the current state of affairs and realization that some action must take place in order to correct the problem. When a small child moves a stool to an appropriate location so that she can use it to get to her toy, she has acted as a designer – of a rudimentary design – by positioning the stool so that she can satisfy her need of playing with the toy.

Design is a purposeful activity, it involves a conscious effort to arrive at a state of affairs in which certain characteristics are evident (Coyne *et al.*, 1990). Suh (1990) defines design as four distinct aspects of engineering and scientific endeavor (see Fig. 5.1):

1. problem definition from fuzzy sets of facts and myths into a coherent statement of the question;
2. creative process of devising a proposed physical embodiment of solutions;
3. analytical process of devising a proposed physical embodiment of solutions; and
4. ultimate check of the fidelity of the design product to the original perceived needs.

Artificial Neural Networks for Intelligent Manufacturing. Edited by Cihan H. Dagli. Published in 1994 by Chapman & Hall, London. ISBN 0 412 48050 6

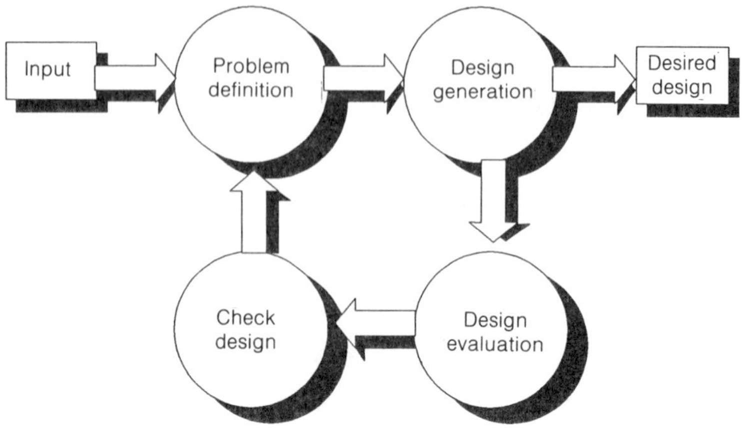

Fig. 5.1 Design as the four distinct aspects of engineering and scientific endeavor.

Every field of engineering involves and depends on the design or synthesis process, which allows us to fulfil needs through the creation of physical and/or informational structures, including machines, software, and organizations (Suh, 1990).

Design is a highly knowledge intensive and ill-structured problem. Most of the practical problems we face in the design are either too complex or too ill-defined to analyze with conventional approaches. By defining functional requirements as a fuzzy set one can perform inexact reasoning during the conceptual or creative phase of the design process.

Design activity usually takes place in an environment in which the functional requirements and constraints are not crisply defined. There tend to be many degrees of freedom in statements describing design specifications. The high degrees of freedom resulting from lack of crisp information could impede the creation of a problem space if not dealt with correctly.

In order to produce a manageable set of data to solve a complex problem, one may choose to ignore factors, including important ones perhaps, that together can contribute to the complete solution of the problem. As a result, we may arrive at completely wrong or incomplete solutions. For example, if we want to design a chair, one of the functional requirements may be 'It must be very comfortable'. Ambiguity results from the use of such natural language which can be meaningful but at the same time vague – what is comfortable to you may not seem so to me. Further, the fuzziness may result from imprecise and incomplete or subjective knowledge of an artifact to be designed. So, being able to mathematically represent the phrase 'very comfortable' greatly simplifies knowledge representation, definition and manipulation of functional requirements, and we use fuzzy sets to provide us with the ability to perform natural language computations and inexact reasoning. A fuzzy set is a class in which the transition from membership to non-membership is gradual rather than abrupt (Zadeh, 1975a; b; 1965). For instance the class of comfortable chairs can be viewed as a fuzzy set. An executive chair is definitely a member of this fuzzy set with a high degree of membership (close to 1) and a classroom chair is more out of the set than in.

In this chapter the feasibility of utilizing fuzzy associative memory (FAM) in design automation is examined. FAM paradigm exploits the relationships between functional requirements and physical structures in order to map fuzzy functional requirements to the crisp design.

5.1.1 Design characteristics and generic steps of design process

Suh (1990) defines design as the culmination of synthesized solutions in the form of product, software, processes or system by the appropriate selection of DPs (design parameters) that satisfy perceived needs through the mapping from FRs (functional requirements) in the functional domain to DPs in the structure domain. This mapping process is not unique and more than one design may result from the generation of DPs that satisfy the FRs. Therefore, there can be an infinite number of feasible design solutions and mapping techniques; see Fig. 5.2. Suh's design axioms provide the principles that the mapping technique must satisfy the input requirements in order to produce a good design, and provide a framework for comparing and selecting design (Suh, 1990). Suh's design axioms relate to Independence and Information (Suh, 1984; 1978), and are defined as follows.

- **Independence Axiom**: In an acceptable design, the DPs and the FRs are related in such a way that specified DP can be adjusted to satisfy its corresponding FR without affecting other functional requirements.
- **Information Axiom**: The best design is a functionally uncoupled design that has minimum information content.

Suh's information axiom coincides with an important principle in design known as 'simplicity', a word that has been used by designer and aestheticians for many centuries and remains a principle of primary concern. So what does

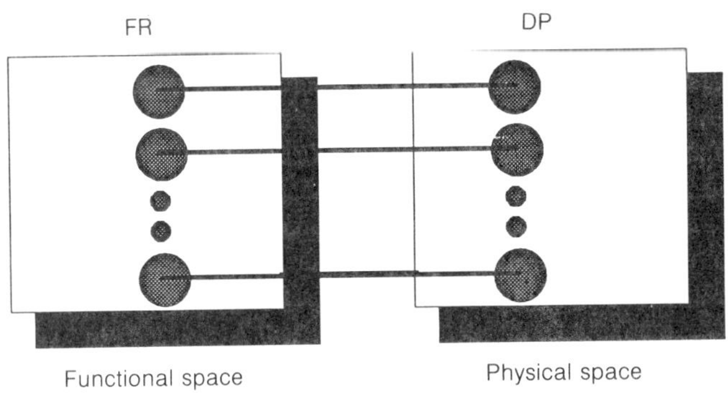

Fig. 5.2 Design mapping paradigm.

the designer mean when he applies the word 'simple' to the work of art? He is not speaking of mental deficiency. Simple in the design means being able to reduce it to the fewest possible lines, shapes and subparts, without reducing its functional requirements or violating its specifications. Simple design will reduce the assembly time and product cost, and increase reliability by many orders of magnitude. A good design is one to which no more can be added, and at the same time, one from which nothing can be subtracted without causing an incomplete design. In other words, good design must not be restrained with the non-essential details.

The independence axiom does not imply that a part has to be broken into two or more separate physical parts, or that a new element has to be added to the existing design. Functional coupling may be achieved without physical separation, although in some cases such physical decomposition may be the best way of solving the problem.

The information axiom implies that to suggest my design is better than yours because it does more than what it was intended is misguided (Suh, 1990). A design should only fulfil the precise needs defined by the FRs – nothing more and nothing less.

It is believed that a designer bases his design on some vague or fuzzy functional requirements. He usually associates sets of fuzzy functional requirements with sets of physical structures, then during the design process, design solutions are retrieved from memory to solve the design problem. Nam Suh (1990) defines the design D as a mapping process from the functional space (FRs) to the physical space (DPs) in order to satisfy the design-specified functional requirements.

Thus, we have

$$D\colon \mathrm{FRs} \rightarrow \mathrm{DPs}$$

$$\mathrm{FR}_i = \sum_j A_{ij}\mathrm{DP}_j \tag{5.1}$$

$$A_{ij} = \frac{\partial \mathrm{FR}_i}{\partial \mathrm{DP}_j}$$

Where A is the design matrix and FR_i is the functional requirement and DP_j is the design parameter. Unfortunately in most cases we do not know the mathematical function between functional requirements and design parameters.

The input to the design process is generally a set of functional requirements and constraints. The output of the process is generally a specification of an artifact. Design is an interactive process. Design involves continuous interplay between what we want to achieve and how we want to achieve it (Suh, 1990). Figure 5.3 shows the six basic phases of design: need assessment, analysis, decomposition, synthesis, integration and evaluation.

Since design is a knowledge intensive activity, the more knowledge and experience the designers have, the better are the chances for generating creative design. In the design process it is assumed that the designer selects a

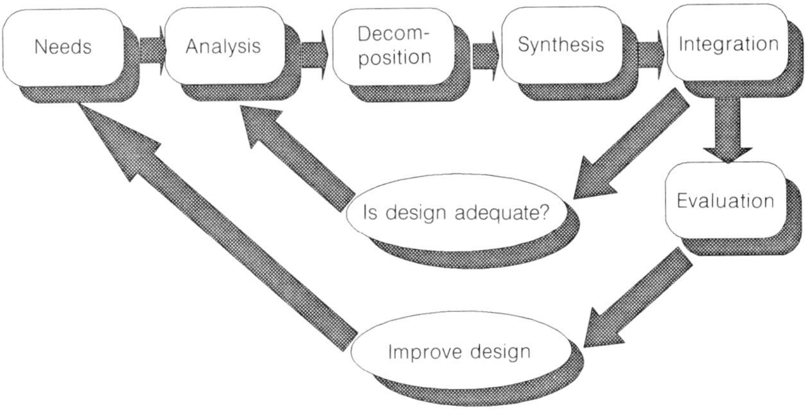

Fig. 5.3 Six basic phases of design – need assessment, analysis, decomposition, synthesis, integration and evaluation.

set of functional requirements which meet all of the perceived needs. Assuming that the functional requirements for certain design are fr_i and constraints are Cs_j where $i, j \in \mathbb{N}_n$, any alternative which satisfies fr_1 and fr_2 and $\cdots$ fr_n and Cs_1 and Cs_2 and $\cdots$ Cs_n, is exactly the desired design.

The view of design as 'generate and test' has received a wide acceptance. During the design process a set of solution alternatives is evaluated, and a solution is suggested based on the degree of satisfaction resulting from selection of alternative and functional requirements. This paradigm of generate and test in design coincides with Suh's (1990) independence axiom since during the design process, as we go from the FRs in the functional domain with DPs in physical domain, the mapping must be such that a perturbation in a particular DP must affect only its referent FR.

The tree structure of design decision (in Fig. 5.4) is described by Suh (1990), where the need is the root and overall design choices make up the branches. Each functional requirement has number of different solutions which are shown as leaves. Each design alternative is evaluated and a solution is suggested. Rarely, one set of solutions satisfies the functional requirements completely while no other set does, but usually each set of objectives offers a certain degree of satisfaction for the decision maker (designer). Yager (1980) defines this process:

> The decision process is now complicated by the fact that acceptability of a particular alternative is determined by the degree to which the alternative satisfies the objective and then modulated by the degree to which the decision maker is satisfied with these objectives. Therefore, what we obtain for each alternative is not a number indicating how well it satisfies the objectives, but a fuzzy set indicating how satisfied the decision maker is with that alternative.

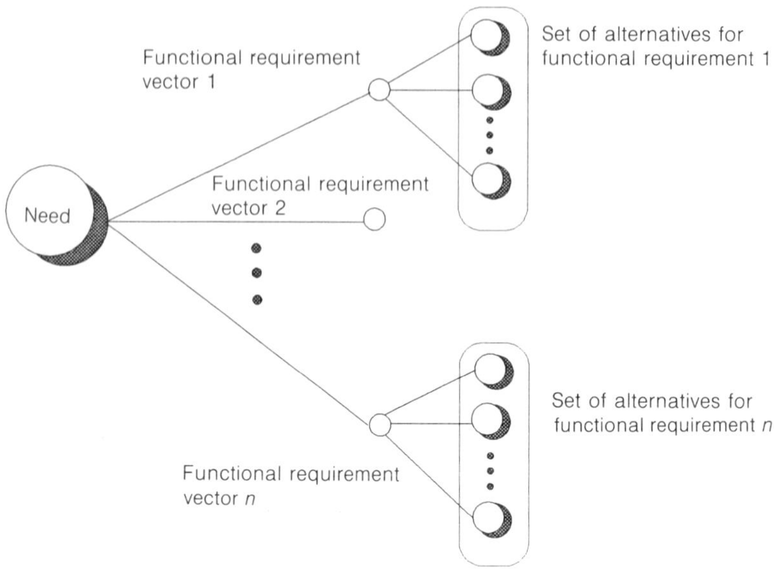

Fig. 5.4 Tree structure of design decision.

5.2 INTRODUCTION TO FUZZY SETS AND BINARY RELATIONSHIPS BETWEEN FUNCTIONS AND STRUCTURES

Fuzzy sets were first introduced by Professor Zadeh (1965), to model the features of uncertainty or vagueness that probability theory could not adequately handle. Decision-making involves perceptions, attitudes, subjection, conceptions and emotions (Zadeh, 1965), and these nomenclatures cannot be modeled or explained by randomness alone. A fuzzy set A in the universal set U is defined by a membership function $\mu_A(x)$ which associates a real number in the interval [0, 1] to each element x in U. The value of $\mu_A(x)$ is the grade of membership $x \in U$ in the fuzzy set A.

In the classical set theory there are sharp boundaries, and an element x in U is either in the set A or not in the set A. The membership function of a **crisp set** has the following form.

$$\mu_A(x) = \begin{cases} 1 & \text{iff } x \in A \\ 0 & \text{iff } x \notin A \end{cases}$$

A fuzzy set is a class in which the transition from membership to non-membership is gradual rather than abrupt (Zadeh, 1975a). A fuzzy set is characterized by its membership function $\mu_A(x)$ which takes on values between and including 0 and 1. If $\mu_A(x) = 0$ then the element is not in A, if $\mu_A(x) = 1$ the element is in A, and if $\mu_A(x)$ is, say, 0.6 then the element is more in A than not in A. As $\mu_A(x)$ approaches the value 1, the more x belongs to A.

Any relation between two sets F and S is known as a binary relation, usually denoted by $R(F, S)$. If we assume that F is a set of functional requirements and S is a set of physical structure, then the binary relation R in $F \times S$ is a fuzzy subset of $F \times S$, i.e. each pair (f, s) where $f \in F$ and $s \in S$. A number between zero and one can indicate how compatible f and s are, so, e.g. if R is the relationship between functional requirements and physical structure then $R(f, s)$ can represent the closeness of certain functional requirements (F) to a physical structure (S).

The binary relation can be represented by a membership matrix or **Sagittal diagram** (Coyne *et al.*, 1990). In matrix representation, each element of the matrix $\mu_{R(f \circ s)}(f, s)$ is represented by the compatibility between f and s. Sagittal diagrams, as shown in Fig. 5.5, resemble a network representation. Each element of the sets F and S is represented by the set of nodes in the diagram; nodes corresponding to one set are clearly distinguished from nodes representing the other set. Elements of $F \times S$ with non-zero membership grades in $R(F, S)$ are represented in the diagram by lines connecting the respective nodes (Coyne *et al.*, 1990). These lines are labeled with the values of the membership grade $\mu_{R(f \circ s)}(f, s)$.

The subscript $R(f \circ S)$ involves the operator $\circ$; this composition is known

	dp1	dp2	dp3	dp4	dp5
fr1	0.8	0.0	0.0	0.0	0.0
fr2	0.0	0.9	0.0	0.0	0.0
fr3	0.9	0.0	0.4	0.0	0.0
fr4	0.0	0.0	0.0	0.7	0.0
fr5	0.0	0.0	0.0	0.0	1.0

Membership matrix

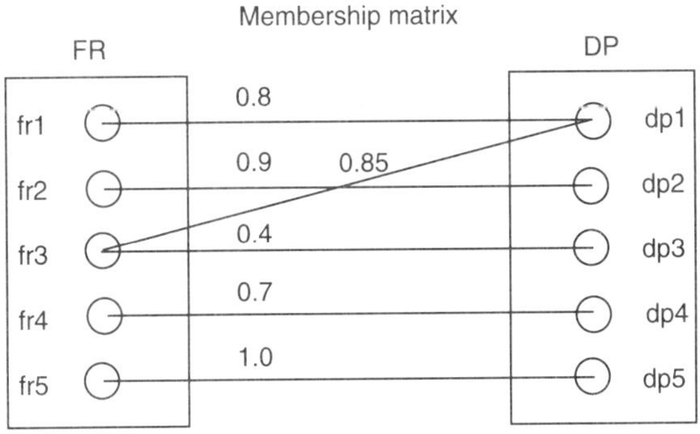

Sagittal diagram

Fig. 5.5 Two representations of the fuzzy binary relationship.

as max–min composition, and the operation is defined as

$$\mu_{(A\circ B)}(U, W) = \max\min[\mu_{(A)}(U, V), \mu_{(B)}(V, W)] \text{ for all } v \in V \tag{5.2}$$

Fuzzy rules are evaluated by means of the compositional rule of inference, and are essentially a fuzzy generalization of the traditional production system (Klir and Folger, 1988). Assume X and A are two fuzzy sets over U and Y, and B is a fuzzy set over V. The statement, 'if X is A then Y is B else C' can be interpreted as $(A \cap B) \cup (A' \cap C)$. Respectively, the rule can be fully captured by a fuzzy relation R defined on $X \times Y$ as follows.

$$R(X, Y) = (A \cap B) \cup (A' \cap C)$$

The fuzzy rules can be generated as A_i (the fuzzy subsets of U) and B_i (the fuzzy subsets of V) where $i \in \mathbb{N}_n$. The above expression can be generalized: if X is A_1 then Y is B_1 else if X is A_2 then Y is B_2 else... or $(A_1 \cap B_1) \cup (A_2 \cap B_2) \cup (A_3 \cap B_3) \cup \cdots \cup (A'_1 \cap A'_2 \cap A'_3 \cap \cdots \cap C)$.

Max–min composition or fuzzy vector–matrix composition resembles vector matrix multiplications, where multiplication is replaced by pair minima and summation is replaced by maxima. For any fuzzy vector pair (F, S), the neural interpretation gives the outer-product correlation matrix (Klir and Folger, 1988) as follows.

$$M = F^{\mathrm{T}} \circ S \tag{5.3}$$

$$m_{ij} = \min(f_i, s_j) \tag{5.4}$$

5.2.1 Fuzzy associative memory

FAM introduced by Kosko (1987), is a two-layer feed forward, hetro-associative, fuzzy classifier that stores an arbitrary fuzzy spatial pattern pair (A_k, B_k) using fuzzy Hebbian learning, where the kth pattern pair is represented by the fuzzy sets (Kosko, 1986; 1987)

$$A_k = \{a_1^k, \ldots, a_n^k\}, \quad B_k = \{b_1^k, \ldots, b_n^k\}$$

$$m_{ij} = A^{\mathrm{T}} \circ B$$

$$m_{ij} = \min(M_A(a_i^k), M_B(b_j^k)) \tag{5.5}$$

where $M_A(a_i^k)$ is a degree of membership of ith member of the fuzzy set A (with the range of 0 to 1) and m_{ij}, as in (5.5), is a fuzzy connection strength from the ith to jth neurons or the max–min composition between the two layers of the network. The recall equations for the first and second layer are

$$M_{A\circ W}(b_j) = \max_{i=1}^{n}[\min(a_i, M_{ij})] \tag{5.6}$$

$$M_{B\circ W^{\mathrm{T}}}(a_i) = \max_{j=1}^{n}[\min(b_j, M_{ij})] \tag{5.7}$$

In this system, FAM is used to synthesize a design from a set of fuzzy functional requirements by exploiting the binary relationships between functional requirements and physical structures.

5.3 SAMPLE PROBLEM DEFINITION

A problem may be expressed as 'given sets of fuzzy functional requirements (FRs) generate the design solution(s) that can satisfy the input requirements'. Assume then that we would like to design a chair, and let the set U be a set of alternative solutions: $U = \{C_i | i \in \mathbb{N}_{11}\}$ as in section 5.4. The input requirement for designing this chair may be stated as 'design a very comfortable chair that can be used more or less in public and should cost in the neighborhood of \$30.00'. Then

F_1^r = The chair must be *very comfortable*
C_1^s = **Use** *more or less in public*
C_2^s = **Cost** should be *in the neighborhood of \$30.00*

Having defined the FRs and design constraints, how do we begin to generate design solution(s)?

Each of the functional requirements and constraints can be represented as a fuzzy set. Therefore it follows that the intersection of functional requirements, F_1^r and the constraints C_1^s, C_2^s forms the desired design of the chair. By adapting the method developed by Bellman and Zadeh (1970) for decision-making in a fuzzy environment and applying it to design, we have the

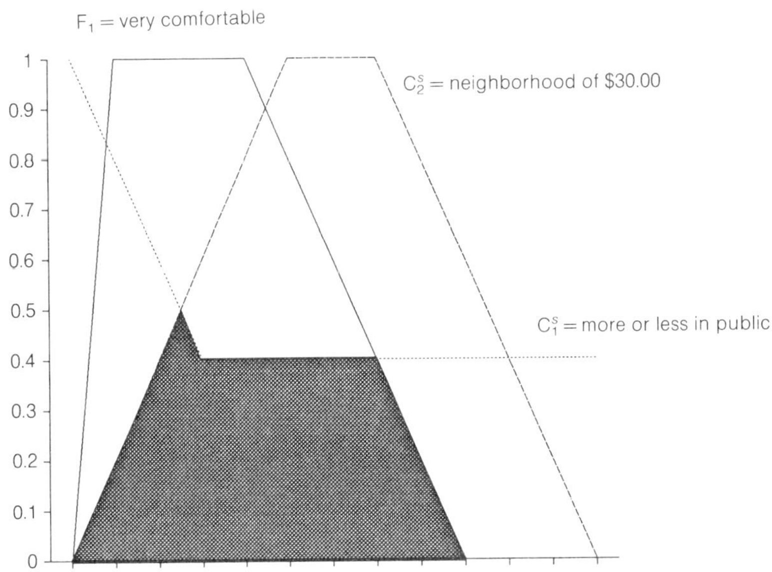

Fig. 5.6. Membership functions for F_1, C_1 and C_2.

following definition: functional requirements can be combined to form a design D, which is a fuzzy set resulting from intersection of all functional requirements. So, we have

$$D=\left(\bigcap_{i=1}^{n} \mathrm{FR}_i\right)\cap\left(\bigcap_{j=1}^{m} \mathrm{CS}_j\right) \tag{5.8}$$

$$\mu_D=\left(\bigwedge_{i=1}^{n} \mu_{\mathrm{FR}_i}\right)\cap\left(\bigwedge_{j=1}^{m} \mu_{\mathrm{CS}_j}\right) \tag{5.9}$$

Therefore to design the required chair our design can be defined as

$$D=F_1\cap C_1^s\cap C_2^s$$
$$\mu_D(c)=\mu_{F_1}(c)\wedge\mu_{C_1^s}(c)\wedge\mu_{C_2^s}(c)$$
$$\mu_D(c)=\min\left[\mu_{F_1}(c),\mu_{C_1^s}(c),\mu_{C_2^s}(c)\right]$$

The relationship between functional requirements and design is given in Fig. 5.6.

5.4 FUZZY KNOWLEDGE REPRESENTATIONS

To experiment with this concept, six generic functional requirements (see Fig. 5.7) and eleven design constraints (see Fig. 5.8) have been generated for designing various classes of chairs. Also eleven different design solutions has been created: C_1, an operational chair; C_2, an executive chair; C_3, contract chair; C_4, a simple office chair; C_5, an office chair; C_6, a stacking chair; C_7, a folding chair, C_8, a long chair; C_9, a classroom chair; C_{10}, an armless chair; and C_{11}, a side chair. Each design solution (chair) satisfies a certain set of functional requirements and design constraints. The fuzzy relationship matrix between functional requirements, design constraints and final solutions are represented in Figs 5.7, 5.8 and 5.9.

We will hypothesize a given set of functional requirements and then use fuzzy associative memory to map a fuzzy set of functional requirements to a crisp set of structures.

Assume we must choose a chair based on the crisp set of 11 pre-designed chairs, where $U=\{c_i \mid i\in\mathbb{N}_{11}\}$ is the set of alternatives (chairs), i.e. the set of possible designs. We also assume that after considerable discussion with marketing groups and analyzing various chairs, the following FAM rule regarding the characteristics of chairs has been indentified:

IF the desired chair must be *highly adjustable*
AND the *cost* is important consideration
AND it must be *used mostly in office*
AND it must be *used for typing*
THEN the chair is an **operational chair**.
or
fr_1 = highly adjustable, $\mu_{\mathrm{fr}_1}=0.8$
Cs_1 = the cost is important consideration, $\mu_{\mathrm{Cs}_1}=0.8$

FRs \ C	1	2	3	4	5	6	7	8	9	10	11
μ_{FR1} (Adjustable)	0	1.0	0	0.8	1.0	0	0	0	0	0	0
μ_{FR2} (Movable)	1.0	1.0	0	0	1.0	0	0	0	0	0	0
μ_{FR3} (Foldable)	0	0	0	0	0	0	1.0	0	0	0	0
μ_{FR4} (Stability)	0	0	1.0	1.0	1.0	1.0	1.0	1.0	1.0	1.0	1.0
μ_{FR5} (Stackable)	0	0	0	0	0	1.0	0	0	0	0	0
μ_{FR6} (Comfortable)	0	1.0	0	0	0	0	0	1.0	0	1.0	0.5

Fig. 5.7 The matrix represents a fuzzy relationship between functional requirements and 11 physical structures. These values are subjective and different designers may select different values.

Cs \ C	1	2	3	4	5	6	7	8	9	10	11
μ_{cs1} (Cost)	0.9	0	0.8	1.0	0	1.0	1.0	0	1.0	0	0
μ_{cs2} (Size)	0	0	0	0	0	1.0	1.0	0	1.0	0	0
μ_{cs3} (Weight)	0	0	1.0	1.0	0	1.0	1.0	0	0	0	0
μ_{cs4} (dining)	0	0	0	0	0	0.8	0	0	0	0	0.9
μ_{cs5} (Office)	1.0	1.0	1.0	1.0	1.0	0	0	0	0	0	0.5
μ_{cs6} (Relaxation)	0	0.7	0	0	0	0	0	1.0	0	0.5	0
μ_{cs7} (Home)	0	0.3	0.5	0	0	0.5	1.0	1.0	0	1.0	1.0
μ_{cs8} (Classroom)	0	0	0	0	0	0	0	1.0	0	0	0
μ_{cs9} (Public)	0	0	0.5	1.0	0	0.5	0.8	1.0	0	0	0
μ_{cs10} (Typing)	1.0	0	0	0	0.8	0	0	0	0	0	0
μ_{cs11} (Aesthetic)	0	1.0	0	0	0	0	0	0.8	0	1.0	0.7

Fig. 5.8 The matrix represents a fuzzy relationship between 11 design constraints and 11 physical structures.

C	1	2	3	4	5	6	7	8	9	10	11
1	1.0	0.3	0	0.7	0	0	0	0	0	0	0
2	0.3	1.0	0	0	0.6	0	0	0	0	0	0
3	0	0	1.0	0.7	0	0	0	0	0	0	0.6
4	0	0	0.7	1.0	0	0	0	0	0	0	0.6
5	0.7	0.5	0	0	1.0	0	0	0	0	0.8	0
6	0	0	0	0.6	0	1.0	0	0	0	0.6	0
7	0	0	0	0	0	0	1.0	0	0	0	0
8	0	0.5	0	0	0	0	0	1.0	0	0.6	0
9	0	0	0	0.6	0	0	0	0	1.0	0	0
10	0	0	0	0	0	0	0	0.6	0	1.0	0
11	0	0	0.6	0	0	0.5	0	0	0	0	1.0

Fig. 5.9 The matrix represents the fuzzy realtionships between 11 physical structures.

Cs_5 = mostly to be used in the office, $\mu_{Cs_5} = 1.0$
Cs_{10} = use for typing $\mu_{Cs_{10}} = 0.5$

Figure 5.10 represents a FAM rule to define an operational chair, the left-hand side as well as the right-hand side is represented as a fuzzy set. If we used the method presented by Bellman and Zadeh for making a design, the design can be stated as $D = fr_1$ and Cs_1 and Cs_2 and Cs_3, i.e.

$$D = fr_1 \cap Cs_1 \cap Cs_2 \cap Cs_3$$

Then one can proceed to select from D the chair which has the largest membership grade (Yager, 1980), since that element (design) satisfies all of our functional requirements and therefore our needs. For instance if our design represented as the fuzzy set D, and contains the following:

$$D = \{0.1, 0.5, 0.4, 0.9, 0.6, 0.8, 0, 0, 0, 0, 0\}$$

Assuming that the rest of the chairs have zero degree of membership in the set. A good design is chair 4 or simple office chair, since its degree of membership is the highest in the fuzzy set D.

In the next section we will explore the possibility of employing the FAM for generating the membership grades for each physical structure based on the given functional requirements and then selecting the one which satisfies all of the functional requirements better than other alternatives.

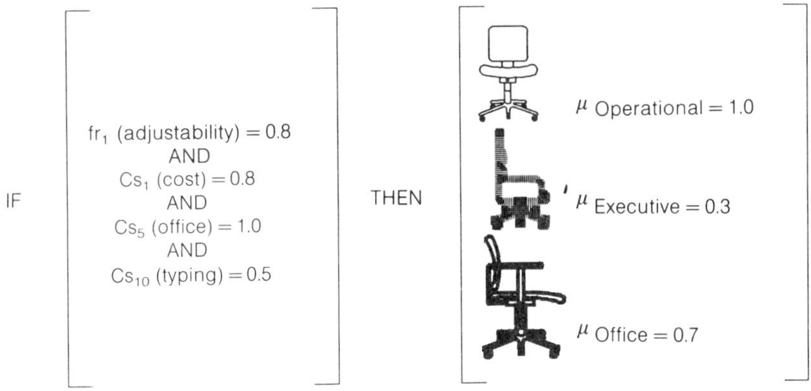

Fig. 5.10 A FAM rule for describing the operational chair.

5.5 MAPPING FUZZY FUNCTIONAL REQUIREMENTS TO DESIGN STRUCTURE BY FAM

Assume that there are n networks, one for each FAM rule as in Fig. 5.11. Then

$$W_i = A_i^{\mathrm{T}} \circ B_i \quad i \in \mathbb{N}_n$$

where A_i and B_i are the antecedent and consequent of the FAM rule i. Now suppose that an l-dimensional input fuzzy vector R is the fuzzy input requirement to the system.

The vector R is fed through the first layer of each network and fires every FAM rule in parallel but to a different degree. Each FAM rule generates an m-dimensional output fuzzy vector P_i, where $i \in \mathbb{N}_n$, and the union of these output fuzzy vectors creates the fuzzy vector D.

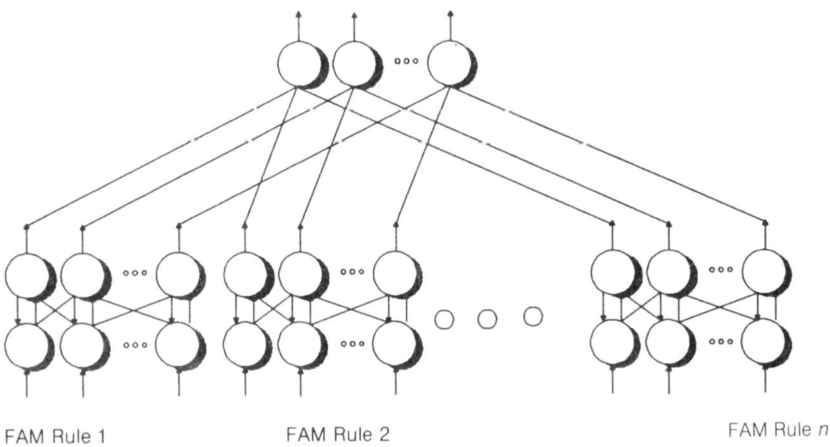

Fig. 5.11 Graphical demonstration of parallel firing and composition of FAM rules.

5.5.1 Defuzzification

Once the output fuzzy vectors P have been generated defuzzification produces a single scalar value $I \in U$, which will be interpreted by the system as the index to the desired design solution.

First, the fuzzy set D must be constructed from composition of all of the output fuzzy vector P_i where $i \in \mathbb{N}_n$.

$$D = \bigcup_{i=1}^{n} P_i \tag{5.10}$$

$$\mu_D(d_k) = \max_{i=1}^{n} \mu_i(P_k) \tag{5.11}$$

The logic operator OR is used to construct the fuzzy set D, since every outcome of FAM rules must be considered as a valid solution to the given design problem. Therefore the fuzzy set D is the union of all output fuzzy vectors P_i (output fuzzy vectors P have been treated as fuzzy sets):

$$D = \{\mu_D(d_k)/C_K \mid i \in \mathbb{N}_n, K \in \mathbb{N}_m, C \in U\}$$

where C_k is the kth design solution with the $\mu_D(d_k)$ degree of membership in the fuzzy set D. Next the technique described in section 5.7 is utilized to select an element in the fuzzy set D with maximum value of membership grade of $\mu_D(d_k)$. The element with maximum grade of membership is the index to the desired design solution.

$$I = \max_{k=1}^{m} \mu_D(d_k)$$

5.6 IMPLEMENTATION AND INPUT/OUTPUT REPRESENTATIONS

The input as we discussed earlier, consists of a fuzzy set FRs. For the purpose of simplicity, the set has been represented as a list. Further, the symbolic representation has been replaced by the positional importance of the degrees of membership in the list. The input layer is designed to have 17 neurons, one neuron for every functional requirement. For example, the functional requirements and design constraints for executive chair (chair 2) are represented as follows:

$$A_2 = (1.0\ \ 1.0\ \ 0\ \ 0\ \ 0\ \ 1.0\ \ 0\ \ 0\ \ 0\ \ 1.0\ \ 0.7\ \ 0.3\ \ 0\ \ 0\ \ 0\ \ 0\ \ 1.0)$$

The first value corresponds to FR_1 (ability to adjust the chair), second to FR_2 (ability to move the chair) and so on. Each degree of membership is assigned to a separate neuron.

The output (second) layer of FAM is designed to have 11 neurons, one for

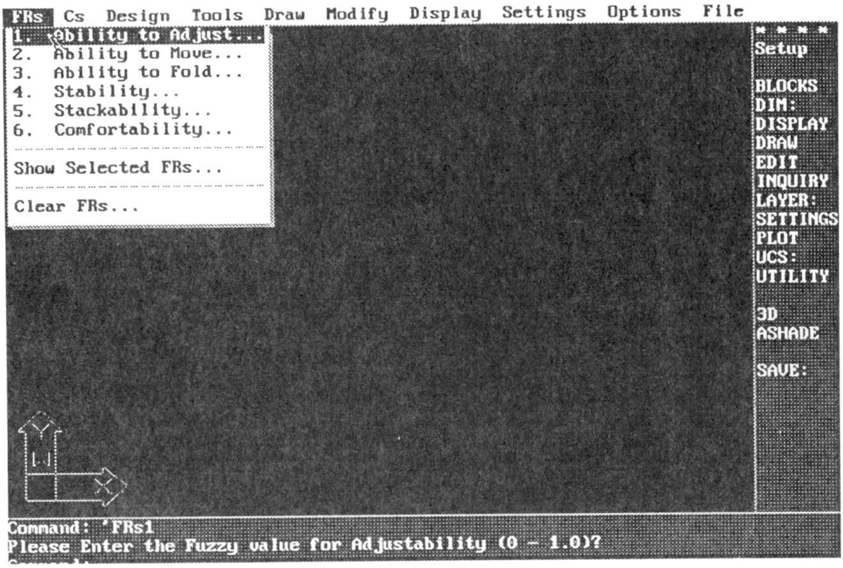

Fig. 5.12 Pull-down menu for selecting the functional requirements. Once the functional requirements are selected, the designer must enter their respective fuzzy values.

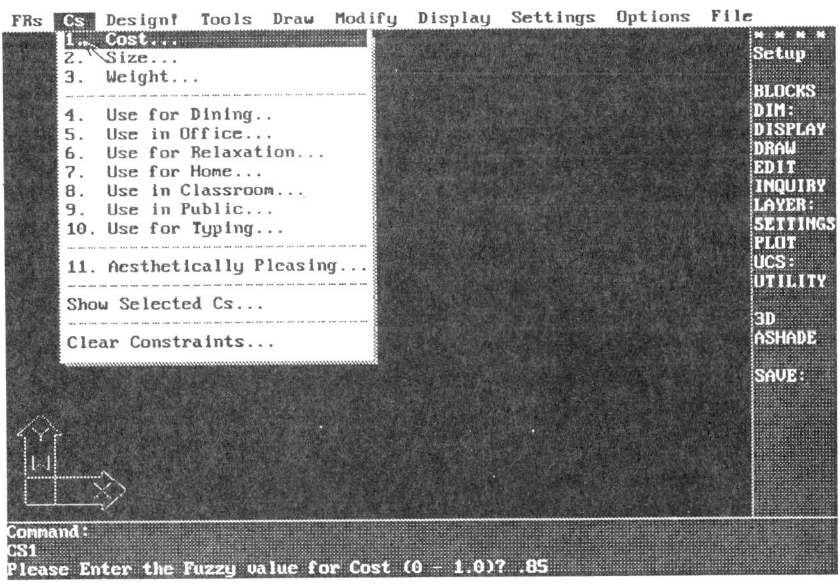

Fig. 5.13 Pull-down menu for selecting the design constraints. Once the design constraints are selected, the designer must enter their respective fuzzy values.

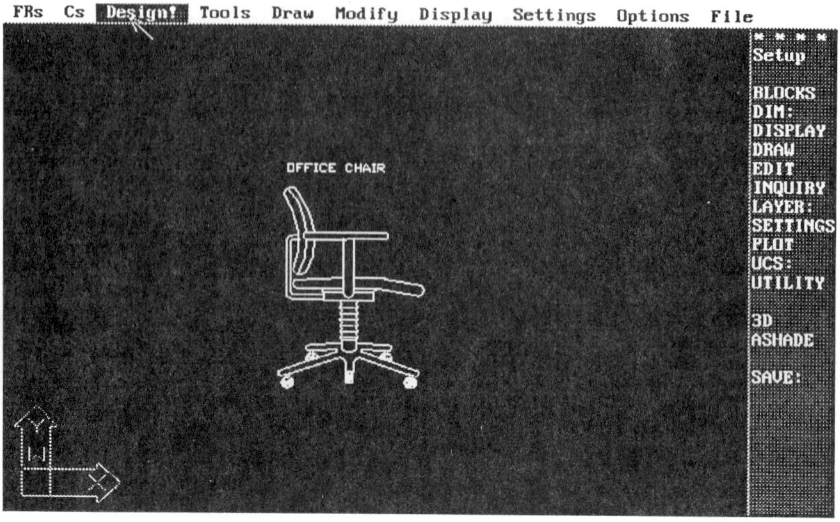

Fig. 5.14 Shows the output of the system after the fuzzy functional requirements have been mapped to the crisp design, in this case, an office chair.

every class of chair. For instance Chair 2 is represented as follows:

$$B_2 = (0.3\ \ 1\ \ 0\ \ 0\ \ 0.6\ \ 0\ \ 0\ \ 0\ \ 0\ \ 0\ \ 0)$$

The AutoCAD, a commercially available computer aided design package has been selected as the basis for the user interface of the system. However, the concepts may be readily adapted and rendered compatible with other CAD systems.

The proposed hybrid system is divided into two separate sections: front end and back end. The front end provides the user interface, and the back end consists of fuzzy associative memory. The front end itself is divided into three sections: pre-processor, post-processor and AutoCAD. The pre-processor communicates with the designer and assists in setting up the functional requirements. The post-processor converts the recall vector of the FAM to graphical representation and displays it; see Figs 5.12, 5.13 and 5.14. The back end does the actual mapping of input requirements to the design solutions. The back end is implemented in C++ and the pre- and post-processors have been written in LISP.

5.7 EXPERIMENTAL RESULTS

To experiment with the system, 11 FAM rules, one for each class of chair, are created. To test the system, 15 design scenarios with known solutions are selected. The system was capable of successfully generating 14 correct solutions; see Table 5.1. For instance, the system was capable of correctly

Table 5.1 Experimental results

No	*Functional requirements (FR_s)*	*Expected design solution(s)*	*Actual design solution(s)*
1	$\{.7/FR_2, .5/FR_{16}\}$	C_1, C_5	C_1
2	$\{.9/FR_1, .9/FR_{12}, .9/FR_{17}\}$	C_2, C_5	C_2
3	$\{.85/FR_6, .9/FR_{17}\}$	C_8, C_{10}	C_2
4	$\{.8/FR_4, 1/FR_7, .9/FR_{11}\}$	C_3, C_4, C_6	C_3
5	$\{.8/FR_1, 1/FR_2, .5/FR_{16}\}$	C_1, C_5	C_5
6	$\{.8/FR_1, .5/FR_4, .6/FR_7, 1/FR_{11}\}$	C_4	C_4
7	$\{.9/FR_5, .8/FR_{10}\}$	C_6	C_6
8	$\{.4/FR_5, 1/FR_{10}\}$	C_{11}	C_{11}
9	$\{.4/FR_5, .5/FR_{10}\}$	C_3, C_6, C_{11}	None
10	$\{.8/FR_4, 1/FR_{12}, .8/FR_{13}\}$	C_3, C_8, C_{10}	C_{10}
11	$\{.9/FR_6, 1/FR_{12}, .9/FR_{13}\}$	C_2, C_8, C_{10}	C_{10}
12	$\{.6/FR_4, 1/FR_{14}\}$	C_4, C_9	C_9
13	$\{.5/FR_6, 1/FR_{14}\}$	C_4, C_9	C_{10}^{**}
14	$\{1/FR_6, 1/FR_{17}\}$	C_2, C_8, C_{10}	C_{10}
15	$\{.8/FR_6, 1/FR_{10}\}$	C_3, C_{11}	C_{11}

**Error

suggesting an executive chair despite the fact that the new input requirements were quite different from the encoded FAM rules. In scenario number 9, it couldn't generate any solution. Overall the system performed with accuracy rate of better than 90%.

5.8 SUMMARY

The chapter describes the conceptual design problem and demonstrates the application of fuzzy associative memory (FAM) in conceptual design by mapping fuzzy functional requirements to crisp design solution.

Computer aided design and manufacturing (CAD/CAM) have become conventional engineering tools. Although these tools can greatly enhance the detailed aspects of design such as analysis of the final design or drafting they are nevertheless incapable of aiding designers during the conceptual or creative phase of the design process.

This chapter has established the basic foundation of true intelligent CAD system, a tool that can aid designers in all phases of the design process. The high accuracy in retrieving a crisp design solution based on the fuzzy input requirements and the simplicity of its practical implementation, makes the system an attractive tool in the idea generation phase of the design process.

REFERENCES

Bahrami, A. and Dagli, Cihan H. (1993) From fuzzy input requirements to crisp design. *International Journal of Advanced Manufacturing Technology*, **8**, 52–60.

Bahrami, A., Dagli, C. and Modarress, B. (1991) Fuzzy associative memory in conceptual design, in *Proceedings of IEEE International Joint Conference on Neural Networks (IJCNN)*, Seattle, WA (8–12 July), pp. I-183, I-188.
Bellman, R.E. and Zadeh, L.A. (1970) Decision making in fuzzy environment. *Management Science*, **17**(4), 144–64.
Coyne, R.D., Rosenman, M.A., Radford, A.D., Balachandran, M. and Gero, J.S. (1990) *Knowledge-based Design Systems*, Addison-Wesley, Reading, Massachusetts.
Klir, J.G. and Folger, A. (1988) *Fuzzy Sets, Uncertainty, And Information*, Prentice Hall, Englewood Cliffs, New Jersey.
Kosko, B. (1986) Fuzzy cognitive maps. *International Journal of Man-Machine Studies*, **24**, 65–75.
Kosko, B. (1967) Fuzzy associative memories, *Fuzzy Expert Systems* (ed. A. Kandel) Addison-Wesley.
Suh, N.P. (1984) Development of the scientific bases for the maufacturing field through the axiomatic approach. *Robotics and Computer Integrated Manufacturing*, **1**(3/4), 399–455.
Suh, N.P. (1990) *The Principles of Design*, Oxford University Press.
Suh, N.P., Bell, A.C. and Gossard , D.C. (1978) On an axiomatic approach to manufacturing systems. *J. Engin. for Industry*, **100**(2), 127–30.
Yager, R.R. (1980) Satisfaction and fuzzy decision functions, in *Fuzzy Sets Theory and Applications to Policy Analysis and Information Systems* (eds P. Wang and S. Chang), pp. 309–20.
Zadeh, L.A. (1975a) The concept of a linguistic variable and its application to approximate reasoning – I, *Information Science*, **8**, 199–249.
Zadeh, L.A. (1975b) The concept of a linguistic variable and its application to approximate reasoning – II, *Information Science*, **8**, 301–57.
Zadeh, A.L. (1965) Fuzzy sets, *Information and Control*, **8**, 338–53.

CHAPTER 6

Machine-part family formation

César O. Malavé and Satheesh Ramachandran

Department of Industrial Engineering, Texas A & M University, College Station, TX 77843-3131.

6.1 CHARACTERISTICS OF GROUP TECHNOLOGY

Group technology is a organizational approach which justifies small and medium batch sized production systems. Group technology based manufacturing systems offer the advantages of flow production as well as the production flexibility of batch manufacturing. By employing clustering techniques, the part-machine spectrum of the manufacturing system and the relevant manufacturing process are analyzed, and arranged according to design, similarity of machining and product flow. This leads to an organization of the production system into self-contained and self-regulating groups of machines called **machine cells**. Each machine cell undertakes a maximal production of a family of parts having similar manufacturing characteristics. The principal advantages of a production system employing group technology lie in a substantially reduced work-in-process inventory and improved delivery performance because of reduced throughput time. The three general techniques of part family–machine group formation are:

1. by inspection,
2. classification and coding, and
3. production flow analysis (PFA).

Inspection

The selection of part families by directly observing the products and the processes used to make them, is generally possible only in the simplest cases.

Artificial Neural Networks for Intelligent Manufacturing. Edited by Cihan H. Dagli. Published in 1994 by Chapman & Hall, London. ISBN 0 412 48050 6

Design drawings, process planning sheets or actual parts are used to identify a family of parts and the corresponding groups of machines. Burbidge (1975) points out that this technique can be used in assembly, process industries and in offices. In the manufacture of parts, however, this technique is only possible when a very small range of parts is involved. Burbidge reported the successful application of this technique in a company manufacturing 500 different parts, but warns that the inspection method is generally inapplicable when the number of parts exceeds 200.

Classification and coding

In this approach, parts are coded by their manufacturing features, so parts having a similar feature code possess similar numbers, letters, etc. The three basic part features that can be classified and coded are:

1. shape or geometry of part,
2. function of part, and
3. manufacturing operations and tooling.

However, the information that can be incorporated into the part codes is not limited to these basic part features, e.g. production planning data can be added to the code systems. Unfortunately, generic classification and coding systems are hard to develop and expensive to maintain. Nevertheless, the overwhelming majority of manufacturing companies that have implemented group technology based production systems, do use this procedure.

Production flow analysis (PFA)

PFA is a method of part-machine grouping which was developed by Burbidge in 1963. PFA is particularly appealing since it does not require a complex part-coding system, is relatively simple to implement, and can be applied to the reorganization of existing production systems as well as to the design of new production systems. The method involves the following stages, which Burbidge (1975) discusses in great detail.

1. In the first stage, machines are classified by numbers according to type of operation that can be performed on them. Machines which can perform similar operations are usually classified with the same type number. The specific need of parts for particular machines within a machine type are considered when the machines are finally allocated to the groups.
2. In the second stage, the parts list and the information in the production route card are carefully checked to identify and ensure correctness of the essential information for analysis. The essential information for each part consist of the operation to be performed on it and the machines necessary to perform each operation.
3. The third stage, factory flow analysis (FFA), involves macro examination of the part flow through the machines. This allows the problem to be

decomposed into major machine-part groups. Although essential to the PFA process, these three stages are merely the necessary preliminaries to provide the required data for the ultimate purpose of the analysis – to determine appropriate machine-part subgrouping for cellular layouts.

4. The last stage, involving the problem of identification of machine cell-part family, is the most difficult part of PFA. The difficulty comes from the amount of information that must be processed accurately to facilitate the formation of manufacturing cells that are dedicated to specific part-family production.

A number of clustering methodologies have been used to tackle the machine-part analysis problem. A brief description of the traditional methods is presented in the next section.

6.1.1 Traditional clustering methodologies

Clustering algorithms are broadly classified into the following five categories: similarity-coefficient based methods, graph theoretic methods, mathematical programming methods, AI based methods and matrix manipulation methods. These are now considered in turn.

Similarity-coefficient based methods

Similarity coefficients can be calculated between different pairs of machines (or different pairs of parts). If machines can be represented as vectors, different distance metrics or other measures can be used to determine similarity between pairs of machines. A simple measure of similarity between two machines could be

$$\frac{\text{the number of parts that require processing on both machines}}{\text{number of parts that require processing on either of the machines}}$$

A matrix of similarity coefficients is developed, and most algorithms use this to construct a tree like diagram called the **dendogram**. Machines high up in the dendogram correspond to machines with a high degree of similarity (primary machines). Machines with lesser degrees of similarity (secondary and tertiary machines) figure in the bottom of the dendogram. Similarity coefficient based algorithms uses this tree to form machine groups. Machine cells and part families are found separately and assigned to each other. McAuley (1972) has used a **single-linkage cluster analysis** and Seiffodini and Wolfe (1986) have used **average-linkage cluster analysis** to construct the part family. The primary disadvantage of this approach lie in the subjectiveness involved in classifying 'similarity'. Arbitrariness exists in deciding thresholds to determine similarity between machines. Nevertheless, these approaches are computationally efficient in the data storage space required.

Graph theoretic methods

These methods treat the machines and parts as nodes, and the processing of components as arcs connecting these nodes. The primary objective of these models is in obtaining disconnected subgroups from the machine-part graph to identify part family and machine cells. Rajagopalan and Batra (1975) have suggested a graph partitioning approach and Lee and Hwang (1991) have proposed a new hierarchical divisive clustering heuristic based on this graph theory to generate clusters.

Mathematical programming methods

Kusiak (1987a) has suggested a 0–1 integer programming formulation called the **p-median** which minimizes the total sum of distances between any two parts, for a prespecified number of part families under the constraint that each part should be assigned to one part family only. Quadratic programming formulation for restriction on number of cells and cell sizes have been developed by Kumar *et al.* (1986). Srinivasan *et al.* (1990) developed an assignment model of identifying preliminary machine cells and part families with the objective of minimizing exceptional elements and voids. Steudel and Ballakur (1987) developed a dynamic programming based heuristic to determine machine groups. Although, these approaches give good solutions, they are computationally inefficient, especially in the case of large scale industrial problems.

Artificial intelligence (AI) based methods

A part family identification problem is based on the knowledge of manufacturing processing requirements. To perform this task, a knowledge-based expert system (Kusiak, 1987b) and the syntactic pattern recognition approach (Wu *et al.*, 1986) have been developed. The problem of AI based techniques is that they are not capable of generalization if a new input does not fall into the existing domain specific knowledge.

Matrix manipulation methods

These methods rearrange the rows and columns of matrix in order to bring the non-zero elements around the main diagonal. McCormick *et al.* (1972) suggested the **bond energy method**. Askin and Subramanian (1987) developed a clustering algorithm that considers the manufacturing cost such as machining cost, set-up cost and material handling cost. The **rank order clustering** (ROC) algorithm and its extension developed by King (1980) and King and Nakornchai (1982) respectively, **direct clustering** algorithm developed by Chan and Milner (1982), and the **cluster identification** algorithm developed by Kusiak and Chow (1987c) are some examples in this category. Matrix manipulation methods have many advantages over the previous approaches as they avoid the subjectiveness involved in the similarity coefficient approaches and the complexity of set theoretic and integer programming approaches. Their popularity stems from these advantages as well as their computational simplicity. These approaches will be dealt with in a bit more detail, as they are often used for comparison with the neural network procedure in the later sections.

Matrix clustering approaches manipulate the rows and columns of the part machine incidence matrix until a block diagonal form is obtained. The part machine incidence matrix contains the process route information of the parts in the form of a matrix

$$X = (x_{i,j})$$

where i = machine number and j = part number. If $x_{i,j} = 1$, machine i is required for processing part j and if $x_{i,j} = 0$ machine i is not required for processing part j. Most of the matrix clustering approaches are able to obtain the block diagonal form in the case of an easily diagonalized matrix, i.e., without the presence of exceptional elements. Exceptional elements are machines or parts which do not permit the arrangement of the incidence matrix in a strictly block diagonal form. In other words, they cause machines to be shared by different families of parts. The shared machine is called a **bottleneck machine** and the part which requires processing on two machine cells is called an **overlapping part**. These approaches do not always lead to a strict block diagonal form as illustrated by the following simple example.

King's ROC is used to diagonalize the initial part machine incidence matrix in Fig. 6.1(a), the diagonalized part machine incidence matrix is shown in Fig. 6.1(b). The arrangement of the final part machine matrix results in two machine cells and two part families.

Machine cell A: Machines 3, 5
Machine cell B: Machines 1, 2, 4

Part family A′: Parts 1, 3, 4
Part family B′: Parts 2, 5, 6

Part families 1 and 2 are the part families associated with machine cells 1 and 2 respectively. Here machine 6 is a bottleneck machine as it has to be shared by both the part families. If machine 6 is associated with the machine cell 1, then part 2 has to visit machine cell 1 for processing and is an overlapping part. Similarly, if machine 6 is associated with machine cell 2, then part 1 has to visit machine cell 2 for processing and is an overlapping part. Thus there are two possible machine cell formations: (i) [3, 5, 6], [1, 2, 4] and (ii) [3, 5], [1, 2, 4, 6], both resulting in inter-cell movement of one part. Easy identification of this information will be useful in the final formation of machine cells.

Although the clustering of the parts is evident for the matrix of this size, such identification for a larger matrix with a reasonable number of bottleneck

	p1	p2	p3	p4	p5	p6
m1		1			1	1
m2		1			1	1
m3	1		1	1		
m4		1			1	1
m5	1			1		
m6	1	1				

(a)

	p1	p2	p3	p4	p5	p6
m6	1	1				
m3	1		1	1		
m5	1		1			
m1		1			1	1
m2		1			1	1
m4					1	1

(b)

Fig. 6.1 Initial and final matrix using ROC algorithm.

machines would be very difficult using the final matrix. This is particularly so when a strictly block diagonal form is not achieved, as in the example shown above. In the clustering approaches, clustering is achieved in two steps: when exceptional elements are present, an initial clustering is done to identify the exceptional elements; and in the second step the exceptional elements are suppressed to arrive at a block diagonal form. This makes identifying the exceptional elements very important.

Sometimes it is desirable to see how possible future parts respond to the obtained classification. Measures exist in the literature of the distance between any two parts. Distance measures between a new part and cells could be useful. Apart from deciding the cell to which the new part is closer, this might help in deciding the assignment of the bottleneck machine. For example, referring to Fig. 6.1, if it is known that there is a possibility of a future part requiring machine 6 and also requiring machines 3 and 5, then it would be better to assign machine 6 to machine cell 1.

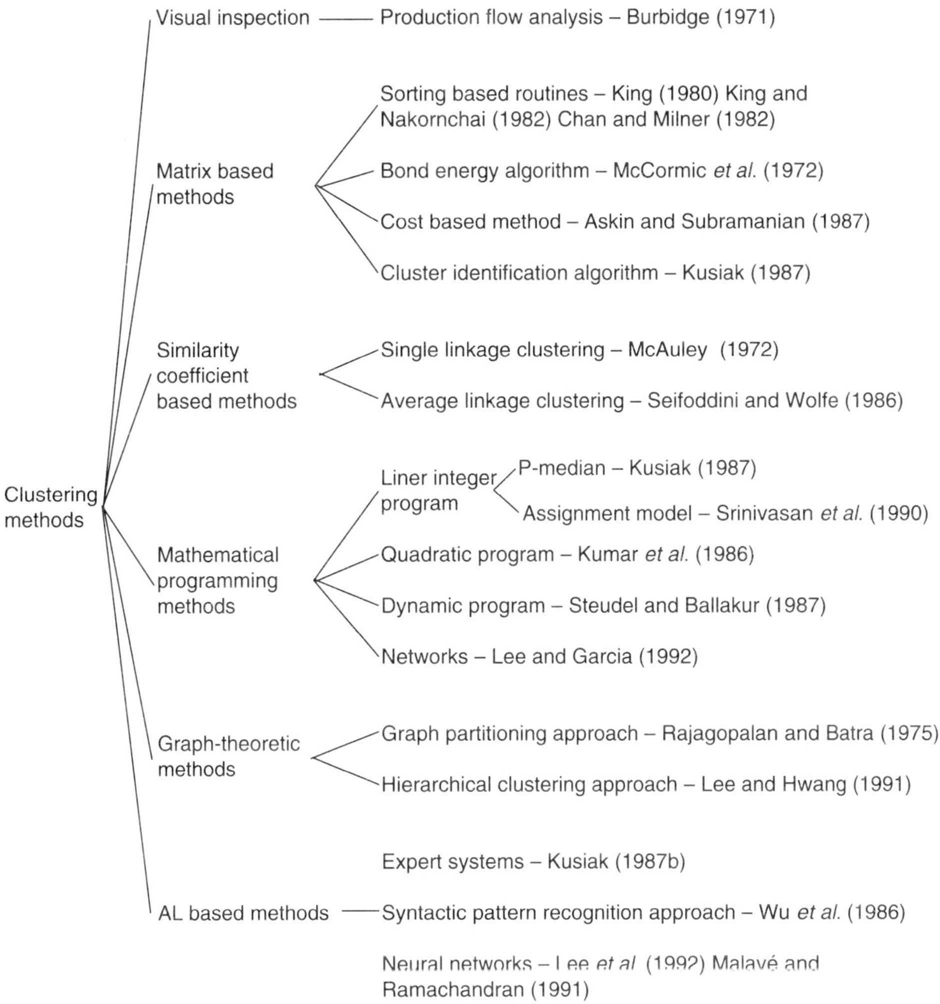

Fig. 6.2 A taxonomy of clustering methodologies in literature.

A brief taxonomy of existing clustering methodologies is shown in Fig. 6.2, and drawbacks and objectives of seven popular algorithms are summarized in Table 6.1.

6.2 NEURAL NETWORK APPROACH

Self-organizing neural networks can be used as classifiers. Most of these self-organizing neural networks are based on competitive learning paradigms. After training with a set of input patterns, these networks have the capability

Table 6.1 A brief summary of a few clustering algorithms and drawbacks

Algorithms	*Main idea*	*Drawbacks*
Rank order clustering	Sorting procedure to obtain a block diagonal form by assigning binary weight to each row and column.	1. The need of storing binary words. 2. Extremely difficult to calculate weights for large size problem. 3. It does not perform well for ill-structured matrices.
Direct clustering	Sorting procedure to obtain a block diagonal form by re-ordering row and columns based on total number of 1s.	1. Human intervention is required to deal with bottle-neck machines. 2. The diagonal in the matrix changes at each iteration.
P-median	A 0–1 integer program formulation to minimize the total sum of distance between any two parts.	1. Excessive computing time is needed (NP-complete). 2. Not a practical approach for large scale industrial problems.
Dynamic programming	Optimal chain of machines for maximizing the sum of similarity along the chain is found by DP and then chain is partitioned into cell by a heuristic approach.	1. Not an efficient procedure for large scale problems because it requires $O(n2^n)$ space and $xO(n^x 2^n)$ time where x is the time to compose one operation.
Graph partion approach	A graph is constructed with machine as nodes and similarity coefficient as arcs. All cliques in the graph are found using Bron-Kresboch's algorithm to form machine cells.	1. Finding all cliques in graph and partitioning a graph is also NP-complete. 2. Limitations on the size of the problem.
Single-linkage clustering	Based on similarity coefficient for grouping machines, hierarchical agglomerative clustering procedure is used.	1. The similarity coefficients used is not appropriate in machine grouping context (biased results). 2. Choosing a different threshhold value in clustering could result in different set of clusters.
Expert systems	Building a knowledge based system consisting of 5 components: 1. Data base 2. Knowledge base 3. Inference engine 4. Request processor 5. Clustering algorithm	1. Existing clustering algorithm should be incorporated into knowledge base model. 2. No generalization ability.

of identifying clusters of similar patterns in the set of input patterns. This property of self-organizing neural networks is used to identify part families and machine cells from a given part machine incidence matrix. The first attempt at the use of neural networks for the part family–machine cell formation problem was by Young (1990). Young used the **interactive activation and competition model** to identify part groups and machine cells from the part machine matrix. The approach does not deal with a lot of issues that usually accompany the group technology (GT) problem. Also, this approach requires a large network architecture, the number of neurons required exceeds the sum of the parts and machines in the part machine matrix.

Later approaches by Malavé and Ramachandran (1991), Lee *et al.* (1992), and Dagli and Huggahalli (1992) are more comprehensive in dealing with the GT problem. Malavé *et al.* used a competitive learning algorithm for part family–machine cell formation. This approach uses the input–output information as well as information from the internal structure of the neural network to form part families and machine cells. Lee *et al.* (1992) extended this with another competitive learning neural network based on the Kohonen's rule, and verified the computational advantages of this approach over other optimal approaches. Dagli and Huggahalli (1992) used the ART-1 neural network, and this approach performs well compared to the conventional approaches. These two approaches will be discussed in detail in the next two sections.

6.2.1 Competitive learning approach

Competitive learning networks inherit their name from the competition that occurs at the output layer of the neural network to respond to an input vector. Most of the networks used in machine cell–part family formation are based on competitive learning algorithms. A good discussion on competitive learning algorithms is found in Malavé and Ramachandran (1991), and will not be considered here. The procedure discussed in this section is based on a simple competitive learning algorithm, as explained in the next section.

Network architecture and learning

Malavé and Ramachandran (1991) used a neural network based on a simple competitive learning algorithm to perform part family and machine cell formation. After training the network with the part vectors, the part family identification is done by observing the steady state input–output information of the neural network. The machine cells are identified from the weight information of the neural network at steady state. The architecture of the neural network used in shown in Fig. 6.3.

The architecture of the neural network consists of two layers, p neurons at the input layer and q neurons at the output layer. Associated with each output layer neuron j is a weight vector $\mathbf{W}_j = (w_{j,1}, w_{j,2}, \ldots, w_{j,p})$. The weight

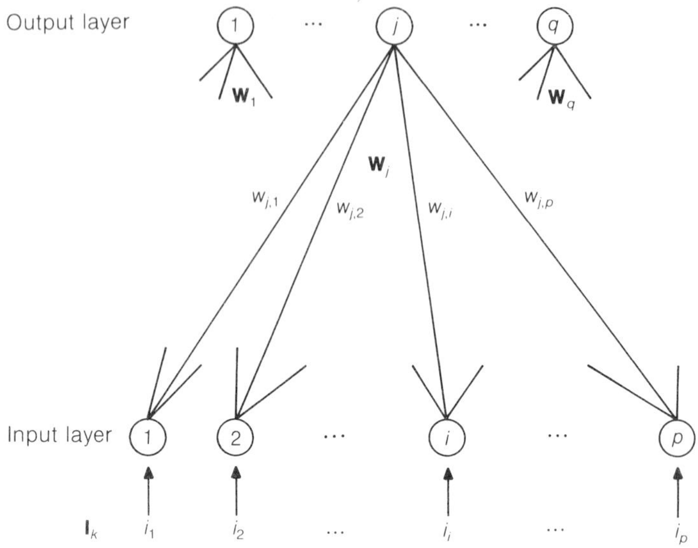

Fig. 6.3 Network architecture for the competitive learning network.

matrix **W** refers to the matrix whose rows are the weight vectors, as shown below.

$$\mathbf{W} = \begin{bmatrix} w_{1,1} & w_{1,2} & \cdots & w_{1,p} \\ w_{2,1} & w_{2,2} & \cdots & w_{2,p} \\ \cdot & & & \\ \cdot & & & \\ w_{q,1} & w_{q,2} & \cdots & w_{q,p} \end{bmatrix}$$

The neural network takes binary patterns as input at the bottom layer. The output from the input layer is the same as the input to it. When an input pattern $\mathbf{I}_k = (i_1, i_2, \ldots, i_p)$ is presented at the input layer, the neurons at the output layer compete among themselves to respond to the input pattern. The output layer neuron k^* which wins the 'right' to respond to the input pattern $\mathbf{I}_k$, is selected as follows:

$$\mathbf{I}_k \cdot \mathbf{W}_{k^*} = \max_{j=1,\ldots,q} (\mathbf{I}_k \cdot \mathbf{W}_j)$$

Thus, the output layer neuron whose weight vector has the highest dot product with $\mathbf{I}_k$ is selected as the winning neuron. Here, the dot product is used as a metric to identify the output neuron whose weight vector is closest to the input pattern. In each learning iteration, only the winning output layer neuron changes its weight vector, and has an output of 1. The rest of the output layer neurons have an output of 0. The winning output neuron moves its weight vector closer to the input pattern by the following weight change equation:

$$w_{k^*,j}(t+1) = \frac{i_j Cr}{N_k} - rw_{k^*,j}(t)$$

where N_k is the number of 1s in the input pattern $\mathbf{I}_k, C$ is a constant and r is another constant called the **learning rate**. Each of the weight vectors $\mathbf{W}_j$ has a fixed quantity of weight distributed among its elements, that is,

$$w_{j,1} + w_{j,2} + \cdots + w_{j,p} = C \quad \text{for } j = 1, 2, \ldots, q$$

During training, the weight vector of the winning neuron moves a part of the weight from the interconnections which receive a 0 at the input layer to the interconnection which receive 1 at the input layer. By this process, the weight vectors shift toward the centroid of the clusters that exist naturally in the input vector space. The learning rate decides the fraction of weight transferred from the interconnections with 0 input (inactive interconnections) to the interconnections which receive 1 input (active interconnections).

Part family identification

From the part machine incidence matrix, a set of binary vectors can be identified in which each vector represents a particular part. These vectors are called characteristic vectors, since they characterize parts. This set of vectors form the training set for the neural network. At steady state, each of these characteristic vectors is associated with a particular output layer neuron. The group of vectors which are associated with a particular output layer neuron form a part family. For example, consider the part machine matrix adopted from Chan and Milner (1982) shown in Fig. 6.4.

		Parts									
		1	2	3	4	5	6	7	8	9	10
	1	0	0	1	1	0	1	0	0	0	0
	2	1	0	0	0	0	0	1	0	0	1
	3	0	1	0	0	1	0	0	1	0	0
	4	0	0	0	1	0	1	0	0	1	0
	5	0	1	0	0	1	0	0	1	0	0
	6	0	0	1	0	0	1	0	0	1	0
	7	0	0	0	0	0	0	1	0	0	1
Machines	8	0	1	0	0	1	0	0	1	0	0
	9	0	0	1	1	0	1	0	0	1	0
	10	1	0	0	0	0	0	1	0	0	1
	11	1	0	0	0	0	0	1	0	0	1
	12	1	0	0	0	0	0	1	0	0	1
	13	0	1	0	0	1	0	0	1	0	0
	14	0	0	1	1	0	1	0	0	1	0
	15	0	1	0	0	1	0	0	1	0	0

Fig. 6.4 Initial part machine matrix, adapted from Chan and Milner (1982).

Part no	Part characteristic vector															Output vector			Part family
1	0	1	0	0	0	0	0	0	0	1	1	1	0	0	0	0	0	1	3
2	0	0	1	0	1	0	0	1	0	0	0	0	1	0	1	0	1	0	2
3	1	0	0	0	0	1	0	0	1	0	0	0	0	1	0	1	0	0	1
4	1	0	0	1	0	0	0	0	1	0	0	0	0	1	0	1	0	0	1
5	0	0	1	0	1	0	0	1	0	0	0	0	1	0	1	0	1	0	2
6	1	0	0	1	0	1	0	0	1	0	0	0	0	1	0	1	0	0	1
7	0	1	0	0	0	0	1	0	0	1	1	1	0	0	0	0	0	1	3
8	0	0	1	0	1	0	0	1	0	0	0	0	1	0	1	0	1	0	2
9	0	0	0	1	0	1	0	0	1	0	0	0	0	1	0	1	0	0	1
10	0	1	0	0	0	0	1	0	0	1	1	1	0	0	0	0	0	1	3

Fig. 6.5 Part vectors, output activations and part families for Example 1.

The matrix in Fig. 6.4 consists of 10 parts and 15 machines. Thus, each part can be represented by a 15 dimensional binary column vector. For example part 1 can be represented by the binary vector

$$[0\ \ 1\ \ 0\ \ 0\ \ 0\ \ 0\ \ 0\ \ 0\ \ 0\ \ 1\ \ 1\ \ 1\ \ 0\ \ 0\ \ 0].$$

The network architecture for this problem consists of 15 neurons at the input layer and 3 neurons at the output layer. (The selection of the number of output layer neurons will be explained in a later section.) The set of 10 part characteristic vectors form the training set, and the network is trained sequentially using this training set. After training, the weight vectors of the neural network stabilize to fixed values with negligible fluctuations. This state of the network is referred to as the **steady state**. At steady state, a group of parts that are associated with a particular output unit form a part family. For the matrix in Fig. 6.4, steady state is achieved after about 100 training iterations, and at steady state the part families formed are shown in Fig. 6.5.

The part families formed are:

Part family 1: Parts 3, 4, 6, 9
Part family 2: Parts 2, 5, 8
Part family 3: Parts 1, 7, 10

Thus, at steady state the part family formation can be done by observing the corresponding output layer activation for each part characteristic vector. The resulting part families formed compare exactly with the part families obtained by classical clustering approaches. The distinct benefit of this approach lies in the ease in identifying the part families as compared to the visual identification of part families of most matrix manipulation approaches.

As a further example, consider a more complicated part machine incidence matrix adopted from King (1980) and shown in Fig. 6.6.

The above part machine matrix consists of 24 parts and 14 machines, each part represented by a column vector. A neural network with 15 input neurons and 4 output neurons was trained with this training set and the resulting part family formation is shown in the Fig. 6.7.

	Parts																							
Machines	01	02	03	04	05	06	07	08	09	10	11	12	13	14	15	16	17	18	19	20	21	22	23	24
1	0	0	0	0	0	1	1	0	0	0	0	0	0	0	0	0	0	0	0	0	0	0	0	0
2	0	0	1	1	0	0	0	0	0	0	0	0	0	0	0	0	0	0	0	0	0	0	0	0
3	0	0	1	1	0	0	0	0	0	0	0	0	0	0	0	0	0	0	0	0	0	0	0	0
4	1	1	0	0	0	0	0	0	0	0	0	0	0	0	0	0	1	0	1	1	0	0	1	0
5	1	1	0	0	0	0	0	0	0	0	0	0	0	0	0	0	1	0	0	1	0	0	1	0
6	0	0	0	0	0	0	0	0	1	1	1	1	0	1	1	1	0	0	0	0	0	1	0	0
7	1	1	0	0	0	0	1	0	0	0	0	0	0	0	0	0	0	0	0	0	0	0	0	0
8	0	0	0	0	1	0	0	0	0	1	1	0	1	0	1	1	1	0	0	0	0	1	0	0
9	0	0	0	0	1	0	0	0	1	0	0	1	1	0	1	0	0	0	0	0	0	0	0	0
10	0	0	1	0	0	0	0	0	0	0	0	0	0	0	0	0	0	0	0	0	0	0	0	1
11	0	0	1	1	0	0	0	0	0	0	0	0	0	0	0	0	0	0	0	0	1	0	0	1
12	0	0	0	0	0	0	1	1	0	0	0	0	0	0	0	0	0	0	0	0	0	0	0	0
13	0	0	0	0	0	1	1	1	0	0	0	0	0	0	0	0	0	1	0	0	0	0	1	0
14	0	0	0	0	0	0	0	0	1	0	1	0	1	0	1	0	0	0	0	0	0	0	0	0

Fig. 6.6 Initial part machine, matrix, adapted from King (1980).

Part	Binary vector														Output activation				Part family
01	0	0	0	1	1	0	1	0	0	0	0	0	0	0	1	0	0	0	1
02	0	0	0	1	1	0	1	0	0	0	0	0	0	0	1	0	0	0	1
03	0	1	1	0	0	0	0	0	0	1	1	0	0	0	0	0	1	0	3
04	0	1	1	0	0	0	0	0	0	0	1	0	0	0	0	0	1	0	3
05	0	0	0	0	0	0	0	1	1	0	0	0	0	0	0	0	0	1	4
06	1	0	0	0	0	0	0	0	0	0	0	0	1	0	0	1	0	0	2
07	1	0	0	0	0	0	1	0	0	0	0	1	1	0	0	1	0	0	2
08	0	0	0	0	0	0	0	0	0	0	0	1	1	0	0	1	0	0	2
09	0	0	0	0	0	1	0	1	1	0	0	0	0	1	0	0	0	1	4
10	0	0	0	0	0	1	0	1	0	0	0	0	0	0	0	0	0	1	4
11	0	0	0	0	0	1	0	0	0	0	0	0	0	1	0	0	0	1	4
12	0	0	0	0	0	1	0	1	1	0	0	0	0	0	0	0	0	1	4
13	0	0	0	0	0	0	0	0	1	0	0	0	0	1	0	0	0	1	4
14	0	0	0	0	0	1	0	1	0	0	0	0	0	0	0	0	0	1	4
15	0	0	0	0	0	1	0	1	1	0	0	0	0	1	0	0	0	1	4
16	0	0	0	0	0	1	0	1	0	0	0	0	0	0	0	0	0	1	4
17	0	0	0	1	1	0	0	0	0	0	0	0	0	0	1	0	0	0	1
18	0	0	0	0	0	0	0	0	0	0	0	0	0	1	0	1	0	0	2
19	0	0	0	1	0	0	0	0	0	0	0	0	0	0	1	0	0	0	1
20	0	0	0	1	1	0	0	0	0	0	0	0	0	0	1	0	0	0	1
21	0	0	1	0	0	0	0	0	0	0	1	0	0	0	0	0	1	0	3
22	0	0	0	0	0	1	0	1	0	0	0	0	0	0	0	0	0	1	4
23	0	0	0	1	1	0	0	0	0	0	0	0	1	0	1	0	0	0	1
24	0	0	0	0	0	0	0	0	0	1	1	0	0	0	0	0	1	0	3

Fig. 6.7 Part vectors, output activations and part families for Example 2.

The four part families formed are

Part family 1: Parts 1, 2, 17, 19, 20, 23
Part family 2: Parts 6, 7, 8, 18
Part family 3: Parts 3, 4, 21, 24
Part family 4: Parts 5, 9, 10, 11, 12, 13, 14, 15, 16, 22

The part families obtained compare exactly with the results by King using the ROC algorithm.

Machine cell formation

Machine cells are formed by assigning the machines among the various part groups formed. The assignment of machine cells can be done by observing the steady state weight matrix, because the steady state weights of the neural network contain information on how the machines are distributed among the various part groups formed. This section discusses the procedure to determine the machine cells from the steady state weights with the aid of examples. The following definitions are required. Let

$T_{j,i}$ = number of parts requiring machine i and associated with output neuron j
T_i = number of parts requiring machine i
T_j = number of parts associated with output neuron j
$PF_{j,i}$ be the set of parts requiring machine i and associated with output neuron j.

Next, define for each machine i and output neuron j the average number of machines required by parts belonging to the set $PF_{j,i}$.

$$\bar{N}_{j,i} = \frac{\sum_{I \in PF_{j,i}} N_I}{n}$$

where N_I refers to the number of machines required by part I and n is the number of parts in the set $PF_{j,i}$. Also,

$$r_{j,i} = \frac{T_{j,i}}{T_j \bar{N}_{j,i}}$$

It can be shown (see Malavé and Ramachandran, 1991) that at steady state, the weight matrix converges approximately to the matrix formed by the $r_{j,i}$. The proof is omitted here, but the essential flavor of the proof is in showing that the weight vectors converge to the centroid of the respective part families formed. Hence, we have

$$r_{j,i} \approx w_{j,i}$$

From the above relationship, we can see that if $w_{j,i} = 0$, then $T_{j,i}/T_j\bar{N}_{j,i} = 0$ and $T_{j,i} = 0$. Thus, if $w_{j,i} = 0$, then machine i is not required by any part belonging to the part family associated with the output neuron j. To understand this better, consider the final weight matrix of the neural network trained with the part characteristic vectors from Example 1.

0.17	0	0	0.17	0	0.17	0	0	0.23	0	0	0	0	0.23	0
0	0	0.19	0	0.19	0	0	0.19	0	0	0	0	0.19	0	0.19
0	0.21	0	0	0	0	0.13	0	0	0.21	0.21	0.21	0	0	0

It can be seen from the steady state weight matrix shown above that for $j = 1$, $w_{j,i}$ is greater than 0 only for $i = 1, 4, 6, 9$ and 14. Thus machines 1, 4, 6, 9 and 14 are the machines required by the parts associated with the output neuron 1, that is, by part family 1. Similarly, machines 3, 5, 8, 13 and 15 are required by part family 2 and machines 2, 7, 10, 11 and 12 are required by part family 3. In this example, each of the columns of the steady state weight matrix has only one positive element. In other words, none of the machines are required by more than one part family. Next, consider the final weight matrix for the second example.

0	0	0	0.45	0.33	0	0.1	0	0	0	0	0	0.06	0
0.18	0	0	0	0	0	0.07	0	0	0	0	0.18	0.44	0
0	0.14	0.27	0	0	0	0	0	0	0.18	0.39	0	0	0
0	0	0	0	0	0.31	0	0.33	0.18	0	0	0	0	0.15

In this case, columns 7 and 13 have two elements positive. This means that machine 7 and machine 14 is required by both the part families 1 and 2. These are the bottleneck machines and are shared by both thc part families.

Most of the applications of neural networks involves a blackbox approach, that is, information is extracted mainly from the input–output relationships. As seen above, the weight matrix which constitutes the internal structure of the neural network has rich information stored in it. The steady state weight matrix offers an easy way to assign machines to the part families formed. By observing the columns, bottleneck machines can also be easily identified.

New part assignment

Neural networks have the ability to 'generalize'. Generalization in a neural network is defined as the network's ability to perform mapping of an input vector not contained in the training set to one of the output units. The generalization ability of the neural network makes it possible to assign new parts to existing machine cells with minimal additional effort. This property of the network is also useful when cell formation could be influenced by possible future new parts. In fact, when bottleneck machines are present and machines are costly to duplicate, the allocation of these bottleneck machines are usually made on the basis of future new parts that could be encountered. Encountering new parts is a distinct possibility in small and medium sized batch manufacturing systems.

The new part vector is fed at the input layer, and the part could be assigned based on the output activation. The neural network behaves as the nearest neighbor classifier. The new part is classified with the output unit whose weight vector has the highest dot product with the characteristic vector of the new part. The primary concern here is that this classification takes place, even if the nearest neighbor is not that close. This concern can be better addressed by the ART-1 model, since the similarity between the new part and the exemplar patterns (weight vector in this case) are compared to a prespecified threshold (called **vigilance**). Part assignment is made only if the similarity is above the threshold.

Kohonen's model

An extension of the previous approach uses a neural network based on the Kohonen's learning rule. The number of input neurons is the same as the dimension of the part characteristic vectors. The number of output neurons can be determined in an algorithmic procedure, and will be shown in this section. The primary difference is in the learning rule. A good exposition on the Kohonen's learning rule and the working of neural network is found in Lee *et al.* (1992), and will not be discussed here. Part families and machine cells are identified in the same manner as before. After training, the part families are formed by observing the output activations at steady state. Machine cells are identified by observing the steady state weight matrix.

Assuming that the number of part families q is unknown, the algorithm starts with an initial small number of output neurons $q = q_0$. Part families and machine cells are identified as before. The procedure is then repeated with q increased in steps of 1, until an unchanged weight vector is found. Thus, the procedure is stopped at $q = q^*$ when an increase in the number of output

		Parts									
		1	2	3	4	5	6	7	8	9	10
	1	0	0	1	1	0	0	0	1	0	0
	2	1	0	0	0	0	1	1	0	0	0
	3	0	0	1	1	0	0	0	1	1	0
	3	0	0	0	1	0	0	0	1	1	0
	4	0	1	0	0	0	0	0	0	0	0
	5	1	0	1	1	0	1	1	1	1	0
	6	1	0	0	0	0	1	0	0	1	0
Machines	7	1	0	0	0	0	1	0	0	1	0
	8	0	0	0	0	1	0	0	0	0	1
	9	0	1	0	0	1	0	0	0	0	1
	10	0	1	0	0	1	0	0	0	0	1
	11	1	0	0	0	0	1	1	1	0	0
	12	1	0	0	0	0	1	1	0	0	0
	13	0	1	0	0	1	0	0	0	0	1
	14	0	0	1	1	0	0	0	1	1	0
	15	0	1	0	0	1	0	0	0	0	1

Fig. 6.8 Initial part machine matrix, adapted from Chan and Milner (1982).

neurons to $q^* + 1$ results in a weight vector that remains unchanged from its initialized value. Also, none of the parts are assigned to the particular neuron which does not change its weight vector from its initialized value.

To make this approach clearer, consider the following example. The initial part machine matrix adopted from Chan and Milner (1982) is shown in Fig. 6.8, and the final part machine matrix achieved by direct clustering algorithm is also shown in Fig. 6.9.

From the final part machine matrix, it is evident that there are three part families in the part machine mix. The steady state weight matrices of the neural network with three and four output neurons respectively are shown in Fig. 6.10. Both times the weight matrices were initialized with the same

	Parts									
	3	4	8	9	2	5	10	1	6	7
14	1	1	1	1	0	0	0	0	0	0
3	1	1	1	1	0	0	0	0	0	0
4	0	1	1	1	0	0	0	0	0	0
1	1	1	1	1	0	0	0	0	0	0
6	1	1	1	1	0	0	0	1	1	1
15	0	0	0	0	1	1	1	0	0	0
13	0	0	0	0	1	1	1	0	0	0
Machines 10	0	0	0	0	1	1	1	0	0	0
9	0	0	0	0	1	1	1	0	0	0
8	0	0	0	0	1	1	1	0	0	0
5	0	0	0	0	0	1	1	0	0	0
12	0	0	0	0	0	0	0	1	1	1
11	0	0	0	0	0	0	0	1	1	1
2	0	0	1	0	0	0	0	1	1	1
7	0	0	0	1	0	0	0	1	1	0

Fig. 6.9 Final part machine matrix achieved by direct clustering algorithm.

0.000	0.476	0.000	0.000	0.000	0.476	0.303	0.000	0.000	0.000	0.476	0.476	0.000	0.000	0.000
0.000	0.000	0.000	0.000	0.178	0.000	0.000	0.325	0.503	0.503	0.000	0.000	0.503	0.000	0.325
0.364	0.000	0.488	0.355	0.000	0.488	0.124	0.000	0.000	0.000	0.111	0.000	0.000	0.488	0.000

(a)

0.000	0.476	0.000	0.000	0.000	0.476	0.303	0.000	0.000	0.000	0.476	0.476	0.000	0.000	0.000
0.125	0.293	0.066	0.399	0.002	0.029	0.349	0.358	0.083	0.393	0.439	0.240	0.219	0.027	0.433
0.000	0.000	0.000	0.000	0.178	0.000	0.000	0.325	0.503	0.503	0.000	0.000	0.503	0.000	0.325
0.364	0.000	0.488	0.355	0.000	0.488	0.124	0.000	0.000	0.000	0.111	0.000	0.000	0.488	0.000

(b)

Fig. 6.10 (a) Steady state weight matrix with three output neurons; (b) with four output neurons.

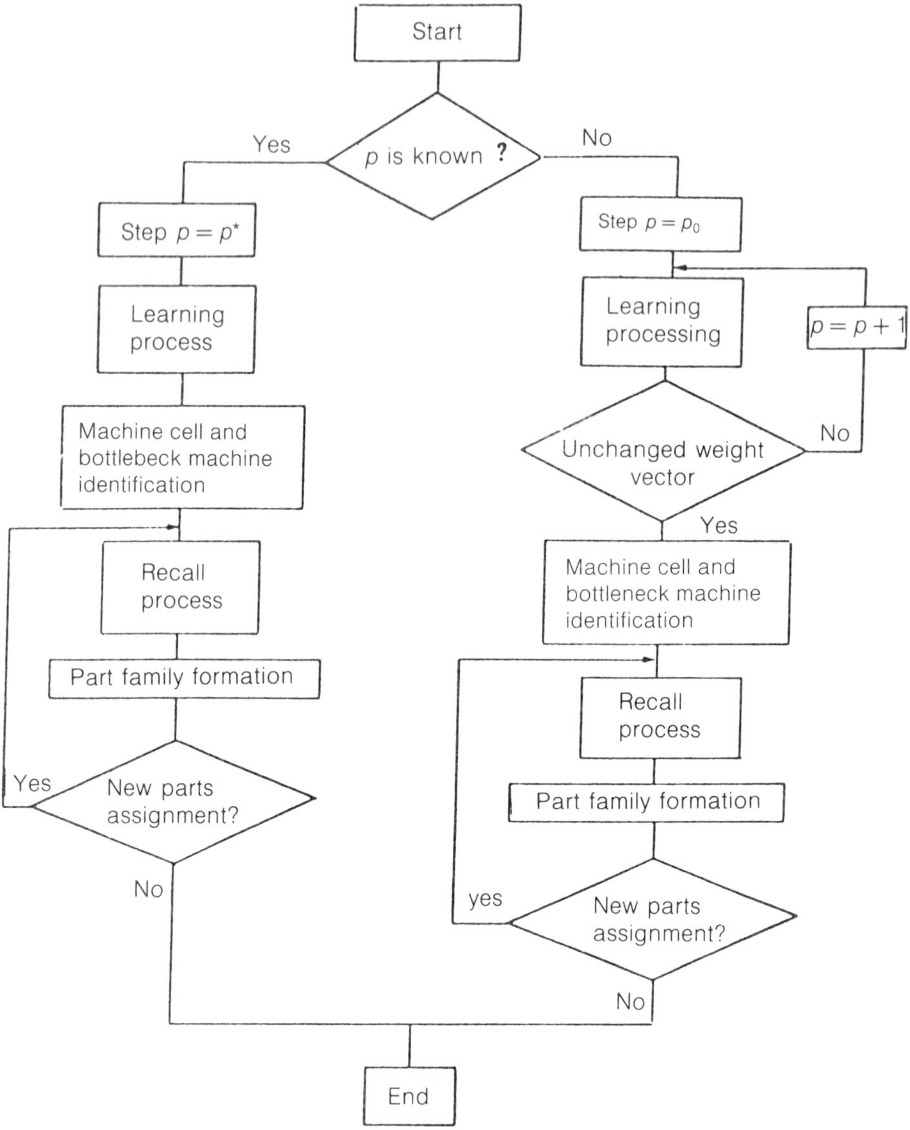

Fig. 6.11 The schematic representation of the GT-based algorithm for part family formation.

values. Notice that, when the number of output neurons are increased to four, three of the weight vectors stabilize to the same final values at steady state. Also, the weight vector associated with the second output neuron does not change during the training process, and no parts are associated with it. Thus, for this process there exists three clusters naturally, and any more output neurons does not create more clusters.

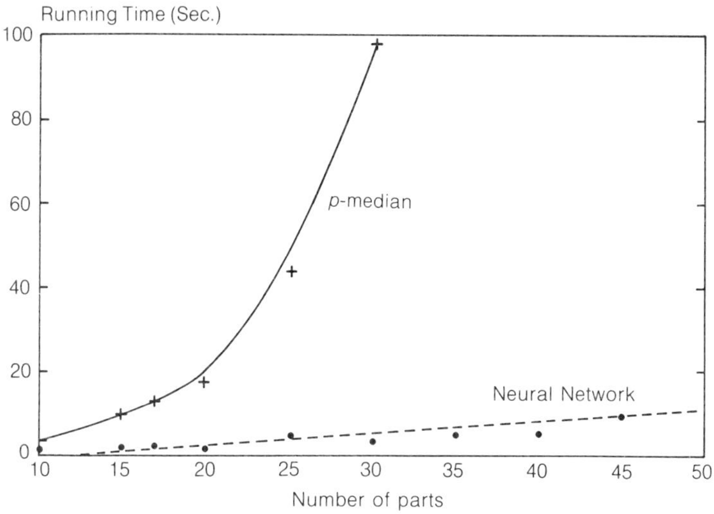

No.	No. of parts	No. of machines	T	E	V	P	Running time (sec.) Neural	-median
1	10	15	41	5	3	3	1.15	3.49
2′	15	16	50	10	23	3	1.54	9.57
3	17	13	62	5	0	4	1.65	13.11
4	20	10	49	0	0	4	1.54	17.21
5	25	25	257	90	0	5	4.12	43.82
6	30	16	83	9	37	5	3.35	97.69
7′	35	20	245	21	12	5	4.56	*
8	40	15	246	14	28	7	4.89	*
9	45	30	592	0	0	6	9.18	*
10	50	40	1032	30	92	8	15.6	*

T: Total number of '1's in matrix
E: Number of exceptional elements
V: Number of voids
P: Number of clusters generated by network
*: Out of memory capacity
/: near-optimal solution matrix

Fig. 6.12 Computational performance of the neural network-based approach versus the *p*-median approach.

The algorithm for the neural network procedure is summarized in Fig. 6.12. The algorithm has been coded in C-language, and the performance as compared to the *p*-median approach is shown in the Fig. 6.13. Comparison to the *p*-median approach was undertaken since *p*-median is an optimal approach for the GT problem. The computational performance of the neural network approach is far superior to that of the *p*-median approach, especially when the number of parts increases.

6.2.2 ART-1 approach

The ART-1 neural network has the capability of classifying a set of binary vectors into groups, each group containing similar vectors based on a

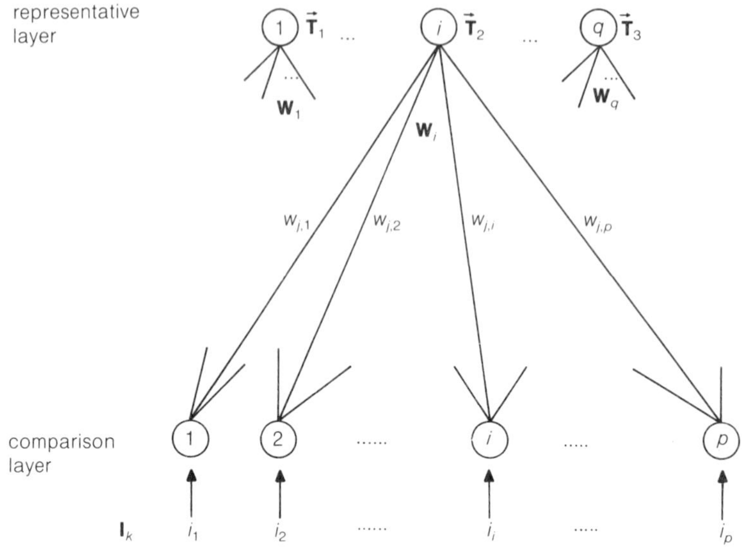

Fig. 6.13 ART-1 neural network architecture.

prespecified degree of similarity. Although ART-1 can be used directly to create part families and machine cells, Dagli and Huggahalli (1992) modified this basic approach to increase the efficiency of the classification procedure. The basic working principle of the ART-1 neural network involves the following two steps:

1. the identification of representative patterns from the set of binary vectors, and storing them,
2. comparing new patterns with the stored representative patterns, and if the new pattern's similarity to a particular stored pattern is above a prespecified similarity coefficient, then the new pattern is classified with the stored pattern.

The ART-1 network

The ART-1 network architecture or topology consists of two layers of neurons, the input layer or comparison layer with p neurons, and the output layer or representative layer with q neurons, as shown in Fig. 6.13. The number of neurons p at the input layer, is usually the dimension of the binary input vectors that are to be classified.

The working of the ART-1 neural network is better understood with the help of Fig. 6.14. The comparison layer receive three inputs (i) an external binary input vector $\mathbf{I}_k$ characterizing a part or a machine (ii) a feedback vector $\mathbf{T}_{k*}$ from the output layer and (iii) a gain vector **G**. The output of the neurons at the comparison layer is 1 if any two of the three inputs to it is 1 (called the two-third rule for activation). Associated with each neuron j in the representative layer are two vectors:

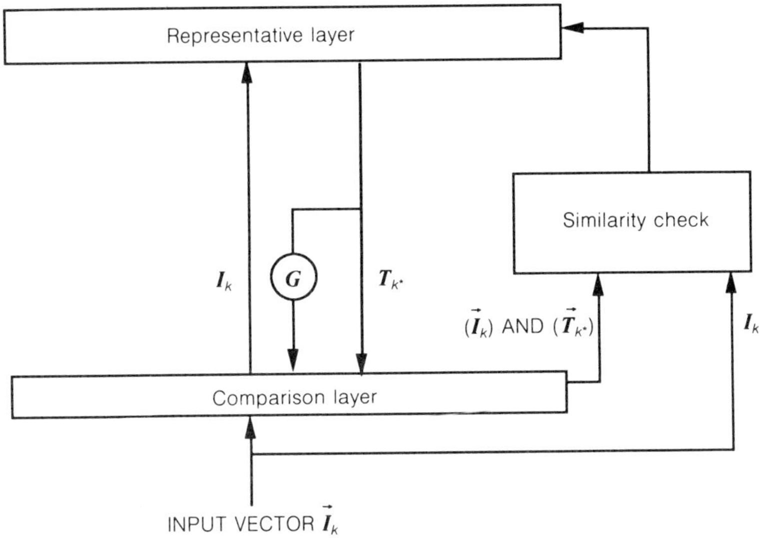

Fig. 6.14 Schematic representation of ART-1 neural network.

1. a weight vector $\mathbf{W}_j = [w_{1,j}, w_{2,j}, \ldots, w_{p,j}]$ and
2. a representative or an exemplar pattern $\mathbf{T}_j = [t_{1,j}, t_{2,j}, \ldots, t_{p,j}]$.

When an input vector $\mathbf{I}_k = [i_1, i_2, \ldots, i_p]$ is presented at the comparison layer of the network, the gain vector $\mathbf{G}$ is initially set to $\vec{\mathbf{G}} = [1, 1, \ldots, 1]$. Thus, by the two-thirds rule of activation at the comparison layer, the output from the input layer is $\mathbf{I}_k$ itself. Next, the output neurons compete with one another to respond to the input vector $\mathbf{I}_k$, and the output neuron k^* which has the closest weight vector to the input vector is identified as follows:

$$W_{k^*} \cdot I_k = \max_{j \in 1,\ldots,q} (\mathbf{W}_j \cdot \mathbf{I}_k)$$

As in the previous approaches, dot product is used as the metric to identify the weight vector which is closest to the input vector. After having identified the output neuron k^*, $\mathbf{T}_{k^*}$ is fed back into the input layer, and if any of the components of T_{k^*} is 1, then $\mathbf{G}$ is set to $\mathbf{G} - [0, 0, \ldots, 0]$. Then, by the two-third rule, the output of the comparison layer is $(\mathbf{I}_k)$ AND $(\mathbf{T}_{k^*})$, a new vector, whose elements are obtained by applying the logical AND on the corresponding elements of the two vectors, giving an estimate of the similarity between $\mathbf{T}_{k^*}$ and $\mathbf{I}_k$. A similarity measure between $\mathbf{T}_{k^*}$ and $\mathbf{I}_k$ is computed as

$$\text{similarity} = \frac{\text{number of 1s in } (\mathbf{I}_k) \text{ AND } (\mathbf{T}_{k^*})}{\text{number of 1s in } \mathbf{I}_k}$$

This similarity measure is compared with a prespecified threshold called the vigilance parameter ρ. If the computed similarity measure is greater than

ρ, then the stored representative pattern associated with the output neuron k^*, is changed to $(\mathbf{I}_k)$ AND $(\mathbf{T}_{k^*})$. Also, $\mathbf{W}_{k^*}$ is changed

$$w_{k^*,i} = \frac{Lc_i}{L-1+\sum_{i=1,\ldots,p} c_i}$$

where c_i is the ith component of $(\mathbf{I}_k)$ AND $(\mathbf{T}_{k^*})$, and L is a constant that is usually set to 2. The weight change rule stated above, moves the weight vector $\mathbf{W}_{k^*}$ closer to the input vector if the similarity of $\mathbf{T}_{k^*}$ and $\mathbf{I}_k$ is greater than the vigilance parameter, thus increasing the possibility of recognizing a similar input vector, at the same time decreasing the chance of recognizing a dissimilar vector. If the similarity is not greater than the prespecified vigilance parameter then the output neuron with the next highest $\mathbf{W}_j \cdot \mathbf{I}_k$ is selected, and the same procedure is repeated.

Thus, unlike the previous algorithms, in ART-1 the winning output neuron changes its weight vector and representative pattern only if the similarity of the input vector and the weight vector of the winning neuron is greater than a prespecified threshold. To show an example iteration, (adopted from Dagli and Huggahalli (1992)), consider the following part machine incidence matrix.

Machines \ parts	1	2	3	4	5	6	7
1		1		1	1	1	
2	1		1				
3	1	1					1
4		1		1		1	
5	1						1

For the classification of the column vectors, column vectors 1 through 7 are presented sequentially at the input layer of the network. Suppose $\mathbf{W}_1$ and $\mathbf{W}_2$ is initialized so that when $\mathbf{I}_1 = [0\ 1\ 1\ 0\ 1]$ is presented, neuron 1 wins, that is $\mathbf{I}_1 \cdot \mathbf{W}_1 > \mathbf{I}_2 \cdot \mathbf{W}_2$. In the first learning iteration there is no exemplar pattern stored at the output neurons, so $\mathbf{I}_1$ is stored as the exemplar pattern at neuron 1, and by using the weight change equations $\mathbf{W}_1 = [0\ 0.5\ 0.5\ 0\ 0.5]$. Now when next column vector $\mathbf{I}_2 = [1\ 0\ 1\ 1\ 0]$ is presented, there are the following two cases depending on the initial value of $\mathbf{W}_2$.

1. Suppose $\mathbf{I}_2 \cdot \mathbf{W}_1 > \mathbf{I}_2 \cdot \mathbf{W}_2$, then $\mathbf{I}_2$ AND $\mathbf{T}_1$ is compared against the stored vector at output neuron 1.

   $$(\mathbf{I}_2) \text{ AND } (\mathbf{T}_1) = [1\ 0\ 1\ 1\ 0] \text{ AND } [[0\ 1\ 1\ 0\ 1]] = [0\ 0\ 1\ 0\ 0]$$

   and the similarity measure between $\mathbf{I}_2$ is computed as

   $$\text{similarity} = \frac{\text{number of 1s in } (\mathbf{I}_2) \text{ AND } (\mathbf{T}_1)}{\text{number of 1s in } \mathbf{I}_2} = \tfrac{1}{3} = 0.333$$

   If the prespecified vigilance parameter ρ is less than 0.333, then the pattern is associated with the first neuron, and $\mathbf{T}_1$ is changed to [0 0 1 0 0] and $\mathbf{W}_1$ is also changed using the weight change equations. If the prespecified vigilance parameter ρ is greater than 0.333, then the input vector is not associated with the neuron 1. $\mathbf{T}_1$ and $\mathbf{W}_1$ are not changed, and the vector $\mathbf{I}_2$ is used to create an initial exemplar at neuron 2.

2. Suppose if $\mathbf{I}_2 \cdot \mathbf{W}_1 < \mathbf{I}_2 \cdot \mathbf{W}_2$, then the vector $\mathbf{I}_2$ is used to create an initial exemplar at neuron 2.

These iterations are repeated by presenting the input vectors one at a time, until the weight vectors show no change with further training iterations. The part groups can be identified by observing the output activations, that is, the winning neurons. In other words, all part vectors which cause a particular output unit to respond or win are classified as a part family. After grouping the columns with a vigilance parameter of $\rho_1 = 0.4$, the resulting matrix is shown below. There are three part families formed:

Part family 1: Parts 1, 3
Part family 2: Parts 2, 4, 5, 6
Part family 3: Part 7.

Parts

Machines	1	3	2	4	5	6	7
1			1	1	1	1	
2	1	1					
3	1		1				1
4			1	1		1	
5	1						1

Notice that a strict block diagonal form is not achieved. In the ART-1 approach proposed by Dagli and Huggahalli (1992), this matrix is called an intermediate matrix. Next, on the intermediate matrix shown above, the row vectors are classified using the ART-1 network with a vigilance of $\rho_2 = 0.4$. The final matrix obtained is shown below.

Parts

Machines	1	3	2	4	5	6	7
1			1	1	1	1	
4			1	1		1	
2	1	1					
3	1		1				1
5	1						1

The 3 machine groups obtained are

Machine group 1: Machines 1, 4
Machine group 2: Machine 2
Machine group 3: Machines 3, 5.

The same procedure is repeated for a different set of vigilances of $\rho_1 = 0.3$ and $\rho_2 = 0.3$. The resultant final matrix obtained is shown below.

$$
\begin{array}{c} \\ \\ \\ \text{Machines} \\ \\ \\ \end{array}
\begin{array}{c} \text{Parts} \\
\left[\begin{array}{c|ccccccc}
 & 1 & 2 & 7 & 4 & 5 & 6 & 3 \\
1 & & 1 & & 1 & 1 & 1 & \\
4 & & 1 & & 1 & & 1 & \\
2 & 1 & & & & & & 1 \\
3 & 1 & 1 & 1 & & & & \\
5 & 1 & & 1 & & & &
\end{array}\right]
\end{array}
$$

The part families obtained are

Part family 1: Parts 1, 2, 7
Part family 2: Parts 4, 5, 6
Part family 3: Part 3

while the machine groups obtained are

Machine group 1: Machines 1, 4
Machine group 2: Machines 2, 3, 5.

A strict block diagonal form is not achieved by this method of classification, although in some cases a better result is possible.

ART-1 in its basic form has many drawbacks, as shown by the above example. The exemplar pattern stored at each output layer neuron gets sparser as more input vectors are applied. For example, suppose that in the first iteration, column vector 1 get stored in output neuron 1, when column vector 2 also gets classified in the same group, the representative pattern stored at neuron 1 is the AND of both these vectors, that is [0 0 0 1 0], which is a very sparse vector. At a later stage, when a vector such as column 4 is presented, it cannot be classified as similar to this group, in spite of its similarity with column vector 2. Thus, bcause of the exemplar pattern getting sparse, the resultant classification could be improper. This effect may be minimized by considering the different solutions obtained for different combinations of vigilance parameters. Also, the obtained classification is very sensitive to the sequence in which the vectors are applied during training. For example, suppose the column vector 1 is applied after vectors 4, 5, 6 and before vectors 3 and 7, then one of the part groups obtained is (2, 4, 5, 6), which is in fact a desirable group. Another short-coming is that the above approach does not offer a way to deal with bottleneck machines and its duplication. Finally, the determination of the optimal vigilance parameters is not always straightforward. Vigilance parameters influence the classification significantly. Higher vigilances result in more groups with lesser number of vectors within a group and greater similarity between the vectors within a group. Lower vigilances result in lesser groups with more vectors within a group. The problem of obtaining an optimal vigilance is not easy, in fact

ART-1 in its basic form shares the same problem as most similarity coefficient approaches, that is, in deciding the similarity threshold.

An improved ART-1 approach

The improved ART-1 approach proposed by Dagli and Huggahalli (1992) incorporates a few changes in the basic approach. To deal with the problem of the representative pattern getting sparse with more number of patterns, instead of storing the AND of the two vectors $\mathbf{T}_{k*}$ and $\mathbf{I}_k$, the vector with the higher number of 1s is stored as the representative vector. This eliminates the possibility of improper classification of later parts due to its comparison with sparse representative patterns. Also, to deal with the problem of sensitivity of the obtained classification to the sequence in which the parts are applied, preprocessing of the input vectors is done prior to training. The vectors are applied in the decreasing order of 1s. This results in the sparser vectors being absorbed into the denser vectors (which are applied first). With these changes, the modified ART-1 algorithm involves the following steps:

1. Arrange the column vectors according to the decreasing number of 1s.
2. Perform grouping, with exemplars created by storing the denser pattern (higher number of 1s).
3. Reorder rows of the intermediate matrix according to the decreasing number of 1s.
4. Group rows, just as in step 2.

The results of the improved procedure adopted on the example problem are shown below in a sequential manner. The figure below shows the columns ordered according to the decreasing number of 1s, followed by the intermediate matrix obtained after training the network with column vectors with a vigilance of $\rho_1 = 0.4$. The part groups obtained are

Part group 1: Parts 1, 7, 3
Part group 2: Parts 2, 4, 6, 5

m/cs \ parts	1	2	4	6	7	3	5
1		1	1	1			1
2	1					1	
3	1	1			1		
4		1	1	1			
5	1				1		

m/cs \ parts	1	7	3	2	4	6	5
1				1	1	1	1
2	1		1				
3	1	1		1			
4				1	1	1	
5	1	1					

The intermediate matrix is rearranged with rows appearing in decreasing order of 1s, and the final matrix obtained after training the network with row vectors with a vigilance of $\rho_2 = 0.4$ is shown below. The machine groups are

Machine group 1: Machines 1, 7, 3
Machine group 2: Machines 2, 4, 6, 5

parts

m/cs	1	7	3	2	4	6	5
1				1	1	1	1
3	1	1		1			
4				1	1	1	
2	1		1				
5	1	1					

parts

m/cs	1	7	3	2	4	6	5
1				1	1	1	1
4				1	1	1	
3	1	1		1			
2	1		1				
5	1	1					

A much more distinct block diagonal form is obtained, with one bottleneck machine in machine 3. Bottleneck machines result in intercellular movements. The decision of duplicating the bottleneck machine is usually an organizational decision, dictated by factors such as cost of the machine, material handling costs, etc.

The modified ART-1 approach presented by Dagli and Huggahalli (1992) also addresses the issue of obtaining optimal vigilances, and detecting and automatically duplicating a bottleneck machine. In order to obtain optimal vigilances ρ_1 and ρ_2, the value of the vigilances for which the number of part groups and number of machine groups formed are the same are found out. This is because a block diagonal form is characterized by an equal number of part families and machine groups. Let ρ_1 and ρ_2 be the vigilance parameters for the column and row vectors classification, and N and M be the number of part groups and number of machine groups formed. N and M increase respectively for increase in ρ_1 and ρ_2, the rate of increase depending on the particular part machine incidence matrix. Since N and M are not very sensitive to small increases in ρ_1 and ρ_2, testing for N and M for very small increments of ρ_1 and ρ_2 is not necessary. Thus, N and M are found for discrete increments of ρ_1 and ρ_2, and the values of ρ_1 and ρ_2 are identified for which $N = M$. The part machine matrix adopted from King and Nakornchai (1982) in Fig. 6.15 is used to explain the approach.

Figure 6.16 shows N and M for classifications obtained with different values of ρ_1 and ρ_2. As seen from Fig. 6.16, the difference between N and M is minimized for $\rho_1 = 0.45$ (5 part groups) and $\rho_2 = 0.4$ (4 machine groups). But, this classification results in an uneven distribution in the number of machines within the various machine groups. But for slightly different values of $\rho_1 = 0.45$ and $\rho_1 = 0.5$, a much more even distribution in the number of machines within the machine groups is obtained. Figure 6.17 shows the classification obtained for $\rho_1 = 0.45$ and $\rho_2 = 0.5$.

The bottleneck machines could be duplicated to achieve a stricter block diagonal form. Figure 6.18 shows the final matrix obtained, after duplicating machine 6, 8 and 10. In the modified ART-1 approach by Dagli and Huggahalli (1992), an algorithm is developed which identifies and duplicates bottleneck

	1	2	3	4	5	6	7	8	9	10	11	12	13	14	15	16	17	18	19	20	21	22	23	24	25	26	27	28	29	30	31	32	33	34	35	36	37	38	39	40	41	42	43
1																																					1					1	
2		1								1																		1				1					1	1		1		1	
3							1										1																	1	1	1							
4					1				1					1					1		1		1						1														
5					1			1	1					1	1	1			1		1		1						1				1								1		1
6	1	1				1	1	1				1	1	1			1		1				1									1	1	1			1		1	1		1	1
7	1												1												1																		
8	1	1	1					1	1		1	1			1				1	1	1		1	1			1	1			1						1	1			1		1
9		1		1						1								1										1				1					1	1		1		1	
10	1											1	1												1	1					1								1				
11			1						1											1				1			1			1													
12											1											1		1			1			1													
13			1																					1																			
14		1				1											1																		1								
15					1									1					1		1												1								1		1
16		1					1			1								1														1					1	1				1	

Fig. 6.15 Initial part machine matrix adapted from King and Nakornchai (1982).

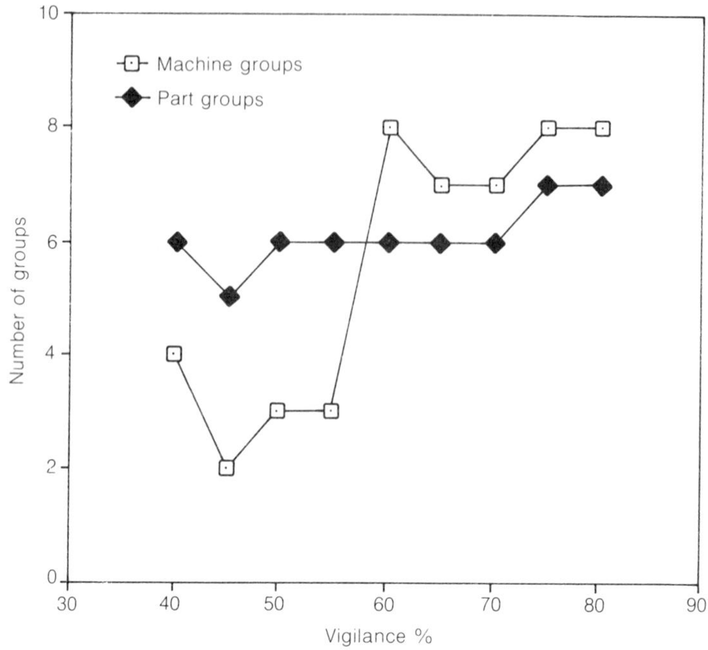

Fig. 6.16 Variation in machine and part groups with vigilance parameters.

machines. The advantage of this algorithm is that a major part of the algorithmic procedure can be performed by the ART-1 network itself, thus providing computational ease. As stated before, duplicating a bottleneck machine is usually a complex management decision dictated by factors such as machine cost, machine performance, cell size constraints, material handling constraints, to name a few. Thus duplicating machines is done only after a performance/cost analysis is carried out.

Dagli and Huggahalli (1992) gives an approach integrating the procedures for part family–machine cell formation, detection and duplicating bottleneck machines, cost performance analysis. This approach leads to an integrated approach for optimal part family–machine cell formation. This approach has been compared on large scale problems against approaches like King ROC algorithm, and the performance compares well in classification obtained, as well as computational time. In addition, the algorithm has been tested on synthetically generated matrices of dimensions 200 × 100, producing reasonably accurate classifications.

6.3 DISCUSSION

The ability of a class of neural networks to categorize binary input vectors into clusters, can be exploited to solve the GT problem. Neural network

	2	37	42	32	38	7	10	17	28	40	6	18	35	4	19	1	9	14	21	23	43	5	8	12	33	41	15	29	16	24	3	27	11	20	30	22	13	25	31	34	39	26	36
8	1	1			1				1						1	1	1		1	1	1		1	1		1	1			1	1	1	1	1					1				
5															1		1	1	1	1	1	1	1		1	1	1	1	1														
4															1		1	1	1	1		1						1															
11																	1													1	1	1		1	1								
12																														1		1	1		1	1							
13																														1	1												
6	1	1	1	1		1		1		1	1				1	1		1		1	1		1	1	1												1			1	1		
10																1								1													1	1	1		1	1	
15															1			1	1		1	1			1	1																	
3						1		1					1																											1			1
14	1							1			1		1																														
7																1																					1	1					
9	1	1	1	1	1		1		1	1		1		1																													
2	1	1	1	1	1		1		1	1																																	
16	1	1	1	1	1	1	1					1																															
1		1	1																																								

Part Groups: Vigilance: 45

Group 1 : 2 37 42 32 38 7 10 17 23 40 6 18 35 4
Group 2 : 19 1 9 14 21 23 43 5 8 12 33 41 15 29 16
Group 3 : 24 3 27 11 20 30 22
Group 4 : 13 25 31 34 39 26
Group 5 : 36

Machine Groups: Vigilance: 50

Group 1 : 8 5 4 11 12 13
Group 2 : 6 10 15 3 14 7
Group 3 : 9 2 16 1

Fig. 6.17 Part machine matrix achieved for $p_1 = 0.45$ and $p_2 = 0.5$.

	2	37	42	32	38	7	10	17	28	40	6	18	35	4	19	1	9	14	21	23	43	5	8	12	33	41	15	29	16	24	3	27	11	20	30	22	13	25	31	34	39	26	36
10																																					1	1	1		1	1	
7																1																					1	1					
6																																					1			1	1		
8																														1	1	1	1	1					1				
11																	1													1	1	1		1	1								
12																														1		1	1		1	1							
13																														1	1												
5															1		1	1	1	1	1	1	1		1	1	1	1	1														
4															1		1	1	1	1		1						1															
8															1	1	1		1	1	1		1	1		1	1																
6															1	1		1		1	1		1	1	1																		
10																1								1																			
15															1			1	1		1	1			1	1																	
8	1	1			1				1																																		
6	1	1	1	1		1		1		1	1																																
3						1		1					1																														
14	1							1			1		1																											1			1
9	1	1	1	1	1		1		1	1		1		1																													
2	1	1	1	1	1		1		1	1																																	
16	1	1	1	1	1	1	1					1																															
1		1	1																																								

Part Groups: Vigilance: 45

Group 1: 2 37 42 32 38 7 10 17 28 40 6 18 35 4
Group 2: 19 1 9 14 21 23 43 5 8 12 33 41 15 29 16
Group 3: 24 3 27 11 20 30 22
Group 4: 13 25 31 34 39 26
Group 5: 36

Machine Groups: Vigilance: 50

Group 1: 10 7 6
Group 2: 8 11 12 13
Group 3: 5 4 8 6 10 15
Group 4: 8 6 3 14 9 2 16 1

Fig. 6.18 Final matrix achieved after duplication machines 6, 8 and 10.

based approaches offer a viable alternative to solve the GT problem, with distinct advantages over the conventional approaches. The principal advantages of neural network based approaches as compared to traditional approaches lie in ease of part family–machine cell identification, computational performance, future new part assignment, ease of bottleneck machine detection, etc. The scope of using neural network to solve the larger GT problem involving scheduling constraints, volume of production, sequencing, etc. looks promising. The inclusion of these additional factors in the design of part families and machine cells is a more realistic approach toward achieving the fuller benefits of group technology.

The implementation of the neural network based approaches as shown in this chapter were done by simulating the neural network on a digital computer. But this methodology does not use the real advantages of neural network, which essentially lie in its parallel architecture. Hardware implementation of neural networks is still a distant reality, and much current research in neural networks is focused on this issue. With hardware implementation, neural network based approaches will be far superior to conventional approaches.

REFERENCES

Askin, R.G. and Subramanian, S.P. (1987) A cost based heuristic for group technology configuration. *International Journal of Production Research*, **25**, 101–13.

Burbidge, J.L. (1971) Production flow analysis. *The Production Engineer* (April/May), 139–52.

Burbidge, J.L. (1975) *The Introduction of Group Technology*, John Wiley & Sons, New York.

Chan, H.M. and Milner, D.A. (1982) Direct clustering algorithm for group formation in cellular manufacturing. *Journal of Manufacturing Systems*, **1**, 65–75.

Dagli, C.H. and Huggahalli, R. (1992) A neural network approach to group technology, in *Neural Networks in Design and Manufacturing* (eds J. Wong and T. Yoshiyasu), World Scientific Pub. Co. (in press).

King, J.R. (1980) Machine-Component grouping in production flow analysis: An approach using a rank order clustering algorithm. *International Journal of Production Research*, **18**(2), 213–32.

King, J.R. and Nakornchai, V. (1982) Machine-Component group formation in group technology: Review and extension. *International Journal of Production Research*, **20**(2), 117–33

Kumar, K.R., Kusiak, A., and Vannelli, A. (1986) Grouping parts and components in flexible manufacturing systems. *European Journal of Operations Research*, **24**, 387–97.

Kusiak, A. (1987a) The generalized group technology concept. *International Journal of Production Research*, **25**(4), 561–9.

Kusiak, A. (1987b) Artificial intelligence and operation research in flexible manufacturing systems. *Information Systems and Operations Research*, **25**(1), 2–12.

Kusiak, A. and Chow, W.S. (1987) An algorithm for cluster identification. *IEEE Transactions on Systems, Man and Cybernetics*, **17**(4), 696–9.

Lee, H. and Garcia, A. (1992) Network flow approach to solve clustering problems in group technology. *IJPR* (in press).

Lee, H. and Hwang, H. (1991) A hierarchical divisive clustering method for machine component grouping problem. *Engineering Optimization*, **17**, 65–78.

Lee, H., Malavé, C.O. and Ramachandran, S. (1992) A self-organizing neural network approach for the design of cellular manufacturing systems. *Journal of Intelligent Manufacturing*, **3**, 325–32.

Malavé, C.O. and Ramachandran, S. (1991) Neural network-based design of cellular manufacturing systems. *Journal of Intelligent Manufacturing*, **2**, 305–14.

McAuley, J. (1972) Machine grouping for efficient production. *The Production Engineer*, **51**, Feb., 53–7.

McCormick, W.T., Schweitzer, P.J. and White, T.W. (1972) Problem decomposition and data reorganization by cluster technique. *Operations Research*, **20**(5), 993–1009.

Rajagopalan, R. and Batra, J.L. (1975) Design of cellular production system: A graph-theoretical approach. *International Journal of Production Research*, **13**(6), 567–79.

Seifoddini, H. and Wolfe, P.M. (1986) Application of the similarity coefficient method in group technology. *IIE Transactions*, **18**(3), 271–7.

Srinivasan, G., Narendran, T.T. and Mahadevan, B. (1990) An assignment model for the part-families problem in group technology. *International Journal of Production Research*, **28**(1), 145–52.

Steudel, H.J. and Ballakur, A. (1987) A dynamic programming based heuristic for machine grouping in manufacturing cell formation. *Computers and Industrial Engineering*, **2**, 215–22.

Wu, H.L., Venugopal, R. and Barash, M.M. (1986) Design of a cellular manufacturing system: A syntactic pattern recognition approach. *Journal of Manufacturing Systems*, **5**, 81–7.

Young, M. (1990) Interactive activation and competition model for machine–part family formation. *Proceedings of the International Joint Conference on Neural Networks at Washington D.C.*, Volume 2, 667–70.

CHAPTER 7

Process planning

Madhusudhan Posani[1] *and Cihan H. Dagli*[2]
[1]*University of Missouri, University Extension, Columbia, Missouri 65211.*
[2]*University of Missouri-Rolla, Missouri 65401.*

7.1 CHARACTERISTICS OF PROCESS PLANNING

Manufacturing planning, process planning, material processing, process engineering and machine routing are some of the names given to the topic referred to here as process planning.

Process planning is the act of preparing detailed operation instructions to transform an engineering design to a final part (Chang, 1982; 1989). It is the critical bridge between design and manufacture. Design information can only be translated through process planning into manufacturing language. The detailed plan contains the route, processes, process parameters, machines, and tools required for production. The process plan provides the instructions for production of the part. These instructions dictate the cost, quality and rate of production. Therefore process planning is of utmost importance to the production system. In general, a process plan is prepared using available design data, and manufacturing knowledge. The process planning function has been described as 'a subsystem that is responsible for the conversion of design data to work instructions.'

Traditionally the process planning activity has been experience based and performed manually. In a conventional production system, a process plan is created by a process planner who examines a new part or engineering drawing, and then determines the appropriate procedures to produce it. The previous experience of the process planner is critical to the success of the plan. Manual process planning has many problems though it is well suited for small firms with a few parts. Variability among the planner's judgment

Artificial Neural Networks for Intelligent Manufacturing. Edited by Cihan H. Dagli.
Published in 1994 by Chapman & Hall, London. ISBN 0 412 48050 6

and experience can lead to differences in the perception of what constitutes the optimal or best method of production. This manifests itself in the fact that most industries have several different process plans for the same part, which leads to inconsistent plans and additional paperwork. In the last decade, there has been a trend to automate process planning, since it increases production efficiency and parts can be produced more economically. To alleviate problems of manual process planning, a computer-aided approach is taken.

Computer-aided process planning (CAPP) serves as a necessary, logical interface between computer-aided design (CAD), and computer-aided manufacture (CAM). CAPP performs process planning automatically by using expert systems and CAD-based part definition as input. It strives to bridge the CAD/CAM gap by producing machine interpretable process plans as input to CAM NC systems from the design CAD model of the part.

Although, during the past few years, CAD and CAM have emerged as the most effective tools for improving productivity and efficiency, the process planning activity has still not been completely integrated in the CAD–CAM cycle. A need, therefore exists to automate the process planning activity and link it with the CAD–CAM cycle for a true computer-integrated manufacturing (CIM) system.

CAPP serves as an interface between computer-aided design and manufacturing. It reduces process planning time and manufacturing cost. It creates consistent and accurate plans.

7.1.1 Summary of the current approaches used

There are three basic approaches to accomplish the task of process planning. These include the traditional manual approach, the computer-assisted variant approach, and the computerized generative approach.

Manual approach

The traditional manual approach begins by examining a two-dimensional engineering drawing that represents a three-dimensional part. Figure 7.1 shows an engineering part drawing that represents a turned shaft with a keyway. The process planner then develops manufacturing process plans and

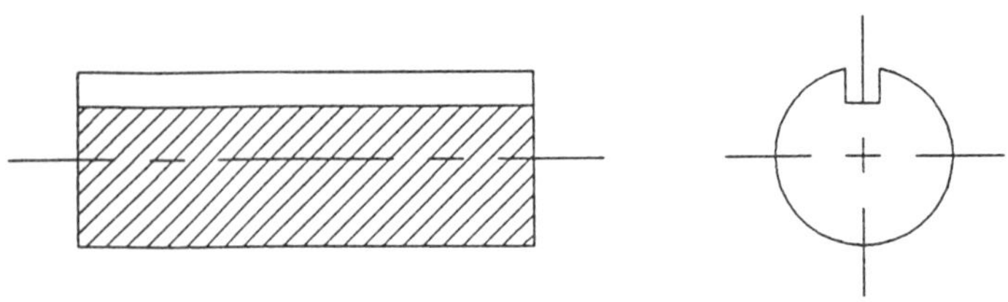

Fig. 7.1 An engineering drawing representing a turned shaft with a keyway.

instructions based upon knowledge of process and machine capabilities, tooling, materials, related costs, and shop practices (Steudel, 1984). This approach draws heavily upon the experience of the manufacturing analyst to develop process plans which are feasible, low cost, and consistent with plans for similar parts. Manual process planning is highly subjective, labor intensive, time consuming, tedious, and often boring. Furthermore, the task requires personnel well trained and experienced in manufacturing shop floor practices.

Despite the many demands and drawbacks, the manual approach often is the best approach for small companies with few process plans to generate. A good process planner can create plans which are accurate, fairly consistent, and cost effective, provided the number of parts are small. Advantages of this approach are good flexibility and low investment costs. The limitations of manual process planning really surface when the number of process plans and revisions to those plans increase. The numerous disadvantages of this method has led to investigation of the automation of the process planning function.

Variant process planning

The variant approach to process planning was the first approach used to computerize planning techniques (Chang, 1989). It is based on the concept that similar parts will have similar process plans. The computer can be used as a tool to assist in identifying similar plans, retrieving them, and editing the plans to suit the requirements for specific parts (Pande and Walvekar, 1989).

In order to implement variant process planning, group technology (GT) based part coding and classification is used as a foundation. Individual parts are coded based upon several characteristics and attributes. The attributes that are significant in the part shown in Fig. 7.1 include its solid cylindrical nature and the milled slot. This classification method is applied to all parts. Parts that have sufficiently common attributes are grouped into a part family. Part family formation is determined by analyzing the codes of the part spectrum. A standard plan consisting of a process plan to manufacture the entire family is created and stored for each part family. An incoming new component is first coded. The code is then sent to a part family search routine to find the family to which it belongs. Since the standard plan is indexed by family number, the standard plan can be easily retrieved from the database. The standard plan is designed for the entire family rather than for a specific component, thus editing the plan is unavoidable. It is edited by the process planner to create a 'variant' to suit the specific requirements of the component being planned. Creation and modification of standard plans are the process planner's responsibility.

Variant process planning systems have several problems associated with them. Setting-up the part family and standard plans database can be extremely time consuming. The components to be planned are limited to similar components previously planned. Experienced process planners are still required to modify the standard plan for the specific component.

Generative approach

Generative process planning is a system which automatically synthesizes a process plan for a new component. The generative approach envisions the creation of a process plan from information available in a manufacturing database without intervention (Wang and Wysk, 1988). Upon receiving the design model, the system is able to generate the required operations and operation sequence for the component. Knowledge of manufacturing has to be captured and encoded into computer programs. By applying decision logic, a process planner's decision-making process can be imitated. Other planning functions such as machine-selection, tool selection, process optimization, etc., can also be automated using generative planning techniques.

A generative process planning system comprises three main components: part description; manufacturing databases; and decision-making logic and algorithms.

Part description forms a major part of the information needed for process planning. The way in which the part description is input to the process planning system has a direct affect on the degree of automation that can be achieved. Some of the methods that have been used for part description include: GT codes; special descriptive languages that describe parts in terms of their geometric and technological elements such as cylinders, cones, chamfers, etc.; 2D and 3D CAD models. Referring to Fig. 7.1, it is possible to describe the part as the removal of a rectangular block representing the milled slot from a solid cylinder. By yet another approach, the same part could be described as a collection of faces and edges that include faces and edges of the cylinder and those of the milled slot. Generative process planning is often constrained by the accuracy and simplicity of part description methods used. Complex parts require a tremendous amount of computer programming and processing.

Generative process planning is regarded as more advanced than variant process planning. Generative process plan design requires long-term investments of people, machinery, and time. Ideally, a generative process planning system is a turn-key system with all the decision logic built in. Existing generative process planning systems can generate process plans for parts with rather simple geometry (Kusiak, 1990). In fact, most existing systems are semi-generative because they require human interaction to describe the part and to obtain information about its features.

Successful implementation of this approach requires that the logic of process planning must be identified and captured. Furthermore, the part to be produced must be clearly and precisely defined in a computer-compatible format, and the captured logic of process planning and the part description data must be incorporated into a a unified manufacturing database.

Artificial intelligence in process planning

Using an artificial intelligence (AI) framework, the process planning problem can be formulated as a sequence of actions (operations) and resources (machines, tools, etc.) that enable the goal state (producing a finished part) to be reached given the initial state (raw material, also called stock). Recent advances demonstrate that practical AI based expert systems are possible (Srihari and Greene, 1987, Srihari, 1989).

There are several advantages in the use of an expert system approach for CAPP function (Gupta, 1990).

- The generation of a process plan requires consideration of a number of factors which influence selection and sequencing decisions for processes, and their important parameters. Normally, a plan begins with the process selection followed by machines and tools selection. The decision process, adequate for producing a complete process plan, requires creation of a hierarchy consisting of several levels. Ideally each level would be required to have a defined domain for deciding specific parameters of the plan. Also, since, a decision at any level can generally influence those at any other level, the system should have the capability of back tracking its partially developed solution from almost any stage. The expert system approach has a potential to provide a good solution to meet these requirements.
- Decision trees and decision tables, often used in traditional generative CAPP systems, work effectively only for simple decision-making processes. These are primarily methods to describe knowledge and are coded line by line in the program. Any modification to the current knowledge would require rewriting of at least some portion of the original program. To automate the process planning function totally, the system must be able to perform a certain level of intelligent reasoning. An expert system organizes the domain knowledge and employs inference mechanism to reason intelligently.
- A productive CAPP system must have access to a tremendous amount of information which include facts about machines in a manufacturing system, rules about selecting and sequencing machining operations, and all necessary machining parameters. Furthermore, the system should be flexible because facts and rules in the database require frequent updating. As new technology, equipment, and processes become available, the most effective way to manufacture a particular part also changes. An expert system stores knowledge in a special manner so that it is possible to add, delete, and modify its knowledge base without re-coding the program.

In spite of the fact that tremendous efforts have been made in developing CAPP systems, the benefits of CAPP in the real industrial environment are still to be seen. This situation has led to some doubts about the current state of the research and implementation of process planning (Alting and Zhang, 1989).

Several different approaches have been used for design/process planning interface. They can produce reasonable results under a very limited domain. The domain is limited to sweep volumes which include $2\frac{1}{2}$D features. Most of the approaches can recognize only separated features. When a feature is taken in as part of an object, these methods may not be efficient or may not even be able to function.

Feature recognition is difficult not only because there is no general theory of geometric reasoning, but also because the definitions of features are imprecise. Although a few features, such as a hole, counterbore, countersink, straight slot, etc., are generally understood, normally the definitions of features are vague. The same geometric feature may be conceived as several different manufacturing features depending on the direction from which they are viewed and the manufacturing processes available.

CAPP summary

In the last decade, process planning has evolved from being a manual approach to one that is highly automated. Although each of the approaches that have been developed have some inherent drawbacks, they have been useful in certain specific applications.

The manual approach requires years of process planning experience. Further it is highly subjective, labor intensive, and time consuming.

The variant process planning approach requires the setting-up of large databases that can be extremely time consuming (Parks, 1989). Like the manual approach, it still requires experienced process planners to modify standard process plans for specific components.

The generative approach is complex, and represents a long-term investment of time, manpower, and computer commitments. Since this is still far from being fully realized, generative systems developed currently provide a wide range of capabilities and can at best be only described as semi-generative. Generative process planning is constrained by the accuracy and simplicity of part description methods used.

7.1.2 Contributions of artificial neural networks

The complexity of CAPP approaches currently in use has caused systems to be developed for specific types of parts and whose implementations are extremely time consuming. The artificial neural networks approach was considered since neural networks are becoming increasingly efficient in image processing and pattern recognition (Rumelhart, *et al.*, 1986). Such a system would consist of an artificial neural network to recognize patterns in an expert systems environment. By simplifying the process of pattern recognition using neural networks, it was felt that CAPP could be made more efficient. An extensive literature survey was conducted to determine if neural networks had been previously used in feature/pattern recognition as applied to CAPP. It was concluded that this study was the first of its kind.

The resurgence of interest in artificial neural networks has been fired by both theoretical and application successes (Wasserman, 1989). Suddenly, it appears possible to apply computation to realms previously restricted to human intelligence – to make machines that learn and remember in ways that bear a striking resemblance to human mental processes and to give a new and significant meaning to artificial intelligence. Features of artificial neural networks that make them attractive to CAPP are discussed below.

Artificial neural networks can modify their behavior in response to their environment. This factor, more than any other, is responsible for the interest they have received in image processing and pattern recognition. Shown a set of inputs, they self-adjust to produce consistent responses (Pao, 1989).

A trained network to some degree is insensitive to minor variations in its input. This ability to see through noise and distortion to the pattern that lies within is vital to pattern recognition in a real-world environment. The artificial neural networks approach to CAPP when compared with conventional CAPP methods has the potential to generalize automatically as a result of its structure and not by using human intelligence embedded in the form of *ad hoc* computer programs as found in the latter.

It was mentioned earlier that the reason for the limited success achieved thus far in generative process planning is the complex system of logic required to describe parts fully. Thus the most difficult task that exists in CAPP as is currently found is the task of part description. The advances made by artificial neural networks in successfully dealing with feature recognition problems greatly simplify this complex task. This means that researchers could place far greater effort on the task of automating process planning for true computer-integrated manufacturing rather than the beginning task of part description. Artificial neural networks thus provide a possible alternative to solving the tedious part description problem as it exists now and help in achieving true manufacturing integration.

7.2 SAMPLE PROBLEM DEFINITION

The purpose of our research is to establish, on a preliminary basis, the utility of neural networks in CAPP. The neural network being designed should be able to read and interpret 2D designs and propose suitable process plans based on previous training and features recognized. In this project 2D designs are represented on $n \times m$ matrices using ASCII format. The edges of parts under consideration are represented by 1s and the rest by 0s. The use of 1s and 0s simplifies the process of part description. A CAD drawing can very easily be converted to a similar matrix by considering the graphics as a matrix of pixels and analyzing each pixel for the presence or the absence of an edge. If an edge is present, the value returned is 1, else it is 0. This method

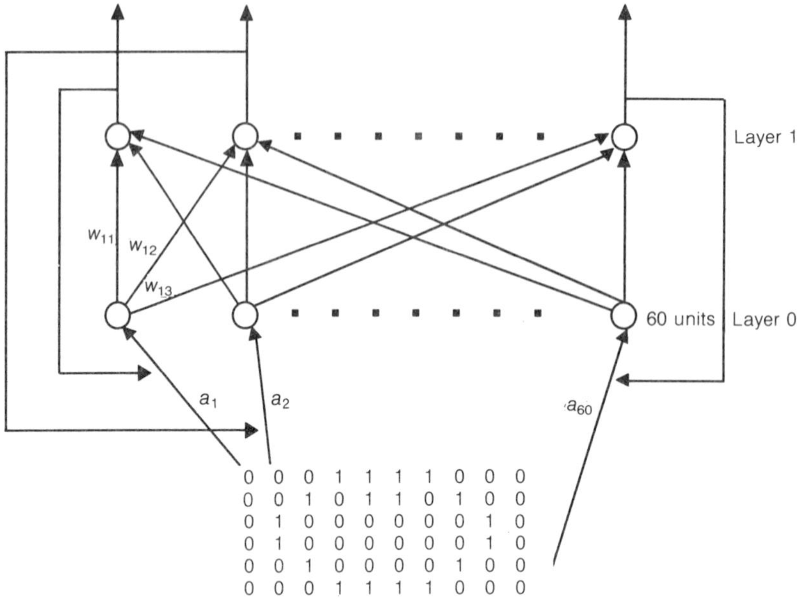

Fig. 7.2 Neural network for part feature learning. This network has lateral and recurrent connections.

of applying artificial neural networks for feature recognition has been successfully used in several other applications.

The model under consideration is first trained using those parts which have existing process plans. Once the network is trained, other 2D parts are applied to the network. Using the nearest neighbor recall, the network will identify these parts and propose process plans based on the features recognized. The methodology is similar to the generative approach since it consists the three main components: part description, manufacturing databases; and decision-making logic and algorithms. The task of part feature identification is performed here by an artificial neural network.

In our research, we used four different features on 6×10 matrices. One such feature on a matrix is shown in Fig. 7.2. The 1s in the matrix represent the edges of a cylinder with a keyway milled on top as seen in the side view of Fig. 7.1.

7.3 DEVELOPMENT OF NETWORK ARCHITECTURE

7.3.1 Network architecture

Since application of neural networks to CAPP was being done for the first time, the following attributes were considered essential in the selection of a suitable artificial neural systems (ANS) paradigm.

- supervised learning
- simplicity in architecture
- quick learning
- improved storage capacity

There are several paradigms which satisfy these requirements. **Backpropagation** was first considered and *BrainMaker V1.9* (©California Scientific Software) was used to identify 2D parts. Since *BrainMaker V1.9* does not allow outside interaction with the DOS shell, this paradigm was shelved.

Despite the demonstrated power of backpropagation, several problems plague its application (Wasserman, 1990). Rumelhart provided a convergence proof cast in terms of partial differential equations making it valid only if network weights are adjusted in infinitesimal steps. Because this implies infinite convergence time, it proves nothing about practical training. In fact, there is no proof that backpropagation will ever converge with a finite step size. Empirical observations show that networks do train, but the duration of the training process is unpredictable and lengthy.

Backpropagation uses gradient descent to adjust the network weights, following the local slope of the error surface toward a minimum (Wasserman, 1990). Gradient descent works well with convex error surfaces that have unique minimum, but it often leads to nonoptimal solutions with the highly convoluted, nonconvex surfaces encountered in practical problems. Under some circumstances, a network can train itself into a state where weight modification comes to a virtual standstill. This network paralysis is a serious problem – once entered it can extend training time by orders of magnitude.

Since one of our objectives was to test the idea of neural networks applied to CAPP, we did not want to spend too much time in training. We therefore avoided using Boltzmann machine, Cauchy machine, adaptive heuristic critic, associative reward-penalty and avalanche matched filter, though each one of them could have been used (Kohonen, 1984). We were fascinated by the simplicity of the one-layer feedback topology **brain-state-in-a-box** (BSB) ANS and also by the fact that despite being first proposed by Anderson, Silverstein, Ritz and Jones (Anderson *et al.*, 1977), this paradigm has been sparingly used in feature recognition and knowledge processing problems. In addition, BSB has excellent storage capacity. We therefore selected BSB for this study.

Brain-state-in-a-box is a good association model that has been applied elsewhere to several sets of experimental data (Anderson *et al.*, 1977). Although it was necessary for Anderson, Silverstein, Ritz, and Jones to oversimplify reality to obtain a model that they could work with easily, the resulting model was sufficiently rich to have a pronounced structure, and, with a very modest amount of manipulation, gave rise to some testable predictions. It was shown that this model could provide a theoretical framework for entities that acted and behaved very much like distinctive features. This is important since feature analysis is a strategy used by the brain. This involves neurons responding when, and only when, a particular feature is encountered. A set of neurons with positive feedback tends to analyze its inputs by most heavily weighing the eigenvectors of the feedback matrix with large positive eigenvalues and by suppressing the rest. These particular eigenvectors are often the most meaningful in terms of the discrimination to be performed, since they contain most of the information allowing discriminations to be made among the stimulus set.

Brain state in-a-box (BSB)

The Brain-state-in-a-box ANS is a one-layer, autoassociative, nearest neighbor classifier that stores arbitrary analog spatial patterns, $A_k = (a_1^k, \ldots, a_n^k)$, using error-correction learning. The BSB learns off-line, operates in discrete time, and displays the one-layer feedback topology shown in Fig. 7.2, where the processing elements (PEs) correspond to components of A_k. The BSB represents an alternative to other Grossberg pattern storage algorithms in that it incorporates the feedback recall dynamics while using the supervised learning procedure. This alternative typically yields improved storage capacity over the unsupervised learning routines at the expense of increased encoding time.

Encoding

The BSB encoding procedure uses the error-correction encoding equation

$$\Delta w_{ij} = \alpha a_i^k d_j \tag{7.1}$$

where Δw_{ij} is the change in the connection strength from the ith to the jth PE (Simpson, 1990), α is a positive constant controlling the learning rate, and

$$d_j = a_j^k - \sum_{i=1}^{n} w_{ij} a_i \tag{7.2}$$

The first term of (7.2) is the desired value of the jth PE and the second term is the computed value of the same PE. During the encoding process (7.2) is employed repeatedly for each pattern until the amount of error d_j produced by each output is sufficiently low.

Recall

BSB recall utilizes a threshold-linear ramp function in an intra-layer feedback operation (Simpson, 1990). When an input is presented, it is processed until all PEs cease to change, typically meaning that they have saturated to either the high or low end of the ramp function using the equation

$$a_i(t+1) = f\left(a_i(t) + \beta \sum_{j=1}^{n} w_{ij} a_j(t) \right) \tag{7.3}$$

where $a_i(t)$ and $a_j(t)$ are the activation values of the ith and jth PEs at time t, β is a positive constant controlling the amount of intra-layer feedback, and $f(\ )$ is the ramp threshold function

$$f(x) = \begin{cases} +\gamma & \text{if } x \geqslant \gamma \\ +x & \text{if } |x| < \gamma \\ -\gamma & \text{if } x \leqslant -\gamma \end{cases} \tag{7.4}$$

Identifying terms in (7.3), the first term in $f(\ )$ is the positive feedback term, and the second term is the lateral feedback term from the other PEs. Equation (7.3) can be rewritten in matrix–vector form as

$$\mathbf{A}(t+1) = f((\mathbf{I} + \mathbf{W})\mathbf{A}(t)) \tag{7.5}$$

where $\mathbf{A}(t)$ and $\mathbf{A}(t+1)$ are the vectors representing the PE activations at time t and $t+1$ respectively, $\mathbf{I}$ is the identity matrix, and $\mathbf{W}$ is the matrix of connection strengths between all the PEs.

During the simulation runs of this algorithm, we experienced slow down in encoding since the saturated high or low values were never reached. This was particularly severe when the input matrix became slightly large. We therefore had to make certain modifications in the encoding and recall procedures, as discussed below.

Modified encoding and recall

Several simulation runs were tried with different modifications. One such run included assigning a value of $+1$ to $+\gamma$. The best results were obtained when input vectors were normalized during encoding and recall to operate in unit space. Thus,

$$a_i^k = \frac{a_i^k}{\| A_k \|} \tag{7.6}$$

for all $i = 1, 2, \ldots, n$ and all $k = 1, 2, \ldots, m$, and where

$$\| A_k \| = \left[\sum_{i-1}^{n} (a_i^k)^2 \right]^{1/2} \tag{7.7}$$

This change helped in a tremendous decrease in training times and the system was able to handle large size input vectors easily.

7.3.2 Input representation

As mentioned earlier, for this study, part features are represented on 10×6 matrices using ASCII format. The edges of features under consideration are represented by 1s and the rest by 0s. Thus the vector $\mathbf{A}_k$ has 60 components being input into as many neurons. This network size was selected due to computer memory and speed constraints. The computer program developed for this study, however, can handle larger matrices.

7.3.3 Output representation

The proposed hybrid system has a modified BSB encoding and recall algorithm that identifies part features and then, based on an expert system, lists simple process planning steps. It is possible to apply this logic to complex features and process plans. This study shows that advanced systems should be able to serve as interfaces between CAD and CAM and provide the required integration.

7.3.4 Training set

To establish optimum neural network performance, several training sets were used. Network training runs were tried with 512 and 1024 input vectors. The modified encoding algorithm showed excellent performance and the system had no difficulty in completing training runs expeditiously. This was in contrast to poor encoding by the standard BSB algorithm.

7.4 ARTIFICIAL NEURAL NETWORK IMPLEMENTATION

A model of an ANN/CAPP system was proposed in our study. Such a system could be further integrated with a visual system to identify new parts and generate suitable process plans. The ultimate objective of this intelligent system will be to read input designs, produce process plans and directly communicate with assigned machine tools by producing NC codes, thus resulting in complete CAD/CAM integration.

The entire system will involve interaction between several databases. Neural network based systems will be much simpler than conventional CAPP systems since complex feature recognition algorithms are avoided. A well-trained ANS/CAPP can be more efficient due to good storage capabilities. The system can be made to be even more powerful if the network is placed on a chip to result in true parallel processing, as opposed to a simulation performed in this study.

7.5 PERFORMANCE OF THE INTELLIGENT SYSTEM ARCHITECTURE

Program code

We mentioned earlier that several versions of the computer program were tested with modifications in the encoding algorithm. Our first major working

Table 7.1 BSB.PAS performance results

Test No.	*Vector size*	*No. of design types*	*No. of training vectors*	*Encoding performance*	*Recall performance*
1	3 × 2	4	512	Excellent	Excellent (100.0%)
2	3 × 5	4	512	Excellent	Good (90.0%)
3	3 × 5	5	1024	Excellent	Excellent (100.0%)
4	3 × 6	4	1024	Poor (stuck on local minima)	—
5	3 × 7	4	1024	Very poor (stuck on first input)	—
6	6 × 10	4	512	—	—

system was based on the algorithm suggested by Simpson (1990). The program written in Pascal was called BSB.PAS.

The most successful version of the algorithm, discussed in section 7.3.1, involves normalizing each input vector. We called this program BSBNORM. PAS.

To understand and appreciate the superior performance of BSBNORM over BSB, we are listing below, the results of several computer runs with different size input vectors in the order that they were executed. The results clearly indicate the reasons for the modifications made by us.

BSB.PAS testing and performance

Several combinations of values for α, β, and γ were tried out in this testing:

α: 0.001, 0.01, 0.10, 0.15, 0.2, 0.25, 0.3

β: 0.001, 0.01, 0.10, 0.15, 0.2, 0.25, 0.3

γ: 0.1, 0.5, 1.0

After extensive testing, it was concluded that the combination of $\alpha = 0.1$, $\beta = 0.1$, and $\gamma = 1.0$ produced optimum results. The test results shown in Table 7.1 are for such a combination of parameters.

BSBNORM.PAS testing and performance

The same combination of parameters were then tried out on BSBNORM.PAS, with similar results. The test results shown in Table 7.2 are with $\alpha = 0.1$, $\beta = 0.1$, and $\gamma = 1.0$.

Table 7.2 BSBNORM.PAS performance results

Test No.	*Vector size*	*No. of design types*	*No. of training vectors*	*Encoding performance*	*Recall performance*
7	3 × 2	4	512	Excellent	Excellent (100.0%)
8	3 × 5	4	512	Excellent	Excellent (100.0%)
9	3 × 5	5	1024	Excellent	Excellent (100.0%)
10	3 × 6	4	1024	Excellent	Excellent (100.0%)
11	3 × 7	4	1024	Excellent	Excellent (100.0%)
12	6 × 10	4	512	Excellent (1 hr. for encoding)	Excellent (100.0%)

7.6 SUMMARY

Process planning is the act of preparing detailed operation instructions to transform an engineering design to a final part. We have proposed and conducted, for the first time, an initial simulation on the feasibility of artificial neural networks as applied to process planning. It was mentioned earlier that the reason for the limited success achieved thus far in generative process planning is the complex system of logic required to fully describe parts. Thus the most difficult task that exists in CAPP as is currently found is the task of part description. The methods that have been traditionally applied to computer-aided process planning involve complex algorithms which transfer designs from one form to another for data manipulation. The advances made by artificial neural networks in successfully dealing with feature recognition problems greatly simplifies this complex task. This means that researchers could place far greater effort on the task of automating process planning for true computer-integrated manufacturing rather than the beginning task of part description. Artificial neural networks thus provide a possible alternative to solving the tedious part description problem as it exists now and help in achieving true manufacturing integration. Neural networks in conjunction with expert systems can be used to generate sophisticated process plans for parts.

The paradigm that was used in this simulation study was brain-state-in-a-box, the selection being based on simplicity of architecture, quick learning, and good storage capacity. Several modifications were tried out in the paradigm encoding algorithm. We found that the best results were obtained when normalized input vectors in unit space were applied.

In this project, 2D designs were represented on $n \times m$ matrices using ASCII format. The edges of parts under consideration were represented by

1s and the rest by 0s. The use of 1s and 0s simplified the process of part description. A CAD drawing can very easily be converted to a similar matrix by considering the graphics as a matrix of pixels and analyzing each pixel for the presence or the absence of an edge. This method of applying artificial neural networks for feature recognition has been successfully used in several other applications.

We have been able to conclude that ANS/CAPP, which is an extremely potent area of research, should be pursued further and have proposed long term outlook for this field.

REFERENCES

Alting, L. and Zhang, H. (1989) Computer aided process planning: the state-of-the-art survey. *International Journal of Production Research*, **27**(4), 553–85.

Anderson, J.A., Silverstein, J.W., Ritz, S.A. and Jones, R.S. (1977) Distinctive features, categorical perception, and probability learning: some applications of a neural model. *Psychological Review*, **84**(5), September 413–51.

Chang, T.C. (1982) TIPPS: a totally integrated process planning system. Doctoral dissertation, Virginia Polytechnic Institute and State University, Blacksburg, Virginia.

Chang, T.C. (1989) *Expert Process Planning for Manufacture*, Prentice Hall, Englewood Cliffs, New Jersey.

Gupta, T. (1990) An expert system approach in process planning: current development and its future. *Computers and Industrial Engineering*, **18**(1), 69–80.

Kohonen, T. (1984) *Self-Organization and Associative Memory*, Springer-Verlag, New York.

Kusiak, A. (1990) *Intelligent Manufacturing Systems*, Prentice Hall, Englewood Cliffs, New Jersey.

Pande, S.S. and Walvekar, M.G. (1989) PC-CAPP – a computer-assisted process planning system for prismatic components. *Computer-Aided Engineering Journal*, **6**(4), August, 133–8.

Pao, Y.H. (1989) *Adaptive Pattern Recognition and Neural Networks*, Addison-Wesley Publishing Company, Inc., New York.

Parks, C.M. (1989) Computer aided process planning with CMPP. *Computers and Industrial Engineering*, **17**(1–4), 246–51.

Rumelhart, D.E., McClelland, J.L. and the PDP Research Group (1986) *Parallel Distributed Processing*, Vol. 1, The MIT Press, Cambridge, Massachusetts .

Simpson, P.K. (1990) *Artificial Neural Systems*, Pergamon Press, Inc., New York.

Srihari, K. and Greene, T.J. (1987) An evaluation and review of computer aided process planning systems. *Proceedings – World Productivity Forum & 1987 International Industrial Engineering Conference*, Washington, DC, May 438–44.

Srihari, K. (1989) A rule-based dynamic prototype CAPP system. *Computers and Industrial Engineering*, **17**(1–4), 240–5.

Steudel, H.J. (1984) Computer-aided process planning: past, present and future. *International Journal of Production Research*, **22**(2), 253–66.

Wang, H.P. and Wysk, R. (1988) A knowledge-based approach for automated process planning. *International Journal of Production Research*, **26**(6), 999–1014.

Wasserman, P.D. (1989) *Neural Computing – Theory and Practice*, Van Nostrand Reinhold, New York.

Wasserman, P.D. (1990) A combined back-propagation/Cauchy machine network. *Journal of Neural Network Computing*, Winter 34–40.

CHAPTER 8

Scheduling

John Y. Cheung
School of Electrical Engineering & Computer Science, The University of Oklahoma, Norman, OK 73019–0361.

An important aspect of automation in manufacturing systems is scheduling. The need for scheduling occurs at every level of the manufacturing process – from scheduling purchase of components and subcomponents, scheduling jobs and machines in the making and asembling process, to scheduling picking, packaging, shipping, etc. Furthermore, scheduling in material handling can occur at more than one level with varying degrees of detail and sophistication, e.g. a month-to-month schedule for orders and component purchase, a week-to-week schedule for components on the assembly line, a day-to-day schedule for jobs to be done in the machine shop, or an hour-by-hour schedule for each machine in the shop. With the emphasis of just-in-time manufacturing and production, scheduling is particularly important to ensure smooth operation of all phases of the entire manufacturing process.

8.1 CHARACTERISTICS OF SCHEDULING PROBLEMS

Scheduling belongs to a larger class of problems called **constraint optimization problems** (COPs). Clearly, scheduling involves looking for a combination that minimizes a particular cost function, subject to the given constraints of the problem. For example, in job shop scheduling and flow shop scheduling, the object is to minimize the work-in-process time or the makespan time, and the constraints would be the maximum capacity of the plant, the number of machines available, etc. There are a large number of manufacturing problems that can be considered as COPs including routing, path planning, etc. There

Artificial Neural Networks for Intelligent Manufacturing. Edited by Cihan H. Dagli. Published in 1994 by Chapman & Hall, London. ISBN 0 412 48050 6

are also other closely related problems such as facilities layout in two dimensions and pallet loading in three dimensions that are also COPs.

A COP may involve discrete (binary) or continuous parameters. For problems dealing with continuous parameters, it is desirable to find an optimal value for the parameters in question such that the cost function can be minimized. For example, in flow shop scheduling, it may be desirable to find the optimal lot size, a continuous variable, such that the completion time of the jobs is minimal. For problems of this type, a number of analytical and heuristic approaches have been reported in the literature. Indeed, there are many applications of COP problems with continuous parameters reported in the literature.

A COP involving discrete (binary) parameters is of particular interest in manufacturing because many assignments of time problems are of this nature. For example, in job shop scheduling, there are n jobs with m machines, it is desirable to find the best job sequence in scheduling the n jobs on the m machines so that the earliest completion time can be achieved. The traveling salesman problem (TSP) where the goal is to find a traveling schedule for a sales person (a drill for example) to visit all given cities (the drill holes) with the shortest path is another example of an assignment problem. All these problems involve making an assignment, a (binary) yes or no decision, on the parameters in question. There are many applications for this type of problem reported in the literature.

Since most of the COPs have been shown to be N-P complete problems, this means that the computation requirement does not increase in a polynomial fashion. The computational requirement of most combinatorial optimization problems in most cases increases as the factorial of the problem size. Hence, complete enumeration to reach the optimal solution is not always possible as the problem size increases beyond the toy problem scenario. But for NP-complete problems, the optimal solution is not always obvious and cannot be guaranteed unless the entire solution base is known. In other words, global optimality can be assured only after the entire solution space has been searched. Since a search of the entire space is often intractable due to the number of possible solutions, finding the optimal solution also becomes intractable. In the light of this, the emphasis of much research has been concentrated on finding a good approximate solution quickly. Experience shows that for many applications, the approximate solutions are often acceptable and are almost as good as the optimal solution, thus the extra computational load to find the optimal solution is not necessary.

8.1.1 Traditional solution approaches

The traditional approach to solving the COPs analytically is similar to other optimization problems using methods such as Newton's method, gradient search, or the method of the steepest descent, etc. These techniques are covered in detail by texts such as Nemhauser and Wolsey (1988). In addition, classical techniques also include dynamic programming, integer programming, or mixed programming. These techniques can be found in most standard undergraduate texts in numerical analysis or operational research such as French's text (1982) on sequencing and scheduling.

8.1.2 Neural network approaches

More recently, a number of novel approaches based on the principles of simulated annealing and genetic search have been reported. Within the last decade, the technology based on neural networks has also been shown to be a viable alternative for some COPs. There are a number of approaches in the use of neural networks for optimization problems.

1. Searching network (Hopfield net)
2. Probabilistic network (simulated annealing)
3. Error-correcting network (multi-layer perceptron)
4. Competing network
5. Self-organizing map

In a **searching type network**, the neural network is set up in such a way that the dynamics of the network is constrained toward a local minimum which represents a possible solution. An example of this type of searching network is the Hopfield net approach. Other networks are based on the **probabilistic outcome** of the network output to improve the search performance. This is true for the simulated annealing approach. In some complex cases where a lot of constraints are present, **error-correcting networks** are used to learn the output patterns of well constructed schedules. This is the popular multi-layer perceptron approach. In a **competing type network,** the solution comes from the winner of the competition within a group of competing output neurons through lateral inhibitions to ensure that only a single choice is made. The self-organizing map approach also allows lateral inhibition among an array of neurons as an additional mean to communicate with one another. The first two methods will be described in detail while the latter three methods will only be briefly reviewed.

8.2 THE HOPFIELD NET APPROACH

The Hopfield net approach was first proposed by Hopfield (1982, 1984), Hopfield and Tank (1985), and Tank and Hopfield (1986). In this approach, the collective computing power of the neural network is used in a parallel fashion to achieve a common goal. Each neuron in the network operates in unison with the available information supplied. Communication between neurons is handled through connection between neurons. The final state of each neuron represents a possible solution to the problem.

8.2.1 Optimization using the Hopfield net

A Hopfield net is a neural network composed of a layer of totally connected artificial neurons as shown in Fig. 8.1. Each neuron of the network is connected to every other neuron in the network. The network is dynamic because the neuronal output is always changing as the network gravitates toward a local minimum of energy level. Hence, the Hopfield net is categorized as a searching type network.

The continuous model of the Hopfield net is analogous to a circuit model of an operational amplifier with resistive connections and additional capacitors as non-linear elements. The dynamics of each neuron in the Hopfield net is governed by the following differential equation

$$\frac{du_m}{dt} = \frac{1}{C}\left(\sum_{n=1}^{N} T_{mn}v_n - \frac{u_m}{R_m} + I_m\right) \tag{8.1}$$

where

$$v_m = g_m(\lambda u_m) = \tfrac{1}{2}(1 + \tanh(\lambda u_m))$$

The output of each neuron is represented by v_m which is related to the u_m, called the **net input**, by the output function g_m. The output function is a non-linear operation and is usually taken to be related to the hyperbolic-tangent function. The parameter λ represents a scaling constant called the **gain constant**, which controls the speed of convergence of the network. When λ is large, the output values of the neuron will settle to the extremal values quickly thus causing the network to settle to a local minimum rapidly.

The net input itself of each neuron is governed by the differential equation. The change in net input is composed of the contribution from the interconnection pattern represented by the weight matrix T_{mn} acting on the output of other

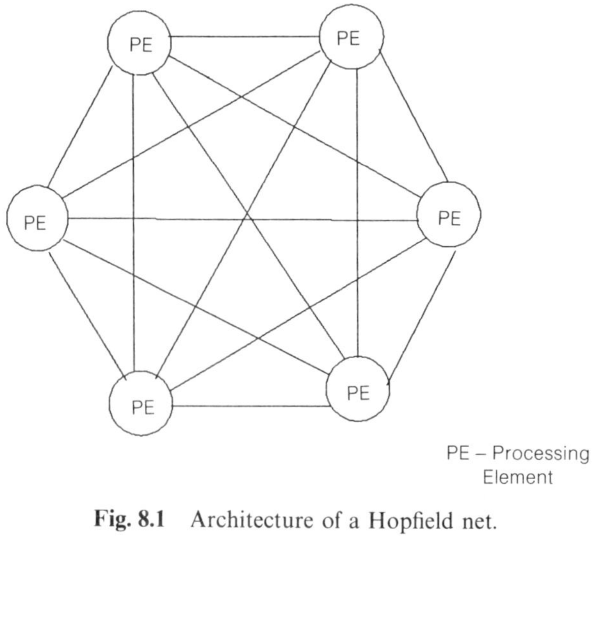

Fig. 8.1 Architecture of a Hopfield net.

neurons, the contribution from a forgetting term ($-u_m/R_m$) and the contribution from the input term (I_m). In the absence of any input, the network would settle at a particular local minimum characterized by the status of the interconnection pattern:

$$u_m^* = R_m \sum_{n=1}^{N} T_{mn} v_n \tag{8.2}$$

For the Hopfield net, there is a continuous energy function E which measures the total energy represented by the network with respect to the network outputs.

$$E = -\frac{1}{2} \sum_{m=1}^{M} \sum_{\substack{n=1 \\ n \neq m}}^{N} T_{mn} v_m v_n + \frac{1}{\lambda} \sum_{m=1}^{M} \frac{1}{R_m} \int_0^{v_m} g_m^{-1}(v)\,\mathrm{d}v - \sum_{m=1}^{M} I_m v_m \tag{8.3}$$

The energy function and the differential equation are related to one another. The differential equation is called an equation of motion corresponding to the energy function E which is called a **Lyapunov function**. A more detailed treatment of the Lyapunov function concept can be found in Goles and Vichniac (1986). It can be shown that the changes in the network output governed by the corresponding equation of motion such as (8.1) will always cause the energy level of the network to either decrease or remain the same. In other words, every change recommended by the equation of motion will always result in a corresponding non-positive change in energy level. In most cases the resultant energy level would be less until the network stabilizes at the local minimum level. At that time there are no further changes in the energy level.

During searching, the Hopfield net is initiated by random values or by values imposed by the inputs (I_m). The input causes a change in the right-hand side of the differential equation thus causing a corresponding change in the net input. This input causes the network to be located to a different part of the energy level space. The dynamics of the network then begin to change the network output toward the closest local minimum.

To apply the Hopfield net approach to COPs, the following five steps are used.

1. Determine the input representation.
2. Determine the cost function in the form of a Lyapunov function.
3. Rewrite the constraints in the form of a Lyapunov function.
4. Determine the equation of motion for each neuron.
5. Randomly start the network.

The first step of applying the Hopfield net to a problem is to find a representation of the problem in such a way that the final solution is indicated by the output of the neurons. For many combinatorial problems, the binary output state of the neurons, i.e. in terms of 0 and 1, is used. This is called the high gain situation where the neuronal output is driven into saturation.

For other problems, a continuous output may be desirable which can be used to indicate the degree of control, for example.

The second step is to determine the proper cost function which must be written in the form of a Lyapunov function. Typically, this energy function is a quadratic function based on the neuronal outputs of the network and other given parameters. The cost function represents the penalty function to be minimized. In order for the Hopfield net to work, one must find an energy function for the penalty function that is a Lyapunov function to the above mentioned equation of motion. If the original penalty function of the COP is not such a Lyapunov function, then it is desirable to modify the penalty function into an equivalent energy function that *is* a Lyapunov function.

Since the Hopfield net has no provision for explicitly coding the constraints of the COP, the third step is to define the constraints and to code the constraints also in the form of a Lyapunov function. There are two sources of constraints. Some constraints arise from the problem itself. Additional constraints come from the representation of the problem defined in step one. These constraints may arise due to the particular way used in representing the problem and may not be constraints imposed by the original problem. All constraints are written in a quadratic form and are then combined with the penalty or cost function to form a generalized energy function. This means that all of the constraints are also constructed in the form of a quadratic.

The fourth step is to derive the equation of motion that is associated with the generalized cost function found from the previous step. If the cost function and the constraints can be put in the standard Hopfield energy function, then the equation of motion for each neuron can be inferred directly by comparing the structure of the two equations. In other words, all the quadratic terms are combined to form the weight matrix T_{mn}, and all the linear terms are also combined as contributions to the input.

The final step is to randomly start the network. If all neurons are initialized to be zero, the correction for each neuron will be identical thus producing a trivial solution. To arrive at a solution, randomly initialize each neuron in the network with a small random number. The equation of motion will drive the network to a local minimum corresponding to the initial output space. If the network is again initialized with another set of neuronal outputs, the equation of motion will cause the network to converge to another local minimum. To get a good solution, typically the network is initialized differently several times to arrive at a variety of solutions and the best solution is selected.

8.2.2 The traveling salesman problem using the Hopfield net approach

The principle of the Hopfield net for scheduling can be illustrated using the traveling salesman problem (TSP). Given the distances, $d_{X,Y}$, between each pair of cities X and Y, the TSP is to find the shortest route that visits each and every city once in the tour and returns to the original city started. Many manufacturing applications can be formulated in terms of the TSP, for example finding the shortest path of a drilling machine that must drill the prescribed number of holes in a metal sheet.

Step 1: Determine the input representation

The first task is to represent the problem from the neural network standpoint. In the Hopfield net, the tour is represented by a permutation matrix. In other words, given N cities, the tour can be represented by an N by N matrix with N^2 elements. Let each row represent a city and each column represent the order of the city to be visited. A valid tour is then a permutation matrix with N ones and $N^2 - N$ zeros. Assume that there are five cities, then a possible matrix would take on the following form.

	1	2	3	4	5
A	0	1	0	0	0
B	1	0	0	0	0
C	0	0	0	1	0
D	0	0	0	0	1
E	0	0	1	0	0

The first row indicates the order of the tour when city A should be visited, in this case the second city. Likewise, the second row indicates what order city B will be visited, in this case first. The rest of the rows are interpreted in the same way. The matrix can also be interpreted by columns. The first column indicates which city the tour begins. The second column indicates

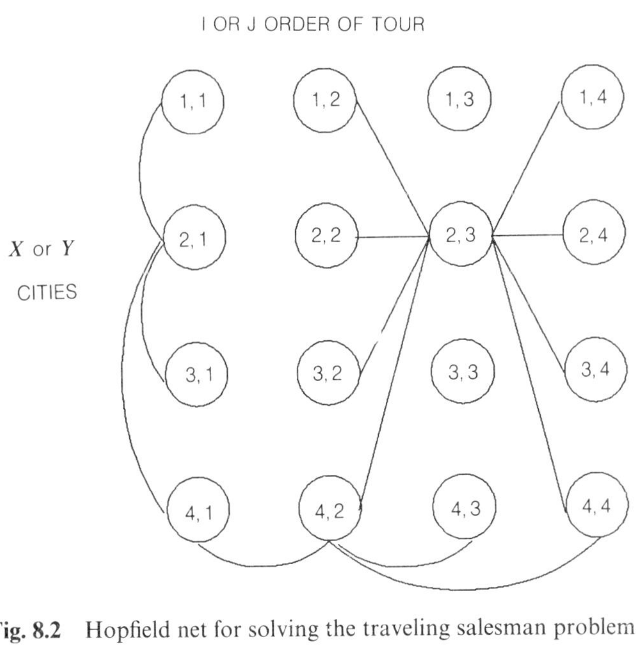

Fig. 8.2 Hopfield net for solving the traveling salesman problem.

which city to visit next, and so on. The matrix above indicates that the tour will be B–A–E–C–D. It is obvious that all valid tours will take on similar form except that the location of the ones will be different according to the order of the tour.

Using the permutation matrix given above, each element of the matrix can now be represented by a neuron thus requiring a total of N^2 neurons and N^4 connections. Each neuron is identified by two subscripts as shown in Fig. 8.2. The first subscript is either X or Y representing the city, while the second subscript is either i or j representing the order. Hence, v_{Xi} refers to the neuron representing the X city being visited at the ith order.

The goal of the neural net algorithm is to cause the network to gravitate towards a local minimum state that is indicative of a tour like the one shown. It is also highly desirable that the tour thus produced will also be one of the shorter tours. To accomplish this, each neuron must be placed in a high gain situation so that the output will asymptotically be very close to 0 or 1, the extremal values, and not to any of the intermediate values. These extremal values then indicate the order that the cities are to be visited.

Step 2: Determine the cost function in the form of a Lyapunov function

For the traveling salesman problem, the cost function is explicitly defined as the shortest distance of the tour from the first city, through all the cities, and returning to the initial city. Since the tour is represented by those elements of the matrix that are one, the cost function can be constructed by attaching a distance measure with those neurons that are on. Though there are N^2 neurons, yet there should be only N on-neurons. The distance of the tour can be calculated directly from the otuput of the neurons.

$$J = \frac{D}{2} \sum_{X=1}^{N} \sum_{i=1}^{N} \sum_{\substack{Y=1 \\ Y \neq X}}^{N} d_{X,Y} v_{X,i} (v_{Y,i+1} + v_{Y,i-1}) \tag{8.4}$$

The last summation in (8.4) gives a measure of the distance of this part of the tour. For a given neuron (X, Y), the summation is over all the neurons in the previous column and the next column. (If the present column is the first column, then the previous column is the last column. Likewise, if the present column is the last column, then the next column is the first column.) The summation pairs the present neuron with each neuron of the previous and next columns. The distance between the previous column and the present column is only summed when the neuron in the previous column is one at the same time that the present neuron in the present column is one. Likewise, the distance between the next column and the present column is only summed when the neuron in the next column is one at the same time that the neuron in the present column is one. The first two summations in (8.4) serve to sum up the distance contributions from all the neurons in the network thus completing the distance calculation for the entire tour. The goal of the problem is of course to minimize the distance measure, J, for the entire tour.

Step 3: Rewrite the constraints in the form of a Lyapunov function

The cost function is often not enough because there are additional constraints that one may further impose on the system. For the TSP, there are no additional constraints to be imposed due to the cost function, though the problem representation does demand some additional constraints to be satisfied in order for the tour to be a valid one. In other words, these constraints are necessary to avoid a trivial solution and to ensure that the solution is valid. Each constraint is now written in the form of a penalty function so that during minimization, those solutions that do not satisfy the constraints are penalized.

For the tour to be valid, it must visit every city only once. This means that there must be and can only be one neuron with output one for every row. This constraint condition can be represented in quadratic form:

$$E_r = \frac{A}{2} \sum_{X=1}^{N} \sum_{i=1}^{N} \sum_{\substack{j=1 \\ j \neq i}}^{N} v_{X,i} v_{X,j} \tag{8.5}$$

The subscripts X and Y refer to the cities while the subscripts i and j refers to the order of their visits. The last summation adds the contribution between the present neuron and all the other neurons in the same row. If both neurons are one, then the multiplication will produce a one, otherwise the multiplication will produce a zero at all other times. Hence, if there are more than one neuron with output one in a single row, a penalty will be imposed because the summation will result in a non-zero but positive value. The first two summations in (8.5) serve to sum the penalty contributions from all the neurons in the network thus completing the penalty calculation for the entire tour.

Likewise, each step of the tour can only be associated with a single city. This means that there can only be one neuron turned on for each column. This condition is indicated as

$$E_c = \frac{B}{2} \sum_{X=1}^{N} \sum_{i=1}^{N} \sum_{\substack{Y=1 \\ Y \neq X}}^{N} v_{X,i} v_{Y,i} \tag{8.6}$$

As before, the last summation adds the contribution between the present neuron and all the other neurons in the same column. If both neurons are one, then the multiplication will produce a one, otherwise the multiplication will produce a zero at all other times. Hence, if there are more than one neuron with output one in single column, a penalty will be imposed because the summation will result in a non-zero but positive value. The first two summations in (8.6) serve to sum the penalty contributions from all the neurons in the network thus completing the penalty calculation for the entire tour.

The two constraints given above are insufficient by themselves to ensure a satisfactory solution because the minimizing process will tend to force all neuronal output to be zero, a trivial solution. In other words, the constraints guard against the situation of having more than one neuron in each row or column, but accept the situation of having none or at most one neuron in each row or column. The feasible solution requires that there should be at least one and no more than one neuron in each row and column. Hence, a third constraint is needed to prevent the trivial solution of having no neuron being active in each row or column.

$$E_t = \frac{C}{2}\left(\sum_{X=1}^{N}\sum_{i=1}^{N} v_{X,i} - N\right)^2 \tag{8.7}$$

The above constraint sums the total number of active neurons through the double summation and compares the total to N. If the total is less or more than N, a penalty is elicited. These three constraints serve to describe a valid tour and penalize those tours that do not satisfy any or all of the constraints.

The total Hopfield energy function is then the sum of the cost function related to the distance measure and the three additional constraints imposed:

$$E = J + E_r + E_c + E_t \tag{8.8}$$

The four constants A, B, C, and D are used to adjust the relative weights given to the penalty contribution of each constraint. During the minimization process, not only is a solution sought that would give the smallest distance, but the same solution must also satisfy, as much as is possible, the given constraints.

Step 4: Determine the equation of motion for each neuron

Since the energy function is in the standard form similar to the Hopfield energy function, the equation of motion can be derived directly from the form of the energy function and the corresponding Hopfield equation of motion. Expanding the double summation of the quadratic term in the energy function, we have:

$$-\sum_{m=1}^{M}\sum_{\substack{n=1\\n\neq m}}^{N} T_{mn}v_m v_n = -\sum_{X=1}^{N}\sum_{i=1}^{N}\sum_{Y=1}^{N}\sum_{\substack{j=1\\j\neq i}}^{N} T_{Xi,Yj}v_{X,i}v_{Y,j} \tag{8.9}$$

Comparing the above equation with the first constraint, first, it is obvious that the summation on Y is not used. The contribution is summed only when Y is the same as X. Secondly, the contribution is summed only when j is not equal to i. From these two points, it is clear that the $T_{Xi,Yj}$ can be written as

$$-A\delta_{XY}(1-\delta_{ij}) \tag{8.10}$$

The first delta function causes the entire term to be zero unless X is equal to Y. The second factor causes the term to be zero when i is equal to j.

Following the same logic as the first constraint, the penalty term in the equation of motion for the second and third constraints can also be written in similar fashion. The distance measure is likewise rewritten in the same format. The final form of the weight matrix is

$$T_{Xi,Yj} = -A\delta_{XY}(1-\delta_{ij}) - B\delta_{ij}(1-\delta_{XY}) - C - Dd_{X,Y}(\delta_{j,i+1} + \delta_{j,i-1}) \tag{8.11}$$

The above equation of motion governs the behavior of every neuron with respect to every other neuron. This is the interconnection matrix for the entire neuronal array.

Step 5: Randomly start the network

Once the equation of motion for each neuron is established, the solution may be found by randomly initializing the network output. The dynamics of the network through the equations of motion would force the network to settle at a local minimum closest to the original initialized point. The solution represents a local minimum. In most cases, the solution also represents a valid tour. Due to the interference of the constraints with the distance measure, occasionally, the solution may represent an invalid tour. For these situations, we simply restart the network with another set of random values. To get a good approximation to the TSP problem, the network is initialized a number of times with different initialization values. The best tour is then selected from the valid solutions obtained.

8.2.3 Hopfield net applications

There are a large number of applications reported in the literature regarding the use of neural networks for COPs. It is of course not possible to list all of them. Some representative works of neural networks for COPs based on the Hopfield approach are listed below.

1. The traveling salesman problem
   Hopfield (1982, 1984)
   Hopfield and Tank (1985)
   Hedge, Sweet, and Levy (1988)
   Szu (1988)
   Van den Bout and Miller (1988)
   Wilson and Pawley (1988)
   Cuykendall and Reese (1989)
   Aiyer, Niranjan, and Fallside (1990a, 1990b)
   De Carvalho and Barbosa (1990)
   Joppe, Cardon, and Bioch (1990)
   Park and Jeong (1990)
   Thomas and Van den Bout (1990)
   Xu and Tsai (1991)
2. The multiple traveling salesman problem
   Wacholder, Han, and Mann (1988, 1989)

3. Graph partitioning problems
   Combinatorial optimization by Fu and Anderson (1986)
   Map and graph coloring by Dahl (1987)
   Combinatorial optimization by Bruck and Goodman (1988)
   Bipartite matching problem by Goldstein, Toomariar and Barhen (1988)
   Graph bisection by Peterson and Anderson (1988)
   General optimization by Peterson and Soderberg (1989)
   Hard problems by Bruck and Goodman (1990)
   Clustering by Kamgar-Parsi and Kamgar-Parsi (1990a, 1990b)
   Multi-constraint zero-one knapsack problem by Lee and Hsu (1990)
   Combinatorial optimization by Peterson (1990)
   Quadratic assignment problem by Wang (1990)
   Combinatorial optimization problems by Yue and Fu (1990)
   Multi-objective 0–1 linear programming problem by Sasaki, Nakahara, Gen, and Ida (1991)
4. Optimization problems
   General optimization by Kamgar-Parsi and Kamgar-Parsi (1987)
   Constraint satisfaction by Tagliarini and Page (1987)
   Constrained optimization by Ramanujam and Sadayappan (1988a, 1988b)
   Constraint satisfaction problems by Johnston and Adorf (1989)
   Dynamic programming problems by Chiu, Maa, and Shanblatt (1990)
   Linear programming by neural network models by Culioli, Protopopescu, Britton, and Ericson (1990a, 1990b)
   A class of quadratic assignment problems by Fang, Wilson, and Li (1990)
   Non-linear optimization by Reklaitis, Tsirukis and Tenorio (1990)
   Structural optimization by Ivezic, Garrett, and Ganeshan (1991)
   Maximum clique problem by Lee, Funabiki, Cho, and Takefuji (1991)
   Optimization for groundwater remediation design by Ranjithan, Garrett and Eheart (1991)
5. Scheduling problems
   Single server scheduling example by Gulati, Iyengar, Toomarian, Protopopescu and Barhen (1987)
   Crew scheduling example by Poliac, Lee, Slage, and Wick (1987)
   Job shop scheduling example by Foo and Takefuji (1988a, 1988b, 1988c)
   Job shop scheduling by Zhou, Cherkassy, Baldwin, and Hong (1990)
   Intelligent scheduling in manufacturing by Dagli, Lammers, and Vellanki (1991)
   Intelligent scheduling for batch manufacturing systems by Jiang (1991)
   Scheduling the Hubble Space Telescope by Johnston and Adorf (1992)
   Multidimensional knapsack problem by Vaithyanathan and Ignizio (1992)

As an example, we will review how Foo and Takefuji (1988a) apply Hopfield type networks for job shop scheduling. The job shop problem is to assign the starting times of a set of jobs for the available machines given the processing time for each task providing that the precedence relationship between the

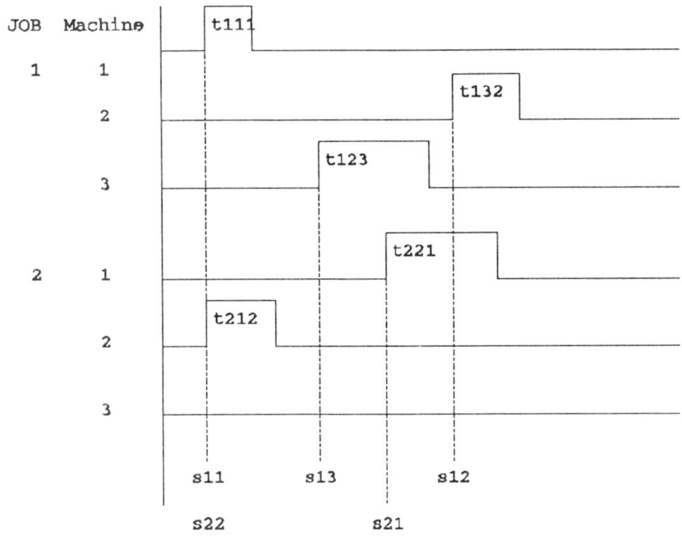

Fig. 8.3 Problem formulation for a job shop scheduling.

tasks have not been violated. The problem can be formulated as a set of integer linear equations. The cost function for minimization is the total starting times of all jobs.

Let the job shop be represented by a triplet (i, j, k) where i denotes the job, j the operation, and k the machine. Let s_{ik} denote the starting time of a job i on machine k and t_{ijk} the processing time for operation j of job i on machine k as shown in Fig. 8.3. All starting times are assumed to be positive. To ensure that two operations are not being assigned to the same machine, we define a binary variable y_{ipk} to indicate the operational sequence. The variable y_{ipk} is 1 if job i precedes job p on machine k or is 0 otherwise.

Assume that there are m machines and n jobs. Then the job shop problem can be formulated by the following cost function and constraints.

$$G = \sum_{i=1}^{n} s_{ik} \tag{8.12}$$

Subject to:

$$\begin{aligned}
&s_{ik} - s_{ih} \geqslant t_{i,j-1,h} && \text{if operation } (i, j-1, h) \text{ precedes } (i, j, k)\\
&s_{pk} - s_{ik} + H(1 - y_{ipk}) \geqslant t_{ijk} && i \geqslant 1, p \leqslant n, 1 \leqslant k \leqslant m\\
&s_{ik} - s_{pk} + H y_{ipk} \geqslant t_{pqk} && i \geqslant 1, p \leqslant n, 1 \leqslant k \leqslant m
\end{aligned}$$

where H represents a positive number that is greater than the longest processing time. The cost function is the sum of all the starting times of the jobs. The smaller this summation is, the sooner the jobs are started.

The solution of course is not to start all the jobs at the same time. Each operation in each job must be performed in order. The precedence order is enforced by three constraints. First, the operation $(i, j-1, h)$ must precede (i, j, k). This precedence relation is represented by the first inequality. The second constraint enforces that operation (p, q, k) starts after the completion of operation (i, j, k) while the third constraint enforces the opposite condition that operation (i, j, k) starts after the completion of operation (p, q, k). The first constraint produces a total of mn inequalities. The second and third constraint each produces a total of $mn(n-1)$ inequalities. Summing additional constraints on s and y yields a total of $mn(n+1)/2$ inequalities.

To implement the problem using the Hopfield net approach with some modifications, let the precedence order y_{ipk} be represented by the output of a set of neurons. Then the cost function and the constraints can be rewritten in terms of a set of neuronal outputs, a vector V of N dimension.

$$G = \sum_{i=1}^{N} A_i V_i, \quad \sum_{i=1}^{N} D_{ij} V_i - B_j \geqslant 0 \tag{8.13}$$

To enforce the bounds and to reinforce satisfaction of the bounds, define

$$f(z) = \begin{cases} 0 & z \geqslant 0 \\ -z & z < 0 \end{cases} \tag{8. 14}$$

Then the equation of motion for the modified Hopfield net can be rewritten as follows:

$$C_i \frac{\mathrm{d}u_i}{\mathrm{d}t} = -A_i - \frac{u_i}{R} - \sum_{j=1}^{N} D_{ji} f\left(\sum_{i=1}^{N} D_{ij} V_i - B_j \right) \tag{8.15}$$

where R is a constant and C_i is a timing constant. Foo and Takefuji showed that the above equation of motion effectively guides the neural net output to a near optimal solution.

8.2.4 Improvements of the Hopfield net approach

The Hopfield net approach has a number of difficulties. Many researchers have reported a slow convergence using the Hopfield net approach. A number of improvements have also been suggested to improve the convergence rate (Abe, 1990; 1991). Kasif (1990) has analyzed the parallel complexity. Bizzarri (1991) also studied the convergence properties of the Hopfield net approach and has suggested numerous improvements.

Another difficulty with the Hopfield net approach is that the constraints are mixed with the cost function. This often results in invalid or unfeasible solutions because the network must compromise between minimizing the cost function and at the same time trying to satisfy all the given constraints. One obvious solution is to tune the weighting constants A, B, C, and D to give different weights to the constraints and the distance measure. At present, the process of tuning these parameters is still not well understood.

A more elegant approach to circumvent the difficulty of getting stuck at a local minimum is to change the form of the cost function and the constraints. A number of such solutions have been proposed in the literature for the TSP. Some proposed a variety of different forms for the energy functions. For example, Brandt *et al.* (1988) suggest using the following energy function.

$$E = J + \frac{\gamma}{2}\left[\sum_{x=1}^{N}\left(1 - \sum_{i=1}^{N} v_{x,i}\right)^2 + \sum_{i=1}^{N}\left(1 - \sum_{x=1}^{N} v_{x,i}\right)^2\right] \tag{8.16}$$

Szu (1988) suggested using a different energy function.

$$\begin{aligned} E = {} & \frac{A}{2}\sum_{X=1}^{N}\sum_{i=1}^{N}\sum_{\substack{j=1 \\ j\neq i}}^{N} v_{X,i}v_{X,j} + \frac{B}{2}\sum_{i=1}^{N}\sum_{X=1}^{N}\sum_{\substack{Y=1 \\ Y\neq X}}^{N} v_{X,i}v_{X,j} \\ & + \frac{C}{2}\sum_{X=1}^{N}\left(\sum_{i=1}^{N} v_{X,i} - 1\right)^2 + \frac{C}{2}\sum_{i=1}^{N}\left(\sum_{X=1}^{N} v_{X,i-1}\right)^2 \\ & + \frac{D}{2}\sum_{X=1}^{N}\sum_{\substack{Y=1 \\ Y\neq X}}^{N}\sum_{i=1}^{N} d_{X,Y}v_{X,i}(v_{Y,i+1} + v_{Y,i-1}) \end{aligned} \tag{8.17}$$

Szu has shown that in conjunction with simulated annealing, the above energy function provides satisfactory results to the TSP. Van den Bout and Miller (1988) also proposed using a modified version of the energy function:

$$E = \frac{d_{\text{mas}}}{2}\sum_{X=1}^{N}\sum_{i=1}^{N}\sum_{\substack{Y=1 \\ Y\neq X}}^{N} v_{X,i}v_{Y,i} + \frac{1}{2}\sum_{X=1}^{N}\sum_{\substack{Y=1 \\ Y\neq X}}^{N}\sum_{i=1} d_{X,Y}v_{X,i}(v_{Y,i+1} + v_{Y,i-1}) \tag{8.18}$$

where d_{mas} is a constant related to the maximum distance. Using simulated annealing (Van den Bout and Miller, 1989, 1990), they have shown that the modified energy function provides improved results for the TSP.

In addition to modifying the energy function, Beyer and Ogier (1991) have proposed using tabu search technique as a replacement of the gradient descent search technique resulting in convergence that was 65 times faster. Similar improvements to the results were also reported by Chakrapani and Skorin-Kapov (1992). Vaithyanathan and Ignizio (1992) have combined the idea of stochastic techniques for a multidimensional knapsack problem and have shown that the modified approach was efficient in producing good approximate solutions.

Another difficulty related to the present Hopfield net formulation is that there are no distinctions between hard and soft constraints. This distinction can be explicitly expressed using the concept of a partition function (Geman and Geman, 1984). The partition function has been reported to be an effective tool in work dealing with the TSP by Peterson and Soderberg (1989). Simic (1990), Yuille (1990), and in work dealing with low level vision by Geiger and Yuille (1989) and Yuille *et al.* (1989). Bozovsky (1990) also analyzed the behavior of the Hopfield net approach. Other improvements to the Hopfield net approach include the use of the elastic net concept for the solution of the TSP by Durbin and Willshaw (1987) and Durbin *et al.* (1989), the use of the elastic net for avoiding obstacles in tour planning by Wong and Funka-Lea (1990), and the use of the guilty net for the TSP by Burke and Damany (1992).

8.3 SIMULATED ANNEALING

The main problem with many search techniques is that the solution often becomes trapped in a local minimum; it is unable to free itself from the local minimum and to move toward the global minimum. To escape the local minimum, the search must be able to break away from the normal search path dictated by the present condition and to follow another search path. This means that the search must move toward a direction that may not appear good at the present time, but on a long run provide a better solution at a later time.

8.3.1 Optimization by simulated annealing

In metallurgy, cracks in metal or other crystalline structures can be cured by heating the material to a high temperature, thus allowing molecular bonds to be easily broken and new bonds to be easily formed. It is commonly accepted that at equilibrium, the probability that the molecules of a material found at a particular energy level can be expressed as

$$\mathrm{P}(E) \propto \mathrm{e}^{-E/k_B T} \tag{8.19}$$

This means that the probability of each energy level is related to the temperature of the material. For a material with two energy states, for example, the probability associated with the change in energy state can be derived.

$$\mathrm{P} = \frac{1}{1 + \mathrm{e}^{\Delta E/k_B T}} \tag{8.20}$$

where $\Delta E = E_{\text{new}} - E_{\text{old}}$ represents the difference or change in energy level. To change the energy state from a higher level to a lower level, ΔE is negative and the probability is high. However to change the energy state from a lower level to a higher level, ΔE is positive and the probaility is low. Hence, the probability shows that the lower energy level is always favored. Equation (8.20) also shows that changes in energy state is related to the temperature T. When the temperature is high, the effects of the energy difference is less pronounced thus giving higher probability to changes to higher energy states. When the temperature is low, the probability of changes to a higher energy state is near zero.

After the material has been heated to a sufficiently high temperature, the material is then allowed to cool gradually. To cure any defects or flaws within a block of metal, the metal block is heated to a high temperature in such a way that any imperfections in the metal crystalline structure are broken loose and reconstructed in the proper manner. It is seen that the crystalline structure is more perfect as the temperature is reduced slowly while the crystalline structure is poor and fraught with imperfections if the temperature is cooled quickly. The latter is due to incorrect bonding frozen into the compound as the temperature has dropped below its mobility range. A perfect crystalline structure represents the lowest energy state. Any remaining imperfection represents a local minimum state but pertaining to a higher energy level.

The principle of simulated annealing follows closely with the real annealing process (Kirkpatrick *et al.*, 1983). The classical simulated annealing algorithm is based on the Metropolis algorithm (Metropolis *et al.*, 1953). The energy state of the material can be considered to be similar to the cost function of the system. The temperature represents the degree of randomness allowable. At the beginning, the temperature is deliberately set high indicating a high degree of randomness. As the temperature begins to cool, the degree of randomness is slowly being reduced.

Aarts and Korst (1989) have shown that when the system is cooled very slowly, the solution tends to converge to the global optimum asymptotically as observed by Geman and Geman (1984) also. However, in general, for simulated annealing to produce the optimal solution, the temperature must be reduced very gradually and slowly.

8.3.2 The Boltzmann machine

The Boltzmann machine is proposed by Hinton *et al.* (1984) and Ackley *et al.* (1985). The Boltzmann machine is a combination of the principle of simulated annealing with the neuronal approach together as a means for solving combinatorial optimization problems and has been shown to have similar behavior

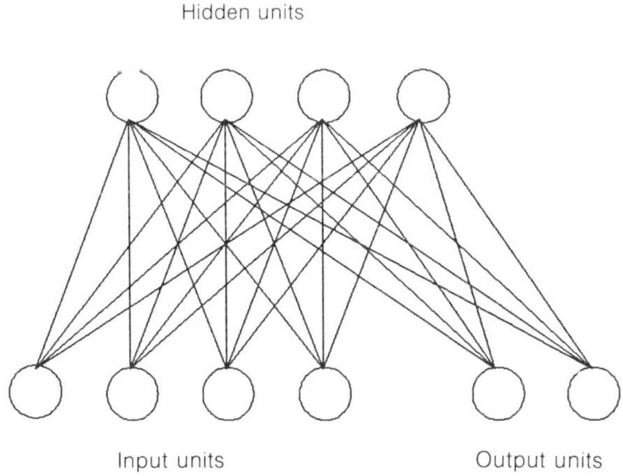

Fig. 8.4 Architecture of a Boltzmann machine.

to the Hopfield net (Derthick and Tebelskis, 1988). The architecture of a Boltzmann machine is similar to the discrete Hopfield net and is usually taken to be an array of fully connected neurons. Part of the array is divided into input/output units while the rest is considered as hidden units as shown in Fig. 8.4. The input to each neuron is binary and the output of each neuron is also binary.

The energy level of the Boltzmann network is similar to the Hopfield energy function.

$$E = -\frac{1}{2}\sum_{i=1}^{N}\sum_{\substack{j=1\\ j\neq i}}^{N} T_{ij}v_i v_j \tag{8.21}$$

For a particular neuron v_k to change its output from 0 to 1, the change in energy can be calculated directly from the energy function as

$$\begin{aligned}\Delta E_k &= E_{v_k=1} - E_{v_k=0}\\ &= -\sum_{\substack{j=1\\ j\neq k}}^{N} T_{kj}v_j \qquad (8.22)\\ &= -\mu_k\end{aligned}$$

The probability of a state change from 0 to 1 can readily be calculated. This probability is implemented by the Boltzmann machine.

To apply the principles of simulated annealing in the Boltzmann machine, there are seven steps in the process.

1. Determine the input representation.
2. Determine the cost function in the form of a Lyapunov function.
3. Rewrite the constraints in the form of a Lyapunov function.
4. Randomly start the network.
5. Select a neuron at random and determine its output stochastically.
6. Repeat Step 5 a number of times until thermal equilibrium.
7. Lower the temperature and repeat Steps 5 and 6 again to a sufficiently low temperature.

These steps show how simulated annealing can be applied to solve COPs. The particular schedule used to lower the temperature is called the annealing schedule and must be determined according to the application on hand.

8.3.3 Application of simulated annealing to the traveling salesman problem

In the previous section, we have shown how simulated annealing can be applied in general for neural networks. In this section, we will give a practical example of using simulated annealing for solving COPs. We will again use the TSP as an example.

Step 1: Determine the input representation

For the TSP, the same input representation used in the Hopfield net can be used again for the simulated annealing approach here. The cities and their respective orders in the tour are represented by an array of n by n neurons. The binary outputs of the array thus indicate the desired order of visit for each city, hence the tour. For optimization problems, since the weights are known, there is no need for the hidden units using the present representation. All units in the array are considered to be input/output units with direct connections between them. Furthermore, the learning procedures of the Boltzmann machine is not needed here. Only the recall procedure of the Boltzmann machine is used.

Step 2: Determine the cost function in the form of a Lyapunov function

The cost function determines the desired measure to be minimized. In the case of the TSP, the desired measure is the total distance of the tour. It is desirable to find a solution that minimizes the distance measure. For the TSP, the network architecture is identical to that for the Hopfield net, hence the same formulation for the cost function given before for the Hopfield net can also be used for this example.

Step 3: Rewrite the constraints in the form of a Lyapunov function

With the same inputs representation as the Hopfield net approach, the same three constraints are also required to avoid the possibility of the network settling into a trivial solution. Hence all three constraints are required. These three constraints and the distance measure are summed to form the total network energy function.

Step 4: Randomly start the network

The output of the array is initialized with random values of 0 and 1. The output may not represent a valid tour at the beginning, but this is not crucial. Up to this point, the problem formulation of the TSP for the simulated annealing approach is exactly identical to that for the Hopfield net approach.

Step 5: Select a neuron k at random and determine its output stochastically

To change the network, select a neuron k at random. It is neither crucial nor desirable to sort the neurons, nor to select them in order. Regardless of the present ouput of the neuron v_k, determine the probability of setting the output of this neuron k to one. This is done by first calculating the energy levels for $v_k = 0$ and $v_k = 1$. The energy levels of this neuron being 0 or 1 is dependent on the entire state of the array as reflected by the energy function.

The probability can then be calculated with equation 8.22 using the difference in energy levels between the two possible outputs for the present temperature. This is the probability of setting this neuron output to one.

To determine the output value of this neuron stochastically, select a random number from a uniform distribution with limits between 0 and 1. This random value is then used to compare with the probability of setting this neuron output to one. If the value of the random number is less than the calculated probability, then set the output of the neuron to be one, otherwise set the neuron output to be zero.

Step 6: Repeat Step 5 until thermal equilibrium

In step 5, the output of a randomly selected neuron is determined stochastically. This process is then repeated a number of times so that every neuron has an equal opportunity of being updated. This does not guarantee that each neuron would be updated or was updated for the same number of times. The neuron array is merely randomly updated for a predetermined number of times. This can be considered as a processing cycle since all neurons should have the chance of being updated.

The processing cycles are then repeated a number of times again until an equivalent thermal equilibrium has been reached for the present temperature. This can be approximated by monitoring the number of state changes in each processing cycle. When the temperature is first changed, a large number of state changes would initially occur. The number of state changes would then gradually decrease as the network settles into the new local minimum condition.

Step 7: Lower the temperature and repeat Steps 5 and 6 again to a sufficiently low temperature

After the network reaches equilibrium at the present artificial temperature, then lower the temperature slightly and repeat Steps 5 and 6 to let the network anneal, i.e. settle into a new thermal equilibrium. It is crucial to lower the temperature slowly so that the network would have chance to settle at the global minimum. The process stops when the number of state changes are reduced to a minimum with further lowering of the temperature.

8.3.4 Applications of the simulated annealing approach

Simulated annealing techniques have been reported in the literature by a number of researchers. We will list only a few of these according to the categories of the applications.

1. Quadratic assignment problems
   Backboard wiring by Steinberg (1961),

Campus facilities planning by Dickey and Hopkins (1972)
Hospital layout by Elshafei (1977)
TSP by Bonomu and Lutton (1984)
Graph partitioning problems by Anderson and Peterson (1988)
Graph bipartite matching problems by Goldstein, Toomarian, and Barhen (1988)
Assignment problems by Akiyama, Yamashita, Kajiura, and Asio (1989)
Assignment problems by Kajiura, Akiyama, and Anzai (1990)
Graph matching problems by Mehta and Fulop (1990)
Graph partitioning problems by Van den Bout and Miller (1990)
Bin packing and pallet loading by Lin *et al.* (1991)
TSP by Yokoi and Kakazu (1991)
Computerized facility layout by Ziai and Sule (1991)

2. Optimization problems
   General optimization problems by Jeffrey and Rosner (1986)
   Global optimization problems using stochastic neural networks by Levy and Adams (1987)
   Non-linear programming by Kennedy and Chua (1988)
   Constrained differential optimization by Plant and Barr (1988)
   General optimization problems by Bibro, Mann, Miller, Synder, Van den Bout, and White (1989)
   Constraint satisfaction problems by Adorf and Johnston (1990)
   Constraint satisfaction problems by Minton, Johnston, Philips, and Laird (1990)
3. Scheduling problems
   Restaurant crew scheduling by Poliac *et al.* (1987)
   Vehicle routing problems by Alfa *et al.* (1991)
   Design of an assembly planning system by Chen and Yan (1991)
   Gate array routing problems by Lu and Thomborson (1991)

As an example, we will review how Poliac *et al.* (1987) proposed to use simulated annealing to solve the restaurant crew scheduling problem. Given the crew's availability times and their skill levels, the problem is to determine the start and stop time of each worker in such a way that all the required tasks of the restaurant are covered. Additional scheduling constraints based on the prevailing policies of the restaurant must also be satisfied.

Let t_j represents the time intervals of a day to a maximum of N intervals, a_{ij} represents the availability of crew member i to a maximum of M crew members on time interval j, p_{ik} represents the proficiency of crew member i on job k to a maximum of Q jobs. The restaurant requirement is represented by r_{jk}, the person proficiency points needed for task k during time interval j, and c_{ijk} the wage of crew member i for working tasking k during time interval j. The desired schedule is then a three-dimensional permutation matrix x_{ijk} for crew member i performing task k during time interval j as shown in Fig. 8.5. To solve this crew scheduling problem, we design a three-

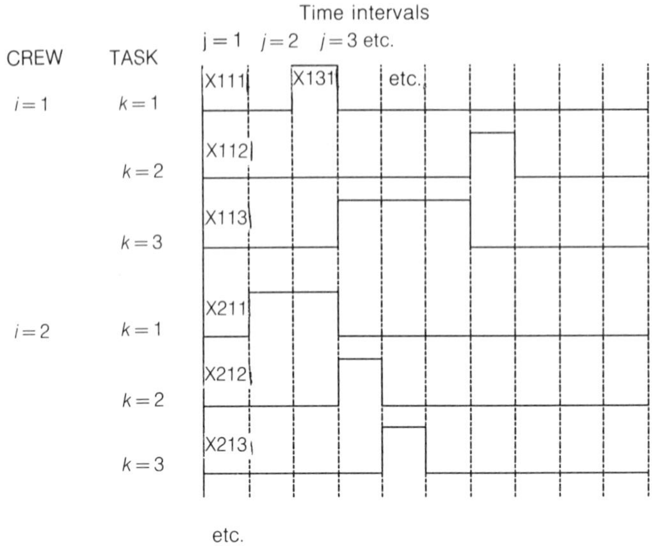

Fig. 8.5 Problem formulation for a crew scheduling problem.

dimensional array of neurons whose binary outputs represent the three dimensional array x_{ijk}. In other words, both cost function and the constraints are coded into the weights so that when the network settles, the output of the network will give the desired schedule for the crew.

The cost function is just the total wages for all crew members using the present schedule. It is of course desirable to minimize this cost function. However, to realize the problem properly, additional constraints must also be specified. The complete crew scheduling problem can be restated in mathematical terms.

$$G = \sum_{i=1}^{M} \sum_{j=1}^{N} \sum_{k=1}^{Q} x_{ijk} C_{ijk} \tag{8.23}$$

Subject to

$$\sum_{i=1}^{M} p_{ik} x_{ijk} \geqslant r_{jk}$$

$$\sum_{k=1}^{Q} x_{ijk} \leqslant a_{ij}$$

$$\sum_{j=1}^{N} |x_{ijk} - x_{i,j+1,k}| \leqslant 2$$

$$w_{\min} \leqslant \sum_{j=1}^{N} \sum_{k=1}^{Q} t_j x_{ijk} \leqslant w_{\max} \quad \text{if} \sum_{j=1}^{N} \sum_{k=1}^{Q} x_{ijk} \neq 0$$

The first constraint is to ensure that the proficiency requirement of each crew member is satisfied. The second constraint is to ensure that the availability of each crew member is satisfied. The third constraint ensures that work shift is contiguous, i.e. there is only one start time and one stop time. The fourth constraint assures that once the crew member starts to work, there is a minimum and maximum time period established by the restaurant management that must be satisfied.

Based on these constraints, the equation of motion for each neuron can be derived.

$$x_{ijk} = f(c_1 T^1_{ijk} - c_2 T^2_{ijk} - c_3 T^3_{ijk} - c_4 t^4_{ijk}) \tag{8.24}$$

where:

$$f(x) = (1 + e^{-K(x + T_{ijk})})^{-1}$$

$$T^1_{ijk} = a_{ij} r_{jk}\left(c_5 p_{ik} + c_6 \sum_{\substack{l=j-d \\ l \neq j}}^{j+d} x_{ilk}(u-1) \right)$$

$$T^2_{ijk} = \sum_{l=1}^{j-c_7} x_{ilk}(u-1) + \sum_{i=j+c_7}^{n} x_{ilk}(u-1)$$

$$T^3_{ijk} = \sum_{\substack{h=1 \\ h \neq k}}^{Q} \sum_{l=1}^{N} x_{ilk}(u-1)$$

$$T^4_{ijk} = c_{ijk}$$

The first weight term provides excitatory input due to labor requirement from the neighboring cells. The second weight term provides the inhibitory input from distant cells (the distance being determined by c_7) so that work-shifts will not be prolonged. The third weight term provides inhibitory input from neighboring cells so that only one task is scheduled for a crew member at any one time. The fourth weight term is the labor cost itself that we wish to minimize. To complete the problem, the labor requirement matrix must also be updated with each iteration.

$$r_{jk}(u) = r_{jk}(0) - \sum_{i}^{m} p_{ik} x_{ijk} \tag{8.25}$$

With the proper annealing schedule, Poliac *et al.* (1987) have shown that the above formulation produced reasonable schedules within a reasonable time.

8.3.5 Improvements to the simulated annealing approach

The simulated annealing approach has been applied either as a technique in itself or as a supplementary method to augment other techniques. The asymptotic behavior of simulated annealing has been analyzed by Matsuba and Masui (1991). A number of improvement has also been suggested in the interest of speeding up the convergence rate of the network. Connolly (1990) has suggested an improved annealing scheme specifically for the quadratic assignment problem (QAP) using a sequential search instead of a random search. Experimental results show that a sequential search is more effective. Lee and Louri (1991) proposed using micro-canonical simulation to improve the speed of simulated annealing. Another improvement has been reported by Galar (1989) using an evolutionary search with soft selection to ensure the global minimization property.

There are a number of variations to the basic simulated annealing technique based on the Metropolis Algorithm (Metropolis *et al.*, 1953) and Kirkpatrick *et al.* (1983). Szu (1986) has proposed a fast simulated annealing (FSA) technique using local search and occasional long jumps coupled with a special cooling schedule. Another technique is the micro-canonical simulated annealing (MSA) method by Creutz (1983). Goldstein *et al.* (1988) have also proposed a variation to the simulated annealing algorithm (VSA) that includes long jumps in the state search space thus providing faster escape from the local minima.

Gutzmann (1987) has shown that the TSP can be solved efficiently using a continuous formulation of the Boltzmann machine discussed so far. In the discrete case, the output of the neurons are simple binary values: zeros and ones. In the continuous case, the output of the neurons may take on a value between the range of 0 and 1. The real output value may be viewed as a fuzzy value among a number of hypotheses thus allowing the continuous machine to evaluate a number of hypotheses at the same time.

8.4 OTHER NEURAL NETWORK TECHNIQUES

In addition to the Hopfield net and the simulated annealing approach described above, scheduling has also been reported in the literature using a number of other neural network paradigms. This include the multi-layer perceptron using the popular backpropagation method, the competition network using the winner-take-all method, and the self-organizing feature map of Kohonen (1989) using the topology preserving method.

8.4.1 Multi-layer perceptron network

A multi-layer perceptron is a feed-forward network with a layer of input nodes. Each input node is connected to a second layer of nodes called the hidden layer. It is possible to have a number of hidden layers although most applications have one or at most two hidden layers. The output of the hidden layer is massively connected to the nodes in the output layer as shown in Fig. 8.6.

During training, sample patterns are presented at the input of the network while desired patterns are simultaneously presented at the output of the

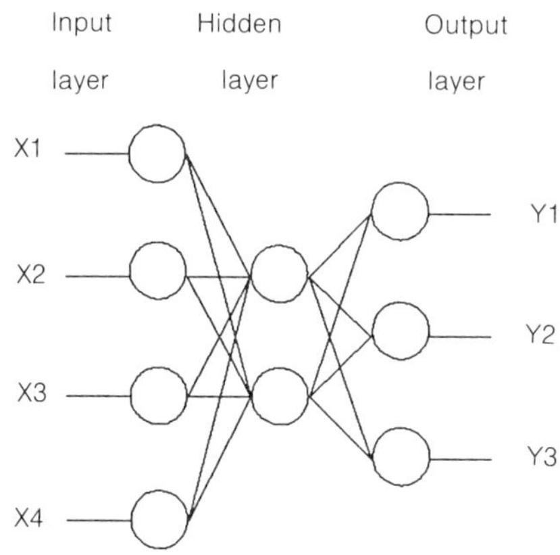

Fig. 8.6 Architecture of a multi-layer perceptron.

network. An error correction approach such as backpropagation based on the generalized delta rule is then used to adjust the weights of the network in such a manner that the network output converges to the training output. During recall, an unknown pattern is presented at the network input, the trained weights cause the network to produce an output pattern closely associated with one of the trained input patterns.

The multi-layer perceptron scheme has been successfully used by Yih *et al.* (1991) to schedule crane operations for a chemical plant. Results showed that the network output is better than a human scheduler. Badami and Parks (1991a, 1991b) used a classifier based approach to solve the flow shop scheduling problem. Mehra and Wah (1991) reported using artificial neural network techniques for load balancing strategies. Potvin *et al.* (1992) also used the same technique for automated vehicle dispatching.

8.4.2 Competition network

Another approach that differs from the Hopfield net approach is based on neighborhood competition using the winner-take-all strategy (Reggia, 1987). The competition network is similar to other feed-forward networks such as the multi-layer perceptron except for the fact that the competition layer contains additional lateral inhibition between the neurons to implement the winner-take-all strategy as shown in Fig. 8.7. The winner-take-all strategy specifies that only one neuron out of the entire group can be active at any one time. When a chosen neuron is active, the rest of the neurons in the same group are inhibited. It is easy to see that the exclusive nature of many constraints can directly take advantage of this property.

In the competition networks, the problem constraints and the heuristic search are designed directly into the equation of motion based on the interactive activation and competition model which was originally proposed by Grossberg

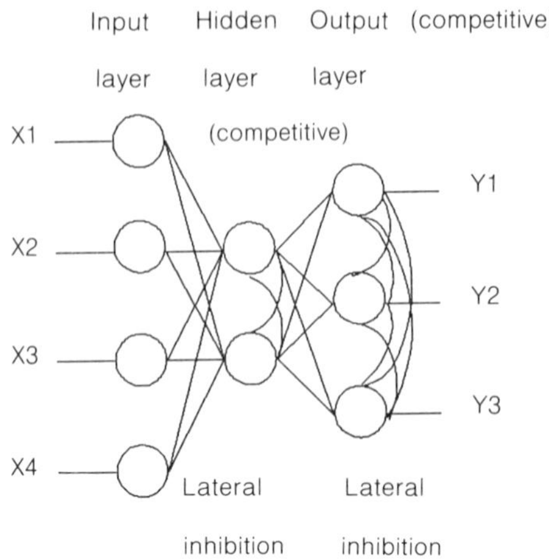

Fig. 8.7 Architecture of a competitive network.

(1978), and by McClelland and Rumelhart (1981). For proof of convergence, one may still have to rely on the Lyapunov energy function. Specific constraint takes on a specific form in the equation of motion (Fang and Li, 1990).

For example, to realize the constraint that only one neuron in a group is on while all the rest is off, the equation of motion is:

$$\frac{\mathrm{d}x_i}{\mathrm{d}t} = \left[(1-\varepsilon) - \sum_{\substack{j=1 \\ j\neq i}}^{N} x_j \right] c_i \tag{8.26}$$

and to realize the constraint that the weighted sum of the neuronal input is less than capacity, i.e.

$$\sum_{i=1}^{N} x_i w_i \leqslant M \tag{8.27}$$

then the corresponding equation of motion would be

$$\frac{\mathrm{d}x_i}{\mathrm{d}t} = \left[M + \varepsilon - \sum_{\substack{j=1 \\ j\neq i}}^{N} w_j f(x_j) - w_i \right] c_i \tag{8.28}$$

Thus, for the TSP, at least two of the constraints are naturally taken care of by the winner-take-all strategy. In other words, it is not necessary to code explicitly that only one neuron is active at one time for each row and only one neuron is active at any one time for each column. The lateral inhibitions from each row and from each column will take care of this situation automatically.

Competitive networks have been proposed for solving the graph bipartitioning problem by Rumelhart and Zipser (1985). Additional information on how competitive networks can be used for COPs can be found in Hertz,

Krogh, and Palmer's book (1991). In addition, Lim and Loe (1991) have also proposed a timetable scheduling algorithm. Lee and Sheu (1990) has proposed the use of competitive Hopfield nets for combinatorial optimization problems and Matsuyama (1990) has proposed the use of competitive self-organization networks for combinatorial problems, more specifically the TSP.

8.4.3 Self-organizing map

The use of a self-organizing map (SOM) for solving COPs have also been proposed by a number of researchers. The architecture of the SOM is shown in Fig. 8.8. There is an array of input units supplying a massive number of connections to the processing layer. An input element in the input layer is connected to the processing layer through a localization function realized by lateral connection weights. In most cases, the form of the localization function is taken to be like a 'Mexican hat' with positive reinforcement for the local clusters, and negative reinforcement to the distance neighbors. When the width of the hat is narrowed, eventually, only a single neuron at the center is active.

The topology-preserving property of the SOM is ideal for the TSP. In essence, we want to map a ring onto the SOM in such a way that the mapping also produces a ring or a closed contour that represents a valid tour through the cities. The weights of output neuron represent the coordinates of the corresponding city while the weights of neighboring neurons also represent the coordinates of the neighboring cities in the tour. This method was shown to be successful in arriving a viable solution by Angeniol *et al.* (1988) and Hueter (1988).

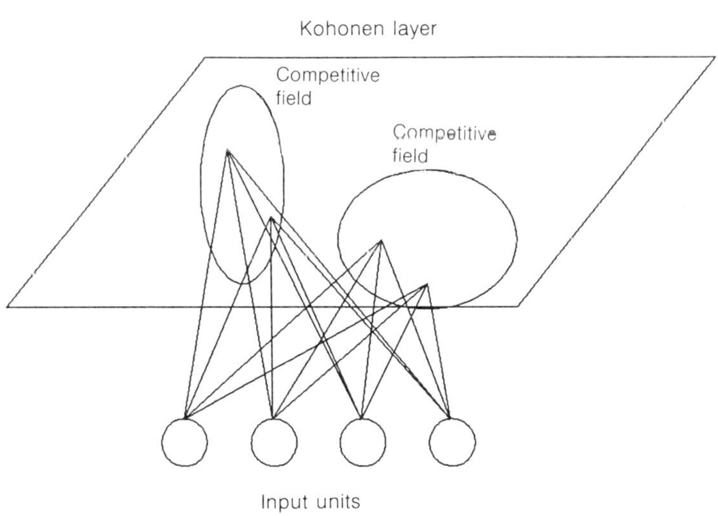

Fig. 8.8 Architecture of a self-organizing map.

A number of researchers have proposed the use of SOM for scheduling and other optimization problems. In addition to Angeniol *et al.* (1988) and Hueter (1988) who proposed using self-organizing maps for solving the TSP, Fort (1988) also proposed the use of SOM for solving combinatorial problems such as the TSP. Fritzke and Wilke (1991) proposed the FLEXMAP based on the SOM also for the TSP. Goldstein (1990) used the SOM for the multiple traveling salesman problem.

8.6 SUMMARY

The use of neural networks for scheduling and other COPs is not only restricted to software simulation. A detailed and comprehensive review can be found by Looi's (1992) article. More recently, special purpose hardware for minimizing Euclidean distance measure has been reported in the literature by Moopenn *et al.* (1988).

It is often difficult to compare the performance of different algorithms. Cervantes and Hildebrant (1987) have compared three neuron-based computation schemes. Huang *et al.* (1991) also compared the implementation and performance of backpropagation technique, simulated annealing technique, and tabu search technique.

For the TSP, Kwon and Lu (1991) have reported a comparative study of four techniques including the Hopfield net and the simulated annealing approaches. For 10 and 30 cities problems, the results from simulated annealing appear much better than that from the Hopfield net. Wang and Chankong (1992) have also compared neural network techniques with simplex methods and have found that the neural net method does not enumerate extreme points, but rather, the massively parallel nature of the network makes it ideal for implementation today.

REFERENCES

Aarts, E. and Korst, J. (1989) *Simulated Annealing and Boltzmann Machines*, Wiley, New York.

Abe, S. (1990) Convergence of the Hopfield neural networks with inequality, *Proceedings of the International Joint Conference on Neural Networks – San Diego*, **III**, 869–74.

Abe, S. (1991) Global convergence and suppression of spurious states of the Hopfield neural networks, *Proceedings of the International Joint Conference on Neural Networks – Singapore*, **I**, 936–40.

Ackley, D.H., Hinton, G.E. and Sejnowski, T.J. (1985) A learning algorithm for Boltzmann machines. *Cognitive Science*, **9**, 147–69.

Adorf, H.M. and Johnston, M.D. (1990) A discrete stochastic neural network algorithm for constraint satisfaction problems, *Proceedings of the International Joint Conference on Neural Networks – San Diego*, **III**, 917–24.

Aiyer, S.V.B., Niranjan, M. and Fallside, F. (1990a) A theoretical investigation into the performacne of the Hopfield model, *IEEE Transactions on Neural Networks*, **1**, 204–15.
Aiyer, S.V.B., Niranjan, M. and Fallside, F. (1990b) On the optimization properties of the Hopfield model, *Proceedings of the International Neural Network Conference – Paris*, **I**, 245–8.
Akiyama, Y., Yamashita, A., Kajiura, M. and Asio, H. (1989) Combinatorial optimization with Gaussian machines, *Proceedings of the International Joint Conference on Neural Networks – Washington, DC.*, **I**, 533–540.
Alfa, A.S., Heragu, S.S. and Chen, M. (1991) A 3-opt based simulated annealing algorithm for vehicle routeing problems. *Computers Industrial Engineering*, **21**, (1–4), 635–9.
Anderson, J. R. and Peterson, C. (1988) Applicability of mean field theory neural network methods to the graph partitioning problems. *MCC-ACA-ST-064-88*.
Angeniol, B., Vaubois, G. de la C. and le Texier, J.Y. (1988) Self-organizing feature maps and the TSP. *Neural Networks*, **3**, 289–93.
Badami, V.S. and Parks, C.M. (1991a) A classifier based approach to flow shop scheduling. *Computers Industrial Engineering*, **21**(1–4), 329–33.
Badami, V.S. and Parks, C.M. (1991b) A classifier based approach to flow shop scheduling. *Computers Industrial Engineering*, **21**(1–4), 401–5.
Beyer, D.A. and Ogier, R.G. (1991) Tabu learning: a neural network search method for solving nonconvex optimization problems, *Proceedings of the International Joint Conference on Neural Networks – Singapore*, **I**, 953–61.
Bibro, G., Mann, R., Miller, T.K., Synder, W.E., Van den Bout, D.E. and While, M. (1989) Optimization by mean field annealing, in *Advances in Neural Information Processing Systems*, (ed. D.S. Touretzky), Kaufman, 91–8.
Bizzarri, A.R. (1991) Convergence properties of a modified Hopfield-tank model. *Biological Cybernetics*, **64**, 293–300.
Bonomi, E. and Lutton, J.L. (1984) The N-city traveling salesman problem: statistical mechanics and the metropolis algorithms. *SIAM Reviews*, **26**(4), 551–568.
Bozovsky, P. (1990) Discrete Hopfield model with graded response (analysis and applications), *Proceedings of the International Joint Conference on Neural Networks – San Diego*, **III**, 651–6.
Brandt, R.D., Wang, Y., Laub, A.J. and Mitra, S.K. (1988) Alternative networks for solving the TSP and the list-matching problem, *Proceedings of the International Conference on Neural Networks – San Diego*, **II**, 333–40.
Bruck, J. and Goodman, J.W. (1988) A generalized convergence theorem for neural networks and its applications in combinatorial optimization. *IEEE Transactions on Information Theory*, **34**, 1089–92.
Bruck, J. and Goodman, J.W. (1990) On the power of neural networks for solving hard problems. *Journal of Complexity*, **6**, 129–35.
Burke, L.I. and Damany, P. (1992) The guilty net for the traveling salesman problem. *Computers Operation Research*, **19**(3/4), 255–65.
Cervantes, J.H. and Hildebrant, R.R. (1987) Comparison of three neuron-based computation schemes, *Proceedings of the International Conference on Neural Networks – San Diego*, **III**, 657–91.
Chakrapani, J. and Skorin-Kapov, J. (1992) A connectionist approach to the quadratic assignment problem. *Computers Operation Research*, **19**(3/4), 287–95.
Chen, C.L.P. and Yan, Q.-W. (1991) Design of a case associative assembly planning system, in *Intelligent Engineering Systems Through Artificial Neural Networks*, (eds Dagli, Kumara, and Shin), *ASME*, 757–62.
Chiu, C., Maa, C.Y. and Shanblatt, M.A. (1990) An artificial neural network algorithm for dynamic programming. *International Journal of Neural Systems*, **1**, 211–20.

Connolly, D.T. (1990) An improved annealing scheme for the QAP. *European Journal of Operational Research*, **46**, 93–100.

Creutz, M. (1983) Microcanonical Monte-Carlo simulation. *Physical Review Letters*, **50**(19), 1411–14.

Culioli, J.-C., Protopopescu, V., Britton, C.L. and Ericson, M.N. (1990a) Neural network models for linear programming, *Proceedings of the International Joint Conference Neural Networks*, I, 293–6.

Culioli, J.-C., Protopopescu, V., Britton, C.L. and Ericson, M.N. (1990b) Neural network for explicitly bounded linear programming, *Proceedings of the International Joint Conference Neural Networks*, I, 381–4.

Cuykendall, R. and R. Reese, (1989) Scaling the neural TSP algorithm. *Biological Cybernetics*, **60**, 365–71.

Dagli, C.H., Lammers, S. and Vellanki, M. (1991) Intelligent scheduling in manufacturing using neural networks. *Journal of Neural Network Computing*, **2**, 4–10.

Dahl, E.D. (1987) Neural Network Algorithm for an NP-complete problem: map and graph coloring, *Proceedings of the International Conference on Neural Networks – San Diego*, **III**, 113–20.

De Carvalho, L.A.V. and Barbosa, V.C. (1990) A TSP objective function that ensures feasibility at stable points, *Proceedings of the International Neural Network Conference – Paris*, **I**, 249–53.

Derthick, M. and Tebelskis, J. (1988) 'Ensemble' Boltzmann units have collective computational properties like those of Hopfield and tank neurons, *Neural Information Processing Systems*, American Institute of Physics, 223–32.

Dickey, J.W. and Hopkins, J.W. (1972) Campus building arrangement using TOPAZ. *Transportation Research*, **6**, 59–68.

Durbin, R. and Willshaw, D. (1987) An analogue approach to the TSP using an elastic net method. *Nature*, **326**, 689–91.

Durbin, R., Szeliski, R. and Yuille, A.L. (1989) An analysis of the elastic net approach to the TSP. *Neural Computing*, **1**, 348–58.

Elshafei, A.N. (1977) Hospital layout as a quadratic assignment problem. *Operational Research Quarterly*, **28**, 167–79.

Fang, L. and Li, T. (1990) Design of competition-based neural networks for combinatorial optimization. *International Journal of Neural Systems*, **I**, 221–35.

Fang, L., Wilson, W.H. and Li, T. (1990) Mean field annealing neural net for quadratic assignment, *Proceedings of the International Neural Network Conference – Paris*, **I**, 282–6.

Foo, Y.P.S. and Takefuji, Y. (1988a) Integer linear programming neural networks for job-shop scheduling, *Proceedings of the International Conference on Neural Networks – San Diego*, **II**, 341–8.

Foo, Y.P.S. and Takefuji, Y. (1988b) Stochastic neural networks for solving job-shop scheduling: part 1. Problem Representation, *Proceedings of the International Conference on Neural Networks – San Diego*, **II**, 275–82.

Foo, Y.P.S. and Takefuji, Y. (1988c) Stochastic neural networks for solving job-shop scheduling: part 2. Architecture and simulations, *Proceedings of the International Conference on Neural Networks – San Diego*, **II**, 283–90.

Fort, J.C. (1988) Solving a combinatorial problem via self-organizing process; an application of the Kohonen algorithm to the TSP. *Biological Cybernetics*, **59**, 33–40.

French, S. (1982) *Sequencing and Scheduling: An Introduction to the Mathematics of the Job-Shop*, John Wiley & Sons.

Fritzke, B. and Wilke, P. (1991) FLEXMAP – A neural networks for the traveling salesman problem with linear time and space complexity, *Proceedings of the International Joint Conference on Neural Networks 91 – Singapore*, **I**, 929–34.

Fu Y. and Anderson, P.W. (1986) Applications of statistical mechanics to NP-complete problems in combinatorial optimization. *Journal of Physics*, **19**, 1605–20.

Galar, R. (1989) Evolutionary search with soft selection. *Biological Cybernetics*, **60**, 357–64.

Geiger, D. and Yuille, A.L. (1989) A common framework for image segmentation, Harvard Robotics Laboratory, Technical Report, No. 89–7.

Geman, S. and Geman, D. (1984) Stochastic relaxation, Gibbs distribution and the Bayesian restoration of images. *IEEE Transactions on Pattern Analysis and Machine Intelligence*, **6**, 721–41.

Goldstein, M. (1990) Self-organizing feature maps for the multiple traveling salesman problem (MTSP), *Proceedings of the International Conference on Neural Network – Paris*, **I**, 258–61.

Goldstein, M., Toomarian, N. and Barhen, J. (1988) A comparison study of optimization methods for the bipartite matching problem (BMP), *Proceedings of the International Conference on Neural Networks*, **II**, 267–73.

Goles, E. and Vichniac, G.Y. (1986) Lyapunov functions for parallel neural networks, *Neural Networks for Computing, AIP Conference Proceedings 151*, (ed. J.S. Denker), 165–81.

Grossberg, S. (1978) A theory of visual coding, memory, and development, in *Formal Theories of Visual Perception*, (eds E.L.J. Leeuwenberg and H.F.J.M. Buffart), John Wiley, New York.

Gulati, S., Iyengar, S.S., Toomarian, N., Protopopescu, V. and Barhen, J. (1987) Nonlinear neural networks for deterministic scheduling, *Proceedings of First International Conference Neural Networks – San Diego*, **IV**, 745–52.

Gutzmann, K.M. (1987) Combinatorial optimization using a continuous state Boltzmann machine, *Proceedings of the International Conference on Neural Networks*, **III**, 721–34.

Hegde, S.U., Sweet, J.L. and Levy, W.B. (1988) Determination of parameters in a Hopfield/Tank computational network, *Proceedings of the Second International Conference Neural Networks – San Diego*, **II**, 291–8.

Hertz, J., Krogh, A. and Palmer, R.G. (1991) *Introduction to the Theory of Neural Computation*, Addison-Wesley.

Hinton, G.E., Sejnowski, T.J. and Ackley, D.H. (1984) Boltzmann machines: constraint satisfaction networks that learn, CMU technical report, CMU-CS-84-119.

Hopfield, J.J. (1982) Neural networks and physical systems with emergent collective computational abilities, *Proceedings of the National Academy of Science*, **79**, 2554–8.

Hopfield, J.J. (1984) Neurons with graded response have collective computational properties like those of two-state neurons, *Proceedings of the National Academy of Science*, **81**, 3088–92.

Hopfield, J.J. and Tank, D.W. (1985) Neural computation of decisions in optimization problems, *Biological Cybernetics*, **52**, 141–52.

Huang, H.H., Zhang, C., Lee, S. and Wang, H.P. (1991) Implementation and comparison of neural network learning paradigms: Back propagation, simulated annealing and Tabu search, in *Intelligent Engineering Systems through Artificial Neural Networks*, (eds Dagli, Kumara, and Shin), 95–100, ASME Press, New York.

Hueter, G.J. (1988) Solution of the travelling salesman problem with an adaptive ring, *Proceedings of the International Conference on Neural Networks – San Diego*, **I**, 85–92.

Ivezic, N., Garrett Jr., J.H. and Ganeshan, R. (1991) Generalized Hopfield network for structural optimization, in *Intelligent Engineering Systems Through Artificial Neural Networks* (eds Dagli, Kumara, and Shin), 849–54, ASME Press, New York.

Jeffrey, W. and Rosner, R. (1986) Optimization algorithms: Simulated annealing and neural network processing, *Astrophysics Journal*, **310**, 473–81.

Jiang, J. (1991) IS: An intelligent scheduler for batch manufacturing systems, *Computers Industrial Engineering*, **21**(1–4), 319–23.

Johnston, M.D. and Adorf, H.-M. (1989) A discrete stochastic neural network algorithm for constraint satisfaction problems, *Proceedings of the NASA Conference on Space and Telerobotics – Pasadena*, **II**, 367–76.

Johnston, M.D. and Adorf, H.M. (1992) Scheduling with neural networks – the case of the Hubble space telescope, *Computers Operation Research*, **19**(3/4), 209–40.

Joppe, A., Cardon, H.R.A. and Bioch, J.C. (1990) A neural network for solving the TSP on the basis of city adjacency in the tour, *Proceedings of the International Conference on Neural Networks – Paris*, **I**, 254–7.

Kajiura, M., Akiyama, Y. and Anzai, Y. (1990) Solving large scale puzzles with neural networks, *Proceedings of the Tools for AI Conference – Fairfax*, **II**, 562–9.

Kamgar-Parsi, B. and Kamgar-Parsi, B. (1987) An efficient model of neural networks for optimization, *Proceedings of the First International Conference on Neural Networks – San Diego*, **III**, 785–890.

Kamgar-Parsi, B. and Kamgar-Parsi, B. (1990a) On problem solving with Hopfield neural networks, *Biological Cybernetics*, **62**, 415–23.

Kamgar-Parsi, B. and Kamgar-Parsi, B. (1990b) Clustering taxonomic data with neural networks, *Proceedings of the International Neural Network Conference – Washington D.C.*, **I**, 277–80.

Kasif, S. (1990) On the parallel complexity of discrete relaxation in constraint satisfaction networks, *Journal of Artificial Intelligence*, **45**, 275–86.

Kennedy, M.P. and Chua, L.O. (1988) Neural networks for nonlinear programming, *I.E.E.E. Transactions on Circuits and Systems*, **35**.

Kirkpatrick, S., Gelatt Jr., C.D. and Vecchi, M.P. (1983) Optimization by simulated annealing, *Science*, **220**, 671–80.

Kohonen, T. (1989) *Self-Organization and Associative Memory*, Springer, New York.

Kwon, T.M. and Lu, Y.Z. (1991) A comparative study of the traveling salesman problem, in *Intelligent Engineering Systems Through Artificial Neural Networks*, (eds Dagli, Kumara, and Shin), pp. 889–94, ASME Press, New York.

Lee, B.W. and Sheu, B.J. (1990) Combinatorial optimization using competitive–Hopfield neural networks, *Proceedings of the International Joint Conference on Neural Networks – Washington D.C.*, **II**, 627–30.

Lee, H.J. and Louri, A. (1991) Microcanonical mean field annealing: A new algorithm for increasing the convergence speed of mean field annealing, *Proceedings of the International Joint Conference on Neural Networks 91 – Singapore*, **II**, 941–6.

Lee, H.-M. and Hsu, C.-C. (1990) Neural network processing through energy minimization with learning ability to the multiconstraint zero-one knapsack problem, *Proceedings of the Tools for AI Conference – Fairfax*, **I**, 548–55.

Lee, K.C., Funabiki, N., Cho, Y.B. and Takefuji, Y. (1991) A parallel neural network computing for the maximum clique problem, *Proceedings of the International Conference on Neural Networks – Singapore*, **II**, 905–10.

Levy, B.C. and Adams, M.B. (1987) Global optimization with stochastic neural networks, *Proceedings of the International Conference on Neural Networks*, **III**, 681–9.

Lim, J-H. and Loe, K.F. (1991) Timetable scheduling using neural networks with parallel implementation on transputers, *Proceedings of the International Joint Conference on Neural Networks – Singapore*, **I**, 122–7.

Lin, Jin-Ling, Foote, B., Chang, C.H. and Cheung, J.Y. (1991) A SMILE for packing and pallet loading in three dimensions, in *Intelligent Engineering Systems Through Artificial Neural Networks*, (eds Dagli, Kumara, and Shin), pp. 763–73, ASME Press, New York.

Looi, C.K. (1992) Neural network methods in combinatorial optimization, *Computers Operation Research*, **19**(3/4), 191–208.

Lu, Y. and Thomborson, C.D. (1991) Gate array global routing using a neural network, in *Intelligent Engineering Systems Through Artificial Neural Networks* (eds Dagli, Kumara, and Shin), 895–900, ASME Press, New York.

Matsuba, I. and Masui, H. (1991) Asymptotic behaviors of simulated annealing and mean-field approximate annealing, *Proceedings of the International Joint Conference on Neural Networks – Singapore*, 923–8.

Matsuyama, Y. (1990) Competitive self-organization and combinatorial optimation: Applications to TSP, *Proceedings of the International Joint Conference on Neural Networks – San Diego*, **III**, 819–24.

McClelland, J.L. and Rumelhart, D.E. (1981) An interactive activation model of context effects in letter perception: Part 1. An account of basic findings, *Psychological Review*, **88**, 375–407.

Mehra, P. and Wah, B.W. (1991) Learning load-balancing strategies using artificial neural networks, in *Intelligent Engineering Systems Through Artificial Neural Networks*, (eds Dagli, Kumara, and Shin), 855–60, ASME Press, New York.

Mehta, S. and Fulop, L. (1990) A neural algorithm to solve the graph matching problem, *Proceedings of the International Conference on Neural Networks – Paris*, **I**, 262–5.

Metropolis, N., Rosenbluth, A.W., Rosenbluth, M.N., Teller, A.H. and Teller, E. (1953) Equations of state calculations by fast computing machines, *Journal of Chemical Physics*, **21**(6), 1087–92.

Minton, S., Johnston, M.D., Philips, A.B. and Laird, P. (1990) A discrete stochastic neural network algorithm for constraint satisfaction problems, *Proceedings of the Eighth National Conference on AI*, **29**, I, 17–24.

Moopenn, A., Thakoor, A.P. and Duong, T. (1988) A neural network for euclidean distance minimization, *Proceedings of the International Joint Conference on Neural Networks – San Diego*, **II**, 349–56.

Nemhauser, G.L. and Wolsey, L.A. (1988) *Integer and Combinatorial Optimization*, Wiley-Interscience, New York.

Park, J.H. and Jeong, H. (1990) Solving the TSP using an effective Hopfield network, *Proceedings of the International Joint Conference on Neural Networks – Paris* **I**, 291–294.

Peterson, C. (1990) Parallel distributed approaches to combinatorial optimization – Benchmark studies on TSP, LU TP 90–2, Department of Theoretical Physics, Lund, Sweden.

Peterson, C. and Anderson, J. (1988) Neural networks and NP-complete optimization problems: A performance study on the graph bisection problem, *Complex Systems*, **2**, 59–89.

Peterson, C. and Soderberg, B. (1989) A new method for mapping optimization problems onto neural networks, *International Journal of Neural System*, **1**.

Plant, J.C. and Barr, A.H. (1988) Constrained differential optimization, in *Neural Information Processing Systems*, American Institute of Physics, 612–21.

Poliac, M.O., Lee, E.B., Slage, J.R. and Wick, M.R. (1987) A crew scheduling problem, Proceedings of the International Joint Conference on Neural Networks – San Diego, **IV**, 779–86.

Potvin, J.Y., Shen, Y. and Rousseau, J.M (1992) Neural networks for automated vehicle dispatching, *Computers Operation Research*, **19**(3/4), 267–76.

Ramanujam, J. and Sadayappan, P. (1988a) Optimization by neural networks, *Proceedings of the International Conference on Neural Networks*, **II**, 325–32.

Ramanujam, J. and Sadayappan, P. (1988b) Parameter identification for constrained optimization using neural networks, *Proceedings of the Connectionist Summer School.*

Ranjithan, S., Garrett Jr., J.H. and Eheart, J.W. (1991) A feedback neural network approach to optimization: An application in groundwater remediation design,

Intelligent Engineering Systems Through Artificial Neural Networks, (eds Dagli, Kumara, and Shin), 861–6, ASME Press, New York.

Reggia, J. (1987) Properties in a competition-based activation mechanism in neuromimetic network models, in *Proceedings of the International Conference on Neural Networks – San Diego*, **II**, 131–8.

Reklaitis, G.V., Tsirukis, A.G. and Tenorio, M.F. (1990) Generalized Hopfield networks and nonlinear optimization, *Advances in Neural Information Processing Systems 2*, (ed. D.S. Touretzky), Kaufmann.

Rumelhart, D. E. and Zipser, D. (1985) Feature discovery by competitive learning, *Cognitive Science*, **9**, 75–112.

Sasaki, M., Nakahara, Y., Gen, M. and Ida, K. (1991) An efficient algorithm for solving fuzzy multiobjective 0–1 Linear programming problem, *Computers Industrial Engineering*, **21**(1–4), 647–51.

Simic, P. (1990) Statistical mechanics as the underlying theory of 'elastic' and 'neural' optimization, *NETWORK: Computers and Neural Systems*, **I**, 89–103.

Steinberg, L. (1961) The backboard wiring problem: A placement algorithm, *SIAM Review*, **3**, 37–50

Szu, H. (1986) Fast simulated annealing, *Proceedings of the AIP Conference on Neural Networks for Computing*, 420–5.

Szu, H. (1988) Fast TSP algorithm based on binary neuron otuput and analog neuron input using the zero-diagonal interconnect matrix and necessary and sufficient constraints of the permutation matrix, *Proceedings of the International Joint Conference on Neural Networks – San Diego*, **II**, 259–66.

Tagliarini, G.A. and Page, E.W. (1987) Solving constraint satisfaction problems with neural network, *Proceedings of the International Joint Conference on Neural Networks*, **III**, 741–7.

Tank, D.W. and Hopfield, J.J. (1986) Simple neural optimization networks: An A/D converter, signal decision circuit, and a linear programming circuit, *I.E.E.E. Transactions on Circuits and Systems*, **33**, 533–41.

Thomas, D.A. and Van den Bout, D.E. (1990) Encoding logical constraints into neural network cost functions, *Proceedings of the International Joint Conference on Neural Networks – San Diego*, **III**, 863–8.

Vaithyanathan, S. and Ignizio, J.P. (1992) A stochastic neural network for resource constrained scheduling, *Computers Operation Research*, **19**(3/4), 241–54.

Van den Bout, D.E. and Miller, T.K. (1988) A travelling salesman objective function that works, *Proceedings of the International Conference on Neural Networks*, **II**, 259–66.

Van den Bout, D.E. and Miller, T.K. (1989) Improving the performance of the Hopfield-Tank neural network through normalization and annealing, *Biological Cybernetics*, **62**, 129–39.

Van den Bout, D.E. and Miller, T.K. (1990) Graph partitioning using annealed neural networks, *I.E.E.E. Transactions on Neural Networks*, **1**, 192–203.

Wacholder, E., Han, J. and Mann, R.C. (1988) An extension of the Hopfield-Tank model for solution of the multiple TSP, *Proceedings of the International Joint Conference on Neural Networks – San Diego*, **II**, 305–24.

Wacholder, E., Han, J. and Mann, R.C. (1988) A neural network algorithm for the multiple TSP, *Biological Cybernetics*, **61**, 11–19.

Wang, J. (1990) A parallel distributed processor for the quadratic assignment problem, *Proceedings of the International Conference on Neural Networks – Paris*, **I**, 262–5.

Wang, J. and V. Chankong (1992) Recurrent neural networks for linear programming: Analysis and design principles, *Computers Operation Research*, **19**(3/4), 297–311.

Wilson, G.V. and Pawley, G.S. (1988) On the stability of the TSP algorithm of Hopfield and Tank, *Biological Cybernetics*, **58**, 63–70.

Wong, W.S. and Funka-Lea, C.A. (1990) An elastic net solution to obstacle avoidance tour planning, *Proceedings of the International Joint Conference on Neural Networks – San Diego*, Paris, **III**, 799–804.

Xu, X. and Tsai, W.T. (1991) Effective neural algorithms for the TSP, *Neural Networks*, **4**, 193–205.

Yih., Y.W., Liang, T.P. and Moskowitz, H. (1991) A hybrid approach for crane scheduling problems, in *Intelligent Engineering Systems Through Artificial Neural Networks* (eds. Dagli, Kumara, and Shin), 867–72, ASME Press, New York.

Yokai, H. and Kakazu, Y. (1991) An approach to traveling salesman problem by a bionic model, in *Intelligent Engineering Systems Through Artificial Neural Networks*, (eds, Dagli, Kumara, and Shin), 883–8, ASME Press, New York.

Yue, T-W. and Fu, L-C. (1990) Ineffectiveness in solving combinatorial optimization problems using a Hopfield network: A new perspective from aliasing effect, *Proceedings of the International Joint Conference on Neural Networks – San Diego*, **III**.

Yuille, A.L., Geiger, D. and Bulthoff, H.H. (1989) Stereo integration, mean field and psychophysics, Harvard Robotics Laboratory, Technical report, 10.

Yuille, A.L. (1990) Generalized deformable models, statistical physics, and matching problems, *Neural Computing*, **2**, 1–24.

Zhou, D.N., Cherkassy, V., Baldwin, T.R. and Hong, D.W. (1990) Scaling neural network for jobs-shop scheduling, *Proceedings of the International Joint Conference on Neural Networks – San Diego*, **III**, 889–94.

Ziai, M.R. and Sule, D.R. (1991) Computerized facility layout design, *Computers Industrial Engineering*, **21**(1–4), 385–9.

CHAPTER 9

Automated assembly systems

Cihan H. Dagli[1] *and Mahesh Kumar Vellanki*[2]

[1] *Engineering Management Department, University of Missouri-Rolla, Rolla, MO 65401.*

[2] *AT&T Bell Laboratories, Engineering Research Center, Hopewell, NJ 08525.*

Assembly is one of the major processes in a manufacturing environment, where many pre-fabricated parts are brought together within specified tolerances to produce a product which has some functional value. The need for sophisticated assembly systems that are intelligent and highly flexible has increased over the years and motivated many researchers to develop innovative techniques in implementing such systems. Today's industry demands more flexibility in assembly automation due to rapid changes in product design imposed by highly competative global markets. Apart from flexibility, it is also desirable to build systems that can quickly adapt to new product designs and market changes. This fact was stated in an editorial by Dagli (1991) as:

> Manufacturing systems of the 21st century will demand more flexibility in product design, process planning, scheduling and process control. This may well be achieved through integrated software and hardware architectures that generate current decisions based on information collected from assembly systems, and execute these decisions by converting them into signals transferred through communication networks. Manufacturing technology has not yet reached this state. However, the urge for achieving this goal is captured by the term 'Intelligent Systems' that has begun to be used more in late 1980s.
>
> Knowledge-based systems, our first efforts to build such intelligent systems.

Artificial Neural Networks for Intelligent Manufacturing. Edited by Cihan H. Dagli.
Published in 1994 by Chapman & Hall, London. ISBN 0 412 48050 6

were not sufficient to generate the intelligence required in automating assembly systems. The limitations imposed by these systems forces researchers to explore new technologies. In the mid-eighties artificial neural networks emerged as a new approach for solving problems that posed difficulties to engineers and scientists and revolutionized the design of intelligent systems. They introduced the power of learning and adaptation and emerged as a solution approach in many manufacturing engineering problems including automated assembly. In this chapter, the use of artificial neural networks in this area is discussed and the concepts are demonstrated through a surface mount assembly of a power supply board.

9.1 AUTOMATED ASSEMBLY

Automated assembly systems have been classified (Arabian, 1989) as: fixed, dedicated, high volume designs; flexible, general-purpose, medium-volume design; and single, flexible, programmable robots.

The **fixed, high volume design** is a customized or special purpose machine which is dedicated to the production of one product. Since all of the fixures and workheads are devoted to this purpose, few, if any, changes in product can be tolerated or accommodated. Tight tolerances are imposed and their feeders and holding fixtures are designed to impose low defect rate on the parts to be assembled. The fixed, high-volume machines are designed to run fast; thus, the cycle time for each operation is fixed, as is the production rate.

The **flexible, general-purpose system** is intended for medium-volume operations, and it is characterized by adaptability to limited forms of product design and style changes. These flexible, medium-volume operations incorporate switchable parts magazines, pallets, cartridges, or reels previously loaded and conveniently available to the workhead; these tools are placed and programmed to be drawn and utilized on a variable basis. In addition, programmable indexed rotary table feeders, automatic tool changers, robotic arms, and end effectors are familiar features of the flexible system. With such flexibility, more than one assembly operation can be completed at one workstation, making it possible to match the system output to a lower demand rate. Also, because of the required extra motions and longer reaches, the assembly time increases, slowing the production rate.

The **single, flexible, programmable robot** is a special case for low-volume, repetitive assembly which requires error-free, accurate operation. It is distinguished from the flexible variety described above only to show that a single robot in a relatively independent or semi-isolated environment, performing a series of assembly steps, is not the same as a robotic arm integrated into a medium-volume operation. The single robot may even be mounted on rails to expand its scope of activity or to extend its reach, at the expense of increasing its cycle time.

The current research is mainly focused on improving the last two methods of assembly: flexible and medium-volume and single flexible/programmable robots. It is necessary to provide intelligence to the robots in these assembly systems in order to make judgments regarding part identification, and/or fault identification. Artificial intelligence systems aided with sensing tools, such as vision systems (Pau, 1990) photoelectric devices, and electron beams, would make it possible to incorporate the desired intelligence into these systems (Bao, 1988). Flexibility in small batch assembly processes is also achieved by examining the routing, as discussed by Taylor and Graves (1990). In this discussion it is observed that, as batch sizes decrease the performance of flexible routing policies improve.

Apart from having intelligence which may contribute to quality and flexibility, the assembly systems have to satisfy another very important criteria of most industries – speed. There are many methods adopted for improving the speed of assembly, such as simultaneous-mounting processes discussed by Chang and Young (1990) and optimal routing and sequence determination discussed by Dagli *et al.* (1991).

Considerable amount of research is done in the area of design of automated assembly systems which satisfy all the three requirements of small/medium scale assembly units: quality, flexibility, and speed (Vellanki and Dagli, 1993; Dagli and Vellanki, 1992; Vellanki and Dagli, 1992; Funk, 1989; Gavish and Seidmann, 1987; Steudel, 1987).

A new approach in the design of intelligent and high precision automated assembly systems is needed. The generic assembly system developed integrates a vision system, artificial neural networks, and a high precision robot. The vision system helps in acquiring visual data from the robot workspace which is later processed to make quantitative deductions. Artificial neural networks provide the central intelligence for interpretation and deduction based on the data obtained as input from the vision system and in turn provide guidance to the robot in performing high precision operations. All three elements, vision system, artificial neural networks, and robot, work together in perfect coherence to form an intelligent system that is not only flexible to product changes but also maintain high precision at reasonable operating speeds. In the following sections details of this generic assembly system are discussed.

A generic assembly system and the theory behind it is presented in section 9.2. The basic components which constitute the assembly system – robot, vision system, and artificial neural network architectures – and the method of integrating them is explained in detail. In section 9.3 the generic assembly system developed is adapted to perform a power supply board assembly. Test cases consisting of five components are executed and performance of the intelligent automated assembly system is evaluated.

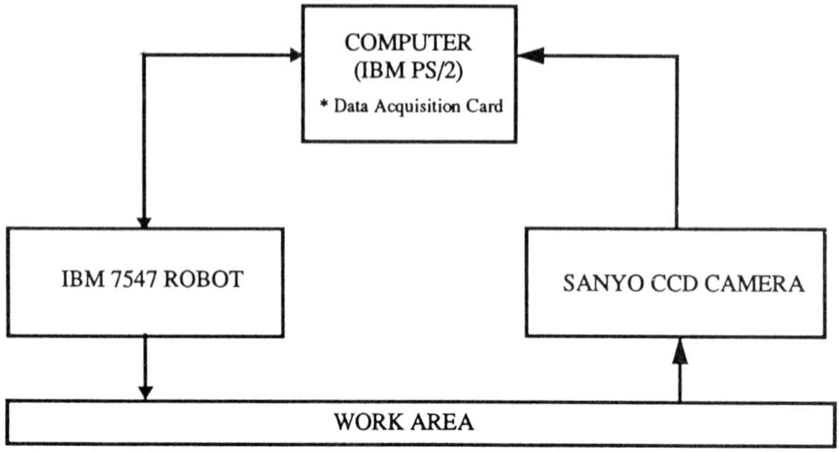

Fig. 9.1 Configuration description of three major components of the assembly cells.

9.2 GENERIC ASSEMBLY CELL

9.2.1 Systems overview

A simple assembly cell is constructed in a computer integrated manufacturing (CIM) environment to demonstrate the proposed architecture for automated precision assembly. The assembly cell consists of three major components: IBM Personal System/2 computer equipped with a data acquisition card, IBM7547 robot, and Sanyo CCD Camera.* Figure 9.1 illustrates the configuration of these components.

The computer is used as the primary processing device to process the information obtained from the camera and the robot. A 'QuickCapture' board from Data Translation† is used as an interface with the camera to perform real-time image processing. One of the serial ports of the computer is configured to communicate with the robot.

The IBM 7547 robot comprises a controller, a manipulator, and a control panel. The manipulator is a part of the robot that actually moves and performs assembly operations in the work area. The controller constitutes the electronics to generate control signals to move the manipulator. It is also connected to a control panel from where commands can be issued manually. Thus the controller serves two purposes: it acts as an interface between a host computer (in this case the IBM PS/2) and the manipulator where the

* IBM®, Personal System/2® are registered trademarks of the International Business Machines Corporation, Sanyo® is a registered trademark of Sanyo Corporation.

† Data Translation® is a registered trademark, and QuickCapture™ is the trademark of Data Translation, Inc.

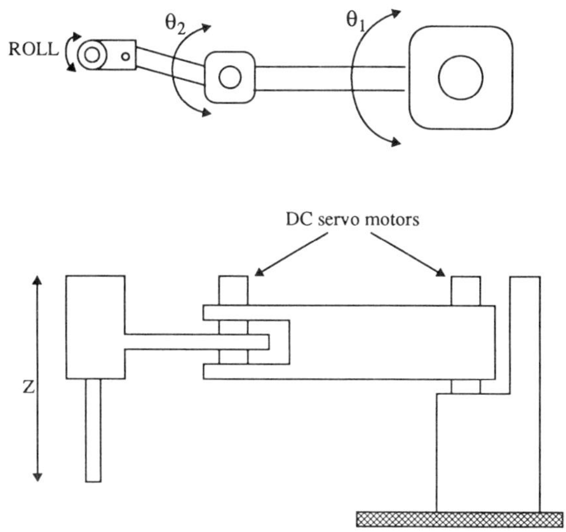

Fig. 9.2 The IBM robot with four degrees of freedom.

computer controls the movement of the manipulator, and it also can be used as a primary control source for the manipulator in a stand-alone system.

The CCD camera is just a passive element in the set-up with its main function being sensing. It is only an output device which sends as its output the live video information of the region of its sight, which is usually some part of the work area. For incorporating flexibility in sensing the camera is mounted on the robot's end effector.

9.2.2 Robot characteristics

The specifications of the IBM 7547 robot and the interface to the host computer is discussed below (IBM 7547 Manufacturing System Hardware Library, 1984). Three main components of the robot are: the manipulator, the controller and the control panel.

The **manipulator** is a two-jointed arm structure with four degrees of freedom (Fig. 9.2). The joints of the arm, called θ_1 axis and θ_2 axis, provide two degrees of freedom through their swivel motion. The end-of-arm rotation, called the roll axis, provides a third degree of freedom. The end-of-arm also provides a fourth degree of freedom through a vertical shaft called the Z-axis. The area spanned by the robot end effector while varying the four parameters – θ_1, θ_2, ROLL, and Z – constitutes the robot workspace.

A planar view of the workspace is shown in Fig. 9.3. The workspace of the robot is limited and is determined by the manipulator arm length(s) and their degrees of freedom.

The **controller** contains most of the electronics for control of the

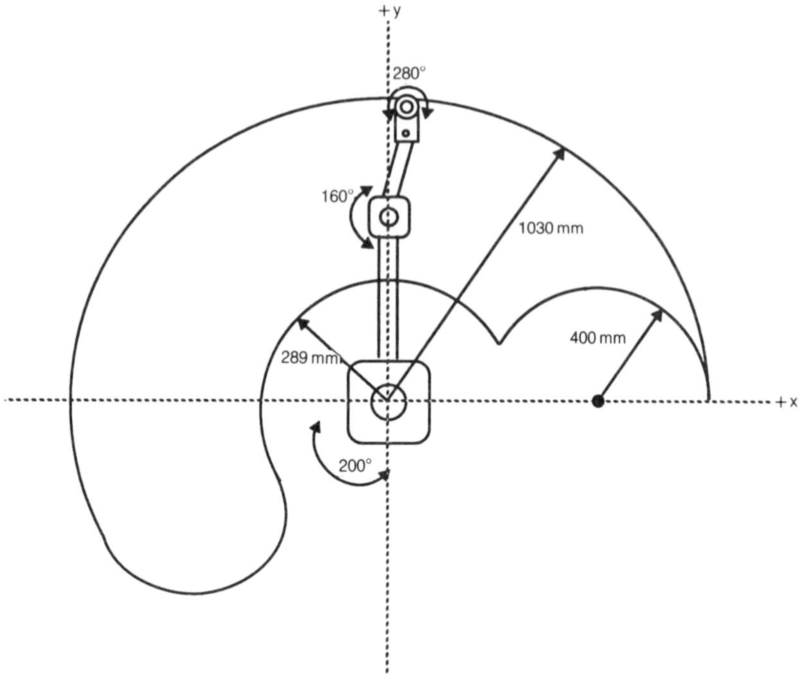

Fig. 9.3 The planar view of the IBM robot workspace.

manipulator operation. A microprocessor coordinates the manipulator's movement and monitors its speed and positioning.

External devices are synchronized with the manipulator through the use of digital input (DI) and digital output (DO) ports. These ports monitor switch closures (DI) external to the system or close relays (DO), allowing events to occur. The controller receives and stores its control program from the programming device and then drives the manipulator by executing the programs. The application programs instruct the controller as to which ports to monitor, the amount of time to wait for an event, and the type of condition to expect.

The **control panel** is the operator interface to the controller and the manipulator.

The robot-manipulator can be controlled through the controller. However, there are three ways to communicate to the controller: through the control panel; the RS-232 serial interface; and the RS-422 serial interface. In the assembly cell the IBM PS/2 is configured to communicate with the controller through the RS-232 interface.

9.2.3 Machine vision system for robot guidance

In the manufacturing industry, tasks requiring the use of eyesight have in the past always been carried out by human operators, and this has restricted the spread of automation mainly to repetitive mechanical tasks. The development of machine vision to the level where it is now a robust and competitively priced sensing technique has opened up new automation possibilities. At present there are two broad areas of application for vision systems in industry: inspection and guidance. **Inspection** is by far the most important area, and there is a great diversity of applications including not only the visual examination of defects which is normally done by an eye but also measurement of dimensions or other characteristics, counting, checking of orientation. The applications in **guidance** include control of robots, vehicles, orientation devices in parts handling, sorting and transport. Automated assembly falls under this category.

Machine vision usually includes the following three elements: image acquisition, feature detection and feature identification.

In generic automated assembly systems, the data capture card is used to carry out the image acquisition and processing operations (The Quick Capture User's Manual, 1989).

The **data capture card** digitizes a video signal, stores an image in the onboard memory buffer, and displays the image in RGB false color or monochrome. The board can digitize images individually or continuously at the same rate cameras produce them – 30 image frames per second. The frame-store memory allows storage of one image onboard. Images are displayed using square-pixels with a 1:1 pixel aspect ratio. Square pixels are free of the geometric distortions that can cause circles to appear as ovals, or squares as rectangles. The QuickCapture 60 Hz model, used in generic automated assembly cell, is compatible with RS-170, RS-330, and NTSC video formats. It outputs RS-170 in RGB format when composite sync is supplied on the green output of the camera. It utilizes 640 × 480 × 8 bits of the frame-store memory buffer. The block diagram of the data capture card is given in Fig. 9.4.

It digitizes an entire video frame in 1/30 second. the video signal is converted into 480 lines by 640 square pixels per line by 8 bits per pixel. The 8-bit information for each pixel represents one of 256 possible gray levels.

As image is acquired, the 8-bit input values are sent through one of four software-selectable input look-up tables stored in onboard RAM. The input look-up tables allow the board to perform real-time image processing on acquired data as it flows into the frame-store memory. The four tables can be loaded with different processing set-ups in order to pass the image unaltered or to perform operations, such as image multiplication and division, intensity correction, and reverse-video.

The memory buffer allows the digitized video input signal to be stored onboard. It can then be accessed over the bus using standard I/O functions. Frame memory operations are controlled by hardware timing circuits and arbitrators, which are fast enough so that video input, video output, and normal memory access can all be done simultaneously.

The data capture card contains a 256 × 24 bit output look-up table for real time manipulation of false color or intensity values assigned to the pixel

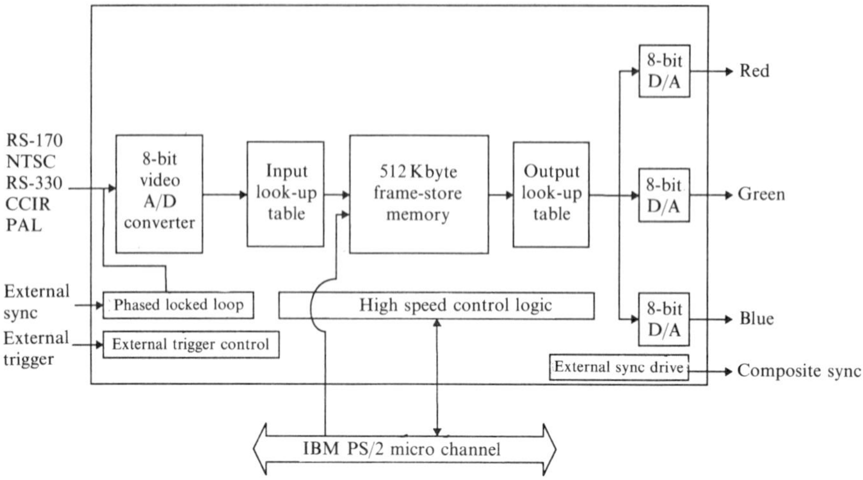

Fig. 9.4 The QuickCapture board block diagram.

data. The 7 by 24 bit overlay output look-up table allows seven graphic overlay attributes to be defined in addition to the 256 display attributes for the image data.

Image pre-processing

Image pre-processing is an essential component of a vision system and it is used to eliminate the noise and reduce the data to manageable proportions. This is an essential step for preparing input data for an artificial neural network. The generic automated assembly system performs image processing through the following sequence of operations: intensity quantification by thresholding, removal of small highlights or dark spots by filtering, object enhancement by filling up missing portions, pattern isolation and centroid computation and resolution reduction.

Thresholding scans the image pixel by pixel and quantizes the intensity on either side of the threshold, ρ, for each of these pixels into two levels, defined as $\text{Pixel}_{\min}$ and $\text{Pixel}_{\max}$ respectively. This process can be expressed mathematically as:

$$P_{xy} = \text{Pixel}_{\min} \quad \text{if } P_{xy} \leqslant \rho$$
$$P_{xy} = \text{Pixel}_{\max} \quad \text{if } P_{xy} > \rho$$
$$\text{for} \quad x = 0, \ldots, 640 \quad \text{and} \quad y = 0, \ldots, 480$$

where

P_{xy} is the intensity of the pixel at (x, y) in the image
ρ is the threshold value

Pixel_{min} is the minimum pixel intensity i.e., black
Pixel_{max} is the maximum pixel intensity i.e., white

It is important to note that the intensities of the actual objects within the image are assumed to be less than the threshold value ρ so that objects after the thresholding process would appear with the intensity Pixel_{min}.

The information from each pixel, whether thresholded to black, white, or analyzed into gray levels, is held in the onboard memory of data capture card for further processing.

The object under examination may have sharp, clearly defined edges, but it is likely that the edges of the image will be quite fuzzy with black pixels where there should be white, and vice versa. Initial pre-processing is carried out on the image for further processing, working on a group of adjacent pixels which form a small rectangular window (Fig. 9.5).

$$P_{ij} = \begin{cases} \text{Pixel}_{max} & \text{if } P_{xy} = P_{x+\delta x,y} = P_{x,y+\delta y} = P_{x+\delta x,y+\delta y} = \text{Pixel}_{max} \\ & \text{and } P_{x+\delta x/2,y+\delta y/2} = \text{Pixel}_{min} \\ \text{Pixel}_{min} & \text{if } P_{xy} = P_{x+\delta x,y} = P_{x,y+\delta y} = P_{x+\delta x,y+\delta y} = \text{Pixel}_{min} \\ & \text{and } P_{x+\delta x/2,y+\delta y/2} = \text{Pixel}_{max} \\ P_{xy} & \text{otherwise} \end{cases}$$

for all

$$i = x, \ldots, x + \delta_x \quad \text{and} \quad x = 0, \ldots, 640$$
$$j = y, \ldots, y + \delta_y \quad \text{and} \quad y = 0, \ldots, 480$$

where P_{ij} is the intensity of the pixel at (i, j) and δ_x, δ_y are the dimensions of the window.

Many more sophisticated operations can be performed to produce fine

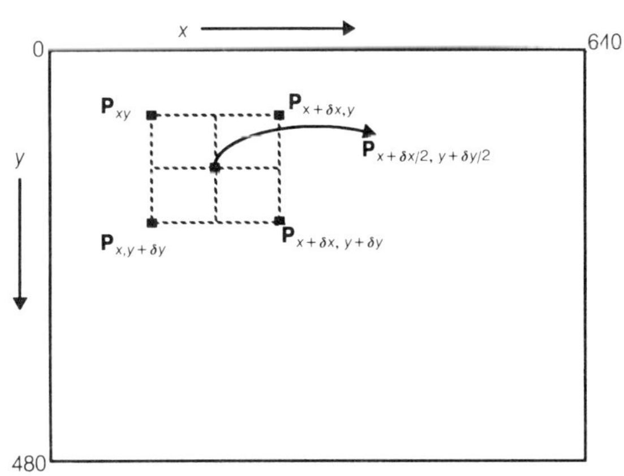

Fig. 9.5 Scanning the image with a rectangular window of size (delta)x by (delta)y pixels.

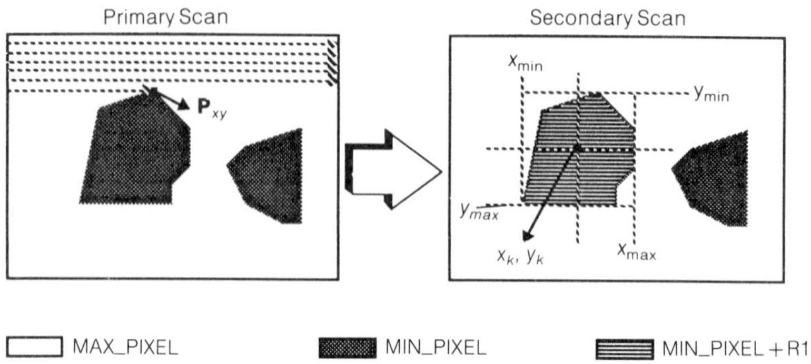

Fig. 9.6 Pattern isolation and center computation.

images and thereby make the subsequent process of pattern identification/recognition easy. Some of them can involve breaking up the picture into meaningful components. It may also include sharpening up the edges between the different areas or to join up lines which are broken in the images. The choice of using these additional techniques depends on the time constraints imposed for pattern identification during automated assembly.

During the final phase of pre-processing the image is scanned pixel by pixel to locate patterns of intensity of $\text{Pixel}_{\text{min}}$ and determine their centroids. A pixel value of $\text{Pixel}_{\text{min}}$ indicates an existence of an object and indicates a secondary scanning process through the object while suspending primary image scanning temporarily (Fig. 9.6).

During this phase the pattern is scanned vertically down line by line while the scan lines are limited by the pattern boundaries on either side forming segments of variable lengths. Each of these lines is scanned starting from a point on that line, which is right below the centre of the previous scan segment, and proceeding on either side until the boundary is detected i.e., when $P_{xy} = \text{Pixel}_{\text{max}}$. While doing this the intensity of all the pixels within the pattern are also changed in order to differentiate between the pattern under scan and other neighboring patterns if any. Thus

$$P_{xy} = \text{Pixel}_{\text{min}} + R_1$$

for all (x, y) within the pattern and

$$\text{Pixel}_{\text{min}} < R_1 < \text{Pixel}_{\text{max}}$$

Also, while scanning the pattern its top line y_{max} the most bottom line, y_{min}, the right-most point x_{max} and the left-most point x_{min} are noted and used to compute the centroid of the object (x_k, y_k) as:

$$x_k = \tfrac{1}{2}(x_{\text{min}} + x_{\text{max}})$$
$$y_k = \tfrac{1}{2}(y_{\text{min}} + y_{\text{max}})$$

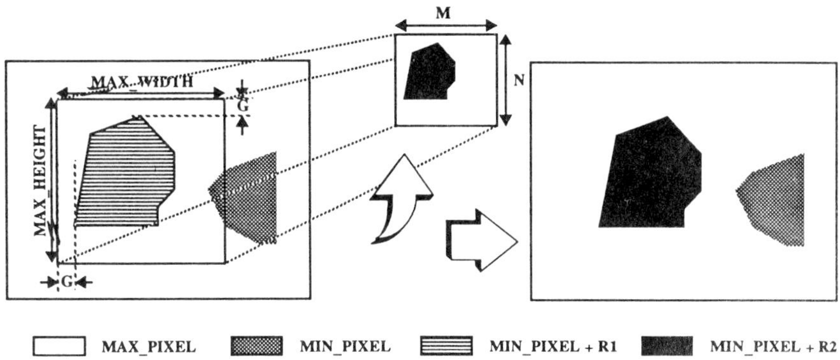

Fig. 9.7 Pattern extraction for recognition.

The next phase involves making preparations for pattern recognition. The pattern under scan is mapped into a binary ($M \times N$) matrix (Fig. 9.7) as an input to the pattern recognition model.

Pattern mapping can be done in two stages as illustrated in Fig. 9.7. First the scanned pattern can be captured into a grid of a standard size determined by the largest object used in an application. Let the size of the grid be $W \times H$ pixels. Imagine that this grid is placed over the pattern as shown in Fig. 9.7 such that its corner coordinates are,

$$\text{top-left:} \quad (x_{\min} - G, y_{\min} - G)$$
$$\text{bottom-right:} \quad ((x_{\min} - G) + W, (y_{\min} - G) + H)$$

Note that $x_{\min}$ and $y_{\max}$ are determined while scanning the object pattern discussed earlier. G, is a constant spacing in pixels at the left and top edges between the pattern and the grid. This offset is introduced to ensure that the pattern is completely enclosed by the grid.

Also, note that this grid is purely an imaginary plane placed over the image and any changes made to the pixels within it do not reflect on the actual image beneath it. The pattern enclosed by the grid is captured into the grid by changing the pixels inside the grid using following equations.

Let D_{lk} be a pixel inside the grid and P_{lk} be a pixel in the actual image, both located at (l, k) w.r.t the actual image. Thus

$$x_{\min} - G \leqslant l \leqslant (x_{\min} - G) + W$$
$$y_{\min} - G \leqslant k \leqslant (Y_{\min} - G) + H$$
$$D_{lk} = \begin{cases} 1 & \text{if } P_{lk} = \text{Pixel}_{\min} + R_1 \\ 0 & \text{otherwise} \end{cases}$$

In other words all the 1s inside the grid constitute the pattern which is just scanned and which consist of pixels with intensity $\text{Pixel}_{\min} + R_1$. It should be noted that by looking for only $\text{Pixel}_{\min} + R_1$ it is possible to ignore other patterns of different intensities which may intrude into the grid. This is also illustrated in Fig. 9.7. Once the pattern is captured into the 'virtual' grid, the pixel intensities of the actual pattern in the image are changed again.

$$P_{lk} = \begin{cases} \text{Pixel}_{\min} + R_2 & \text{if } P_{lk} = \text{Pixel}_{\min} + R_1 \\ P_{lk} & \text{otherwise} \end{cases}$$

where,

$x_{\min} - G \leqslant l \leqslant (x_{\min} - G) + W$
$y_{\min} - G \leqslant k \leqslant (y_{\min} - G) + H$
R_2 is a constant and $R_2 \neq R_1$
$\text{Pixel}_{\min} < R_2 < \text{Pixel}_{\max}$

The change made to the pixel intensities of the scanned pattern helps in distinguishing it from the rest of the unscanned patterns. At any instance during scanning there can exist patterns with only three different pixel intensities – $\text{Pixel}_{\min}$, $\text{Pixel}_{\min} + R_1$, and $\text{Pixel}_{\min} + R_2$ – each representing unscanned, being scanned, and already scanned patterns respectively.

The second step in pattern mapping involves reduction in resolution of the grid from $W \times H$ to $M \times N$. The amount of reduction is dependent on the computing speed used while executing pattern recognition algorithms and time constraint imposed by the automated assembly on pattern recognition. In any case, the resolution reduction factor y is determined after deciding the size $M \times N$ matrix as

$$y = W/M = H/N$$

The matrix is used as an input to the pattern recognition model that identifies the components.

Pattern recognition

A prototype neural network based vision system is developed to perform pattern recognition. Adaptive resonance theory (ART) neural network paradigm is selected for this purpose. This architecture has the capability of updating its memory with any new input pattern without corrupting the existing information. This is especially important in flexible automated assembly where the product designs change with time. The possibility of an infinite number of views of the same pattern could pose a serious problem. This possibility is eliminated by assuming only three defined geometric views of each component (length–breadth, breadth–height, and height–length) viewed from a fixed distance. This assumption is appropriate for many automated assembly systems as most of the components and tools usually possess definite geometric shapes.

As discussed in Chapter 3 the ART1 operation mainly involves three stages – initialization, search, and learning. The initialization involves resetting the weights in the neural network. Fig. 9.8 which shows a typical ART1 architecture (Simpson, 1990). The feedback connections between the

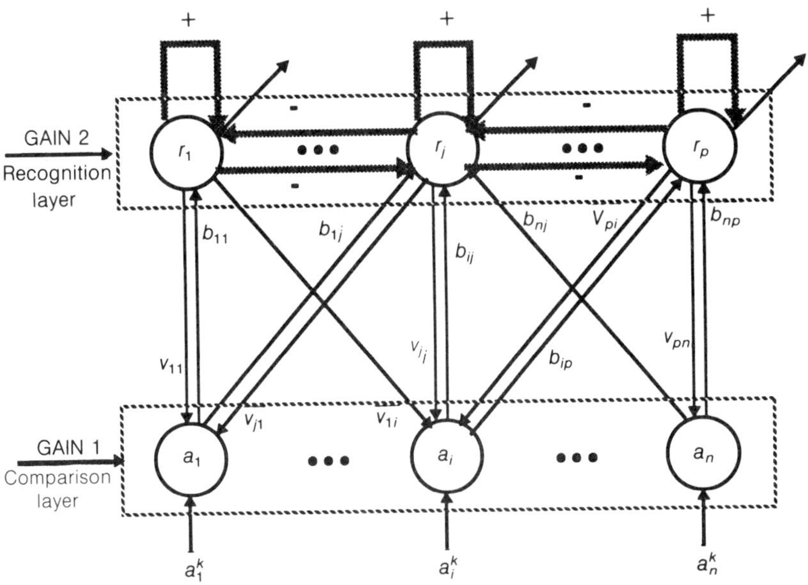

Fig. 9.8 Topology of the binary adaptive resonance theory (ART-1) system.

comparison and recognition layers facilitate a resonance upon proper match between encoded and input patterns. The comparison layer elements accept inputs from the environment and the recognition layer elements each represent a pattern class. Each layer's activations are controlled by the gain outputs. During the operation the recognition layer elements compete with each other to determine the proper class of the presented input. This process is achieved through lateral inhibition. These later interactions are represented as shaded self-exciting/neighbor-inhibiting connections in Fig. 9.8.

The ART1 network can operate in two learning modes: slow learning and fast learning. In **slow learning** only a small portion of the input pattern is used to change the corrresponding weights. After repeated presentations of the entire training set, the most prominent characteristics of each pattern will emerge and become committed to long-term memory. **Fast learning** immediately encodes the input pattern directly onto the weights and does not require more than one or a few passes through the training set. Thus, the pattern classes formed using fast learning are not as statistically representative as those found using slow learning. Fast learning is well suited to applications that have crisp data sets and require immediate learning. In a typical assembly process the components are usually well defined, and the images can be obtained fairly noise free. It is also necessary to learn new components quickly to reduce set-up time. Thus fast learning methodology is adopted in generic automated assembly model.

Camera calibration

Pattern recognition component of the generic automated assembly system provide information related to the part location and identification. However, this is not sufficient to guide the robot to the exact location of the parts as their locations are expressed in terms of camera coordinates, namely, pixels in x and y directions. These co-ordinates do not represent fixed points in space because the region of the work space scanned changes based on the location of the robot arm. However, it is possible to identify a workspace with known boundaries in which all components that need to be assembled are placed. Each time before the start of the assembly the robot is directed to move to a particular location represented by constant location and roll coordinates x, y, z and ROLL respectively so that the onboard camera can capture the image of the workspace and later the components within it can be identified.

Now, there is a need to map the camera coordinates in a fixed region of space to the corresponding robot coordinates in the same region. Due to several reasons (such as lens distortion and misalignment of the camera with respect to the workspace) a significant amount of non-linearity is introduced between the camera and robot coordinates. In other words, they cannot be mapped on a one-to-one basis. Even if an expensive set-up is used, there will still be some non-linearity in their relationship. It is not feasible to estimate an n-polynomial non-linear equation from the data or the graph representing the camera versus robot coordinates. Many methods are presently adopted by the industry, such as non-linear approximations and coordinate matrix multiplication, which are probably well suited when an accurate and expensive set-up is used. However, it is possible through the use of neural networks to capture the non-linearity between the data sets accurately. A backpropagation neural network is identified for this purpose.

Generic automated assembly systems uses four-layer backpropagation for camera calibration; the network architecture is depicted in Fig. 9.9. The training set of the neural network, comprises pairs of camera coordinates and corresponding robot coordinates. The training data can be acquired by placing a virtual grid over the work space and locating each crosspoint within the grid in both coordinates to form training pairs. This process is illustrated in Fig. 9.10 where the camera coordinate of each crosspoint is represented in pixels and the corresponding robot coordinate in millimeters. It should be noted that the camera coordinate and the robot coordinate refer to the same physical point in space. Also, the camera is fixed to the robot arm, and the workspace is fully captured in the camera viewfield while it forms a small portion in the whole robot workspace. Thus, each crosspoint within the camera viewfield is located with respect to its top-left corner $(0, 0)$ as origin while the same point in the robot workspace is located with respect to a fixed point (I_x, I_y). The top-left corner $(0, 0)$, which is a reference point in the camera viewfield, corresponds to (I_x, I_y) which is a reference point in the robot workspace.

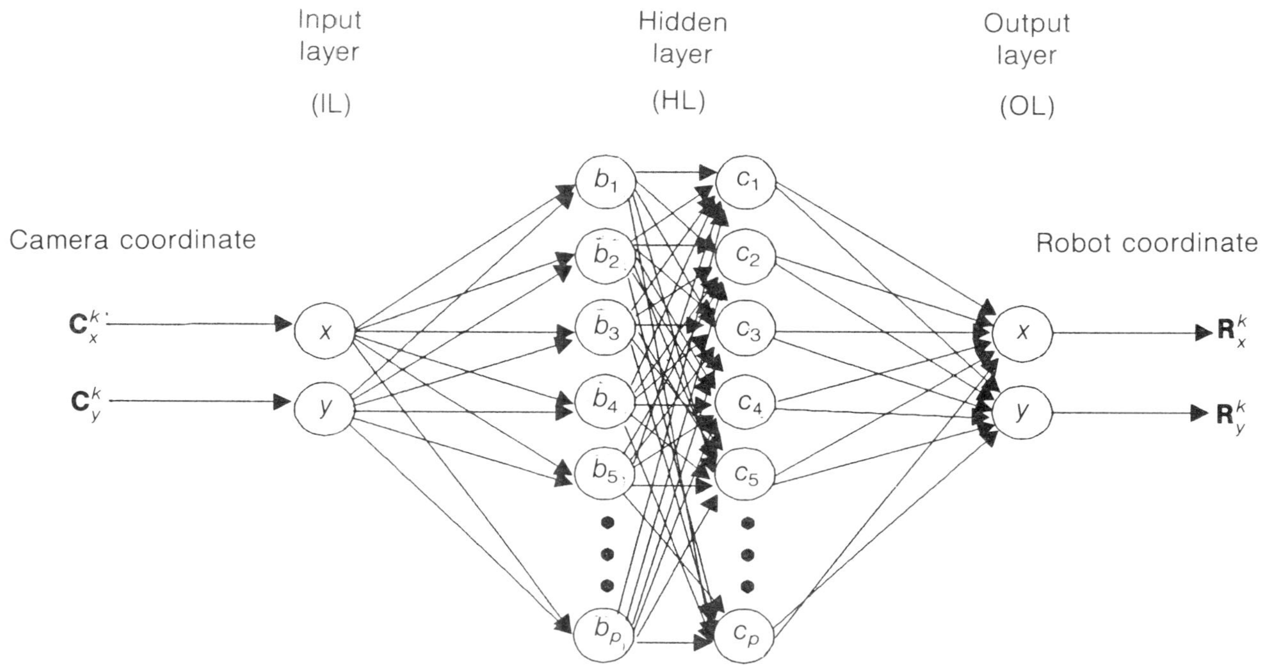

Fig. 9.9 Backpropagation network topology.

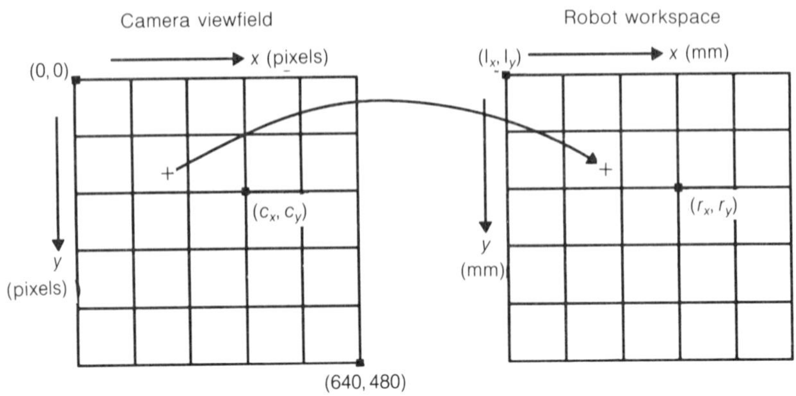

Fig. 9.10 Camera: robot coordinate mapping process.

The next step in preparing the training data involves normalization of each of these coordinates such that they range between 0 through 1. Normalized robot coordinates R_x and R_y can be computed as:

$$R_x = \frac{I_x + r_x}{\mathrm{NR}_x} \quad R_y = \frac{I_y + r_y}{\mathrm{NR}_y}$$

where NR_x, NR_y are normalization factors and $0 \leqslant R_x \leqslant 1$, $0 \leqslant R_y \leqslant 1$. Normalized camera coordinates are

$$C_x = \frac{c_x}{\mathrm{NC}_x} \quad C_y = \frac{c_y}{\mathrm{NC}_y}$$

where NC_x, NC_y are normalization factors and $0 \leqslant C_x \leqslant 1$, $0 \leqslant C_y \leqslant 1$.

The objective is to train the neural network with coordinate pairs obtained from each crosspoint until the maximum mapping error is reduced to meet the tolerance level. The network is later used passively to compute the robot coordinate of a point, for instance '+' marked in the figure, which does not belong to the training set and whose camera coordinate is known. Presumably the mapping of the new point should occur with the same accuracy as obtained while training the network. After the network has been fully trained it is used to recall a corresponding robot coordinate in a normalized form. The location of an object in the camera viewfield is always identified by its center the value of which is computed in pixels at the pre-processing stages. The value C_x, C_y is processed through the backpropagation neural network and the values of R_x, and R_y determined are converted back to original robot coordinates for locating the object. At this point it is important to mention that the height at which an object needs to be picked up is determined by its identity, determined by ART1 neural network. However, there is a limitation to the system – it is not rotation invariant. In other words, it is ensured that all the objects placed in the inventory area are approximately aligned vertically. Thus the robot always picks them up either at 0, or 90, or 180 degrees.

To summarize, the vision system is useful only to an extent of providing the information of where the object is located in a two-dimensional plane while the information related to the height and orientation is obtained from the properties of the object. Thus, with the available information it is possible to direct the robot to move to an exact point in space represented in four demensions (x, y, height, and roll) in order to pick up an object.

9.3 POWER SUPPLY BOARD ASSEMBLY: A CASE STUDY

As a case study a surface mount assembly of a power supply board is considered. The simple power supply board circuit used consists of five electronic components: transformer, two capacitors, rectifier, and voltage regulator. These five components are of different shapes and sizes and are placed in the area of the robot workspace in order to be assembled.

Figure 9.11 shows the schematic of the assembly cell layout consisting of the robot, its workspace, and the regions of operation within the workspace: the inventory area where the parts are placed, mounting area where the parts are assembled over a printed circuit board (PCB) (Fig. 9.12), and the gripper station from where the appropriate gripper for handling a component is selected (Fig. 9.13). The actual assembly cell is shown in Fig. 9.14. The circuit

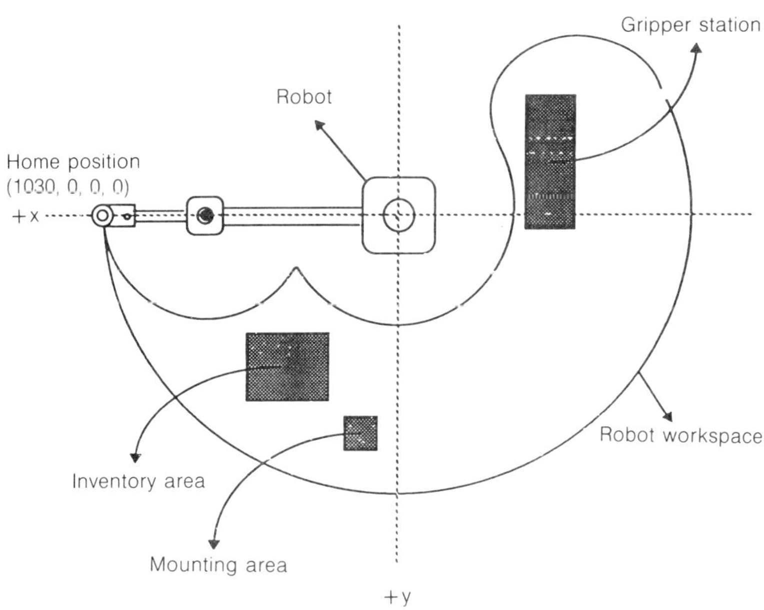

Fig. 9.11 A schematic of power supply board assembly cell.

Fig. 9.12 The parts of the inventory area, and the PCB on the mounting area.

Fig. 9.13 The gripper station.

Fig. 9.14 The actual assembly cell.

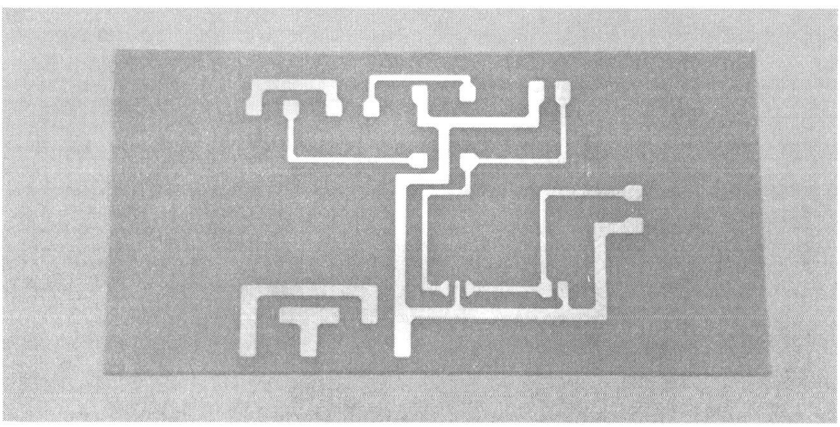

Fig. 9.15 The printed circuit board (PCB) and the circuit layout.

layout on the PCB, as shown in Fig. 9.15, is used to determine the destination coordinates of all the above five components in the inventory area, shown in Fig. 9.16. Before initiating the assembly process the sequence of placing the components on the printed circuit board is determined so that collisions can be avoided during component placement.

A software utility program has been developed exclusively for this purpose. The input to the software consists of the dimensions of all the parts and their corresponding destination coordinates on the circuit layout. Based on

Fig. 9.16 The five components on the inventory area.

the layout description and component dimensions, the program determines an optimum placement sequence to avoid collisions. If, according to the software, collisions cannot be avoided, the user is prompted to redesign the layout to meet the specifications.

9.3.1 Assembly process

The assembly process takes place in a sequence of nine steps: system reset, speech synthesizer activation, image acquisition, image pre-processing, coordinate mapping, preparation for final assembly, assembly process initiation and final assembly.

System Reset

In this initial phase of program execution, the main components of the assembly cell are reset and configured to the required specifications. These components involve the communication ports, the data acquisition card inside the computer, and the robot. The communication ports of the computer are configured to meet the specifications of the communication protocol between the computer and the robot controller and between the computer and the speech synthesizer. The communication ports in the IBM PS/2 are configured to meet the above specification, namely, full duplex transmission with a half-duplex end-to-end user protocol, baud rate of 4800 even parity 7 data and 3 stop bits.

The data acquisition card is allowed to synchronize with the camera signals and is set initially to a continuous display mode. The robot, irrespective of its present position, is moved to its home position, and the speed is set within the limits of safety.

Speech synthesizer activation

The user is prompted to choose or disregard this feature in the assembly cell. The purpose of developing the speech synthesizer is to provide the user with a status report, in 'audio', at every stage during the assembly operation. The system simulates a human commentary over the events which take place during the assembly. Evidently, this feature introduces more delays in the assembly process because it is driven by the same computer which takes care of all other components in the cell.

Image acquisition

During this phase the robot moves over the inventory area to a predetermined coordinate and allows the onboard camera to view the inventory area in which the components are placed. Immediately after the inventory area is brought into camera viewfield the data acquisition card, which was originally set to continuous display mode, is now made to freeze the frame. This operation captures the image as seen by the camera (Fig. 9.17) into the local memory at the card.

The image occupies 640 × 480 bytes of memory inside the card. Each of the 640 × 480 bytes represents the intensity of the corresponding pixel which ranges from 0 to 255.

The ambient light over the inventory area strongly influences the quality of the image. If the lighting conditions are poor there can occur many dark spots in the image which can subsequently be interpreted as valid components. On the other hand if there is excessive ambient light, the objects/patterns within the image will be distorted with bright spots inside them. Hence, to ensure an appropriate number of lumens to strike the inventory area, directional bulbs fixed adjacent to the camera, as shown in Fig. 9.17, are used as light sources.

Figure 9.18 shows the frozen image as it appears in the monitor which is connected to the output of the data acquisition card. The image consisting of patterns of all five components is now ready for processing.

Image pre-processing

This operation mainly involves noise elimination and pattern enhancement. Every pixel in the image is processed during this phase to generate a fairly clean image. These changes, while being made to the image, can be observed in the on-line monitor.

At the beginning of the pre-processing every pixel in the image is quantized into two intensity levels, Pixel_{min} and Pixel_{max}, based on the threshold, ρ,

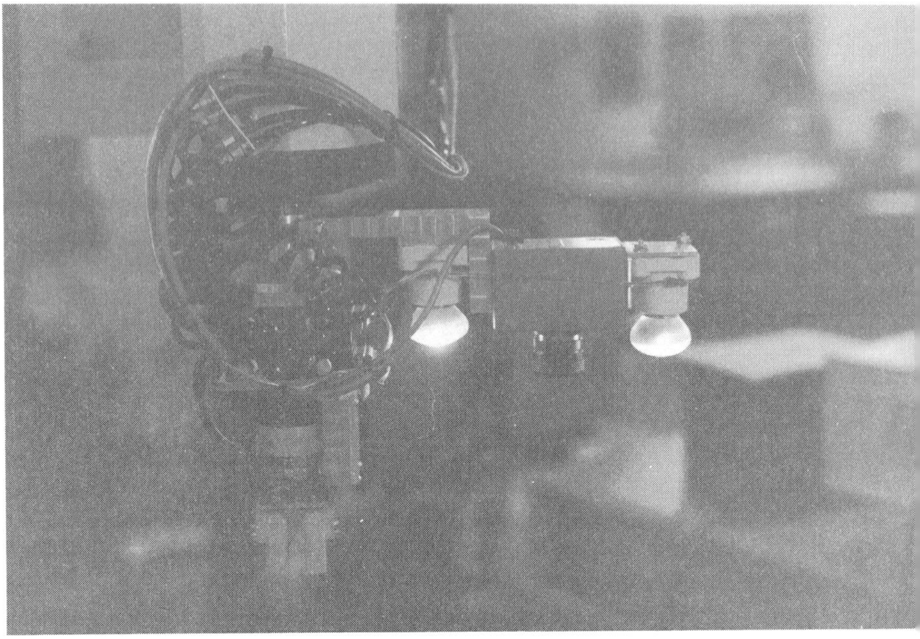

Fig. 9.17 The camera and lights fixed to the end-effector of the robot.

Fig. 9.18 Frozen image of the inventory area.

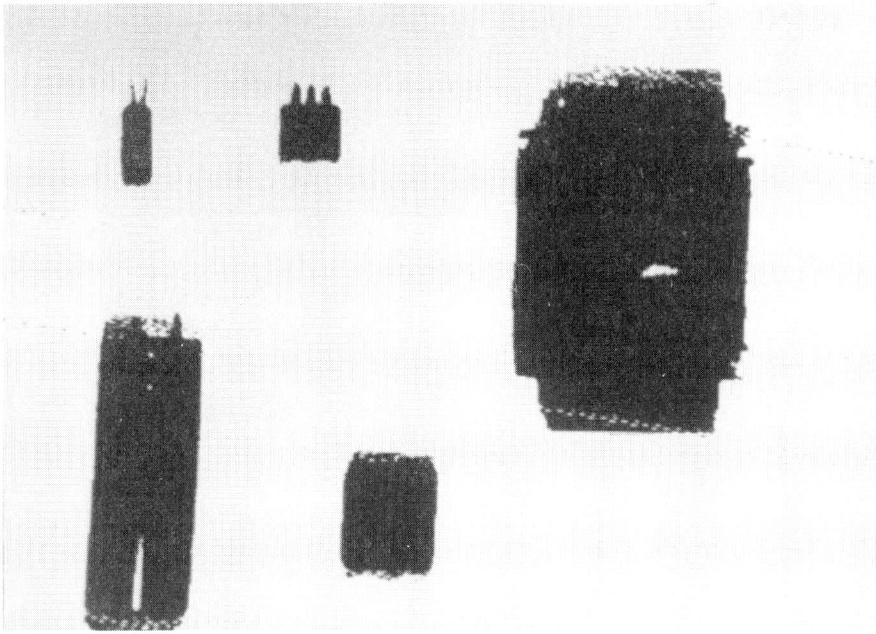

Fig. 9.19 The image after performing a threshold operation.

(Fig. 9.19). For this application Pixel_{min} is set to zero and Pixel_{max} is set to 250; ρ is selected based on the lighting conditions at the time of image acquisition and usually ranges between 95 and 105. After filtering the image with the desired threshold the next step involves noise elimination and pattern enhancement. Since the background intensity is Pixel_{max}, white in color, all the small chunks of undefined patterns with intensity Pixel_{min}, black in color, are changed to Pixel_{max} as part of noise elimination procedure. Similarly all the small chunks of white patterns encapsulated within black patterns are changed to black as part of pattern enhancement. Necessary care is taken not to fill these white holes when they represent an actual hole inside the object. This is determined based on the dimensions of the hole. Noise elimination and pattern enhancement, as explained earlier, is done using window scanning techniques. This is done in three stages: initially, black holes are filled using a window of 3 pixels ($\delta_x = \delta_y = 3$), then white holes are filled with a window of 20 pixels ($\delta_x = \delta_y = 20$) and finally white holes are filled again with a window of 10 pixels ($\delta_x = \delta_y = 10$). The final result of the processing is displayed in Fig. 9.20.

Image analysis

Every pattern in the image is isolated and is passed through the ART1 neural network after performing the necessary reduction in resolution. Just before

Fig. 9.20 The final outcome after image processing.

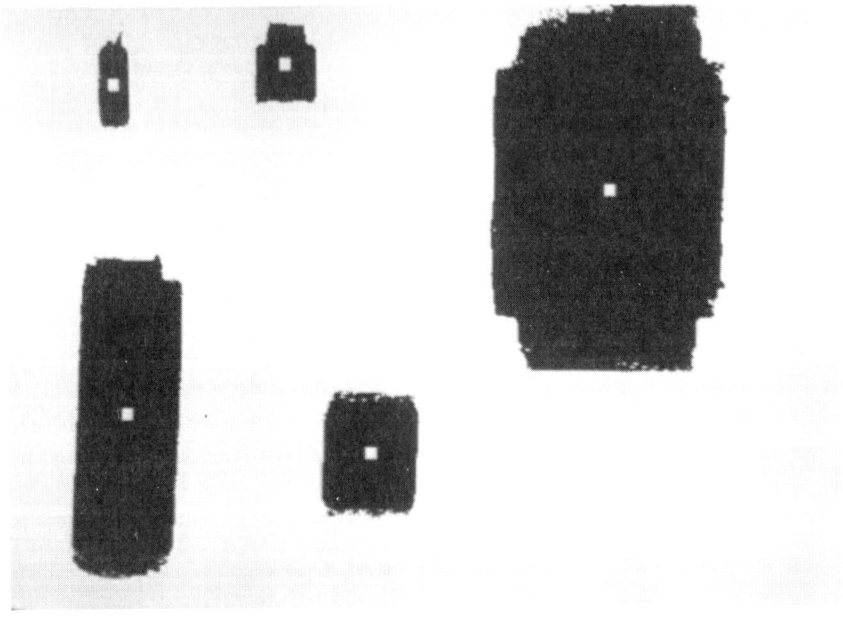

Fig. 9.21 The image after the analysis.

doing the inquiry on the pattern's identity its center is computed and recorded in pixels. The maximum width W and maximum height H (for any pattern in the image) are assigned the value of 200 pixels.

Due to the limitations imposed by the dynamic memory of the computer, the resolution of each pattern is reduced by a factor of two from $W \times H$ and of 40 000 pixels to a grid of $M \times N$ of 10 000 pixels. The binary matrix of $M \times N$ is then presented to the ART1 neural network paradigm to determine the part identity.

The parameters of the ART1 structure used are: 10 000 comparison layer neurons, initially 10 recognition layer neurons, vigilance parameter of 0.89 and deformation parameter of 0.4. The bottom-up b_{ij} and top-down v_{ij} weights within the network are initialized to satisfy the following inequalities.

$$0 < b_{ij} < \frac{L}{L - 1 + (M \times N)} \qquad L > 1 \quad \tfrac{1}{2} < v_{ji} \leqslant 1$$

for all $i = 1, \ldots, 10\,000$ and $j = 1, \ldots, 10$.

Then, the values of b_{ij} for $L = 2$ become

$$b_{ij} = \frac{L}{L - 1 + (M \times N) + 1}$$

$$= \frac{2}{2 - 1 + 10\,000 + 1} = 1.9996 \times 10^{-4}$$

Upper bound of 1 is used for top-down weights v_{ji}.

Figure 9.21, shows the image after performing the analysis. It can be seen that the center of every pattern is marked by a white block. The output of this stage is a database of components represented by their name and location in pixel coordinates.

Coordinate mapping

The center coordinates of every pattern obtained from the previous stage are now ready to be passed through the backpropogation neural network to determine the actual robot coordinates in the workspace. Before passing through the neural network, the camera coordinates are normalized, and the output of the network, which is also in the normalized form, is converted back to the original form to obtain a valid robot coordinate. A two-hidden-layer backpropagation network with two neurons in both input and output layer and 20 neurons in each hidden layer is used for mapping.

The normalization factors for robot and camera coordinates are dertermined such that values are well spread over the range zero through one. This will allow the backpropagation network to converge faster into a minimum error while training. Pixel coordinate values are normalized by $NC_x = 640$ and $NC_y = 480$ and robot cordinates by $NR_x = 150$ and $NR_y = 200$ in X and Y

directions respectively. The *y*-robot coordinate is reduced by 550 units before normalization to achieve well spread of values over the range zero through one. An accuracy of 99.98% is achieved in estimating a robot coordinate given a camera coordinate. The output of this stage is a text file consisting of the names and codes of the components and their corresponding locations in the inventory area.

Preparation for final assembly

At this stage preparation is made to initiate the final assembly process which is a sequence of pick-and-place operations. A component is picked up from the inventory area and placed at its corresponding destination coordinate on the printed circuit board. The information regarding the component coordinates in the inventory area is obtained after the execution of several steps discussed above. The information regarding the destination coordinates of each of these components on the PCB is fixed for a specific circuit layout and assembly sequence is determined prior to the start of the assembly stored as text files.

Assembly process initiation

The robot is set with an appropriate speed and three text files mentioned earlier are opened and the pick-and-place sequence is initiated beginning with the first component listed in the sequence file. The appropriate gripper to handle the first component is loaded unless it is already present. The gripper is then used to pick a component from inventory area with high precision.

Figure 9.22 shows a voltage regulator being picked by using a parallel gripper. Note the precision required in coordinate mapping in order to pick the part, when the clearance between the part and the gripper is less than a millimeter. This part is later placed at the appropriate destination coordinate on the PCB as shown in Fig. 9.23.

Also, note the accuracy in placing the component such that its leads perfectly coincide with the contact terminals on the surface of the PCB. Figure 9.24 shows the importance of determining the assembly sequence to ensure collision free placement of components.

Assembly

The final assembly finishes after performing a sequence of pick-and-place operations. All the components are surface mounted with an acceptable tolerance (Fig. 9.25). The product, is now ready to undergo the soldering process, where the actual connections between the component leads and the metallic conductors on the PCB are made – a process not considered in this case study.

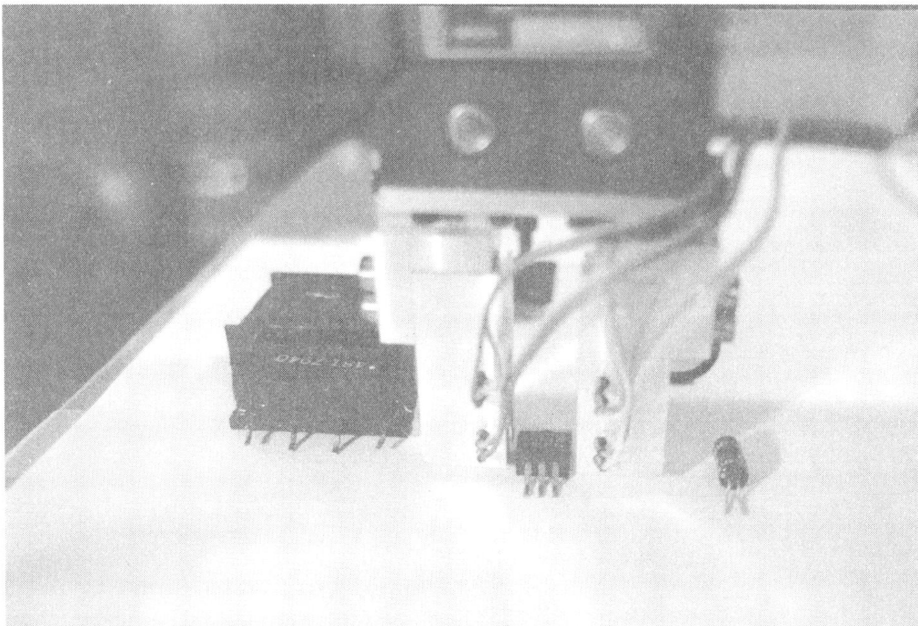

Fig. 9.22 A pick operation on a voltage regulator using a parallel gripper.

Fig. 9.23 The place operation on a voltage regulator using a parallel gripper.

Fig. 9.24 An example of a collision-free component placement.

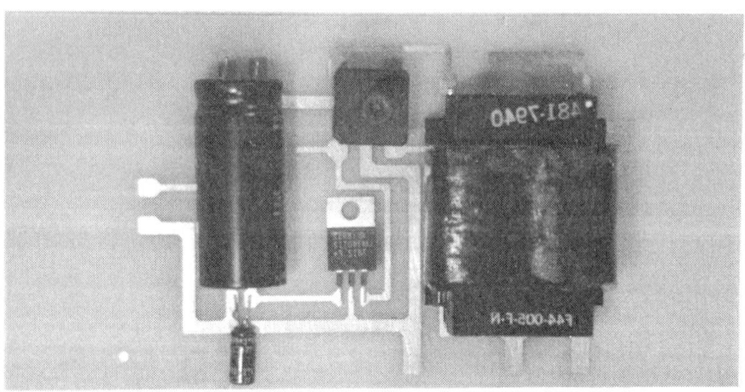

Fig. 9.25 The final product, a fully-assembled power supply board.

9.3.2 Software organization

The software for running the system is developed in C++ using object oriented programming principles. It is structured based on the three key elements of the systems: robotics, vision and artificial neural network simulation. The development of an object-oriented software requires a detailed study of each of these processes and their interactions. After studying the processes thoroughly, it is possible to identify the sub-processes in each of them which are repeatable and may perform the same function. All such processes and functions can then be placed under a common class from where they can be accessed with appropriate access permissions. The property of inheritance in object-oriented programming can also be used to inherit all the necessary functions with the appropriate access control from one class to another. This allows a well-defined and structured flow in the program development and is also useful in preparing comprehensible documentation. However, there are certain disadvantages in programming using an object-oriented methodology. The penalty is execution time, which will be high compared to the relatively unstructured 'C' programming. But the loss in execution time for following a structured approach is worthwhile when considering the huge amounts of time spent in debugging an unstructured program and also the efforts, which are sometimes futile, to expand the software to provide more functionality.

For the current application, the software is developed by creating three main classes (Fig. 9.26): robot control, ART1 neural network, and backpropagation neural network, each consisting of related functions and data. A fourth class, which performs all the machine vision operations, is derived from the ART1 neural network class. This is because the machine vision operations include the ART1 neural network simulations for pattern recognition. The declaration of classes and the functions within the class is done at a higher level, and the header files consist of these declarations. The next stage in the hierarchy is to define all the functions declared within each class.

The definitions strictly limit their operations on the data specified within the class, and the reverse is also true. That is, all the data declared and defined

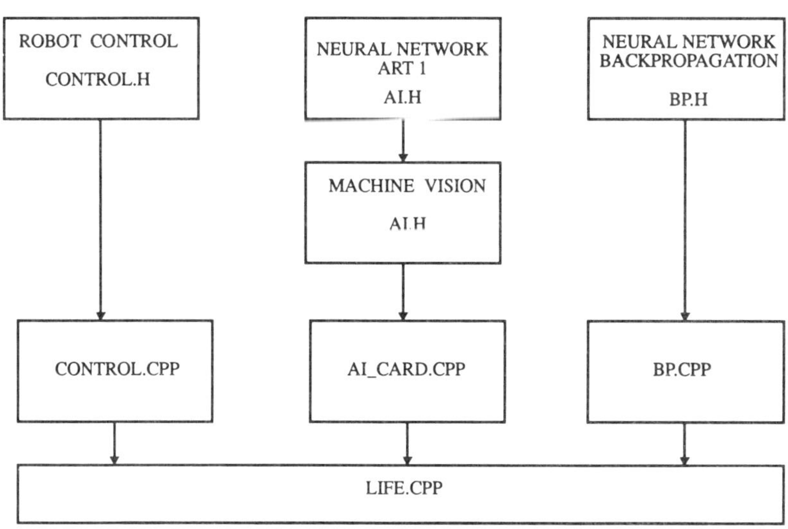

Fig. 9.26 The software organization.

within the class can be accessed only through the functions declared and defined in that class and also from other sources which are given access rights. The class declarations and definitions of functions within each class provide the basic infrastructure for developing any application program – in this case, to perform a high precision and flexible printed circuit board assembly. It is achieved by declaring objects under these classes and accessing all the functions through these objects to perform intelligent assembly tasks. An application specific software can be provided with 'life' by including these classes, which enable the software to 'see' and 'understand' objects through the vision system, to 'move' the robot, and also to 'speak' through the use of an external speech synthesizer (not shown in the figure).

Apart from the main software, other tools are available to assist the user in configuring the assembly system for performing certain key operations. **Calibration tool** is used to switch the vision system into a frozen state and allow the user to move a floating pointer over the frozen image and make note of the pixel coordinates of each crosspoint and later use them for training the backpropagation neural network. Another tool converts the raw data obtained from each crosspoint on the workspace, in pixel coordinates and robot coordinates, into the corresponding normalized data and in accordance with the specifications of the backpropagation neural network training program. A **route optimizer** generates an optimum component placement sequence based on certain heuristics. Also, there are many other utility programs developed by using the functions declared in all three classes. These allow the user to conduct and test various robotic operations like movement in the x, y, z, and roll axes and the fixure controls, lights on/off, and gripper close/open, directly from DOS commands. The tools also allow the user to access the vision system, to take snapshots of certain regions on the workspace, to compress the images, and to store them as ASCII files. Similarly, they allow the user to view any ASCII image file, whose size is specified by the compress factor, and view them on the screen. The tools also enable the user to select one of the eight possible cameras connected to the multiplexer card of the vision system and display the corresponding visual data on the monitor.

9.3.3 Execution and results

The execution of the assembly process is determined by the sequence of commands issued in the final application program. As mentioned earlier, the assembly process is executed in nine stages, but at an abstract level some of the processes can be grouped together, and therefore the assembly process can be considered to mainly consist of five stages: initialization, image acquisition, image pre-processing, image understanding, and assembly process initiation and completion. Also, before using the software for performing the assembly tasks, the sequence of component placement on the printed circuit board

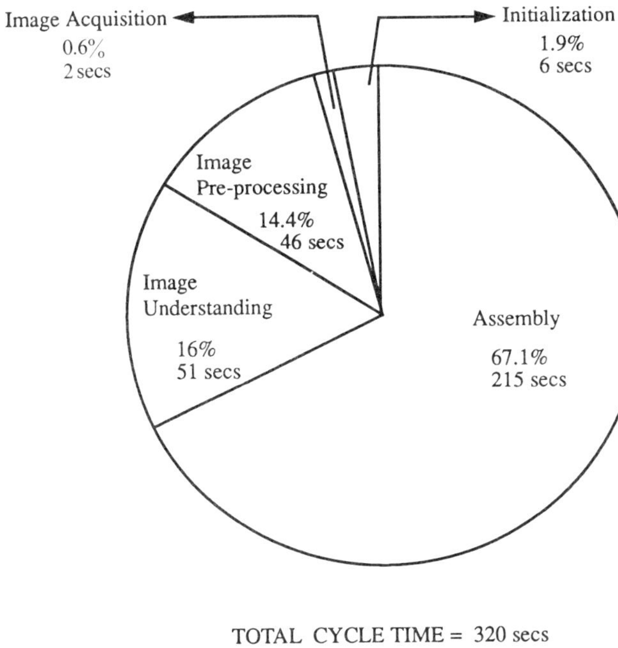

Fig. 9.27 Pie chart representing the cycle time distribution.

should be known. This information is usually supplied during product design stage.

The results of execution are obtained in the form of timing data which are analyzed to determine the bottlenecks in the system. The timing data are obtained by polling the system clock at the beginning and end of every subprocess. Figure 9.27 shows the time taken by each subprocess during the execution of the assembly process of a power supply board. However, the route determination process, being a non-repeatable process which occurs only once when the product design changes, is not shown in the table and accounts for one or more seconds depending on the number of components. In Fig. 9.27, system reset and speech synthesizer activation operations are grouped as initialization, image analysis and coordinate mapping operations are classified as image understanding, preparation for final assembly, assembly process initiation and assembly operation are combined under assembly. Total cycle time of assembly is 320 seconds.

9.3.4 Performance evaluation and improvement

The strategy employed to improve the performance of the system is divided into three stages. First, the processes which contribute to a major share in the cycle time are identified. Second, the tools used to execute these processes are discussed. Finally, at the third stage, the means of improving the performance of these tools to improve the overall performance is discussed.

According to the timing data shown in Fig. 9.27, the final assembly contributes to a fairly large portion of the cycle time. This operation is mainly conducted by the robot. The large operational time is due mainly to the delays introduced by the mechanical motions of the robot in performing the pick-and-place operations of the components during assembly. Due to the laboratory safety regulations, the robot is operated at almost 1/14th of its maximum speed. Assuming that the robot is operated at its maximum speed, the operational time of assembly can be reduced from 215 seconds to almost 15 seconds. However, the 15 seconds operational time may still not be acceptable by many industry standards. In order to reduce this time further, the only alternative is to use a more sophisticated and customized robot for the purpose of assembly. Also, the work area of the robot can be reconfigured such that the components, tools and fixtures can be easily and quickly accessed.

The next major processes which take most of the cycle time are the image pre-processing and image understanding. The image pre-processing mainly consists of changing pixel intensities in an image of 640×480 pixels. This operation results in a clean image which can be analyzed later for distinct patterns. Assuming that a parallel computer consisting of 480 PEs is used to process the image such that each PE operates on 640 pixels sequentially and is synchronized with the rest of the elements, it would take for a 30 MHz system about 21.3 μsecs to complete this process as compared to 46 seconds for a personal computer. Depending on the necessity, more PEs can be used to reduce the total processing time. However, there is a limit of 640×480 PEs that may be used to process the image. The number of PEs can be determined based on cost performance as beyond a certain optimum number any further increase cannot provide an economical increase in performance. During the image understanding phase, ART1 and backpropagation neural network paradigms used for pattern recognition and coordinate mapping tasks, respectively. Though there is an additional time taken for file-handling purposes, most of the execution time during this phase can be contributed by the neural network simulations. Neural networks are another area of application for massively parallel computers. In the case of ART1 the maximum number of processing nodes required at any instant is at the comparison layer and equals 10 000. Similarly, for the backpropagation neural network the maximum number of processing nodes required at any instant is at the two hidden layers, which is 40. Thus, assuming a 30 MHz massively parallel computer which implements both of these neural networks and performs the two operations, pattern recognition and coordinate mapping sequentially, and with an average number of iterations in ART1 being 4, and the number of layers in an ART1 neural network being 2, and that in a backpropagation neural network being 4, the operational time in executing the neural networks can be computed as

$$((4 \times 2) + 4) \times 0.03 \times 10^{-6} = 0.36\,\mu\text{seconds}$$

Assuming that the additional processes which follow or precede the neural network operation during this phase account for 500 μseconds, the total operational time during this phase can be reduced to 500.36 μseconds which is fairly negligible compared to the other processes. Initialization, and image acquisition, are system dependent and cannot be improved further unless a more sophisticated vision system and computer is used. Thus by following the proposed improvements on the system it is possible to reduce the cycle time to 23 seconds.

Apart from the high execution times there are other limitations to the system. The pattern recognition is not rotation invariant. The components in the inventory area have to be placed with either a vertical or horizontal orientation to perform the assembly. This problem can be alleviated by adopting other methods of pattern vector generation, such as feature vectors. The ART1 network used is capable of processing only binary input vectors. The image in this case is represented by a vector consisting of 1s and 0s. For more reliable pattern recognition the input vector should be an analog vector which can represent more information. The ART2 network which *can* accommodate analog input vectors may be used for this purpose. During image acquisition, a homogeneous background is required to produce an image which can be processed for identifying distinct patterns within it. An analog pattern vector generation and the use of ART2 network may also help in overcoming this limitation.

9.4 SUMMARY

The generic automated assembly system described in this chapter clearly demonstrates the fact that it is feasible to achieve high precision and low set-up times in a typical assembly process for small volume production. The case of printed circuit board assembly discussed proves the fact in a limited sense; it is possible to build autonomous systems for manufacturing that can adapt to changes in manufacturing environment through the use of artificial neural networks. The results, though strongly influenced by laboratory limitations, show that the precision in assembly is very high. All the components of the system – vision, artificial neural networks, and the robot – work together with perfect coherence and can reduce the set-up time imposed due to design changes. Through the use of more sophisticated and high speed equipment it is possible to assemble parts for high volume production. The model developed and tested demonstrates the power of integrating neural networks, vision, and robotics in achieving high-precision assembly with low set-up times which is an important competitive edge in today's global economy.

REFERENCES

Aranian, J. (1989) *Computer Integrated Electronics Manufacturing and Testing*, (1st edn) New York: Marcel Dekker, Inc.

Bao, H.P. (1988) An expert system for SMT printed circuit board design for assembly, *Manufacturing Review*, **1**(4), 275–80.

Chang, C. and Young, L. (1990) A simultaneous-mounting process for automated printed circuit board assembly, *International Journal of Production Research*, **28**(11), 2051–64.

Dagli, C.H. (1991) Editorial, *Journal of Intelligent Manufacturing*, **2**, i.

Dagli, C.H., Lammers, S. and Vellanki, M. (1991) Intelligent scheduling in manufacturing using neural networks, *Journal of Neural Network Computing*, Spring, 4.

Dagli, C.H., and Vellanki, M. (1992) A Hybrid Intelligent Architecture for Automated Printed Circuit Board Assembly, in *Lecture Notes in Economic and Mathematical Systems*, (eds A. Jones, T. Gulledge and G. Fadel) Springerverlang.

Funk, J. (1989) Design for assembly of electrical products, *Manufacturing Review* **2**(1), 53–9.

Gavish and Seidmann (1987) Printed circuit board assembly automation – Formulations and algorithms, *Proceedings of the IXth ICPR*, Cincinnati, Ohio, 662–73.

IBM 7547 Manufacturing System Hardware Library: For Authors, Editors, and Copywriters, 1st edition (1984) USA: International Business Machines.

IBM Manufacturing System Software Library – AML/Entry Version 4 User's Guide: For Authors, Editors, and Copywriters. 2nd edition (1985) USA: International Business Machines.

Lynch, M. and Dagli, C.H. (1991) Backpropagation neural network for stereoscopic vision calibration, *SPIE Volume 1615. Machine Vision Architectures, Integration and Applications*, Boston, Massachusetts, 289–98.

Pau, L.F. (1990) *Computer Vision For Electronic Manufacturing*, Plenum Press, New York.

The QuickCapture User's Manual: For Authors, Editors, and Copywriters, 2nd ed. (1989) Marlboro: Data Translation, Inc.

Simpson, P.K. (1990) *Artificial Neural Systems – Foundations, Paradigms, Applications, and Implementations*. Pergamon Press, Inc., New York.

Steudel, M. (1987) Fully automatic flexible assembly of an electromechanical timer group, *International Journal of Production Research*, **25**(7), 1029–44.

Taylor, D. and Graves, R.J. (1990) An examination of routing flexibility for small batch assembly of printed circuit boards, *International Journal of Production Research*, **28**(11), 2117–35.

Vellanki, M., and Dagli, C.H. (1993) Artificial neural network approach in printed circuit board assembly, *Journal of Intelligent Manufacturing*, **4**, 109–19.

Vellanki, M., and Dagli, C.H. (1992) Automated precision assembly through neuro-vision, *Proceedings of Applications of Neural Networks III*, SPIE, Orlando, Florida, p. 49–504.

CHAPTER 10

Manufacturing feature identification

Mark R. Henderson
Associate Director, Computer-Integrated Manufacturing Systems Research Center, Associate Professor of Mechanical/Aerospace Engineering, Arizona State University, Tempe, Arizona 85287-5106.

10.1 CHARACTERISTICS OF MANUFACTURING FEATURES

Integration of CAD/CAM involves the use of a design description, design database model and applications while ensuring data consistency. The data types needed in a variety of applications may be different from the data type stored in a design database. In order to automate the CAD/CAM process, computer understanding of the meta-knowledge in the database or description and automatic data conversion to the next analysis stage are necessary (Woodwark, 1988).

Engineering considerations arise quite frequently from geometric factors. **Geometrically complete models** have been available for some time, but they always lack some explicitly represented data which, for some engineering applications, are only implicitly embedded in the model. It is because these application considerations depend on how we want to use the geometry, that the geometry does not necessarily incorporate the pattern properties. The computer, therefore, must figure out the implicitly embedded patterns. For example, in manufacturing analysis, the machining patterns for material removal and the face types for surface finishing must be recognized. Also, if we want to partition a part for finite element analysis, the concave and convex corners and edges matter. For display of wire-frames and calculation of mass

Artificial Neural Networks for Intelligent Manufacturing. Edited by Cihan H. Dagli. Published in 1994 by Chapman & Hall, London. ISBN 0 412 48050 6

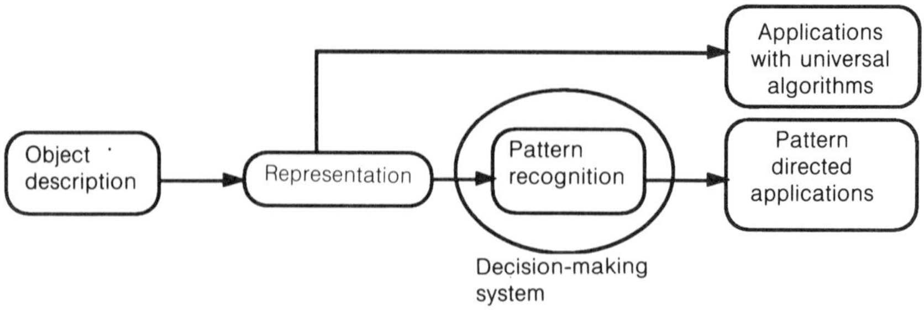

Fig. 10.1 The role of pattern recognition in CAD/CAM integration.

properties, the recognition process may not be needed. In general, if there is a universal method to deal with the object as a whole, no pattern recognition is needed. However, if the designed object is considered as a collection of some specific process-related features, pattern recognition is useful. For an automatic link between the geometric data and certain engineering considerations, shape pattern recognition plays a crucial role in a decision-making system. An illustration of a CAD/CAM integration schema is shown in Fig. 10.1.

Several modeling schema can be used for representation of three-dimensional solid geometry in computers. The solid modeling database is a design database which is converted from a designer's description with complete geometric information for a solid object. The two predominant solid modeling representations are:

1. **constructive solid geometry** (CSG), stored as a tree of primitive volumes with their respective boolean operators used to construct the solid geometry; and
2. **boundary representation** (B-rep) containing the boundary elements of the resultant object.

The B-rep database stores low level primitives such as faces, edges and vertices (FEV), and their topological relationships. For the explicit representation of adjacency relationships of boundary elements, the robustness of a B-rep justifies its usefulness in industrial applications and, consequently, this kind of solid modeling is popular in today's industry.

Current representations in product modeling of 3D objects for manufacturing are based on wireframe, surface, and solid models. The biggest difference between solid representations and surface or wire-frame representations is that solid modelers have sufficient topological information. The database used in the recognition typically is the boundary representation (B-rep). B-rep data structures contain two kinds of entities: geometric and topological. The geometry represents information about the geometric shape of an object such as the exact geometric location. The topology describes the relationships among geometric entities. Some general problems exist in representing the product model completely and succinctly.

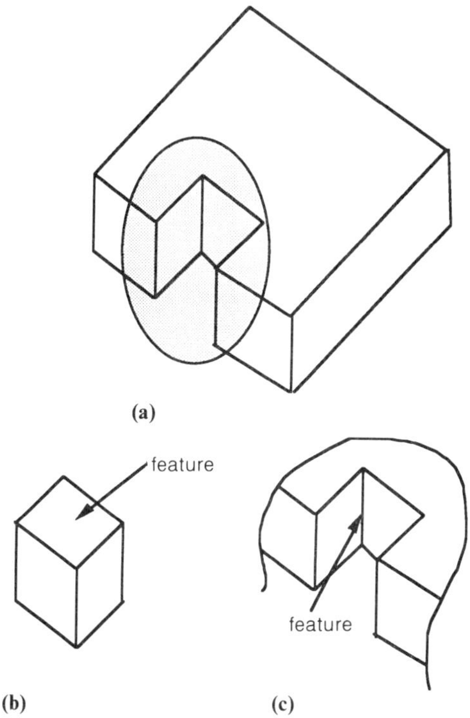

Fig. 10.2 Different types of features: (a) shaded area in an object; (b) manufacturing feature; (c) finite element feature.

A feature-based product description is one proposal to solve the problem, where features may be defined as a labeling scheme for part components or sets of components. A feature is typically defined as a set of faces, edges, and vertices. Features are useful because they represent a semantic interpretation of the model by adding connotations to the sets of entities (Fig. 10.2). A pocket, for example, consists of a set of faces usually defining a depression with straight sides. Attributes of a pocket include the number of sides, the depth and the dimensions of the pocket opening. A pocket, or other feature, can be mapped directly to an application such as process planning, N/C machining or fixture design. Having features in a model allows automation of these tasks.

10.1.1 Traditional feature identification methods

All previous work in feature recognition can be broadly classsified into human-assisted or automatic feature recognition. In **human-assisted feature recognition** systems, there is considerable human intervention in all stages of the recognition process. In **automatic feature recognition** systems, the recognition and extraction stages are completely automated. Automatic feature recognition algorithms can be further classified into

- machining region
- rule-based
- graph-based
- CSG-based
- application-based algorithms.

In order to recognize features from a B-rep model automatically, many different techniques have been used. They can be grouped into three general approaches: rule-based, graph-based, and neural net recognition. Four measures of recognition efficacy are given.

- The patterns are easy to define to the recognizer.
- Recognition is fast.
- The correct features are recognized with the correct labels.
- The recognizer can handle partial or interfering features.

The **rule-based approach** (Kyprianou,1980; Stately *et al.*, 1983; Henderson and Anderson, 1984; Hwang,1988; Vandenbrande, 1990) uses AI techniques to develop a set of feature rules. Rules for features are written by determining the necessary and sufficient conditions for the feature of choice and expressing them in a logic statement. The basic format for a rule-based system, a hole in this case, is

IF an inner loop is found, and
  the inner face is a closed concave cylinder, and
  ⋮
THEN these faces comprise a flat-bottom-straight-hole

Rule-based algorithms identify a feature based on certain prespecified rules that are characteristic to the feature. Disadvantages of the rule-based approach are:

- rules are non-unique to a feature;
- rules cannot be devised for every conceivable feature; and
- recognition involves repeated exhaustive searches of the solid model which takes time if a part is complicated.

The work of Kyprianou (1980), Jakubowski (1982), Henderson (1984), Choi, Barash and Anderson (Choi *et al.*, 1984), and Hwang (1988) are representative examples of this category.

Graph-based systems (Joshi and Chang, 1988; Sakurai and Gossard,1988; Chuang, 1991) require matching the feature graph to the appropriate subgraphs in a solid modeler database. An object and its graph are shown in Figs 10.3 and 10.4, respectively. The diagram on the right-hand side of Fig. 10.4 is the graph of an object and the one on the left-hand side is the graph of a feature pattern. It is said a feature is recognized if the pattern $(F_2, F_7, F_8, e_1, e_2, e_5)$ matches the feature pattern.

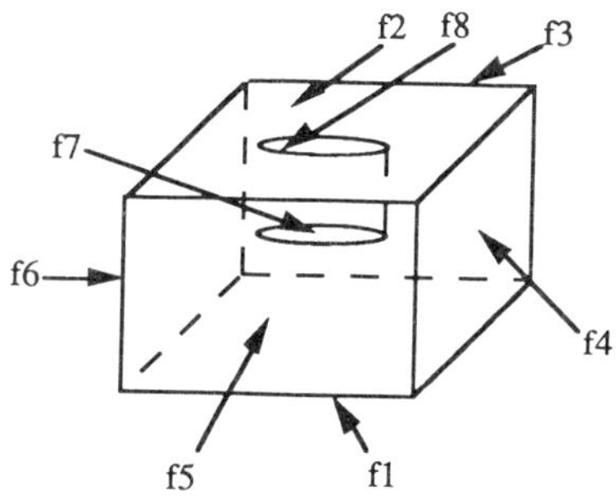

Fig. 10.3 A labeled object.

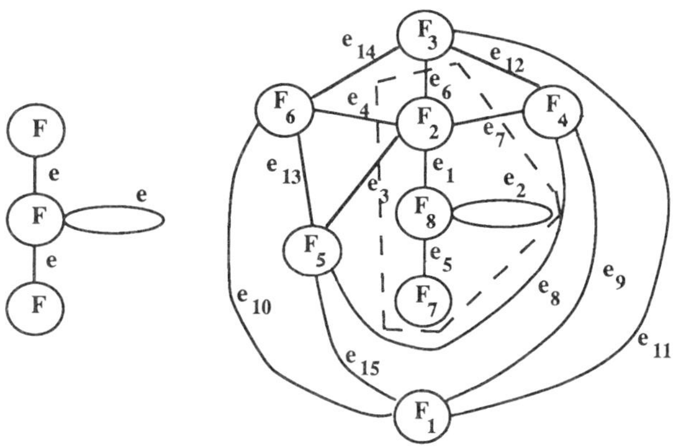

Fig. 10.4 An example for a graph-based system.

Graph-based algorithms organize the B-rep of an object into a graph structure. These graphs can have either faces, edges or vertices as nodes and the other two entities as arcs. During the recognition process, these graphs are split into subgraphs using well-developed graph manipulation algorithms. The main disadvantage of graph-based techniques is that extensive preprocessing is necessary to construct the graphs and additional computation is necessary to extract the feature subgraph from the rest of the graph. The algorithms of Ansaldi (1985), Joshi and Chang (1988), Sakurai and Gossard (1988), DeFloriani (1989), Ganu (1989), Chuang (1991) and Gavankar (Gavankar and Henderson, 1990) are a few examples of this class.

Problems in the above-mentioned recognition techniques are described as follows.

- Rule-based systems have no formal mechanism for rule definition. The rules are typically *ad hoc* and depend on the user's view of the feature.
- Graph matching is an exact process and, therefore, graph-based systems do not have the ability to look for interfering features.

- Because they rely on graph topology, graph-based systems ensure only topological equality.
- Both graph-based and rule-based systems perform exhaustive searches for matching their patterns. The execution time will increase rapidly if an object is very complicated.
- Neither graph-based nor rule-based systems have the ability to learn or to improve their behavior dynamically.
- None of these techniques can recognize partial features or incomplete features.

10.1.2 Contributions of artificial neural networks

It is only recently that neural nets have been applied to recognition of features in product models. The results are limited, therefore, but the potential has been demonstrated. Three research approaches will be discussed: one approach for 2D features from engineering drawings and two approaches for 3D features from solid models.

Product models are usually represented by either engineering drawings or solid models. Feature-based models can be created by two methods: feature-based design and feature recognition. **Feature-based modeling**, otherwise called 'design by feature' uses primitive features to create or modify objects. Automated **feature recognition** labels features by reasoning about the geometry and topology information. It has been shown that 'design by feature' is not sufficient for complete automation of product analysis (Shah, 1991). Feature recognition is important for linking design and downstream applications, and even with feature-based modeling, it is necessary for the system to recognize features (Cutkosky *et al.*, 1988). Feature recognition enables users to generate many different application-specific views, addressing applications that the designer could not originally foresee (Ames, 1991). Neural nets have the advantage in recognition that they can be taught using exemplars (which makes setting up the recognition patterns straightforward) and they can recognize partial features using floating point outputs. A block diagram is shown in Fig. 10.5.

2D feature identification

Recognition of features from engineering drawings is important because most product models have sometime along their life existed as 2D drawings. Millions of these drawings are in drawers in design and manufacturing

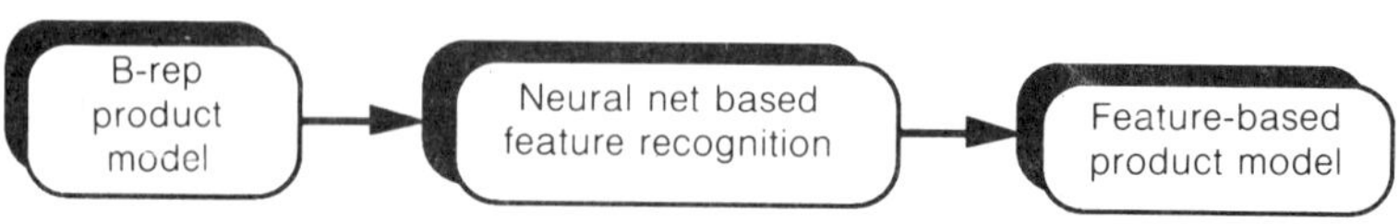

Fig. 10.5 The simple model for the neural net.

Fig. 10.6 A two-dimensional multi-segment profile.

locations and could be directly converted to computer-based models if they can be processed into feature-based or solid models in 3D. Or, if they are to remain two-dimensional, the features such as dimensions, holes, slots, chamfers, etc., can be recognized and used for automated process planning or other applications. As Peters (1991) mentions, recognition is essential for:

- feature based analysis over assemblies containing wireframe and solid modeled components; and
- transfer to design knowledge captured in existing wireframe data bases into emerging solid models of next generation products.

Peters was one of the first to attempt recognition of 2D features from engineering drawings using neural nets. Two-dimensional profiles of shapes are represented as sets of lines and curves. These profiles are then subdivided into a connected loop of edges, E*i* (Fig. 10.6). Features are characterized by the triplet (C*i*, A*i*, L*i*) where C*i* is the curvature of the *i*th element, A*i* is the interior angle of the *i*th curved segment and L*i* is the arclength of the *i*th segment. The input to the neural net is a vector of triplets for a given profile.

Peters' system schematic is shown in Fig. 10.7. The net used in this case is from Nestor, Inc. Future work will include more complex profiles than straight lines and arcs and still also investigate whether these types of measures will be extensible into 3D.

3D feature identification

Three-dimensional feature recognition adds surfaces and three more degrees of freedom to the primitives and their connectivities. The essence of 3D recognition tries to solve the problem of finding a pattern of faces and edges

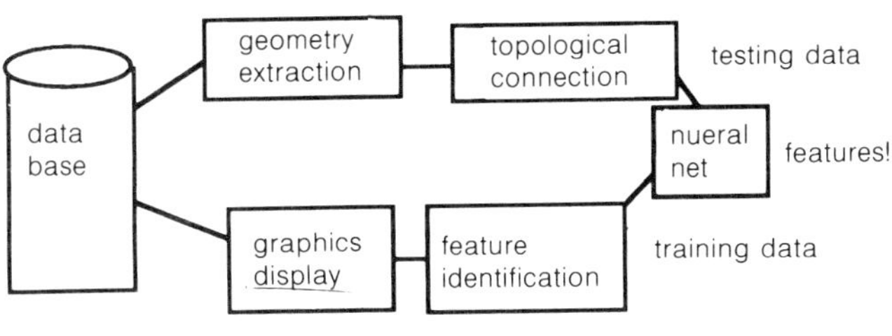

Fig. 10.7 Two-dimensional neural net recognition of features.

that satisfy some geometry and topology properties. The results so far from two research efforts (Hwang, 1991; Prabhakar, 1990) show that neural nets have the advantage that the exemplars are simple to construct, the nets are easy to train and the recognition speed is much better than other methods. The following section discusses briefly the first known effort in 3D feature recognition from solid models using neural nets. Later, the more detailed discussion will be based on work by Hwang using a perceptron network.

Prabhakar (1990), in the first attempt known to use neural net techniques for solid model feature recognition, developed a system called FRENET (*f*eature *r*ecognizer and *e*xtractor based on a *n*eural net *e*valuation *t*echnique), and proposed a five-layer quasi-neural net. A face-adjacency matrix containing 8 digit codes for all faces in the part was input and the codes were calculated based on the morphological content of the faces. The training method of this net is not receptive to traditional neural net training – during one training session, the trained feature was presented only once, and the weights and other parameters set at that time. Recognition was done for holes, slots and pockets, but limitations were found. For example, the output was binary, so no partial features could be recognized. Also, training was explicit for a particular specific feature; the net could not generalize a feature pattern.

The adjacency matrix, which was described in the previous section, is a two-dimensional array of integer vectors with each row giving information about one face of the part. Each element of the array, i.e. one of these vectors, gives information about a face (according to the row) and its relationship to another face. It is assumed that each vector has a length of 8, and edge type and face type occupy positions 1 and 2 respectively. (In this example, a plane face is denoted by the number 1, a cylindrical face by 2 and a circular edge by 4.) It is also assumed that another position, say 3, indicates if a particular vector is a diagonal element of the adjacency matrix (1 if it is a diagonal element, 0 if not) – in which case it gives information about the face itself. The other (non-diagonal) vectors in the same row give information about its relationship to the other faces in the part.

For the example being considered, the algorithm must search for a cylindrical face adjacent to two plane faces. This means that it must search for a row that contains the three vectors corresponding to the three feature rules. The first vector must have a 1 in position 3 (diagonal element) while the other two must have a 0 in this position. The first vector must also have a 2 in position 2 (cylindrical face), while the other two must have a 1 in this position (plane face). Also, the second vector must have a $+2$ in position 1 (convex circular common edge) while the third vector must have a -2 in this position (concave circular common edge). So the three vectors to look for are $(1,2,1,x,x,x,x,x)$, $(2,1,0,x,x,x,x,x)$ and $(-2,1,0,x,x,x,x,x)$ where x indicates that the values in these positions do not matter in the recognition of the feature. It is assumed that the part will be modeled on the ROMULUS solid modeler in which a cylindrical face is adjacent to itself, with a straight smooth common edge. A smooth edge is assumed to be convex by the program. Therefore the first vector must have a $+1$ in position 1. If a row is found that has three vectors satisfying these criteria, then a flatbottomed blind hole has been recognized. It must be noted that the two faces need not be adjacent to each other. For example, a rectangular slot can be defined such that one of the parallel faces is the primary feature face, and the other is a secondary feature face. A primary face is defined as a face which itself 'defines' the feature and is usually one to which all other feature faces are adjacent; these secondarily adjacent faces are hence known as secondary feature faces.

A feature is recognized if the recognizer can detect the presence of the sequence of codes corresponding to the feature in any row. Each row represents information about one face of the part and its relationship to all the other faces in the part, but this information is lost when the next row is processed. So the recognizer can recognize only those features which can be defined with respect to one face, the primary feature face. The other constituent faces of the feature are called secondary feature faces. A primary feature face and a secondary feature face can be formally defined as follows.

- **Primary or principal feature face** is the face that characterizes the feature. In other words it is the one face with respect to which the topological and geometric relationships of all other feature faces are described. With the present implementation, there can be only one principal feature face in a feature type.
- **Secondary feature face** is a face which comprises a feature but is not the principal feature face. Its topological and geometric relationships are described with respect to the principal feature face. A secondary feature face need not be adjacent to the primary feature face.

Relationships among the constituent faces of a feature must be restricted to relationships between the principal feature face and the secondary feature faces. In other words, the codes that indicate the presence of a feature must all lie in a single row of the adjacency matrix. This restriction was imposed in order to simplify the design of the net. This is not a fatal restriction, as most of the well-known features can still be recognized. Table 10.1 gives a list of some commonly used features and the capacity of the net to recognize these features. The terms 'easy' and 'difficult' to identify pertain to the degree of difficulty of identifying the primary and secondary feature faces, and the relationships between them that needs to be communicated to the recognition system. In the case of features characterized as 'easy to identify', the choice of the primary feature face is obvious. In the case of 'difficult to identify' features, the choice is difficult, but not impossible. In most cases of such features, there is more than one choice, and the performance depends on the choice. Features which have identical topological relationships between their feature faces but different in their relative sizes also fall into this category. The table indicates the reason for the characterization.

Table 10.1 Domain of neural net-based recognition system

S. No.	*Major Category*	*Feature*	*Comments*	*E*	*D*
1	Macroholes	Countersink		•	
		Counterbore	Cannot dist. from spotface	•	
		Spotface	Cannot dist. from counter-bore	•	
		Straight through hole		•	
		Slant through hole		•	
		Straight flat bottomed hole		•	
		Slant FB hole		•	
		Straight conical bottomed hole		•	
		Slant CB hole		•	
		Centerhole		•	
2.	Slots	Long narrow through going slot		•	
				•	
		Through slot		•	
		T-slot		•	
		Dovetail slot		•	
		V-slot		•	
		Circular slot			•
3.	Pockets	Rectangular pocket		•	
		Tringle pocket w/o rounded edges		•	
		Triangle pocket with rounded edges	Will not extract all faces		•
		Cylindrical pocket	Will not dist. from FB hole	•	
		Complex pocket	Will not extract all faces		•
4.	Cutouts	Circular cutout			•
		Triangular cutout		•	
		Rectangular cutout			•
		Irregular cutout		•	
		D-shaped cutout		•	
5.	Grooves	Seal groove	Will not dist. from circular slot		•
		Non-closing angular channel groove		•	
6.	Keyways	External keyway	Will recognize as straight slot		•
		Internal keyway	„	•	

Table 10.1 (*Contd.*)

S. No.	*Major Category*	*Feature*	*Comments*	*E*	*D*
7.	Bevels	Corner bevel	Cannot dist. from other bevels	*	•
		Side bevel 1	"	*	•
		Side bevel 2	"	*	•
8.	Interfering features	Intersecting slots (without merge)	Will recognize as 4 slots	•	
		Intersecting slots (after merge)		•	
		Hole through corner		•	
		Hole through edge		•	

*Easy to detect bevels but not the particular type of bevel
E Easy to identify feature
D Difficult to identify feature
dist. = 'distinguish'

Intersecting features can be recognized to some extent by the neural net. The output and performance of the net depends on the representation of the feature in the solid modeler and the user's definition of the feature.

Recognition of a feature requires creation of an input format to a neural net which conveys information about faces, edges and vertices to the net in a convenient and useful way. Nets typically accept integer values, especially when they are used to recognize pixel patterns in image analysis. The solid model is a complex graph structure. The challenge here is to convert that structure to a format suitable for net input. The solution described here is a vector of floating point face scores which summarize the important properties of the faces in the model and in the feature.

The most important information for recognition is the face geometry and topology. Kyprianou (1980) used morphological content of a face to analyze the characteristics of a geometric pattern.

10.2 SAMPLE PROBLEM DEFINITION

A feature can be defined in terms of the topology and geometry of its faces. This means that if a set of faces that satisfy some predefined relationships and possess some predefined characteristics is found in the test part, then the feature type defined by these relationships and characteristics has been recognized. This is best explained with an example. If a cylindrical face is found in the part that is adjacent to two plane faces and the common edges are circular, one concave and the other convex, then a flatbottomed blind hole

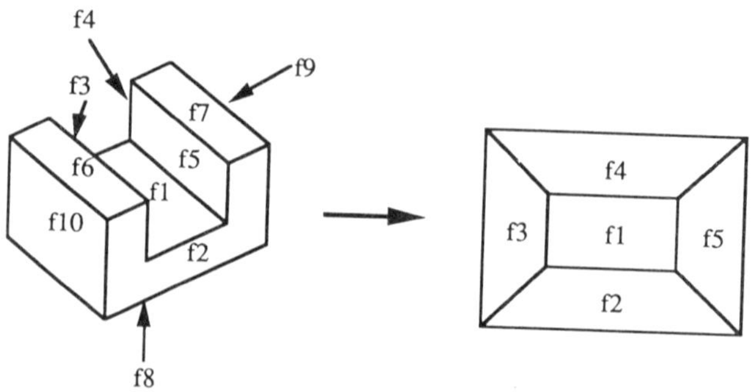

Fig. 10.8 A two-dimensional viewpoint.

has been recognized. Here the characteristics are the face types and the relationships are the adjacency and the common edge information. The above conditions are the feature rules for a flat-bottomed blind hole.

Hwang used a set of floating point face attributes to evaluate the topological and geometric pattern in an object. Both an object and a feature in a B-rep data structure consist of a set of faces. Each face has n neighboring faces. In order to understand the relationships between each face and its n neighboring faces, it is possible to convert a three-dimensional object into a two-dimensional face set. A simple example is illustrated in Fig. 10.8. The two-dimensional face set is based on the 'primary feature face' $f1$. Each face in this object has the same structure as $f1$.

A face definition consists of a face geometry and a set of bounding edges and vertices. If a value is assigned to edges and vertices based on their geometric and topological information, then these values can be converted to a face score which includes face geometry information together with the information about the edges and vertices on the face. The evaluation formula can be written as

$$F_s = f(F_g, E_g, V_g, A_t) \tag{10.1}$$

where F_s is the face score, F_g is the face geometry information, E_g is edge geometry information, V_g is vertex geometry information, and A_t is the adjacency among faces, edges, and vertices.

The face score is a measure of face complexity based upon the convexity or concavity of the surrounding region. A high measure of complexity, that is, a high face score is assumed to reflect the presence of a 'region of potential interest' or a 'feature'. The basic idea is that the number of concave edges and concave vertices a region (one or more faces) has, the higher the score because concavity tends to indicate material removed from the region or material added to the area surrounding the region. Therefore, the reasoning goes that a higher face score indicates a likely feature face which in turn indicates removal or addition of material. It is not always true that a feature is the presence of material addition or deletion. However, in this first try at neural

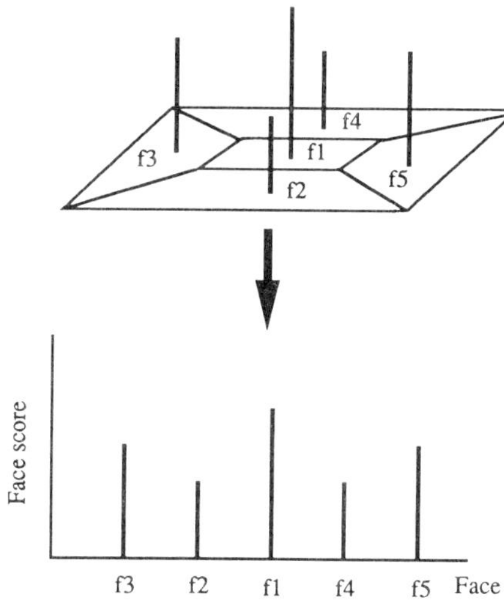

Fig. 10.9 The formation of a face score vector.

net recognition it seemed that the face score calculated this way may help the net to recognize our concept of a feature.

After face scores are evaluated for all faces in the example object, a face score graph can be drawn as in Fig. 10.9, representing the relationships of face scores between a face and its neighboring faces. The collection of face scores (the lower figure in Fig. 10.9) describes the local (the surrounding region of $f1$) behavior and can form a face score vector.

Since each face in an object has a certain face score, a non-zero difference between a face score and its neighboring face score indicates a geometric or topological change between these faces, which form a region, and the region may be defined as a feature. Therefore, based on the viewpoint of the face score, a feature may be defined as the region, comprised by faces of an object, characterized by a specific face score gradient. In other words, a feature is based on a set of face score *changes*. The feature definition depends on identification of the primary face of a feature which is defined as the face with the highest score in the defined region.

10.3 DEVELOPMENT OF NETWORK ARCHITECTURE

A schematic of the feature recognizer is shown in Fig. 10.10. The idea for neural net feature recognition is an attempt to represent a feature by using a numerical form of the relationship:

$$\text{Feature} = f(F_{GT}, W) \tag{10.2}$$

where F_{GT} is the geometry and topology information for a face, and W is the weighting function for the net.

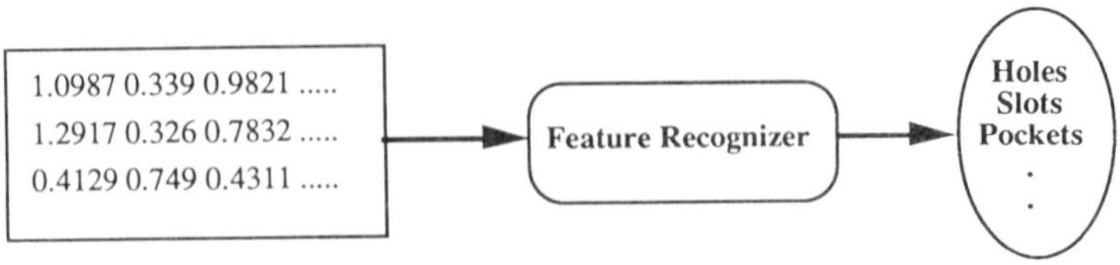

Fig. 10.10 The status of feature recognizer.

The weighting function is the vector of weights assigned to the nodes of the net and can be obtained through training the neural net. Two factors, gain (η) and threshold (θ), affect the convergence and the result of a training algorithm. Formula (10.2), then, may be modified as

$$\text{Feature} = f(F_{\text{GT}}, W, \theta, \eta) \tag{10.3}$$

10.3.1 Network architecture

The net structure of the perceptron is shown in Fig. 10.11.

Overall, the training algorithm attempts to find a suitable weight vector which gives the appropriate output feature value. If such a weight vector exists, the trained patterns are said to be **linearly separable**. The detailed training steps are described as follows.

During training, several sets of face scores are repeatedly input to the neural net. The user can use as many trained patterns as she/he wants. In the trained patterns, a partial feature is optional, but a feature pattern and a non-feature pattern are required. A user might want to use one partial feature as the trained pattern and then try to recognize a different partial feature with a different confidence factor, where the confidence factor represents the distance between the feature and trained pattern. The training procedure for the perceptron can be described in five steps.

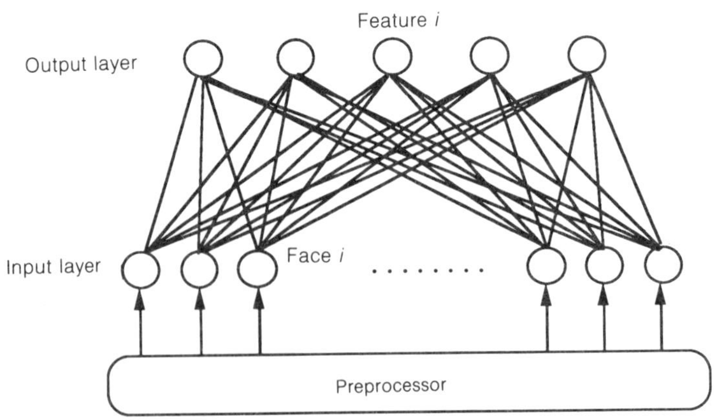

Fig. 10.11 The net structure of a perceptron.

1. Initialize all weights.
2. Calculate the actual net output using the formula

$$\text{Feature}_j(t) = \sum_{i=0}^{N-1} W_{ij}(t)F_i - \theta \tag{10.4}$$

   where W_{ij} is the weight from face$_i$ to feature$_j$, F_i is the score of face$_i$, θ is the threshold in the output neuron, and t is the number of trained exemplars.
3. Adjust weights between features and faces.

$$W_{ij}(t+1) = W_{ij}(t) + \eta[d(t) - \text{Feature}_j(t)]F_i(t) \tag{10.5}$$

   where η is a gain term and $d(t)$ is the desired correct output at the t times.
4. If all patterns are presented, go to Step 5 – if not, go to Step 2.
5. If the actual output is not close to the desired output, go to Step 2; if it is, stop training. In other words, stop adjusting weights and store the necessary information, such as weights and a gain factor, into a database because the weights have converged.

The perceptron training procedure is a kind of gradient descent algorithm (Duda and Hart, 1973). The basic gradient algorithm starts with arbitrarily chosen weight vector $\mathbf{W}_0$ and computes the gradient vector of $\mathbf{J}(\mathbf{W}_0)$, where $\mathbf{J}(\mathbf{W}) = \sum(-\mathbf{W}^T\mathbf{F})$ in the perceptron algorithm. The next value $\mathbf{W}_1$ is obtained by moving some distance from $\mathbf{W}_0$ in the direction of steepest descent, i.e., along the negative of the gradient. In general, $\mathbf{W}_{i+1}$ is obtained from $\mathbf{W}_i$ by the algorithm

$$\mathbf{W}_{i+1} = \mathbf{W}_i - \eta\nabla\mathbf{J}(\mathbf{W})$$

where η is a positive scale factor that sets the step size. If the size is small enough, a weight vector will converge to a solution minimizing $\mathbf{J}(\mathbf{W})$.

The overall purpose for this training algorithm is to find an appropriate weight vector which has the desired projection on the trained patterns. A simple example is described in Fig. 10.12. Figure 10.12(a) contains an initial weight vector ($\mathbf{W}_i$) and two trained pattern vectors ($\mathbf{F}_1, \mathbf{F}_2$). After it is trained for the first time, the weight vector is adjusted to $\mathbf{W}_1$. $\mathbf{W}_2$ is produced when it is trained for the second time.

An output can be thought of as the length of the projection from the weight vector to a trained pattern vector. In other words, the desired output has the same meaning as the desired projection. The training is terminated when each actual projection is almost equal to its desired projection from a trained pattern.

The perceptron learning scheme was proposed by Frank Rosenblatt (1958). The general proof of its convergence is known as the **Perceptron Convergence Theorem**. The difference between adjusted output and desired output is the error before training, and the difference between actual output and desired output is the error after training. To make sure the training is convergent, the

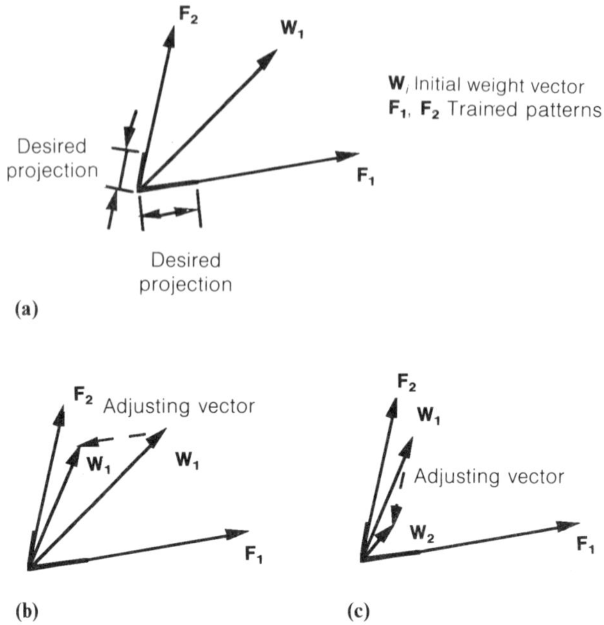

Fig. 10.12 Training steps: (a) initial weight before training; (b) the first training; (c) the second training.

gain factor η must fit the condition.

$$\left|\frac{\text{the error before trained}}{\text{the error after trained}}\right| = \left|\frac{\text{adjusted output} - \text{desired output}}{\text{actual output} - \text{desired output}}\right| < 1$$

or

$$\left|\frac{\mathbf{W}^{\mathrm{T}}_{\text{new}}\mathbf{F} - \theta - d}{\mathbf{W}^{\mathrm{T}}_{\text{old}}\mathbf{F} - \theta - d}\right| < 1 \tag{10.6}$$

where $\mathbf{W}_{\text{new}}$ and $\mathbf{W}_{\text{old}}$ are the new and old weight vector, respectively, and $\mathbf{F}$ is the face score vector of the trained pattern. From the formula above

$$\mathbf{W}_{\text{new}} = \mathbf{W}_{\text{old}} + \eta(d + \theta - \mathbf{W}^{\mathrm{T}}_{\text{old}}\mathbf{F})\mathbf{F}$$

Therefore,

$$\begin{aligned}\left|\frac{\mathbf{W}^{\mathrm{T}}_{\text{new}}\mathbf{F} - \theta - d}{\mathbf{W}^{\mathrm{T}}_{\text{old}}\mathbf{F} - \theta - d}\right| &= \left|\frac{\mathbf{W}^{\mathrm{T}}_{\text{old}}\mathbf{F} + \eta(d + \theta - \mathbf{W}^{\mathrm{T}}_{\text{old}}\mathbf{F})\mathbf{F}^{\mathrm{T}}\mathbf{F} - \theta - d}{\mathbf{W}^{\mathrm{T}}_{\text{old}}\mathbf{F} - \theta - d}\right| \\ &= \left|\frac{\mathbf{W}^{\mathrm{T}}_{\text{old}}\mathbf{F} - \theta - d - \eta(\mathbf{W}^{\mathrm{T}}_{\text{old}}\mathbf{F} - \theta - d)\mathbf{F}^{\mathrm{T}}\mathbf{F}}{\mathbf{W}^{\mathrm{T}}_{\text{old}}\mathbf{F} - \theta - d}\right| = |1 - \eta|\mathbf{F}|^2| < 1\end{aligned} \tag{10.7}$$

If there are n trained patterns, then for cyclical training, the equation can be rewritten as

$$|(1-\eta|\mathbf{F}_1|^2)\cdot(1-\eta|\mathbf{F}_2|^2)\cdot\cdots\cdot(1-\eta|\mathbf{F}_{n-1}|^2)\cdot(1-\eta|\mathbf{F}_n|^2)|<1$$

or

$$\left|\prod_{i=1}^{n}(1-\eta|\mathbf{F}_i|^2)\right|<1 \tag{10.8}$$

After the sample patterns are trained, the information needed for further recognition will be stored in a feature database. The information stored in a feature database should include at least the feature name, the size of the weight vector, the weight vector values, the threshold, and the gain factor of the trained feature. For instance, the information of a through hole, which has to be stored in the feature database, is the following.

[slot 8 0.418946 0.144826 0.138723 0.087950 −0.052162 −0.102934 −0.421723 −0.472496 0.2 0.099136 60]

The recognition procedures are described as follows:

1. Read the weight vectors, threshold, and gain factors for all pre-trained features.
2. Read the face scores of the object to be recognized.
3. Calculate the inner products between a face score vector in the object and all the weight vectors from the trained pattern.
4. Compare the result with the desired projection from Step 3 and determine whether a feature and/or partial feature is recognized. If a projection from a weight vector to a face vector matches its desired projection within a designated tolerance, then a feature is recognized. The recognition of partial features is based on how close a projection to its desired projection is. For example, a designer may define a partial feature to require a 50% confidence level to its desired projection. If the difference between these two projections is less than 50% of its desired projection, then a partial feature is recognized.
5. Repeat Steps 3 and 4 for each face in the object.

10.3.2 Input representation

A problem that must be addressed in the conceptual design of the neural net recognizer, is how to represent the solid model of a part so that it can be understood by a neural net. The most important constraint that must be kept in mind is that the net must not be expected to perform any logical operations explicitly. The net will perform only simple arithmetic operations. This apparently goes against the grain of conventional feature recognizers which check to see whether a potential feature satisfies a prespecified set of rules. This process almost certainly involves a series of logical decisions made by the recognizer. A feature is considered recognized if and only if all of the set of rules for the feature are satisfied. A solution to this apparent impasse between the two technologies is to code the solid model and use the principles of pattern recognition.

To code the solid model, the elements and attributes required to recognize any feature must first be identified. Perusal of literature (Henderson, 1984; Hwang, 1988; Sakurai and Gossard, 1988) revealed that the answer lay in the characteristics mentioned in our definition of a feature. These characteristics can be broadly classified into topology and geometry. Topological characteristics consist of face, edge and vertex adjacency information. Geometry may include characteristics such as face type, number of internal loops, etc., and relationships between faces such as parallelism, perpendicularity, etc.

The face score vector is comprised of a set of face scores which can be subdivided into several sections (Fig. 10.13). The main face score represents the geometric and topological information of the face under consideration. The second section describes the characteristics of the neighboring faces immediately adjacent to the primary face. The third section represents the face scores of the faces secondarily adjacent to the main face, that is the faces which are separated from the main face by at one intervening face. There is a potential that the face score vector may have many elements, in fact too many elements for the neural net input. Because the input to the net may be truncated, the first element in each section summarizes the section with the average face score of that section. The following elements list the face scores of the first neighboring faces sorted in decreasing order.

The following example describes an 8-element vector which contains only the first and second sections of the scheme in Fig. 10.13. The reason for choosing the 8-element vector is that most faces have between 3 and 7 adjacent neighbors. The scheme of the 8-element vector is similar to Fig. 10.13. The first element is the score of the main face input to the net. The second element is the average score of the neighboring faces of the main face. The third through the last number are the scores of the neighboring faces with the highest face scores listed first (the higher score of the node indicates the more likelihood of a feature). If there are more than six adjacent faces, only the six adjacent faces with the highest scores are listed. In this example, the last two elements are assigned to 1.0 because the main face has only four neighboring faces.

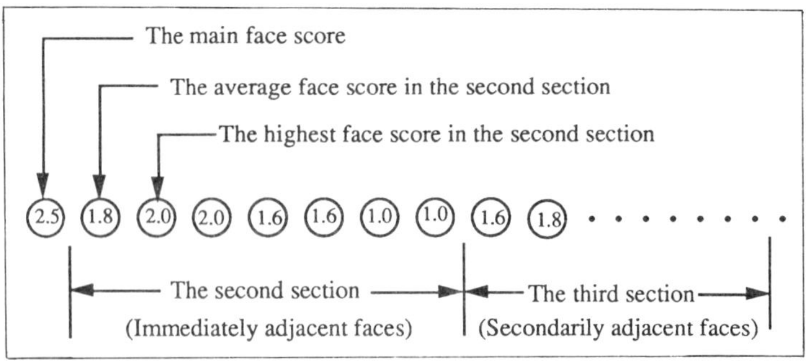

Fig. 10.13 The scheme of a face score vector.

Information on adjacent faces 7 through n would be included in the average adjacent face score, the second element in the vector. For example, a face score vector in this case is

$$[2.5 \quad 1.875 \quad 2.0 \quad 2.0 \quad 1.75 \quad 1.75 \quad 1.0 \quad 1.0]$$

10.4. ARTIFICIAL NEURAL NETWORK IMPLEMENTATION

This section describes the topics involved in implementing an experimental neural net feature recognizer. This recognizer uses the ACIS™ solid modeler (1990) interface on a Silicon Graphics Personal IRIS 4D/20 workstation which runs the Unix™ operating system. ACIS is a recently released, three-dimensional, boundary-representation, geometric solid modeler. The ACIS geometric solid modeler consists of the ACIS kernel, which is a core solid modeler, and ACIS parametric surface, which extends the kernel to handle sculptured surfaces. ACIS is written in C++, an object-oriented language. The implementation of this neural net system is built on the ACIS kernel.

The implementation can be divided into four categories (see Fig. 10.14):

1. evaluate face score vectors from a B-rep solid model;
2. train exemplar patterns and update feature databases;
3. classify and recognize features; and
4. output results.

The implementation code for Step 1 was written in C++. Steps 2, 3 and 4 were implemented in C. To check the net speed, a starting timer is net before Step 1 and an ending timer is set right after Step 3. The total execution time is equal to the difference between the ending timer and the starting timer.

There are two ways to create wireframes, surfaces, and solids when programming ACIS. One uses the applications procedural interface (API)

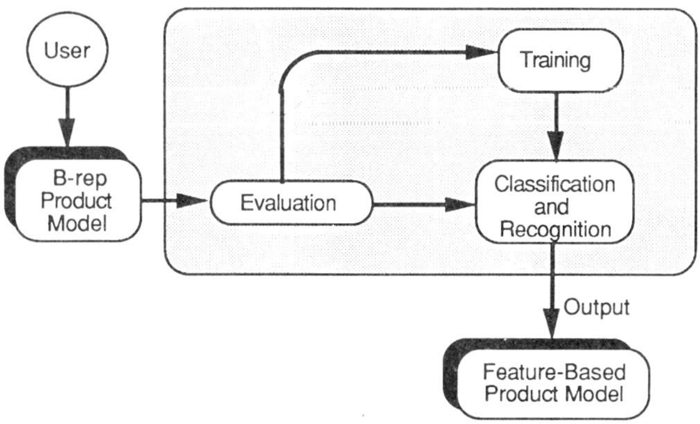

Fig. 10.14 Four categories of implementation.

which provides interfaces to create, modify and access the data structure of objects. For example, the C command *api_make_cuboid* creates a block centered at the origin with given width, depth, and height; *api-subtract* performs a boolean subtract operation; *api_body_mass_pr* calculates mass properties of a body. The other method uses the ACIS test harness, which is a simple application program written in C++. It supports English-like commands, entered using the keyboard. For example, 'BLOCK bodyname Width × Depth y Height z' has the same performance as the C command *api_make_cuboid* mentioned above. The test harness also provides simple graphic output, and several different forms of file input and output.

Although the test harness is not a supported product, it was chosen as a tool to create objects for a further use in evaluating face score vectors.

The data structure of the ACIS solid model uses the half-edge data structure. For a solid body, the topology entities include BODY, LUMP, SHELL, SUBSHELL, FACE, LOOP, COEDGE, and EDGE, VERTEX. The geometry entities include POINT, CURVE (of types STRAIGHT, ELLIPSE, and

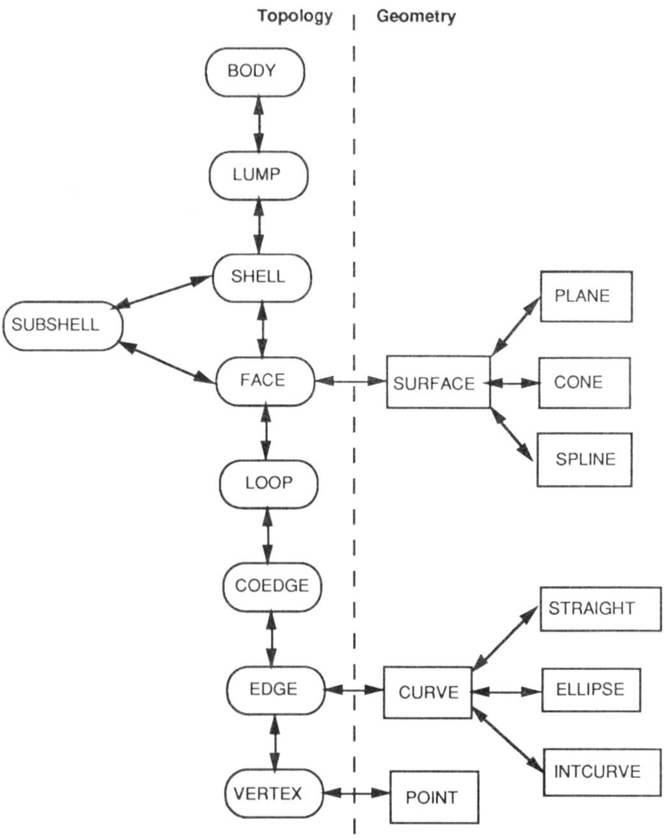

Fig. 10.15 Hierarchical view of ACIS data structure for a solid.

INTCURVE), and SURFACE (of types PLANE, CONE, and SPLINE). In addition, the data structure contains a general item ATTRIB and includes all system- and user-defined attributes, such as color. The relationships between all entities for a solid body are shown in Fig. 10.15. The definitions of the topological entities are described as follows:

- A lump represents a connected part of space, bounded by one or more shells.
- A shell is one of a separable part of a body's boundary.
- A subshell represents a subdivision of a shell or subshell.
- A face is bounded part of a single geometric surface in space. The boundary is represented by loops.
- A loop is a connected part of the boundary of a face.
- A coedge stores the relationships of the edge with adjacent edges and with the entities which own the edge.
- An edge represents a bounded part of a space curve.
- A vertex is a corner of a face and an end of an edge.

More detailed definition of the geometric and topological entities can be found in the ACIS Kernel manual (1990).

After an object is created in the Test Harness, a text file is stored by using the command 'SAVE bodyname TO filename'. A text file, which contains sufficient information to describe an object, is used in the preprocessor to evaluate face score vectors.

The procedures of the preprocessor for evaluating face score vectors of an object are described as follows:

1. Extract data from ACIS data structures

The topology of an object is accessed, using the *next( )* method. The *next( )* function allows the program to trace the next entity in a linked list of entities. For example, the folllowing routines visit all faces in a shell of the object named 'body', and call the function *access_geometry( )* for each.

```
FACE *this_face;
this_face = body - > shell() - > face();
while (this_face! = NULL)
{
   access_ geometry (this_face);
   this_face = this_face - > next();
} // while.
```

The geometry of the object is accessed using the *geometry( )* and *identity( )* method. For example, from the example above, the function *access_geometry( )* may be written as

```
void access_geometry (FACE *face_ptr)
{
   SURFACE *the_surface = face_ptr - > geometry();
```

```
int the_type=the_surface->identity();
if(the_type==PLANE_TYPE)
{
      PLANE *the_plane=(PLANE *)the_surface;
      }
      else if (the_type==CONE_TYPE)
      {
            CONE *the_cone=(CONE *) the_surface;

      }
}//the end of access_geometry
```

2. *Evaluate implicit geometric data*

The evaluation of some geometric data such as edge direction and angles between adjacent faces is not straightforward. These must be evaluated from explict topological and other geometric data. See Hwang's (1988) thesis for details on the algorithm.

3. *Calculate face scores*

First, set an edge score $E = f(\phi) = c + \phi/180$, where ϕ is the adjacent face angle of the edge, and $c = 0$ when the edge is convex and $c = 1$ when the edge is concave. E is assigned to be 1 when the edge is a smooth edge. Second, compute a vertex score which is equal to the sum of the scores of the edges sharing the vertex. Third, calculate a face score, based on the convexity/concavity of the face and on the average score of the vertices on the face.

4. *Store face score vectors*

All face scores of the object are storted in a face structure (see Fig. 10.16). A pointer connects the main face to its neighboring faces. After the face score vectors are generated, the scores of the neighboring faces are sorted in descending order.

```
typedef struct face_struct
{
 FACE *pointer;
 char name[8];
 int num_neighbor;
 float score;
 struct face_struct *neighbor[12];
 struct face_struct *next;
} face_struct;
```

Fig. 10.16 A face structure written in C.

5. *Output all score vectors to disk for further recognition use*

The sequence of faces in the output file is based on the main face score. The higher face score comes out earlier because the basic assumption is that a higher score indicates a likely feature face.

10.4.1. Training – the perceptron

The implementation of this algorithm is discussed as follows. All trained patterns are stored in a feature database.

1. Read the face score vector of trained patterns.
2. Check to see if the trained patterm has been presented before by comparing the names of the patterns in the database.
   If the pattern appeared before, then copy the data (see Fig. 10.17) from the feature database to an accessible data structure.
   Else, assign the gain factor η to be $0.9 \times 2/\mathrm{Max}(|\mathbf{F}|^2)$, where $|\mathbf{F}|^2$ is the norm of a trained pattern, the threshold is 0.2, and the initial weight vector ($\mathbf{W}$) is

$$[0.8 \quad 0.7 \quad 0.6 \quad 0.5 \quad 0.4 \quad 0.3 \quad 0.2 \quad 0.1]$$

3. Calculate the actual outputs of the trained patterns and check the difference between the actual and desired output.
   If the differences for all trained patterns are less than 0.01 (the designated tolerance), the convergence has been achieved, and then exist.
   Else calculate a new weight vector and then go to Step 2.

The recognizer was divided into two parts. The first part attempts to find simple features. The second part matches compound feature rules to find compound features. A simple feature is a feature in which the main face must directly connect to all of its associated faces. A compound feature is the collection of two or more non-intersecting simple features. The confidence factor of a compound feature is defined as the lowest confidence factor of the simple features which form the compound feature. To recognize features, an object is provided to the recognizer face by face.

Simple features

To recognize simple features after the net classifies potential features, three simple feature rules must be applied.

Pattern Name	Number of Weights	Weight Vector	Threshold	Gain Factor	Number of Training

Fig. 10.17 The structure in the feature database.

1. The face must not be one of the faces in any feature already found.
2. To be the main face of a simple feature, the face must have the highest face score in the feature.
3. To be the main face of a partial feature, the score of the face must be greater than 1.5 which is the face score of a planner face with 90° convex angles with its adjacent faces since it is assumed that the boundary of a planar face with only convex edges does not form a feature.

For step 2 we calculate the value (Feature_j) of the face score vector of the face and the weight vector of trained patterns from the feature database, using the formula

$$\text{Feature}_j = \mathbf{W}_j^{\mathrm{T}}\mathbf{F}_i - \theta$$

where $1 \leqslant i \leqslant$ number of faces and $1 \leqslant j \leqslant$ number of trained patterns.

For step 3, we determine if each face value (Feature_i) drops within the designed tolerance (0.01 for a feature and 0.5 for a partial feature), and pick one which has the highest confidence factor. The designated tolerance is arbitrarily chosen since the recognizer picks the highest confidence factor. A face is the main face of a feature when the face has the highest confidence factor which is greater than 98% for the trained pattern since the designated tolerance for a simple feature is assigned to be 2%. A face is the main face of a partial feature when the highest confidence factor is between 98% and 50% since the designated tolerance for recognizing a partial feature is assigned to be 50%.

Then we store the main face information of a feature or partial feature into a feature structure, which contains a feature code, as shown in Fig. 10.18. If a simple feature is merged into a compound feature then the feature code is set to be 0. The feature code is 1 for a simple feature, 2 for a partial feature, and 3 for a compound feature.

```
typedef struct feature_struct
{
        char name[20];
        char main[8];
        int num_associate;
        char associated[30][8];
        // feature code -- 0: deleted feature  1: simple feature
        //    2: partial feature  3:compound feature
        int code;
        // the main vector (normal for a plane axis direction for a
        // cylinder or cone
        unit_vector vect;
        float factor;
        struct feature_struct *next;
} feature_struct;
```

Fig. 10.18 A feature structure.

Compound features

From the definition, a compound feature is a combination of simple features and/or partial features. The main face of a compound feature is the main face of the simple feature which has the highest confidence factor since the highest confidence factor represents the closest to the trained pattern. Each compound feature must have a rule to define it. Two compound features are considered: a counter-bored hole and a nested-slot. The rule for recognizing a counter-bored hole is defined as:

> *if* one through-hole and one blind-hole share an associated face and the axes of their main face are collinear,
> *then* one counter-bored hole is found and is comprised by the through-hole and blind-hole.

The rule for recognizing a nested-slot is defined as:

> *if* two steps and one slot share at least two associated faces and the normals of their main faces are either parallel or perpendicular each other,
> *then* one nested-slot is found and comprised by the steps and the slot.

Output of feature recognizer

The output of the recognizer contains six pieces of information. The data structure is shown in Fig. 10.18. The first part shows the kind of feature found currently, there are three kinds considered: simple, partial and compound features. The second part is the name of the feature, and the third part is the confidence factor of the feature found. The fourth part is the name of the main face of the feature, and the fifth part is the list of associated faces. The sixth part is the total execution time to go through the evaluation and recognition of an object. For example, the output is shown as follows.

!! Recognized partial feature
⇒ Feature Name: step
  confidence factor = 95.15%
⇒ Main Face f26
⇒ Associated Faces f28 f10 f29 f4
Total execution time = 0.610 seconds.

10.5 PERFORMANCE OF THE INTELLIGENT SYSTEM ARCHITECTURE

Before the examples are introduced to be recognized, three patterns (see Fig. 10.19) were trained and stored in the feature database, shown in Table 10.2.

Table 10.2 The Feature database

pocket	8	.198326	.145909	.084153	.022397	−.039359	−.101116	−.139524	−.201280	.2	.026462	238
slot	8	.418946	.144826	.138723	.087950	−.052162	−.102934	−.421723	−.472496	.2	.099136	60
through_hole	8	.049791	.180217	.117344	.054472	−.002361	−.065233	−.128106	−.190978	.2	.006086	242

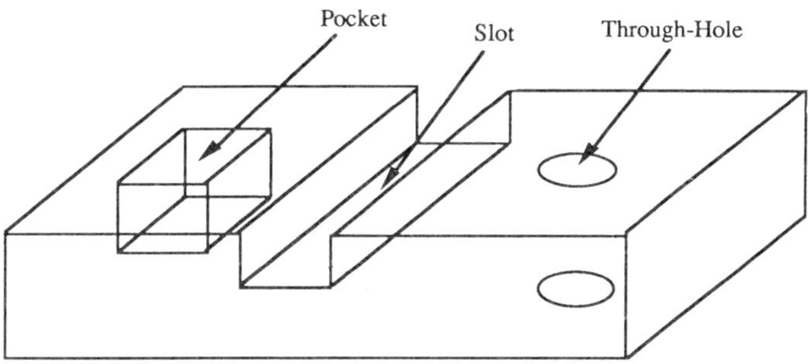

Fig. 10.19 Three trained patterns.

10.5.1 An object with a simple (partial) feature

The example shown in Fig.10.20 is a simple part with a half hole which was not trained in the net. After the object is created, the recognizer reads the text file from the ACIS solid model and the recognition process is shown as follows.

%output
Is this a training feature?(y/n)n
Input the file name ⇒ object1
!! Recognized partial feature
⇒ Feature Name: through_hole
confidence factor = 75.52%
⇒ Main Face f0
⇒ Associated Faces f4 f3 f2 f1

Total execution time = 0.230 seconds
Quit the program?? (y/n)y

The result shows that the recognizer makes a reasonable guess that the feature has some characteristics of the trained hole patterns, although the feature (a half-hole) is not in the feature database.

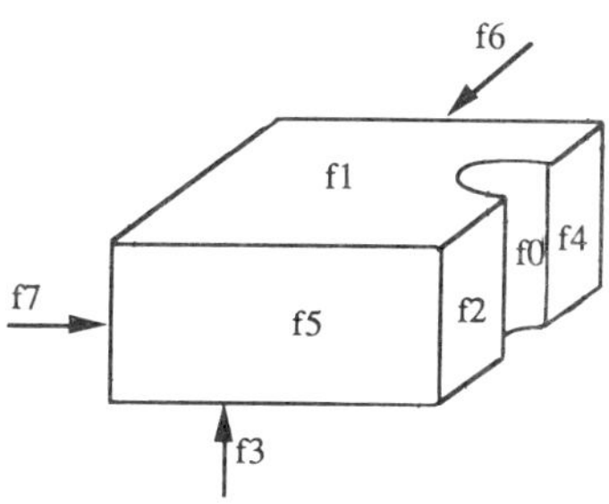

Fig. 10.20 An object with a partial feature.

10.5.2 Objects with intersecting features

The examples in this section show the recognition of intersecting features. The intersecting features are difficult to recognize because the intersection destroys part or all of the characteristics of the original features.

The first example, two intersecting slots, is from Ames (1988) and is shown in Fig. 10.21. The output from the recognizer is:

!! Recognized simple feature
⇒Feature Name: slot
confidence factor = 99.01%
⇒Main Face f0
⇒Associated Faces f4 f1 f6 f2
!! Recognized simple feature
⇒Feature Name: slot
confidence factor = 99.41%
⇒Main Face f8
⇒Associated Faces f9 f4 f11 f7

Total execution time = 0.300 seconds

Since the intersection destroys only a slight part of the chracteristics (f4 in Fig. 10.21), the result (two slots) is expected. From the confidence factor of the

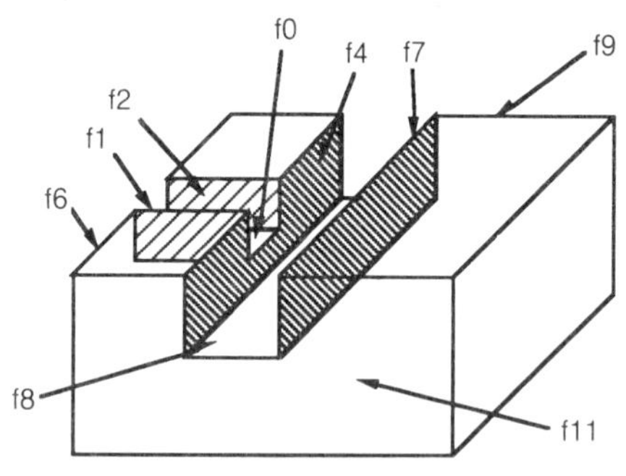

Fig. 10.21 Two intersecting slots.

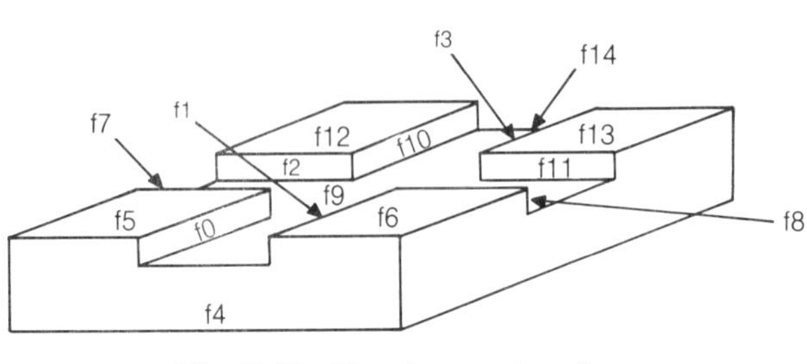

Fig. 10.22 Two intersecting slots.

above result, it is easy to see that both slots are close to the trained pattern in Fig. 10.19.

The second example shows that if the intersection destroys a complete face then the result is unpredictable. An example mentioned in Sakurai's paper (Sakurai and Gossard, 1988) is shown in Fig. 10.22, and the output is:

!! Recognized partial feature
⇒ Feature Name: slot
confidence factor = 91.72%
⇒ Main Face f0
⇒ Associated Faces f4 f5 f7 f9

!! Recognized partial feature
⇒ Feature Name: slot
confidence factor = 91.72%
⇒ Main Face f1
⇒ Associated Faces f4 f9 f8 f6

!! Recognized partial feature
⇒ Feature Name: slot
confidence factor = 91.72%
⇒ Main Face f2
⇒ Associated Faces f10 f12 f14 f9

!! Recognized partial feature
⇒ Feature Name: slot
confidence factor = 91.72%
⇒ Main Face f3
⇒ Associated Faces f11 f9 f14 f13

Total execution time = 0.400 seconds

In this example, several features go unrecognized and none of the recognized slots has face $f9$ as the slot face because $f9$ (in Fig. 10.22) has many associated faces (12 associated faces) when comparing the trained patterns (4 associated faces). Those associated faces contribute to the face score and increase its value over the designated tolerance of the trained pattern (slot), where the face score vector is

$$[2.833 \quad 2.083 \quad 2.250 \quad 2.250 \quad 2.250 \quad 2.250 \quad 2.250 \quad 2.250]$$

The unpredictable results may be due to the fact that after a face is recognized as the main face of a feature or partial feature it is eliminated from further consideration as a main face for other features. This example shows that the decision to incorporate this face elimination step may be a major obstacle to recognizing this particular type of feature. Figure 10.22 has potentially 15 logical slots or partial slots (steps). Only four oddly-configured illogical face sets are found (trace the faces on Fig. 10.22)

If the net is trained with a cross-slot configured as shown in Fig. 10.22, then the result is as expected. A cross-slot was trained and added to the data in the feature data base as:

cross_slot 8 1.082612 −.938331 −.321034 −.184203 −.047371 .089460 .226292 .363123 0.2 .060178 912.

The result, after the cross-slot was trained, is:

!! Recognized simple feature
⇒Feature Name: cross_slot
confidence factor = 99.42%
⇒Main Face f9
⇒Associated Faces f7 f15 f10 f2 f14 f3 f11 f16 f8 f1 f4 f0

Total execution time = 0.410 seconds

The third example shown in Fig. 10.23 is similar to the cross slot but the through-slot destroys almost four side faces of the original pocket. The result is:

!! Recognized partial feature
⇒Feature Name: cross_slot
confidence factor = 78.16%
⇒Main Face f9
⇒Associated Faces f7 f11 f2 f13 f14 f3 f12 f10 f1 f4 f5 f0

Total execution time = 0.360 seconds

The result seems logical. The net recognizes the intersection of a slot and a pocket as a partial cross-slot because of its similarly in face scores to the cross-slot pattern trained in the previous example. From the human point of view, the geometric shape in Fig. 10.23 'looks like' a cross-slot. The neural net seems to have generalized appropriately in this case.

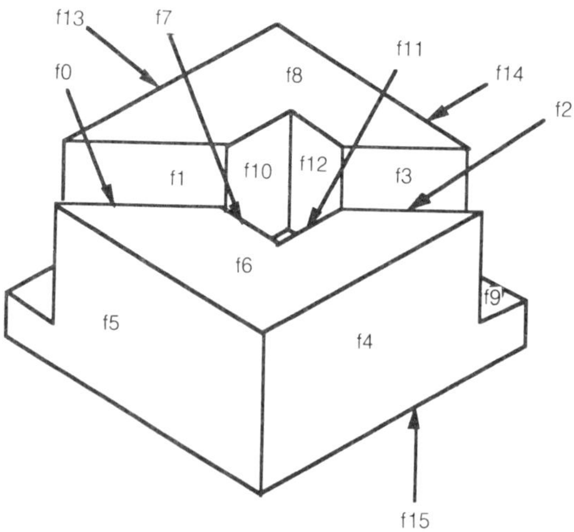

Fig. 10.23 A slot and a pocket intersection.

10.5.3 A complex object

The example shown in Fig. 10.24 is a complex object which is similar to the object used for demonstration in several references (Sakurai and Gossard, 1988; Hummel, 1989; Vandenbrande, 1990; Chuang, 1991). A blind-hole and a step are trained before recognition. If the blind-hole has not been trained, then four partial through-holes would be recognized.

The data for a blind-hole in the feature database is

blind_hole 8 .036697 .180799 .110938 .060527 .003634 −.059743 −.123121 −.186498 0.2 0.005668 268

The result is shown as follows:

!! Recognized compound feature
⇒ Feature Name: counter-bored
confidence factor = 89.73%
⇒ Main Face f1
⇒ Associated Faces f4 f2 f0 f3

!! Recognized compound feature
⇒ Feature Name: counter-bored
confidence factor = 89.73%
⇒ Main Face f5
⇒ Associated Faces f4 f6 f0 f7

!! Recognized partial feature
⇒ Feature Name: blind_hole
confidence factor = 94.45%
⇒ Main Face f8
⇒ Associated Faces f10 f9

!! Recognized partial feature
⇒ Feature Name: blind_hole
confidence factor = 94.45%
⇒ Main Face f11
⇒ Associated Faces f10 f12

!! Recognized partial feature
⇒ Feature Name: blind_hole
confidence factor = 94.45%
⇒ Main Face f13
⇒ Associated Faces f10 f14

!! Recognized partial feature
⇒ Feature Name: blind_hole
confidence factor = 94.45%
⇒ Main Face f15
⇒ Associated Faces f10 f16

!! Recognized partial feature
⇒ Feature Name: pocket
confidence factor = 95.52%
⇒ Main Face f25
⇒ Associated Faces f23 f22 f21 f24

!! Recognized partial feature
⇒ Feature Name: pocket
confidence factor = 86.11%
⇒ Main Face f17
⇒ Associated Faces f10 f18 f20 f19

!! Recognized partial feature
⇒ Feature Name: step
confidence factor = 95.15%
⇒ Main Faces f26
⇒ Associated Faces f28 f10 f29 f4

Total execution time = 0.610 seconds

In the references which use this example, the faces f17, f10, f18, f19, and f20 form a notch feature. However, in this neural net recognizer those faces form a partial pocket feature with confidence factor of 86.11%. Since a notch is not defined in the feature database, the recognizer made its best guess. The recognition speed is very fast. The complete process (to convert the object to

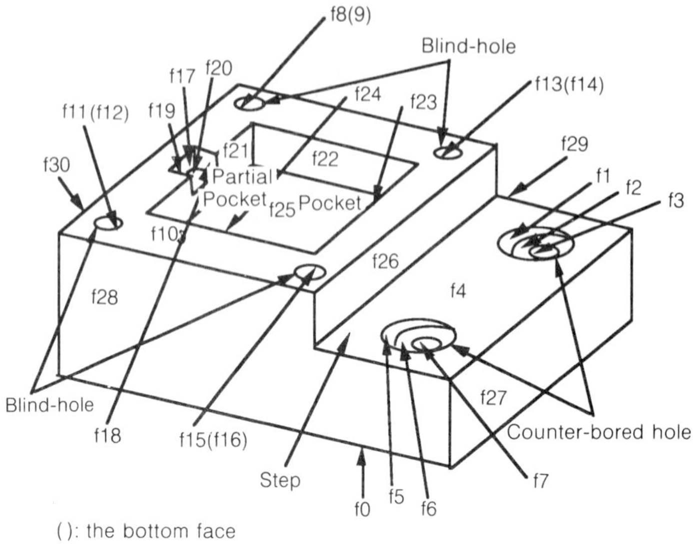

Fig. 10.24 A complex object.

face score vectors and recognize features) for this complex object took 0.61 seconds to complete. Recognition using graph matching took two and half minutes (Chuang, 1991) on the same CPU.

10.6 SUMMARY

Recently, neural nets have been applied to diverse fields and have many successful examples. Unlike other techniques, the neural network matches patterns by using only mathematical calculations which take less time than other approaches. When the problem is outside the range of learning, neural nets have reasonable behavior and accuracy. The conclusions for this chapter are listed below.

1. The recognition speed is very fast. Even a somewhat complicated object took less than one second to process.
2. To achieve recognition of compound features, some feature rules were applied.
3. The net can recognize partial features. A confidence factor is computed to represent the distance (similar to the Hamming distance) from trained patterns. Multiple feature classifications can occur for each feature face, each with different correlation coefficients to an exemplar. In this research, a feature is selected because it has the highest confidence factor. That is not always the best way to classify a feature, however. For example, from the manufacturing point of view, a partial feature in Fig. 10.20 might be

recognized perferably as a partial slot instead of a partial through-hole because of manufacturing methods and reduced costs.

4. The recognition in the research allows feature intersections as long as the intersecting features have been trained or the intersections do not destroy the identifiable shape of the feature.
5. Both memory storage and computational complexities are linear and are based on the number of faces in an object.
6. A factor, gain function η, affects the convergence of training and the results of recognition. A small gain function to a very slow convergence, but a large gain function can cause overshoot and divergence. This research has shown that a mathematical function either too small or too large gain should be avoided. The gain function is determined by the norm ($|\mathbf{F}_i|^2$) of the face score vectors of trained patterns. The inequality equation is

$$\left| \prod_{i=1}^{n} (1 - \eta |\mathbf{F}_i|^2) \right| < 1$$

where n is the number of trained patterns. For example, for two trained patterns the inequality equation may be simplified as

$$0 < \eta < \frac{a + b - \sqrt{a^2 + b^2 - 6ab}}{2ab}$$

or

$$\frac{a + b + \sqrt{a^2 + b^2 - 6ab}}{2ab} < \eta < \frac{a + b}{ab}$$

where a and b is norm of two face score vectors. Therefore, it is inappropriate if the gain is less than

$$\frac{a + b + \sqrt{a^2 + b^2 - 6ab}}{2ab}$$

(too small) or greater than $a + b/ab$ (too large).

7. The initial weight vector does not affect the training convergence. The final weight vector, which depends on the initial weight vector, can lie anywhere within the valid region and is, therefore, not unique. Since the training procedure is attempting to solve a under-determined problem (the unknowns–weights exceed equations–trained patterns), the number of trained exemplars is limited to the number of unknown weights.
8. The order of face scores in the vector is important for recognition and training. The assumption of the recognition is that the higher face score indicates more likelihood to form a feature.
9. A key to successful manufacturing automation requires a feature-based viewpoint. This neural net approach shows the potential to allow a user to automatically generate multiple viewpoint feature models for downstream applications.

It is apparent from the results of this research that further investigation in neural net and feature recognition is necessary. To complete the integration between CAD and its downstream applications the following capabilities may be added.

- *A graphical pattern trainer* would allow users to select a region (formed by one or more faces) and name it. Then, the neural net would automatically be trained for further recognition.
- *A compound feature recognizer* would require extending the scheme of a face score vector in this research to the third section or even the fourth section for recognizing more complex compound-features. It would not then be

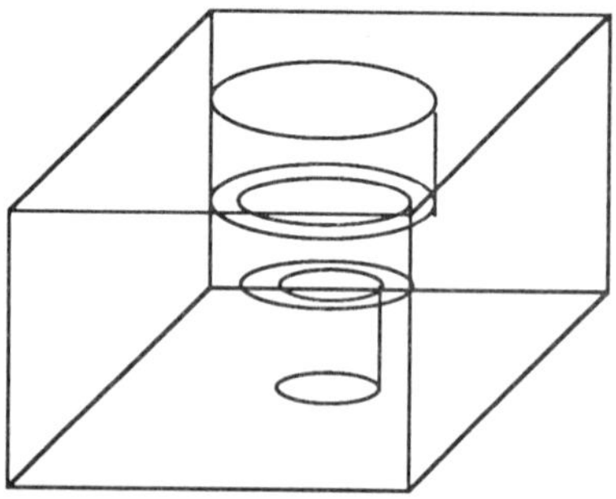

Fig. 10.25 A compound hole.

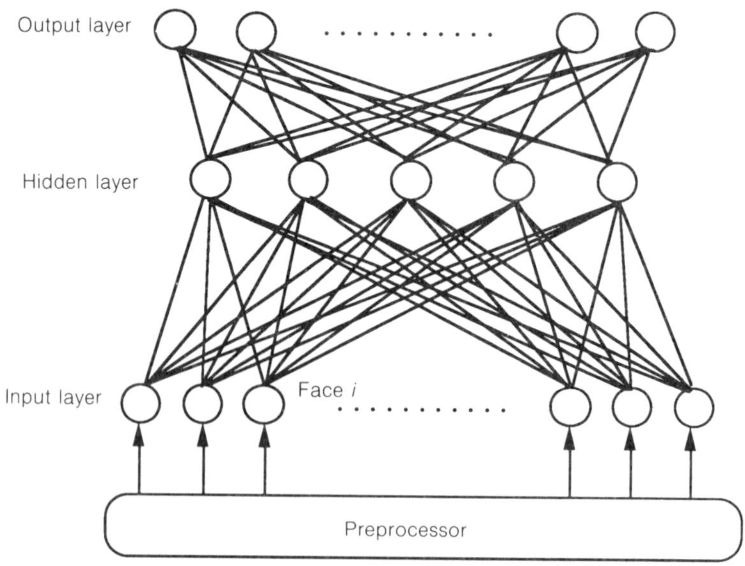

Fig. 10.26 A modified net structure.

necessary to apply compound feature rules. If a feature shown in Fig. 10.25 is attempted to be recognized as a compound hole, then the rules set in this research do not work. In addition, the net structure may be modified as in Fig. 10.26, where the hidden layer collects simple features and outputs the value to the output layer which collects compound features as well as simple features.

ACKNOWLEDGEMENTS

The author gratefully acknowledges the financial support provided for this research by the Department of Mechanical and Aerospace Engineering at Arizona State University and the National Science Foundation under Presidential Young Investigator Grant 86-58021.

REFERENCES

ACIS Kernel Manual No. Sapati Technologies, Boulder, Co. (1990).

Ames, A. (1988) Automated generation of uniform group technology part codes from solid model data, in *ASME-Computers in Engineering Conference*, San Francisco.

Ames, A. (1991) Production ready feature recognition based automatic group technology part coding, in *ACM Symposium on Solid Modeling Foundations and CAD/CAM Applications*, Austin, Texas.

Ansaldi, S. (1985) An edge-face relational scheme for boundary representations. *Computer Graphics Forum*, **4**, 319–32.

Choi, B.K., Barash, M. and Anderson, D.C. (1984) Automatic recognition of machined surfaces from a 3D solid model. *Computer-Aided Design*, **16**(2), 81–6.

Chuang, S.-H. (1991) *Feature recognition from solid models using conceptual shape graphs*, Ph.D., Arizona State University.

Cutkosky, M., Tenenbaum, J. and Muller, D. (1988) Features in process-based design, in *ASME Computers in Enginearing*, San Francisco, CA.

DeFloriani, L. (1989) Feature extraction from boundary models of 3D objects *I.E.E.E. Pattern Analysis and Machine Intelligence*, **11**(8), 785–98.

Duda, R. and Hart, P. (1973) *Pattern Classification and Scene Analysis*. New York.

Ganu, P. (1989) *Graph-based extraction of one-and two-connected features*. M.S., Arizona State University.

Gavankar, P. and Henderson, M.R. (1990) Graph-based extraction of protrusions and depression from solid models. *Computer-Aided Design* (June).

Henderson, M.R. (1984) *Extraction of feature information from three-dimensional CAD Data*, Ph.D., Purdue University, West Lafayette, IN.

Henderson, M.R. and Anderson, D. (1984) Computer recognition and extraction of form features. *Computers in Industry*, **5**, 329–39.

Hummel, K, (1989) The role of features in the implementation of concurrent product and process design, in *ASME Winter Annual Meeting*, San Francisco, CA.

Hwang, J. (1988) *Rule-based feature recognition: Concepts primitives and implementation*, M.S., Arizona State University.

Hwang, J. (1991) *Applying the perceptron to 3-D feature recognition*, Ph.D., Arizona State University.

Jakubowski, R. (1982) Syntactic chracterization of machine part regions. *Cybernetics and Systems*, **13**(1) 1–24.

Joshi, S. and Chang, T.-C. (1988) Graphi-based heuristics for recognition of machined features from a 3D solid model. *Computer-Aided Design*, **20**(2).

Kyprianou, L. (1980) *Shape classification in computer-aided design*, Ph.D., Cambridge, UK.

Peters, T.J. (1991) *Encoding mechanical design features for recognition via neural nets* (Technical report No. CSE-TR-91-20), Department of Computer Science, University of Connecticut.

Prabhakar, S. (1990) *An experiment on the use of neural nets in form feature recognition*, M.S., Arizona State University.

Rosenblatt, F. (1958). The perceptron: A probabilistic model for information storage and organization in the brain. *Psychological Review*, **65**(6), 386–408.

Sakurai, H. and Gossard, D. (1988) Shape feature recognition from 3D solid models, in *ASME Computers in Engineering*, San Francisco, CA.

Shah, J.J. (1991) Assessment of features technology. *Computer-Aided Design*, **23**(5), 331–43.

Staley, S.M., Henderson, M.R. and Anderson, D.C. (1983). Using syntactic pattern recognition to extract feature information from a solid geometric model data base. *Computers in Mechanical Engineering*, **9** (September), 61–6.

Vandenbrande, J. (1990) *Automatic recognition of machinable features in solid models*, PhD. thesis published as Technical report No. IRIS #260. Computer Science Dept. and Institute for Robotics and Intelligent Systems, University of Southern California.

Woodwark, J. (1988) Some speculations on feature recognition. *Computer-Aided Design*, **20**(4), 189–96.

CHAPTER 11

Vision based inspection

Joydeep Ghosh
Department of Electrical and Computer Engineering, The University of Texas at Austin, Austin, Texas 78712-1084.

11.1 INTRODUCTION

In industrial manufacturing, product inspection is an important step in the production process. Since product reliability and quality management is of utmost importance in most mass-production facilities, 100% inspection of all parts, subassemblies, and finished products is often attempted. As a result, the inspection process is often the most costly stage in manufacturing.

A central task for machine-based inspection is inspecting for visual appearance. Visual inspection seeks to identify both functional and cosmetic defects. At present, visual inspection in many manufacturing processes is carried out by human inspectors whose performance is generally inadequate and variable. The human visual system is adapted to perform in a world of variety and change; the visual inspection process, on the other hand, requires observing the same type of image repeatedly to detect anomalies. Moreover, the accuracy of human visual inspection declines with dull, endlessly routine jobs. So it is not surprising that over the years, substantial research and development efforts have been directed at vision based inspection and in general 3D object recognition, and considerable progress has been achieved (Arman and Aggarwal, 1992, Besl and Jain, 1985; Chin and Harlow, 1982; Chin and Dyer, 1986; Freeman, 1989).

There are several advantages to automating visual inspection. One obvious advantage is the elimination of human labor, which is increasingly expensive. Human inspectors are slow compared to modern production rates, and they make many errors. By substituting for humans, an automated vision based

Artificial Neural Networks for Intelligent Manufacturing. Edited by Cihan H. Dagli. Published in 1994 by Chapman & Hall, London. ISBN 0 412 48050 6

inspection system in fact frees them from dull and routine tasks. Automated inspection also facilitates

- performing inspection in unfavorable environments
- analyzing statistics on test information and keeping records for management decisions and
- matching high-speed production with high-speed inspection.

In light of all these advantages, why is it that versatile, automated vision systems are not widespread, and why are we interested in investigating neural networks for this task? To answer these questions, one must first realize the complexity of the tasks involved. First, in order to visually reconstruct the world that projects itself as an image on a 2D surface such as the retina, one needs to not only analyze the image but also have a model of the nature of that world and of the imaging process. A plausible model might assume, for example, that objects are rigid, surfaces are smooth and boundaries are continuous. Humans are almost always unaware not only of these internal models but also of the tremendous amount of computation needed for preprocessing and recognition of images, so that vision seems effortless. On the other hand, even the basic step of detecting edges poses a problem for machine vision. To quote (Marr, 1982, p. 16):

> ... Edges that are of critical importance from a three-dimensional point of view often cannot be found at all by looking at the intensity changes in an image. Any kind of textured image gives a multitude of noisy edge segments; variations in reflectance and illumination cause no end of trouble; and even if an edge has a clear existence at one point, it is as likely as not to fade out quite soon, appearing only in patches along its length in the image.

The tremendous amount of research in machine vision over the past three decades has been influenced not only by physiological or psychophysical discoveries and psychological observations about perception among living beings, but also by advances in signal processing, computational mathematics, pattern recognition and artificial intelligence. Some researchers in the recently rejuvenated field of neural networks are also attempting to develop useful models of biological and machine vision. With the human visual system serving as a common source of inspiration, it is not surprising that neural network approaches to image processing/understanding often have commonalities with more traditional techniques. However, they also bring fundamentally different elements of adaptation and learning, and promise breakthroughs through massively parallel and distributed implementations in VLSI (Ghosh and Hwang, 1989).

In this chapter, we present both conventional and artificial neural network (ANN) approaches to vision-based inspection, and to the larger problem of 3D object recognition. The goal is not only to compare different methodologies, but also to highlight the relationships among them so that more comprehensive techniques incorporating the best of the diverse approaches can be developed.

For our purposes, an ANN is a collection of computing cells (artificial neurons) interconnected through weight links (artificial synapses with varying strengths). The cells compute asynchronously and in parallel using information available locally or from topologically adjacent cells through the weighted links. The knowledge of the system is embodied in the pattern of interconnects and their strengths which vary as the system learns or adapts itself. This learning can be supervised by an external input (the teacher) or through some local rule such as Hebb's rule (Hebb, 1949). It should be mentioned that techniques such as relaxation can be cast in the above framework without the incorporation of learning.

In section 11.2, we outline the main steps involved in vision based inspection. Sections 11.3 and 11.4 summarize key techniques of traditional machine vision that have been used for representing and recognizing 3D objects respectively. The emphasis in these sections is on model-based recognition of 3-dimensional (3D) objects. Section 11.5 explores neural-like techniques for early visual processing such as edge detection and segmentation. The next section covers neural network approaches or recognition that are based on a biologically motivated viewer-centered representation of objects. Section 11.7 visits more direct neural network applications to object recognition.

11.2 CHARACTERISTICS OF VISION BASED INSPECTION SYSTEMS

Figure 11.1 shows a simple pictorial diagram of an automated visual inspection system. This installation can replace and improve present quality control operations involving manual or visual measurements. The transport moves the objects to be inspected into the scanning station. The scanner collects visual data describing each object and sends them to the processor to be analyzed. After analysis, decisions are made, and the processor directs the sorter to reject a defective item, if any.

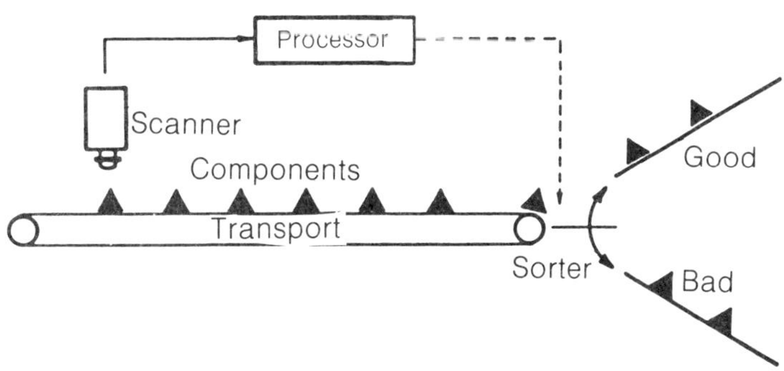

Fig. 11.1 Simple pictoral diagram of an automatic visual inspection system, adapted from Chin and Harlow (1982).

The goal of any vision-based recognition system is to identify and locate a specific object in the scene. The system should have adequate knowledge of the shape of the desired object. Such a priori knowledge of the object is provided through a model of the object, and in most cases the model contains information regarding the geometry of the objects. Therefore, the key issues involved in machine vision based inspection include data collection, representation, model construction and matching strategies (Arman and Aggarwal, 1992).

The goal of data acquisition is to yield information about the surface characteristics and dimensions of objects in the scene. As we shall soon see, such information is not directly obtainable from input data, and some preprocessing is warranted. Two general approaches exist to collect the data. In the first approach, passive sensing modalities such as visible light intensity cameras or thermal cameras are used, and are possibly augmented by depth cues to recover the necessary 3D information. In the second approach, active sensors needing external energy sources, such as laser range scanners and sonar sensors, are used to obtain range or depth data. Though the first approach is preferred because of its non-intrusive nature, it lacks the necessary resolution and precision for many common tasks. While the best choice of the sensor is application dependent, the second approach is usually preferred nowadays for 3D object recognition (Arman and Aggarwal, 1992).

The 2D array of numbers (representing depth or intensity) provided by a sensor is not of much use in its raw form. A suitable representation scheme must be used to describe both the data and the model. In fact, representation is the most critical issue that determines the effectiveness of the inspection system. Unlike quantifiable entities like motion and intensity, a full description of even reasonably complicated 3D shapes requires hundreds of parameters, and an object-model matching based on such descriptions is not computationally feasible. A representation is feasible if it is unambiguous, unique, not sensitive to noise, easy to store and convenient to use in the matching stage. In a large number of cases, local surface properties such as surface normal and surface curvature are used to describe 3D shapes. Various surface-based and volumetric-based representation schemes are described in the following section.

The construction of object models is the third issue which must be addressed in any model based vision system. There are two main approaches to model construction. In the first approach, the actual objects are used to generate a model, i.e., data points obtained from several viewpoints of the object are integrated in a coherent fashion to provide information from all the viewing angles (Vemuri and Aggarwal, 1986). The resulting model can be object-centered (i.e. independent of the viewer coordinates) or viewer-centered (Chakravarthy *et al.*, 1991). Recently, CAD-based systems are being increasingly used where a set of predefined primitives allows the user to interactively construct the model of an object (Bhanu and Ho, 1986).

Once the appropriate descriptions are derived from the data and models, the vision system attempts to match the two descriptions. We note that the matching process is not trivial. For example, since range scanners sample points on the surface, the vision system based on such sensors must rely on information derived from the surfaces. Moreover an ideal vision system must be able to recognize arbitrarily shaped objects from any viewing direction. It should also be able to cope with partial occlusion from other objects in the scene and with cases of self-occlusion. Noise in the input images and subjectivity in edge detection and segmentation can also lead to ambiguity in the extracted features that further complicate the matching process.

11.3 REPRESENTATION OF 3D OBJECTS

The representations used for describing shapes of 3D objects can be divided into surface-based, discontinuity-based and volumetric-based schemes.

11.3.1 Surface-based representation

In this representation, surface properties like the surface normal and surface curvature are estimated from the collected range data points. This is facilitated by the much studied theory of surface reconstruction, which uses a priori knowledge of surface characteristics such as smoothness and discontinuities to interpolate between the range data points, and to perform surface based segmentation (Besl, 1988). We outline below some key techniques in surface reconstruction and representation, see (Bolle and Vemuri, 1991) for a recent review.

Gaussian sphere

In this representation, the orientation of the surface normal at each point in the surface is mapped onto corresponding points on the unit sphere. The mapping is not local in plane regions, partially local in parabolic regions, and strictly local at elliptic and hyperbolic regions, allowing for unambiguous representations of convex objects only. In all cases the representation is rotation invariant; however translation and information on the dimensions of the objects (scaling) are not preserved.

Several modifications to the gaussian sphere exist which overcome, to an extent, the limitations of this representation. In the extended gaussian sphere, each point on the sphere is assigned a weight equivalent to the corresponding surface patch area on the object. In a second extension, Nalwa (1988) defines a support function for a surface to be assigned the distance from the origin of the sphere to the tangent plane at the surface. Preserving similar information, Kang and Ikeuchi (1991) associate with each surface normal a complex number where the magnitude of the number is the corresponding surface area and the phase is the distance of the surface from a predefined origin of the coordinate system. This information is essential in determining the translation component of any geometric transformation (pose determination). Nalwa further represents the surface normal discontinuities as symbolic links, where arcs connect the surfaces on either side of the discontinuity. Bhanu and Ho (1986), and Korn and Dyer (1987) implement various tessellation schemes of the sphere to obtain a multi-resolution representation. Another extension is the **dual spherical representation** in which planar surfaces are used to approximate smooth and polyhedral objects, and the properties of the gaussian sphere are combined with the dual space to further cover the types of objects that can be represented. The dual space is based on the duality that exists between planes and points. In dual representation, planes are represented as points and their orientation is signified by the position of the dual space point. Edges are defined as lines connecting the points in dual space. Regularization can be employed so that no dangling 2D or 1D artifacts result from the boolean operations.

The common problem with such schemes is that in complex scenes, proper object partitioning must be achieved beforehand to avoid the interaction among the representations of each object in the sphere.

Moment based

Shape can be defined as a function of mass distribution, and properties of shape as moments of functions. Using such a definition, a **shape polynomial** can be used to measure similarities between shapes (Taubin *et al.*, 1989). As with most moment dependent representations, this representation suffers from sensitivity to occlusion.

11.3.2 Discontinuity-based representation

Instead of storing information about the surfaces, this method preserves the information about the points where the characteristics of an area on the surface change. This results in curves embedded in 3D space, lowering storage requirements and increasing the efficiency in the higher level algorithms. The detected curves are represented by mathematical means such as parametric polynomial curves and rational B-splines. However in most cases the information in not complete because information about the surface is lost.

Space curve

Space curves are used to describe bounding contours and the spine of the sweep representations. Mokhtarain (1988) uses the general forms of expression for curvature and torsion of space curves to describe the curves over several scales. Following the smoothing process with gaussian smoothing filters of increasing widths(s), the zero crossing of the torsion and similarly the curvature over the scales are extracted.

Surface primal sketch

Several types of discontinuities are detected and modeled in the representation

- *steps* – where the depth map is discontinuous;
- *roofs* – where the surface normal is discontinuous;
- *smooth joints* – where the principle curvature is discontinuous; and finally
- *shoulder* – the combination of two roofs.

After gaussian smoothing at a set of scales, the principle directions and curvatures are computed everywhere for each smoothed image. Next, the zero crossings of the gaussian curvature are marked, and, finally the descriptions of the four types of discontinuities are matched to detect the discontinuity points on the surface. Mathematical models of each of the discontinuities are presented and analyzed in (Ponce and Brady 1987).

Aspect graph

Aspect graphs are graphs in which each node represents a topologically distinct 2D viewpoint of a 3D object, and two nodes are connected if one can transit from one view to the other by a small displacement that brings out (or occludes) a new feature such as an edge. Gigus and Malik (1988) have studied aspect graphs of polyhedral objects, while Eggert and Bowyer (1989) create aspect graphs by using the symmetry of the objects represented by generalizing cylinders and dividing the gaussian sphere into bands so that each view of the object within each division is the same. Aspect graphs are a manifestation of a biologically motivated viewer-centered paradigm of object representation and recognition and are therefore discussed in more detail in section 11.6.

11.3.3 Volumetric representations

This class of representations describe volumes rather than surfaces or discontinuities of surfaces. While most such representation schemes are efficient in describing shapes, their disadvantage is that first, in most cases, the objects have to be symmetrical and simple in shape. Second, these representations cannot be used directly in the matching stage, i.e., other properties must be derived from this representation. Hence this class has been used most often for CAD modeling rather than for recognition tasks.

Superquadrics

Superquadrics are an extension of the basic quadric surfaces and solids. Several parameters are used to determine the object's squareness, size, bending, tapering, and twisting. Minimization techniques are used to recover these parameters from a set of data. However since the minimization problem is non-linear, good initial estimates are required for numerical stability. In addition, if object symmetry is not assumed, data points from multiple viewpoints are necessary. The disadvantage of this representation is its lack of uniqueness, i.e., the same shape may be approximated by more than one set of parameters. This representation has been limited to the modeling of shapes.

Octrees

The octree, a tree structure, is constructed by recursively decomposing a cubical volume into eight subcubes until each resultant sub-cube is homogeneous with respect to some criterion. This representation is mostly useful for modeling (Chein *et al.*, 1988).

Constructive solid geometry

In this representation, solids are represented by the boolean combination of primitive solids. Each object is represented by a tree whose leaves are primitive solids (such as block, wedge, cylinder, etc.) which may be transformed, and other nodes are the boolean operations used to construct the solid; these operations have to be regularized to assure the validity of the constructed objects. The major drawback of using this approach for recognition is that the knowledge of primitive solids used during the construction may not be effective to the recognition system, since only surfaces are visible and the sub-patterns may not be.

11.4 MODELING AND MATCHING STRATEGIES

Models are based on a priori geometrical and topological knowledge that the vision system has about a set of objects. There are two main approaches to model building. In the first 'manual' approach, multiple viewpoints of the object are integrated in a coherent fashion to provide a 3D description of the object (Vemuri and Aggarwal, 1986). In the second approach, a CAD/CAM system is used, where a set of predefined primitives allow the user to interactively construct the CAD model of the object (Bhanu and Ho, 1986). For example, ALPHA-1, a CAD system developed at the University of Utah, uses B-spline surface patches and subdivided polygons. A surface is represented by its control mesh and the related parameters, and a polygon is represented using its contour. ALPHA-1 also allows for high level shape operators, such as bending and twisting, and set operations on surface patches (Bhanu and Ho, 1986).

Once the appropriate descriptions are derived from the data and a suitable model is developed, the vision system tries to match the two descriptions and complete the task of object recognition. In the first step, a correspondence is established between the two sets of descriptions to provide 'an interpretation' of the input data (Ballard and Brown, 1982). In the second step, using the established correspondences, a geometrical transformation is derived such that the model may be transformed to the orientation of the object in the scene.

Several approaches have been applied to finding a corresponding match in the model for each feature found in the scene. In tree search, the data from the sensor is arbitrarily ordered and the first data feature is hypothesized to correspond with extracted model feature. This would represent the first level of the tree. Given each such hypothesis, the second data feature is hypothesized to correspond to each of the model features accordingly. Each path in this tree from the root to a leaf represents a possible solution to the correspondence problem. Instead of searching the entire tree, various methods are followed to prune the search tree and thereby cutting the search space.

The other popular approach to establishing correspondence has been to represent the sensory data and the model data as graphs in which the nodes of the graphs represent surface patches and the arcs represent the relations among surfaces. The problem of establishing correspondence may then be approached as a (sub)graph isomorphism problem. Several of the ANN techniques for object matching that are considered in sections 11.6 and 11.7 are based on this approach.

11.5. ARTIFICIAL NEURAL NETWORKS (ANNs) FOR VISION-BASED INSPECTION

The major use of ANNs in engineering applications have been (i) to serve as distributed associative memories, and (ii) for approximating an unknown function of input features given a training set consisting of (possibly noisy) input–output sample. Classification can be considered as a special case of function approximation where the desired mapping is to a 'one out of M' function. Sometimes the input vectors, representing features of interest, may themselves be extraced from raw inputs using a neural network. In the area of vision-based inspection, and object recognition in general, both approaches have been explored. It is interesting to note that a major goal of the classical perceptrons, as can be guessed from their very name, was to recognize objects whose images were projected on an artificial 'retina'. With the notable exception of the neocognitron and related work, current efforts are usually not purely ANN based, but rather employ a hybrid approach incorporating techniques from statistical pattern recognition as well as computer vision. For example, both input preprocessing/feature extraction as well as object modeling can be done using some conventional technique, and then the matching of the extracted features to the object models be performed using an ANN.

In this subsection, we shall consider ANNs proposed for two key tasks in early vision that every vision-based inspection system has to address, namely, edge detection and image segmentation. However, the thrust of this chapter is to present a more biologically motivated approach, namely recognition of objects based on a viewer-centered approach. In section 11.6 we introduce this approach and its biological plausibility, and compare it with object-centered modeling/recognition that is the dominat paradigm in conventional machine vision. Then, we consider ANN based techniques that use a viewer-centered basis for object recognition. In section 11.7, we examine some hybrid approaches based on object-centered models that use ANNs for the matching or recognition process.

11.5.1 Detection of oriented edges in computer and biological vision

An image is a distribution of light intensity values. An edge in an image is characterized by a rapid, sustained change in intensity as one traverses in a direction orthogonal to it. Thus an edge can be described by its location, its direction (orthogonal to the local intensity gradient), and its magnitude which is a measure of the intensity change. Edges are useful in defining boundaries formed by combining local edges and for detecting shapes. A texture can be distinguished by a particular pattern of edges. The detection of edges is central to image pre-processing in biological systems, wherein the retina, the lateral geniculate nucleus (LGN) and visual cortex seem to be actively involved in some forms of contrast enhancement or edge detection. In this section, we look at a typcial computational scheme for detecting edges, and then introduce self-organizing neural network models that lead to development of cells that are sensitive to particular orientation and position of edges. These neural networks use simple local learning rules to yield cell behavior similar to those found in the striate cortex of mammals.

Most computational schemes for edge detection use operators for detecting gradient maxima of the intensity surface (Marr and Hildreth, 1980) or for estimating the parameters of an assumed edge model. Since edges in actual images deviate substantially from ideal step discontinuities, some form of post-processing is required to eliminate the edges caused by noise and other effects. In natural images, edges can occur over a wide range of scales or resolutions. Marr and Hilderth showed that filtering an image with a gaussian filter restricts the range of resolutions over which intensity changes can occur while maintaining the spatial location of detected discontinuities in a near-optimal fashion. After an image has been gaussian filtered at several resolutions by adjusting the variance of the gaussian function, the laplacian operator is applied to detect intensity change maxima, indicated by the zero crossings in the output. The effect of the gaussian function and laplacian operator can be combined to yield a single laplacian of gaussian or LOG filter. An edge is indicated by a segment of zero-crossings that occur at the same positions for more than one resolution. Another popular filter is the 'difference of gaussian' or DOG filter, which is remarkably similar to the LOG filter. In an attempt to relate his model to the human visual system, Marr postulated that the LGN computes the zero-crossings of a LOG filter while simple cells detect lines of zero-crossings that indicate an edge.

In the human eye, each ganglion cell is known to collect the output of a localized group of photoreceptors that form its receptive field. Interestingly, ganglion cell receptive fields tend to have center-surround profiles that are qualitatively similar to DOG or LOG filters. A uniform illumination on the receptive field elicits no response from a ganglion, but some form of contrast does. The long axons from all the ganglion cells are bundled together to form the optic nerve, which follows two separate pathways to the brain. One leads to the superior colliculus in the midbrain, presumably for controlling eye movements, while the other leads to the visual cortex via the LGN.

The visual cortex has been subdivided into several areas according to function and/or neural structure. Each area is essentially a 2D layered sheet of neurons. Interconnections among the areas are invariably bi-directional. Area V1, also known as area 17 or the striate cortex, contains a more or less topographic map of the visual field, just like the LGN. It contains simple cells, complex cells and hypercomplex cells. A simple cell fires at maximum frequency in response to a small edge with a specific angular orientation and position in the visual field. Its response falls dramatically for small shifts from this position, or for a change of more than 20 degrees in orientation. Complex cells, while remaining orientation sensitive, are more tolerant to position changes. Hypercomplex cells have been observed to respond to line terminations and corners. These three types of cells are believed to be organized in a roughly hierarchical fashion. A simple cell receives input from a group of LGN cells responding to the same edge; the receptive field of a complex cell is a group of neighboring simple cells with the same orientation preference, and hypercomplex cell receives excitatory input from some complex cells and inhibitory inputs from others. Further details of the human visual system can be found in Van Essen (1985); Arbib and Hanson (1987).

Through their studies of the visual striate cortex (area 17) of the cat and monkey, Hubel and Wiesel (1962) found neurons that are selectively sensitive to light bars and edges of a certain orientation. They also detected 'functional columns' in which neurons within a column are sensitive to the same orientation, and neighboring columns tend to respond to stimuli of similar orientation. Van der Malsburg (73) succeeded in developing a neural network model that could evolve to exhibit behavior similar to those found experimentally by Hubel and Weisel. This model consisted of a 2D 'cortical plane', containing excitatory (E) and inhibitory (I) cells of uniform and equal density. The E cells had excitatory connections to nearby E and I cells, with connection strengths decreasing with distance, while the I cells would inhibit E cells over a wider area. Bell-shaped functions were used to describe the decreasing intra-cortical strengths with cell distance for both cell types. The input image was projected onto the 'retina', a plane of A cells that were initially connected to the cells in the cortical plane through synapses of random strengths. Only the connections afferent to the E cells were modified using Hebb's (1949) rule and renormalized so that the net strength of afferent connections to a cell remained constant. Malsburg's results showed that after learning, the cells in the artificial 'cortex' would form groups that maximally responded to edges of specific orientation, thus indicating that cortical organization is not genetically predetermined. Since this pioneering work, several other self-organizing approaches to feature detection have been discussed, including those by Grossberg (1987) and Kohonen (1989).

11.5.2 Cooperative processes and boundary completion

Several biologically motivated techniques have been proposed for connecting edges in order to form more complex curves and boundaries. In fact, cooperative processes such as relaxation labeling have been explored by the vision community for over a decade, without explicitly casting them in a neural network framework. These mechanisms are very akin to neural network models such as the Boundary Contour System (Grossberg and Mingolla, 1986).

Local intensity edges typically form a part of a global line or boundary rather than occurring in isolation. Thus the presence or absence of a nearby edge of similar angular orientation would tend to reinforce the hypothesis of the existence of an edge at a given point in the intensity field. This idea is the basis of a cooperative process called **relaxation labelling** (Davis and Rosenfeld, 1981) that can be used to enhance lines and curve in an image. First, a bank of filters that detect small lines at different orientations are used to generate an array of nodes, with each node corresponding to a location in the image, and having a line-orientation label assigned to it. These labels are updated by a relaxation process, such that they become more compatible with neighboring labels. Thus adjacent 'no-line' labels support one another, and so do lines with similar orientation, while two adjacent labels corresponding to orthogonal orientations antagonize each other. A refined relaxation-labeling algorithm with associated convergence conditions is given in (Hummel and Zucker, 1983). Cooperative processes have also been used to obtain curvature estimates, for detecting corners in dotted line drawings, and for enhancing perceptually significant features (Walters, 1987).

A cooperative-competitive scheme called the **boundary contour system (BCS)** has been proposed by Grossberg and Mingolla (1985) to explain how edges are filled in when part of a boundary is missing, and how illusory contours can emerge from appropriately positioned line-terminations. The BCS consists of a hierarchy of locally-tuned interactions that controls the emergence of image segmentation and also detects, enhances and completes boundaries. The interaction of BCS with a feature contour system and an object recognition system, as developed in Grossberg (1987), attempts to attain a unifying precept for form, color and brightness. While the BCS is largely preattentive, the model does allow feedback from the object recognition system to guide the segmentation process.

The BCS consists of several stages arranged in an approximately hierarchical organization. The image to be processed forms the input to the earliest stage. Here, elongated and oriented receptive fields called **masks** are employed for local contrast detection at each image position and each orientation. Thus there is a family of masks centered at each location, and responding to a prescribed region around that location. These elliptical masks respond to the amount of luminance contrast over their elongated axis of symmetry, regardless of whether image contrasts are due to differences in textural distribution, a step change in luminance or a smoother intensity gradient. The elongated receptive field makes the masks less sensitive to differences in average contrast in a direction orthogonal to the major axis. However, the penalty for making them sensitive to contrasts in the preferred orientation is the increased uncertainty in the exact locations of contrast. This positional uncertainty becomes acute during the processing of image line ends and corners.

The authors assert that all line-ends are illusory in the sense that they are not directly extracted from the retinal image, but are created by some process that generates line terminations, and propose a mechanism based on two short-range competitive stages followed by long-range cooperation, to mimic such phenomena. They also propose a boundary completion process that involves long-range cooperation between similarly oriented pairs of input groupings. This mechanism is able to complete boundaries across regions that receive no bottom-up inputs from the oriented receptive fields, and thus accounts for illusory line phenomena such as the completion of the square edges in a reverse-contrast Kanisza square. The process of boundary completion occurs discontinuously across space, using the gating properties of the cooperative cells successively to interpolate boundaries within progressively finer intervals. Unlike a low spatial frequency filter, this process does not sacrifice spatial resolution to achieve a broad spatial range. The co-operative cells used in this stage also provide positive feedback to the cells of the second competitive stage so as to increase the activity of cells of favored orientation and position, thereby providing them with a competitive edge over other orientations and positions. This feedback helps in reducing the fuzziness of boundaries. The detailed architecture, equations and simulation results can be found in Grossberg and Mingolla (1985; 1986).

11.5.3 Image segmentation

To segment an image into relatively homogeneous components, some indicator of uniformity (or lack thereof) is first required. Typical cues are intensity, orientation, location of boundaries or edges, and texture. A hierarchical neural network for texture segmentation and labeling has been proposed by Dupaguntla and Vemuri (1989). The underlying premise of their approach is that textural segmentation can be achieved by recognizing local differences in texels. The architecture consists of a feature extraction network whose outputs are used by a texture discrimination network. The feature extraction network is a multilayer hierarchical network governed by the BCS theory. The image intensities input is first pre-processed by an array of cells whose receptive fields correspond to a difference of gaussian filter, and which follow the feed-forward shunting equations of Grossberg. The output of this array of cells form the input to a BCS system and are processed by oriented masks which are then fed into.the two competitive stages of the BCS theory. However, the long-range cooperative processes described above are not used. Instead, the outputs of the second competitive stage activate region encoding (RE) cells at the next level. Each RE cell gathers its activity from a region of orientation masks of the previous layer, as well as from a neighborhood of adjacent RE nodes of the same orientation.

The RE cells are functionally analogous to the complex cells in the visual cortex, with the intra-layer connections helping to propagate orientation information across this layer of cells. The outputs of the feature extraction network are used by a texture discrimination network which is essentially Kohonen's single-layered self-organizing feature map (Kohonen, 1989). At each position, there are T outputs, one for each possible texture type, which is assumed to be known a priori. Model (known) textures are passed through the feature extraction network. For a randomly selected position, the output cell of the texture discrimination network that responds maximally is given the known texture-type label. The weights in the texture discrimination network for that position are adapted according to the feature-map equations. Since these weights are the same for all positions, one can simply replicate the updated weights for all positions. The hierarchical scheme described above has been applied to natural images with good results. However, it is computationally very intensive, since there are cells corresponding to each orientation and position at every hierarchical level.

The SAWTA mechanism

A cooperative-competitive feedback network called smoothing, adaptive winner-take-all network (SAWTA) has been developed recently for performing texture-based segmentation (Ghosh *et al.*, 1990). It uses the multichannel texture model of Bovik, Clark and Geisler (1990) which characterizes textures as superpositions of localized narrow band frequencies of light intensity band variations. These variations are detected by a bank of Gabor filters which are known to be most responsive to localized spatial frequencies. The network consists of n layers of cells, with each layer corresponding to one Gabor filter, as shown in Fig. 11.2. On the presentation of an image, a feedforward network using local receptive fields enables each cell plane to reach an activation level corresponding to the amplitude envelope of the Gabor filter that it represents, as outlined in the preceding paragraphs. Let $m_i(x, y)$, $1 \leqslant i \leqslant n$, be the activation of the cell in the ith layer with retinotopic coordinates (x, y). Initially, the n cell activations at each point (x, y) are set proportional to the amplitude responses of n Gabor filters.

To implement the SAWTA mechanism, each cell receives constant inhibition from all other cells in the same column, along with excitatory inputs from

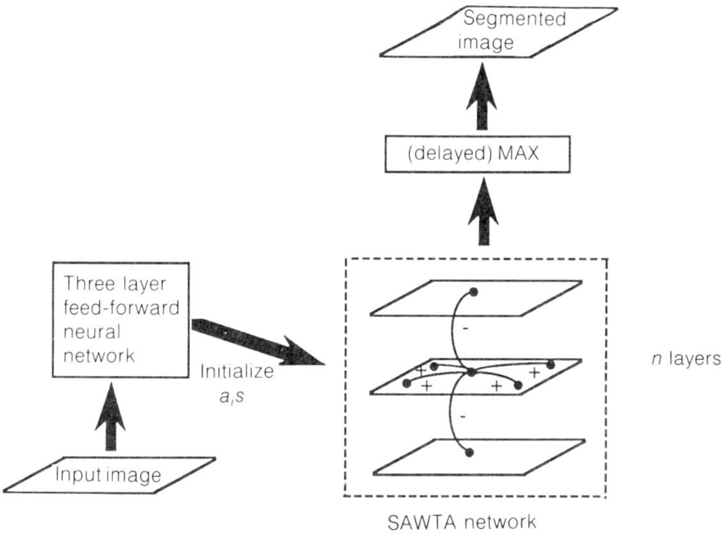

Fig. 11.2 The SAWTA network for segmentation of textured images.

neighboring cells in the same row or plane. The synaptic strengths of the excitatory connections exhibit a 2D gaussian profile centered at (x, y). The network is mathematically characterized by shunting cooperative-competitive dynamics (Grossberg and Mingolla, 1985) that model on-center off-surround interactions among cells which obey membrane equations (Kandel and Schwartz, 1985). Thus, at each point (x, y), the evolution of the cell in the ith layer is governed by:

$$\tau \frac{\mathrm{d}}{\mathrm{d}t}(m_i) = -m_i + (A - m_i)J^+ - (B + Cm_i)J^- \tag{11.1}$$

Where J^+, J^- are the net excitatory and inhibitory inputs respectively, and are given by

$$J^+ = \alpha \sum_{(x_n, y_n)\in R} m_i(x_n, y_n) \exp\left\{\frac{-[(x - x_n)^2 + (y - y_n)^2]}{2\sigma^2}\right\}$$

and

$$J^- = \sum_{j=i} f(m_j(x, y))$$

Here, R is the neighboring region of support and f is a sigmoidal transfer function. The convergence of a system described by (11.1) has been shown for the case when the region of support R consists of the single point (x, y) (Cohen and Grossberg, 1983). The network is allowed to run for 10–20 iterations before segmentation is performed by assigning to each point (x, y) the label i

corresponding to the subscript of m whose value is maximum. Several results using SAWTA are given in Ghosh and Bovik (1991).

The SAWTA network does not require a feature extraction stage as in Dupaguntla and Vemuri (1989) or computationally expensive masking fields. The incremental and adaptive nature of the SAWTA network enables it to avoid making early decisions about texture boundaries. The dynamics of each cell is affected by the image characteristics in its neighborhood as well as by the formation of more global hypotheses. It has been observed that usually four spatial frequencies are dominant in human visual systems. This suggests the use of a mechanism for post-inhibitory response that suppresses cells with activation below a threshold and speeds up the convergence of a SAWTA network. The adaptive learning network of Kohonen (1989) can be used to change both excitatory and inhibitory synaptic strengths (J^+, J^-), in response to a teaching input. Also, the SAWTA network can be easily extended to allow for multiple 'winners'. Then, it can cater to multicomponent textures, since a region that contains two predominant frequencies of comparable amplitude will not be segmented but rather viewed as a whole.

11.6 VIEWER-CENTERED OBJECT RECOGNITION

Underlying all the methods presented in sections 11.2–11.4 is an object-centered philosophy that aims to construct models of objects independent of the viewer position. Object-centered approaches are typically targeted to specific practical applications and not presented as plausible models of biological vision. In fact, there is substantial evidence to suggest that biological visual systems are viewer-centered in which an object is internally represented as well as recognized in terms of its image or view obtained from different directions or vantage points, as well as the relationships among these views. The internal models of the world generated by an organism are tailored to its anticipated sensory experience. The argument that this internal construction takes the form of a function that enables the organism to react appropriately to an inflow of sensory data was first put forth by Koenderink and Van Doorn (1979). For instance, the *trompe-l'oeil* paintings must be regarded as mere flat surfaces covered with pigments. When observed from a single vantage point, through a peep-hole for instance, an illusion of 3D perception may occur, but on changing the viewing position the observer would discover that the view is only a flat surface. Hence, solid shape is an invariant extracted from the visual input depending on the organism's position, orientation and motion relative to the object. In particular, we are perceptive of the transitions from one view of an object to another view as that object moves naturally in our field of view, and also notice the key features that seem to generalize across several vantage points.

Several important *physiological* studies support a viewer-centered representation. For example, cells have been discovered in the fundus of the anterior superior temporal sulcus of macaque monkey brain which selectively respond to 2D or 3D images of familiar heads from characteristic viewpoints (Perett *et al.*, 1985). Different views maximally activate different classes of cells. Moreover, certain cell groups respond to view transitions, say from a frontal to a lateral view of the same face. Such cells are deemed to enforce relational constraints on the different views or aspects of the same object. Finally, cell groups exist that respond to a particular object regardless of the view, and thus integrate the information provided by the first two types of cell groups.

Psychological evidence supporting viewer-centered recognition include several studies that show that familiar views are recognized much more quickly that unusual ones, even though the former might be simpler to compute (Edelman *et al.*, 1989). Another aspect of train of thought is that in the case of wire-frame objects with depth cues or multiple orthogonal views which theoretically provide sufficient information to construct an unambiguous object-centered representation, human subjects seem unable to recognize the wireframes from unusual vantage points. Finally, from a *computational* perspective, Rosenfeld (1987) notes that there does not seem to be sufficient time to compute the 3D volumes involved in object-centered volumetric models, nor to hypothesize a 3D object, including its position and orientation, and generate the expected 2D appearance, and finally match the stimulus with the imagined object.

A viewer-centered object representation scheme depicts a 3D object as a network of connected visual events encountered as the observer explores the space surrounding the object. A view of the object, consisting of contours, hillocks and other singularities, is termed as an 'aspect'. For small changes in viewer position this structure of singularities does not change; one still sees the same aspect. For larger displacements, however, when a contour becomes invisible or a hillock appears in view, a visual event is said to have occurred (Koenderink and Van Doorn, 1979). A full description of the object aspects can be presented in the form of tables, or simple diagrams, or as a connected graph with the aspects as nodes. A connected graph is preferred because it naturally expresses the relationship among various visual events. This is the aspect graph of the object.

Similar graph-based approaches have been suggested by Freeman and Chakravarty (1980). Their method consists of dividing the viewing space of an object with planar surfaces into a set of vantage domains, each uniquely corresponding to a characteristic view (CV). More recently Platinga and Dyer (1986) developed an algorithm to construct the aspect graph of an object. Their method also estimates the total number of CVs of an object and addresses the problem of equivalence between two given views of an object. Recent efforts in this area include that of Eggert and Bowyer (1989). All these realizations of the original idea suggested in Koenderink and Van Doorn (1979) require a rigorous geometric characterization of the object under study, a stipulation that is not granted even in very simple cases of biological vision. Moreover, the size of the aspect graph becomes prohibitively large even for simple non-convex polyhedral objects (Wang and Freeman, 1990). A strict geometric treatment is altogether infeasible in case of large scale problems of spatial representation like route learning or architectural walkthrough.

A novel approach to spatial representation that does not involve stringent geometric treatment of the 3D scene is suggested in the next section. The technique exploits the ability of artificial neural networks to 'learn' an arbitrary function from a number of instances. Aspects of an object, involving distinct features, not necessarily geometric entities, are encoded as weights of a neural network. A procedure to construct the aspect graph from the neural network representation is given. That is, the aspect graph is encoded in a distributed manner as weights of the neural network. This approach leads to great compactness in representation, a feature that becomes most useful for treatment of larger problems.

11.6.1 Robust viewer-centered object representation using RBFs

In this section, we use radial basis functions networks (RBFs) for representing objects such that they can estimate the viewer position given an object view, and conversely, given a vantage point, predict what features should be seen.

RBF networks are a class of single hidden-layer feedforward networks in which radially symmetric basis functions are used as the activation functions for the hidden layer units (Broomhead and Lowe, 1988). The output of the jth hidden node, $R_j(x)$, and that of ith output node, $f_i(x)$ are given by

$$f_i(x) = \sum_j w_{ij} R_j(x) \tag{11.2}$$

$$R_j(x) = R(\|x - x_j\| / \sigma_j) \tag{11.3}$$

where $R(x)$ is a radially symmetric function such as a gaussian. We call x_j the jth centroid and σ_j its width. In earlier applications, the centroids in an RBF network were determined using heuristics such as performing k-means clustering on the input set, and widths were held fixed during training. Alternatively, we can vary both the centroid locations and associated widths of the receptive fields by performing gradient descent on the mean square error of the output. This leads to a variant of the RBF, which is called the adaptive kernel classifier (AKC).

Consider a quadratic error function

$$E_p = \tfrac{1}{2} \sum_i (t_i - f_i(x_p))^2$$

where t_i is the target function and f_i is as defined in (11.2). Then, the update rules for various network parameters using gradient descent on the expected mean square error (MSE) are given by:

$$\Delta w_{ij} = \eta_1 (t_i - f_i(x)) R_j(x_p) \tag{11.4}$$

$$\Delta x_{jk} = \eta_2 R_j(x_p)(x_{pk} - x_{jk})/\sigma_j^2 \left(\sum_i (t_i - f_i(x_p)) \cdot w_{ij} \right) \tag{11.5}$$

$$\Delta \sigma_j = \eta_3 (\| x_p - x_j \|^2) R_j(x_p)/\sigma_j^3 \left(\left(\sum_i (t_i - f_i(x_p) \right) \cdot w_{ij} \right) \tag{11.6}$$

where $\|\cdot\|$ denotes the Euclidean norm, and $R(\cdot)$ is a gaussian. These constitute the learning scheme for the adaptive kernel classifier. The AKC is able to perform the same level of approximation as RBF using fewer hidden units. However, training time is increased since centroids and center widths are also adapted using the generalized delta rule which is a slow procedure.

A similar scheme, called the **gaussian potential function network (GPFN)**, which involves segmentation of the input domain into several potential fields in form of gaussians was proposed by Lee and Kil (1988). The gaussian potential functions of this scheme are not radial functions. Instead, the size of these potential fields is determined by a correlation matrix. The network parameters are computed by gradient descent as in the case of AKC. More general schemes like the regularization networks are studied by Poggio and Girosi (1990). Though more complex decision regions can be shaped out of potential fields that are not radially symmetric, often receptive fields of radial functions would suffice for most decision tasks.

Robust object representation

A simple polyhedral solid studied in Wang and Freeman (1990) is selected for our experiments. (The particular object is chosen only for the sake of simplicity; the technique used is applicable regardless of the geometry of the object.) The aim is to construct a map of the view space of the object onto the corresponding CV. For this, each CV consisting of several features or singularities must be expressed numerically by an appropriate encoding technique. Sufficient number of these features are located on the surface of the object and identified by numbers. The simulations are based on two different sets of features viz., edges and faces. For the chosen object in Fig. 11.3 these features are labeled with numerals. Each view is represented as a binary vector

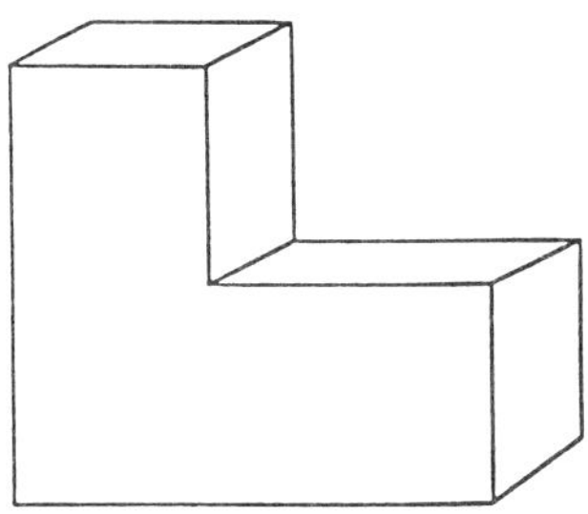

Fig. 11.3 A simple plane-faced solid.

Fig. 11.4 Characteristic views of the object in Fig. 11.3, from Wang and Freeman (1990).

with one bit allotted to each feature (edge or face). The size of this vector equals the total number of edges or faces of the object depending on the situation. The presence or absence of a feature in a view is encoded by setting the corresponding bit in the binary vector to 1 or 0 respectively. Bits corresponding to partially occluded features are arbitrarily set to 1 or 0 depending upon the extent of their visibility. The object considered has 8 faces and 18 edges. For this object, Wang and Freeman (1990) have determined 71 different characteristic views (see Fig. 11.4). Based on this data, a complete list of binary vectors and the corresponding viewer coordinates is prepared. A portion of this data is used to train the RBF network. Performance is tested on significant tasks like (1) estimating viewer position from a new view and (2) predicting expected features in a view from a given viewer position.

Estimating viewer position from object view

Estimating camera location from distinct images of an object is a well-studied problem (Mitchie *et al.*, 1985). It is related to the problem of estimating the position and motion of an object in space by observing the correspondence between a small number of points from two distinct images of the object. In the situation studied here, the question can be formulated as follows: given sufficient number of views of the object and the corresponding viewer coordinates, is it possible to estimate the viewer coordinates that correspond to an object view not learnt before? To address this question, a partial set of view–viewer position pairs is selected and the network is trained to map the set of binary vectors of the views onto the corresponding set of viewer coordinates. Assessing the performance of the network consists of feeding a binary vector that is absent in the training set and comparing the estimated coordinates with the actual values. The training set has at most one binary vector of edge features per CV, for a total of 50 vectors. Table 11.1 compares the estimated viewer positions with actual positions for representative samples, and shows that even the coarse binary coding of the input yields a fairly accurate mapping. It is also observed that even for positions corresponding to CVs for which there is not even a single training example, the error in a coordinate value is large only if the absolute value of the coordinate is high. This is due to a sensitive dependence of the view on the viewer position at some critical distance from the object center. Note that the training set is extremely sparse, and more training samples will surely increase the position estimation accuracy.

Predicting expected features in a view

An interesting question that can be posed in many situations of spatial representation is: if I see certain features from various vantage points, what could I expect to see from a new viewing point? The success with which this problem can be solved clearly depends on how much of the scene is already

Table 11.1 Actual and estimated viewer positions for training set of edge features

Correct viewer position			Estimated viewer position		
x	*y*	*z*	*x*	*y*	*z*
−24.3	−1.8	−1.8	−24.2	−2.5	−2.4
24.3	−1.8	−1.8	24.3	−4.0	−3.6
0.0	−24.9	0.0	1.1	−11.0	−3.4
0.0	25.5	−1.8	−0.2	25.2	0.9
−20.8	−1.8	−21.6	−23.0	−2.6	−22.3
15.3	1.8	26.7	15.6	2.3	25.3
−20.8	1.8	21.6	−21.9	2.7	24.8
20.8	1.8	21.6	21.1	4.3	21.9
0.0	0.0	−24.9	−1.3	0.5	−24.8
0.0	1.8	24.9	−1.4	0.4	24.5
0.0	2.2	2.2	0.7	4.4	2.1
0.0	21.4	21.4	−1.1	23.2	19.6
0.0	−21.4	−21.4	0.0	−18.2	−21.9
25.3	1.8	1.8	24.1	3.4	−0.8
−25.3	−1.7	1.8	−22.4	−2.7	0.9

Table 11.2 Number of incorrect features (faces) in the test views

Faces			
View no.	*No. of errors*	*View no.*	*No. of errors*
9	3	43	0
11	3	44	1
12	3	45	1
21	0	46	0
22	0	51	0
23	0	52	0
29	0	54	0
35	0	56	0
39	2	59	0
41	2	60	0
42	2		

seen. The question can be formulated in case of our polyhedral object as follows. A partial set of view-viewer position pairs are given. From a viewer position that is absent in the partial list the corresponding binary vector of the object view must be estimated. Clearly, this is a more difficult mapping problem since it is a map from a low dimensional space to one of high dimensionality. Table 11.2 gives the list of test views and the number of incorrectly guessed features in the feature vectors for face chosen as features. The same procedure can be applied to recognize edges, but typically the percentage of errors will increase because of the greater number of edges than faces.

Aspect graph construction

The aspect graph gives the relational structure of the object views seen from various points in the viewer space. Several methods for constructing the aspect graph have been investigated. In one approach, the viewing space of a plane faced solid is divided into a finite set of vantage point regions each corresponding to a CV. These regions are taken as nodes of the aspect graph and nodes corresponding to adjacent regions in the viewer space are connected (Wang and Freeman, 1990). The present approach to construction of aspect graphs essentially consists of relating the obvious adjacency structure of the points in the viewer space to that of the points in the abstract space of the object views. The network is trained on the complete set of view/viewer position pairs for the purpose of constructing the aspect graph. The observer co-ordinates are the input to the network in this case. Once the network is trained on the complete data set, the aspect graph can be systematically generated as follows.

1. The object considered is imagined to lie wholly within a sphere whose size is comparable to the size of the object.
2. A sufficient number of points uniformly distributed on the surface of the sphere are chosen as viewer positions for the object and the corresponding views are generated by the neural network trained on the complete data set. Since the exact location of the points on the sphere surface is not important, symmetry can be exploited to decrease computation.
3. A graph is drawn with the CVs as nodes. CVs corresponding to neighboring points on the sphere are connected.

The aspect graph constructed using exhaustive search is given in Fig. 11.5(a). The aspect graphs constructed for the cases of faces shown in Fig. 11.5(b), using a set of 84 coordinates. It will be noted that the aspect graph fails to include some views that appear in the complete aspect graph shown in Fig. 11.5(a), but is remarkably informative considering the sparseness of the training data.

Comments

The work described above is primarily motivated by the observation that internal representations of the 3D world consist of intricately mixed topological and metric information of the objects. The aspect graph is an explicit representation of how the topological features of an object, as viewed by an observer moving around the object, are interconnected. But the graph itself does not provide any quantitative information related to the object. On the other hand, object-centered models like the octree contain numerical description but give no clue as to how various visual events recorded by a viewer are ordered. This ordering is the central feature of all internal representations of 3D objects generated by biological vision systems as argued in Koenderink and Van Doorn (1979). A single scheme that handles both metric and symbolic data is clearly lacking in the existing literature.

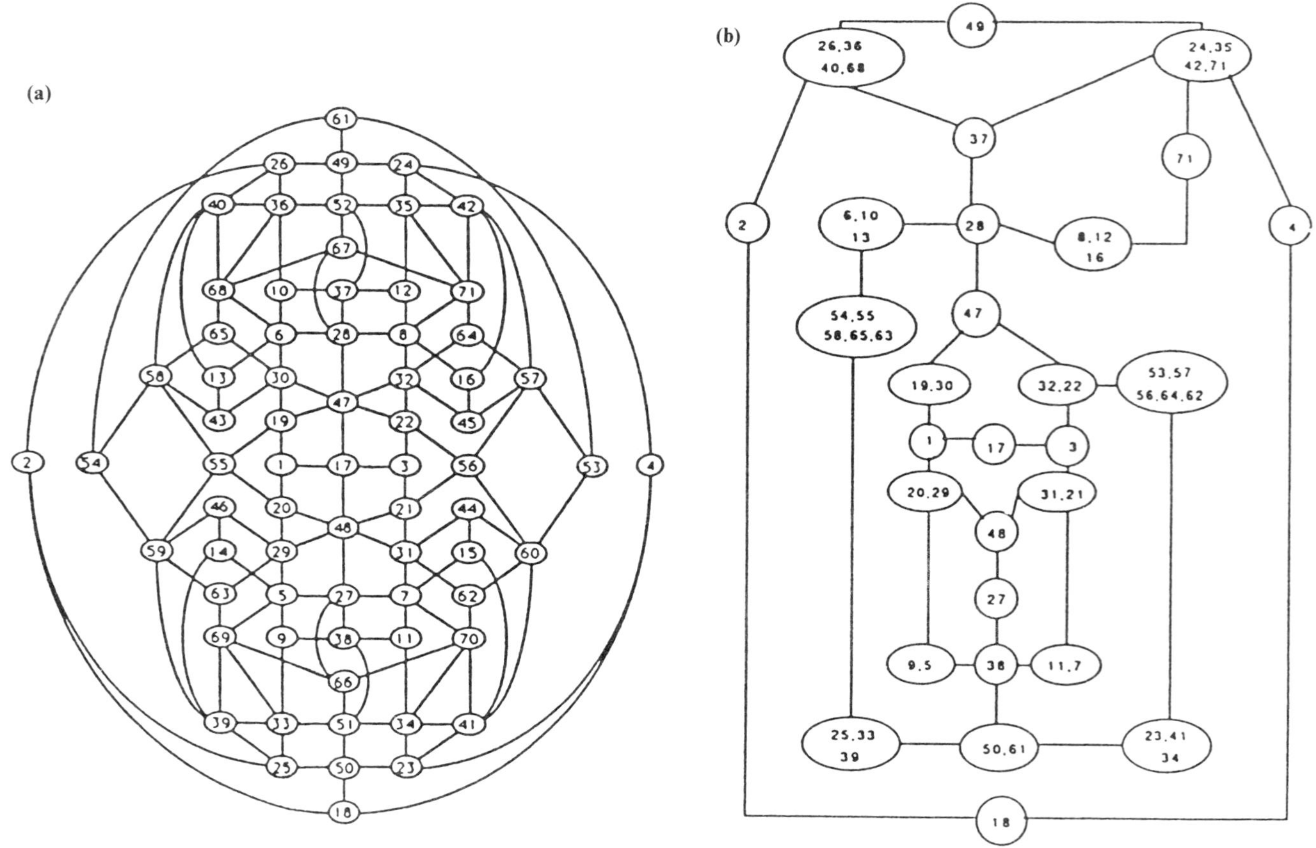

Fig. 11.5 (a) The complete aspect graph of the object in Fig. 11.3; (b) aspect graph connecting 58 views with faces as object features.

Like many problems of computational vision, object representation belongs to the class of ill-posed problems. There is a shift of paradigm recently from exact and rigorous approaches to 'soft' approaches using neural network-based techniques to these problems. Similar approaches to the representation problem are introduced in the pioneering work of Poggio and others (Poggio and Edelman, 1990). Several neural network realizations have appeared in subsequent literature (Lin *et al.*, 1991). The line of investigation pursued in Poggio and Edelman (1990) assumes the following:

- for each object there exists a smooth function mapping any perspective view onto a 'standard' view, and
- this multivariate function can be constructed from a small number of views of the object.

It is interesting to see how such a formulation falls in line with the argument presented in the original paper on the topic of the aspect graph (Koenderink and Van Doorn, 1979).

Koenderink and Van Doorn have made it convincingly clear that a biologically plausible, viewer-centered model of an object must bring out the graph structure inherent in an organism's internal model of the object. According to this argument the organism learns the relations between views as an outcome of its displacement relative to the object. Then the relative position of the organism with respect to the object becomes an important parameter of the model. In other words, the views are 'remembered' in terms of the relative position. This feature is not prominently brought out by a model where all object views are mapped onto a special view (Poggio and Edelman, 1990).

Mapping the viewer space onto the abstract space of the object views is a natural step toward construction of the aspect graph. Neural network based approaches to the representation problem have considered maps from various object views to a special view. It is not apparent how the aspect graph can be constructed from such a map. In this paper we have constructed a map from the viewer coordinates to the corresponding object view. It is shown how the aspect graph can be directly constructed from this map. The simulations described above are based on a sparse training data set. Only one viewing point is considered for each CV. In an experimental setup, however, data acquisition is straightforward and the representation is expected to be more robust.

As a further step, one would like to have a comprehensive system that can extract a string of aspects from sequences of images and then associate them with a previously experienced object if there is sufficient similarity or form an internal representation of a new object. Promising work in this direction has been recently reported by Seibert and Waxman (1992). The overall architecture of the neural system for 3D object learning and recognition is shown in Fig. 11.6. The early modules of the vision system are arranged primarily in a feedforward manner to extract features, determine attentional cues that drive

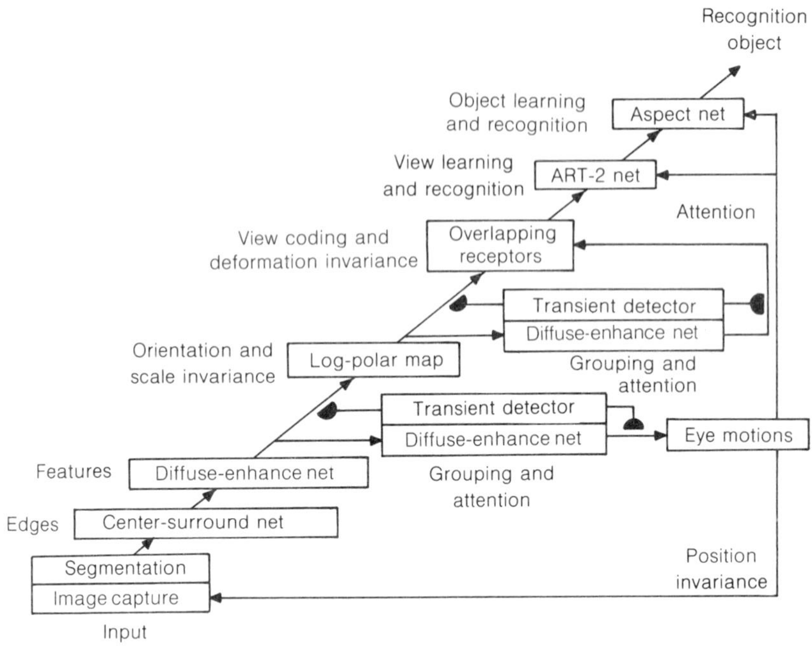

Fig. 11.6 Neural system architecture for three-dimensional learning and recognition, adapted from Seibert and Waxman (1992).

saccadic and pursuit camera-motions (Seibert and Waxman, 1989), and subsequently learn 2D aspects using an ART2 network. In the aspect network, each object is represented by a node whose synapses correspond to the edges of its aspect graph. Thus, it accumulates evidence every time its associated aspects and allowable aspect transitions are re-experienced. A winner-take-all network is implemented among the object nodes, which recognizes an object as soon as its node's accumulated evidence exceeds a threshold. If none of the accumulated evidences are high, then a novel object is suggested, and therefore a new object node is allocated. This system has been used to recognize various types of military aircrafts.

11.6.2 Hopfield networks for recognition

Despite the strong biological motivations for considering aspect-graph representations, this approach at present suffers from high construction time, excessive storage requirements, and the limited nature of objects that have been modeled so far. Hierarchical techniques provide a way for tackling complexity, and have also been applied for aspect graph representations. An innovative hierarchical approach recently proposed by Dickinson *et al.* (1992) aims to provide the indexing power of complex 3D object-centered primitives while exploiting the convenience of 2D viewer-centered aspect matching. An aspect hierarchy of boundary groups, faces and aspects are used to represent a fixed set of 3D volumetric primitives. An object of interest is modeled in terms of these 3D primitives, and probabilities inherent in the aspect hierarchy are used to distinguish it from other objects of interest, given any 2D view (aspect). This technique has been applied to both synthetic line drawings and real images and is notable in that the size of the aspect is fixed, independent of the size of the object database.

As noted earlier, the main part of a recognition process is to establish the correspondence relationships between the contents of the image and the object model, and can be posed as a graph matching problem. This in turn is often formulated as an optimization problem where an energy function is minimized. In Pravin and Medioni (1989), a Hopfield network realizes a constraint satisfaction process to match visible surfaces of 3D objects. In Mjolsness *et al.* (1989) object recognition is posed as an inexact graph matching problem and then formulated in terms of constrained optimization. In Mohan (1989) the problem of constraint satisfaction in computer vision is mapped to a network where the nodes are the hypotheses and the links are the constraints. The network is then employed to select the optimal subset of hypotheses which satisfy the given constraints.

To convey the flavor of such optimization frameworks, we consider here the two-level hierarchical approach of Lin *et al.* (1991), where Hopfield networks are used to recognize objects by first matching surfaces and then matching vertices of the likely object candidates as indicated by the surface matching network. It is assumed that the unknown object is in the form of line drawings obtained by segmenting the original image and each model is a 2D projection of a 3D object whose identity and pose are to be determined. Surface correspondence between an unknown object and an object model in the database is established by designing a Hopfield network in the form of a 2D array, wherein the rows represent the features of an input image while the columns represent the features of an object model (see Fig. 11.7). The output of a neuron reflects the degree of similarity between two nodes, one from the image and one from the object model.

The proposed Hopfield net for surface matching has a flexible structure and is able to solve the surface correspondence problem even if the numbers of polygons in the input image and the object model are different. Also, based on the outputs of the neurons in the network, a similarity measure between any object model and the input image can be derived even if they contain a different number of surfaces. This similarity measure is used to reduce the search space. Based on the proposed simple features, a set of object models most similar to the input image is selected from the database. This process can be considered as a coarse search step because 'good' candidates as well as some 'bad' candidates are selected due to roughness of the feature set.

The next step is to apply Hopfield networks to establish vertex correspondences. First, the order of all the vertices in each polygon are determined.

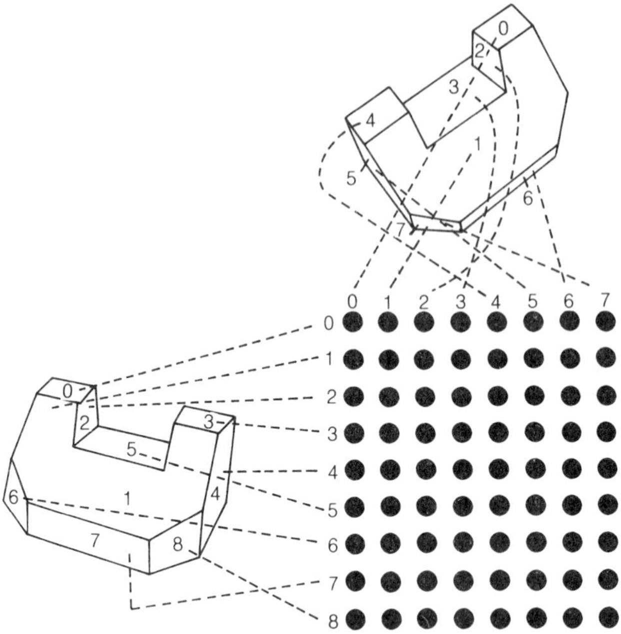

Fig. 11.7 Row-colum assignment for surface correspondence establishment using the continuous network, adapted from Lin et al (1991).

This is usually achieved at the pre-processing stage by selecting the vertex of a polygon with minimum x coordinate as the starting vertex. If two vertices happen to possess the same minimum x coordinate, the one with smaller y coordinate is selected as the starting vertex. The subsequent vertices are numbered sequentially in a clockwise direction. A useful feature for vertex correspondence establishment is the shape number (Underwood and Coates, 1985). This is used to form another 2D array which is similar to the one used to match surfaces. Both the networks use the continuous Hopfield model, which is found to be more suitable than using a binary Hopfield model, as was done in Nasrabadi and Li (1991).

11.7 DIRECT, OBJECT-BASED ANN APPROACHES

Since ANNs can be used to realize an unknown mapping given sample points, it is not surprising that they have also been directly applied to match object models. For example, Hwang and Henderson (1991) convert the topological and geometric information in a boundary representation model of an object to a set of real numbers corresponding to 'face scores'. These vectors are then mapped onto features by using a single-layered perceptron. This approach has been used to recognize slots with an associated confidence factor which is computed from the distance between the input and trained features. However, this method is limited to simple objects, and the extraction and coding of inputs is not automated.

Another use of the mapping capabilities of feed-forward networks is given by Smith and Jepsen (1991) for defect location using non-destructive evaluation techniques. Their two-layer network locates defects by deriving inputs from a window that is sliding over input image and, at each step, labeling the center pixel with the thresholded network output. This is motivated by the observation that human inspectors rely primarily on local characteristics of the defect itself, rather than on the context of an entire part for defect recognition. Their results show less sensitivity to noise than statistical novelty detection techniques applied to the same images.

Khotanzad and Liou (1991) have developed a robust and fast system for recognition as well as pose estimation of a 3D object from a single 2D perspective of it taken from an arbitrary viewpoint. The silhouette of the object is first normalized with respect to scale and location. Then, a set of rotation invariant features called **Zernike moments** of the image are extracted. The Zernike moments are only rotation invariant. To achieve location invariance, the origin is moved to the centroid of the image. Scale invariancy is achieved by enlarging or reducing each object such that its zeroth order moment is set to a constant. A bank of multi-layer feed-forward neural networks are used to classify the extracted features. These networks have the same topology and map the same input to one of C classes representing C object types. The main difference between them is in the initial condition of their weights and the sequence of presentation of training samples to them. Since gradient descent converges to a local minimum in weight space, if initial conditions and sequence of presentations are altered, the weights may converge to a different set of values which sometimes differ greatly. Another interpretation is that these different weights correspond to different ways of forming generalizations about the patterns inherent in the training set. Thus, one can argue that the collective decision produced by the bank of ANNs is less likely to be in error than by any single ANN. After the object is classified, its pose (aspect and elevation angles) is estimated by another bank of neural networks which are trained for the particular class to which the object is found to belong. A similar approach to object identification and pose estimation using 3D moment invariants has been proposed by Lu, Lo and Don (1991). Such approaches do well if the objects are unoccluded and silhouette-based features are sufficient to describe an object.

11.8 CONCLUDING REMARKS

Object recognition for vision-based inspection demands robust mechanisms for image processing and feature extraction coupled with powerful object representation and matching paradigms. For inspection, often the type of objects and views expected may be known a priori, thus simplifying the search space of possible object-view combinations. However, one may be interested not only in identifying an object but also determining whether it deviates significantly from a standard representative of that object. Besides such visual clues, other sensing modalities such as X-ray or infra-red imagery may also be used for detection of damaged or deformed parts. System-level specifications and performance metrics then include minimum identifiable defect size, detection probability, false alarm rate and throughput. Neural networks complement statistical pattern recognition techniques for this purpose, and are often found to be more robust to noise and parameter variations.

In this chapter we noted that most of the ANN techniques have been applied at the object matching stage, and also gave a few examples where neural networks have been used for feature extraction. A significant recent development is the emergence of 'silicon retinas' (Mead, 1989) which are analog VLSI chips that use networks of resistive grids and operational amplifiers to perform edge detection, smoothing, segmentation, compute optic flow and so on. The possibility of implementing sensors, early vision algorithms as well as object matching routines on single chips can spell the emergence of low-cost, widely available systems for vision-based inspection. To realize this promise, it is imperative that more versatile object representations and matching techniques, that are amenable to VLSI implementation using neural-like circuitry, are developed.

ACKNOWLEDGEMENTS

I would like to thank Jake Aggarwal, Farshid Arman, Srinivasa Chakravarthy and Madhu Katta for useful discussions at various stages of this work.

REFERENCES

Arbib, M.A. and Hanson, A.R. (1987) *Vision, Brain and Cooperative Computation*, MIT Press, Boston, MA.

Arman, F. and Aggarwal, J.K. (1993) Model based object recognition in dense range images – A review. *ACM Computing Surveys*, **25**(1), 5–43.

Ballard, D.H. and Brown, C.M. (1982) *Computer Vision*, Prentice-Hall, Englewood Cliffs, NJ.

Besl, P.J. and Jain, R.C. (1985) Three-dimensional object recognition. *ACM Computing Surveys*, **17**(1), 75–145.

Besl, P.J. (1988) *Surfaces in Range Image Understanding*, Springer-Verlag, New York.

Bhanu, B. and Ho, C.C. (1986) Geometric design based 3D models for machine vision, in *8th International Conference on Pattern Recognition*, Paris, France, (Oct 27–31), pp. 107–10.

Bolle, R.M. and Vemuri, B.C. (1991) On 3D surface reconstruction methods, in *I.E.E.E. Trans. Pattern Analysis and Machine Intelligence*, **13**(1), pp. 1–13.

Bovik, A.C., Clark, M. and Geisler, W.S. (1990) Multichannel texture analysis using

localized spatial filters. *I.E.E.E. Trans. Pattern Analysis and Machine Intelligence*, **PAMI-12**, 55–73.
Broomhead, D.S. and Lowe, D. (1988) Multivariable functional interpolation and adaptive networks. *Complex Systems*, **2**, 321–55.
Chakravarthy, S.V., Ghosh, J. and Jaikumar, S. (1991) Aspect graph construction using a neural network of radial basis functions, in *Intelligent Engineering Systems Through Artificial Neural Networks* (eds C.H. Dagli, S.R.T. Kumara and Y.C. Shin), 465–72, ASME Press, New York.
Chein, C.H., Sim, Y.B. and Aggarwal, J.K. (1988) Generation of volume/surface octree from range data, in *Proc. of Computer Vision and Pattern Recognition*, Ann Arbor, MI (Jun 5–8), pp. 254–60.
Chin, R.T. and Harlow, C.A. (1982) Automated visual inspection: A survey, in *I.E.E.E. Trans. on Pattern Analysis and Machine Intelligence*, **PAMI-4**, (6), Nov.-82, pp. 557–73.
Chin, R.T. and Dyer, C.R. (1986) Model-based recognition in robot vision. *ACM Computing Surveys*, **18**(1), 67–108.
Cohen, M.A. and Grossberg, S. (1983) Absolute stability of global pattern formation and parallel memory storage by competitive neural networks. *IEEE Trans. Systems, Man and Cybernetics*, **13** (5), 815–26.
Davis, L.S. and Rosenfeld, A. (1981) Cooperating processes for low-level vision: A survey. *Artificial Intelligence*, **3**, 245–64.
Dickinson, S.J., Pentland, A.P. and Rosenfeld, A. (1992) 3D shape recovery using distributed aspect matching. *I.E.E.E. Trans. PAMI*, **14**(12), 174–98.
Dupaguntla, N.R. and Vemuri, V. (1989) A neural network architecture for texture segmentation and labeling, in *Proc. International Joint Conference on Neural Networks*, pp. 127–44 (I).
Edelman, S., Bulthoff, H. and Weinshall, D. (1989) Stimulus familiarity determines recognition strategy for novel 3D objects. *A.I. memo no. 1138.*, AI Lab, MIT.
Eggert, D. and Bowyer, K. (1989) Computing the orthographic projection aspect graph of solids of revolution, in *Proc. of I.E.E.E. Workshop on Interpretation of 3D Scenes*, Austin, Texas (Nov 27–29), pp. 102–8.
Freeman, H. (ed.) (1989) *Machine Vision for Inspection and Measurement*, Academic Press.
Freeman, H. and Chakravarty, I. (1980) The use of characteristic views in the recognition of three-dimensional objects, in *Pat. Rec. in Practice* (eds E. Gelsema and L. Kanal), North Holland Publishing Co., Amsterdam.
Freeman, H. and Wang, R. (1990) The use of characteristic view classes for 3D object recognition, in *Machine Vision for Three-Dimensional Scenes* (ed. H. Freeman), Academic Press.
Ghosh, J. and Hwang, K. (1989) Mapping neural networks onto message-passing multicomputers. *Journal of Parallel and Distributed Computing*, **6**, 291–330.
Ghosh, J., Gopal, N. and Bovik, A.C. (1990) Textured image segmentation using localized receptive fields, in *Proc. Int'l Joint Conf. Neural Networks*, Washington, DC, Jan 15–19, pp. 283–6.
Ghosh, J. and Bovik, A.C. (1991) Processing of textured images using neural networks, in *Artificial Neural Networks and Statisticial Pattern Recognition* (eds I.K. Sethi and A. Jain), Elsevier Science, Amsterdam, pp. 133–54.
Gigus, Z. and Malik, J. (1988) Computing the Aspect Graph for Line Drawing of Polyhedral Objects, in *Proc. of Computer Vision and Pattern Recognition*, Ann Arbor, MI, Jun 5–8, pp. 654–61.
Grossberg, S. and Mingolla, E. (1985) Neural dynamics of perceptual grouping: Textures, boundaries and emergent segmentations. *Perception and Psychophysics*, **38**, 141–71.
Grossberg, S. and Mingolla, E. (1986) Computer simulations of neural networks for

perceptual psychology. *Behavior Research: Methods, Instruments and Computers*, **18**, 601–7.

Grossberg, S. (1987) Cortical dynamics of three-dimensional form, color and brightness perception, I: Monocular theory. *Preception and Psychophysics*, **41**, 87–116.

Hebb, D.O. (1949) *Organization of Behavior*, Wiley.

Hubel, D.H. and Wiesel, T.N. (1962) Receptive fields, binocular interaction, and functional architecture in the cat's visual cortex. *J. Physiol, London*, **160**, 106–54.

Hummel, R.A. and Zucker, S.W. (1983) On the foundations of relaxation labelling processes. *I.E.E.E. Trans. on Pattern Analysis and Machine Intelligence*, **PAMI-5**, 267–87.

Hwang, J.L. and Henderson, M. (1991) Applying the perceptron to 3D feature recognition, in *Intelligent Engineering Systems Through Artificial Neural Networks* (eds C.H. Dagli, S.R.T. Kumara, Y.C. Shin), 485–94, ASME Press, New York.

Kandel, E. and Schwartz, J. (1985) *Principles of Neuroscience*, Elsevier, New York.

Kang, S.B. and Ikeuchi, K. (1991) Determining 3D object pose using the complex extended Gaussian image, in *Proc. of Computer Vision and Pattern Recognition*, Maui, Hawaii, Jun 3–6, pp. 580–5.

Khotanzad, A. and Liou, J. (1991) Recognition and pose estimation of 3D objects from a single 2D perspective view by banks of neural networks, in *Intelligent Engineering Systems Through Artificial Neural Networks* (eds C.H. Dagli, S.R.T. Kumara, Y.C. Shin), 479–84, ASME Press, New York.

Koenderink, J.J. and Van Doorn, A. (1979) The internal representation of solid shape with respect to vision, in *Biological Cybernetics*, **32**, 211–16.

Kohonen, T. (1989) *Self-Organization and Associative Memory*, 3rd edn, Berlin, Heidelberg: Springer-Verlag.

Korn, M.R. and Dyer, C.R. (1987) 3D multiview object representations for model based object recognition. *Pattern Recognition*, **20**(1), 91–104.

Lee, S. and Kil, R. (1988) Multilayer feedforward potential function network, in *Proc. IJCNN*, pp. 161–71.

Lin, W.C., Liao, F.Y., Tsao, C.K. and Lingutla, T. (1991) A heirarchical multiple view approach to 3D object recognition, in *I.E.E.E. Trans. Neural Networks*, **2**(1), 84–92.

Lu, M.-C., Lo, C.-H. and Don, H.-S. (1991) 3D object identification and pose estimation, in *Proc. of ANNIE'91* (Nov 10–13), St. Louis, Missouri, pp. 473–8.

Marr, D. and Hildreth, E. (1980) Theory of edge detection. *Proc. Roy. Soc. London B*, **207**, 187–217.

Marr, D. (1982) *Vision*, W. H. Freeman Publishing Co., San Francisco.

Mead, C.A. (1989) *Analog VLSI and Neural Systems*, Addison-Wesley: Reading.

Mitchie, A., Seida, S. and Aggarwal, J.K. (1985) Determining position and displacement in space from images, in *Proc. I.E.E.E. CVPR*, Jun 19–23, pp. 504–9.

Mjolsness, E., Gindi, G. and Anandan, P. (1989) Optimization in model matching and perceptual organization. *Neural Computation*, **1**, 218–29.

Mohan, R. (1989) Application of neural constraint satisfaction network to vision, in *Int. Joint Conf. on Neural Networks*, Washington, DC, Jun 18–22, pp. II: 619–20.

Mokhtarian, F. (1988) Multi-scale description of space curves and three-dimensional objects, in *Proc. of Computer Vision and Pattern Recognition*, Ann Arbor, Michigan, Jun 5–9, pp. 298–303.

Nalwa, V.S. (1988) Representing oriented piecewise C^2 surfaces, in *Proc. of Int. Conf. on Computer Vision*, Tampa, Florida, Dec 5–8, pp. 40–51.

Nasrabadi, N.M. and Li, W. (1991) Object recognition by a Hopfield neural network. *I.E.E.E. Trans. Systems, Man and Cybernetics*, **21**(6), 1523–35.

Perett, D.I., Smith, P.A.J, Potter, D.D. *et al.*, (1985) Visual cells in the temporal cortex sensitive to face view and gaze direction. *Proc. of Royal Society of London, Series B*, **223**, 293–317.

Plantinga, W.H. and Dyer, C.R. (1986) An algorithm for construction of aspect graph, in *Proc. IEEE Symp. on Foundation of Computer Sciences*, pp. 123–31.

Poggio, T. and Girosi, F. (1990) Networks for approximation and learning. *Proc. I.E.E.E.*, **78**(9), 1481–97.

Poggio, T. and Edelman, S. (1990) A network that learns to recognize three-dimensional objects. *Nature*, **343**, 263–6.

Ponce, J. and Brady, M. (1987) Toward a surface primal sketch, in *Three-dimensional Machine Vision* (ed. T. Kanade), Kluwer Academic Publishers, Boston, Massachusetts, pp. 195–240.

Pravin, B. and Medioni, G. (1989) A constraint satisfaction network for matching 3D objects, in *Int. Joint Conf. on Neural Networks*, Washington, DC, Jun 18–22, pp. II: 281–6.

Rosenfeld, A. (1987) Recognizing unexpected objects: A proposed approach. *Intl. J. of Pattern Recognition and Artificial Intelligence*, **1**(1), 71–84.

Seibert, M. and Waxman, A.M. (1989) Spreading activation layers, visual saccades, and invariant representations for neural pattern recognition systems. *Neural Networks*, **2**(1) 9–27.

Seibert, M. and Waxman, A.W. (1992) Learning and recognizing 3D objects from multiple views in a neural system, in *Neural Networks for Perception*, Vol. 1 (ed. H. Wechsler), Academic Press, pp. 426–44.

Smith, F.G. and Jepsen, K.R. (1991) Case study of image processing techniques for nondestructive evaluation, in *Intelligent Engineering Systems Through Artificial Neural Networks* (eds C.H. Dagli, S.R.T. Kumara, Y.C. Shin), 429–34, ASME Press, New York.

Taubin, G., Bolle, R.M. and Cooper, D.B. (1989) Representation and comparing shapes using shape polynomials, in *Proc. of Computer Vision and Pattern Recognition*, San Diego, CA, Jun 4–8, pp. 510–16.

Underwood, S.A. and Coates Jr., C.L. (1985) Visual learning from multiple views. *I.E.E.E. Trans. Computers*, **C-24**(6), 651–61.

Van der Malsburg, C. (1973) Self-organization of orientation sensitive cells in the striate cortex. *Kybernetik*, **14**, 85–100.

Van Essen, D.C. (1985) Functional organization of primate visual cortex. *Cerebral Cortex* **3**, 259–329, Plenum Press.

Vemuri, B.C. and Aggarwal, J.K. (1986) 3D model construction from multiple views using range and intensity data, in *Proc. of Conference on Computer Vision and Pattern Recognition*, Miami Beach, Florida, Jun 22–26, pp. 435–7.

Walters, D. (1987) Selection of image primitives for general purpose computing. *Computer Vision, Graphics and Image Processing*, **37**, 261–98.

CHAPTER 12

Performance analysis of artificial neural network methods

Benito Fernández R., Ph.D.

Neuro-Engineering Research and Development (NERD) Laboratory, Departments of Mechanical Engineering, The University of Texas at Austin, Austin, Texas, 78712-1063.

12.1 INTRODUCTION

The complex demands of many modern applications cry out for flexible, adaptable devices that can quickly react to changing circumstances as solutions rather than mechanical brute force. Artificial neural systems (ANS) promise this capability. It is the intent of this chapter to realize ways to evaluate this accomplishment and its performance.

As a result of the study conducted by the DARPA's (1988) System Application Panel, the following conclusion was offered:

> Neural networks provide significant advantages in solving processing problems that require real-time encoding and interpretation of relationships among variables of high-dimensional spaces.

Performance analysis of neural networks (NN) is by no means an exact science, just as the same holds for other technologies. In general authors will use different sets of measures to evaluate their specific networks and applications against the others available and hence justify/defend their selection/design.

Artificial Neural Networks for Intelligent Manufacturing. Edited by Cihan H. Dagli. Published in 1994 by Chapman & Hall, London. ISBN 0 412 48050 6

Performance evaluation is in general a weighted mutliple criteria. The selection of the variables to measure and the way the variables are to be measured present interesting dilemmas. Due to the broad spectra of artificial neural systems (ANS) and the myriad of applications, a consistent performance measure is not realistic and comparison among different ANS architectures might be futile.

The essential components defining an ANN are (see Fig. 12.1):

- network topology (i.e., MLP, ART, CABAM, etc.);
- network size (number of layers, number of nodes per layer);
- network functionality (nodes input operator, $f(\cdot)$, transfer function, $H(s)$, and discriminatory function, $\Pi(\cdot)$);
- learning procedure (learning algorithm, on/off-line, learning parameters, sequential/batch updating, etc.);
- training/validation (training size, quality, data format, etc.);
- implementation/realization (analog, discrete, software, etc.).

The performance measure will involve the selection of these features and quantifying, in some form, how successful the selection was. Another facet of performance measure, not covered here, is how well ANS compare against alternative approaches. As specified before, the result of the performance evaluation will depend significantly on the application. We should always strive for the application of the right tool for the required job.

12.2 ARTIFICIAL NEURAL SYSTEMS IN MANUFACTURING

Today's complicated machine tools, manufacturing cells and systems require systematic methods of control, monitoring and fault diagnostics. In intelligent manufacturing systems, unprecedented and unforeseen situations are expected to be solved, within certain limits, even on the basis of incomplete and imprecise information.

The manufacturing studies board (1986) contends that manufacturing is not limited to the material transformations performed in the factory. It is a system encompassing design, engineering, purchasing, quality control, marketing and customer service as well as materials transformation. The system approach is a key principle not only for manufacturing technology, but also for organizational structure, supplier relations, and human resource management.

In manufacturing, neural networks have been used in the diagnosis and control of suspension systems of automobiles; in the discovery of process problems in the injection molding of plastic parts (Smith and Dagli, 1991); in the detection of defects in rolling element bearings in large rotating machines (Stevenson *et al.*, 1991); in the processing of picture images in cameras (McAulay *et al.*, 1990); in the estimating of bore tolerance in circular end milling; in the design of complex adaptive controllers for continuous processes (Dornfeld, 1990; Sutton, 1992).

Neural nets help manufacturers to pursue CIM by offering a new level of understanding for the relationships between design decisions, manufacturing and inspection methods and capabilities, and profitability (Liu and Liu, 1990; Wang and Malakooti, 1992). This level of understanding enables significant planning, operational and strategic opportunities (Din, 1990). The application area of neural networks in CIM is broad; it covers nearly all of the fields spreading from the design phase through simulation, control, monitoring and quality assurance to the maintenance.

12.3 THE POWER OF NEURAL NETWORKS

Neural networks are computational structures modeled on biological processes. Some of its advantages are: adaptation and learning, ease of implementation, and self-organization. As a general definition (see Fig. 12.1), we can say that a neural network is a directed graph and Fernández (1991):

- it has arbitrary topology that connect nodes (neurons) and links (axon/synapse/dendrite);
- each node has an input operator, $f(\cdot)$, a transfer function, $H(s)$, and a discriminatory function, $\Pi(\cdot)$;
- the internal node variable is part of the state variables (called STM–short-term memory) – the only state of the network once it learns;
- the rest of the state variables while learning is represented by the links – the links values change during learning (called LTM–long-term memory);
- the network functionality/capability[1] is given by the network topology, the nodes functionality, and the network weights;
- networks are (in general) organized into layers – an input layer, an output layer, and in general one or more hidden layers; and
- neural networks have a metric in Hilbert space with linear transformations between layers, and (in general) non-linear operators within layers.

The neurodesigner must specify the appropriate elements (architecture, node's functionality, learning rule, and training set) for an efficient design. He/she must decide on the following details.

- Are we emulating a biological system? Is it necessary to reproduce the neuron firing rate curve?
- What are the speed requirements? Can learning and decision making take place within real-time constraints?
- Should the node's functionality be invertible, monotonic, or continuous?
- What are the nature of our interconnecting weights? Will they be binary, continuous, or discrete?
- And finally, what type of output do we want/need?

[1] 'cognitive task'

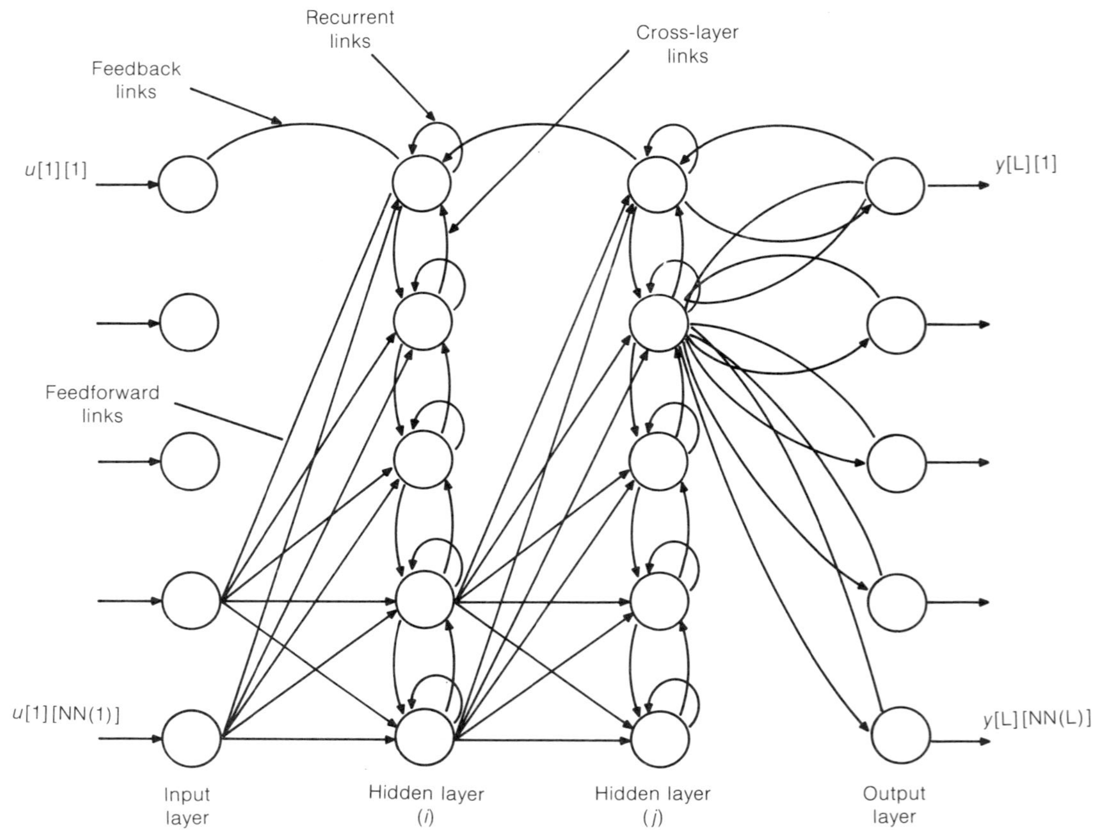

Fig. 12.1 General artificial neural network architecture.

The behavior of a neural net depends upon two things:

- the information processing properties of the individual network elements (**network model**); and
- the manner in which those elements are organized (**network topology**).

The neural network learning is accepted from large amount of data in the form of training sets. The power of neural networks lie in the model-free approximation capabilities:

> Given a finite-dimensional, time-invariant, deterministic system from which input–output data has been drawn, and given reasonable amount of data and time, there exist an ANN that will adjust its parameters to reproduce the system in an input–output sense (McLelland *et al.*, 1986; Rumelhart *et al.*, 1986; Hecht-Nielsen, 1987; Cybenko, 1989).

12.3.1 Some basic ANN mechanisms

ANNs are a low-level information-processing mechanism. There are several fundamental ANN building-block functions (Cruz, 1988):

- **encoder** (instar): a circuit to recognize, or generate a response to the presence of, a particular static activation pattern;
- **decoder** (outstar): a circuit to produce a desired static activation pattern on command;
- **sequence encoder:** a circuit to recognize a particular spatiotemporal input activation pattern; and
- **sequence decoder** (avalanche): a circuit to produce a desired output spatiotemporal activation pattern on command.

Both static and adaptive versions of these are possible. In addition to the four circuits cited above, there are some housekeeping functions that recur across numerous ANN applications.

- **Competitive mechanisms** – a pervasive mechanism where the neurons compete against each other inhibiting laterally the activation level of the other neurons of the same cluster.
- **Cooperative mechanisms** – another pervasive mechanism where the neurons cooperate among each other exciting laterally the activation level of the others neurons of the same cluster.
- **Normalization mechanisms** – or scaling – is a mechanism by which, the net tries to adjust the neurons weights of the same cluster such that the activation level does not saturate or never fires.

Cooperation and competition usually work together through lateral voting (casting yeas and nays) to neighboring neurons. The information in an activation pattern resides in the relative activation levels of the pattern's constitutive nodes, rather than in the absolute levels. Normalization acts as an automatic gain control that regulates the total activation in a pattern, while preserving the ratios of this levels.

12.3.2 Network topology and classification

An artificial neural network (ANN) is defined mainly by three features; topology, functionality, and learning. **Topology** refers to the number of layers, number of nodes in each layer, and the way nodes are connected (see Fig. 12.1). **Functionality** refers to the transfer function and discriminatory function (if any) of each node, and the cost function of the network outputs. **Learning** refers to the learning algorithm and the values of learning parameters (e.g., learning rates, vigilance coefficient, momentum).

ANNs can be classified differently according different criteria: the type of learning (supervised, unsupervised, self-supervised), the mode of operation (on-line, off-line), the network functionality (pattern associator, auto associator, classification builder, regularity detector), the network interconnection architecture (feedforward, feedback, recurrent), the number of layers, the learning algorithm (Hebbian, backpropagation, resonant, Widrow–Hoff, simulated annealing), and the type of data processed (binary, discrete, continuous, fuzzy).

The most popular neural network topology is feedforward networks, e.g., multilayer-perceptron (MLP). But networks may also have feedback connections, 'cross-talk' (intra-layer connections, for winner-takes-all functionality), and recurrent connections. Recurrent networks are used for dynamic memory encoding and for dynamic system identification, as well as for the encoding of sequences (spatio-temporal networks, i.e., state machines).

12.3.3 Node functionality

The node's functionality provides the network with the capability of representing the knowledge associated with the task at hand. Node functionality is composed of three parts (see Fig. 12.2):

- an **input operator**, $f(\cdot,\cdot)$, that combines the inputs and interconnection weights $\mathbf{W}$ to form a single value s (e.g. $s = f(w, y) = \mathbf{w}^{\mathrm{T}} \cdot \mathbf{y}$) ready for the transfer function;
- a **transfer function**, $H(s)$, that operates on the node's input operator output (e.g. integration) to form a value ready for the discriminatory action, x; and
- a **discriminatory function**, $\Pi(\cdot)$, that operates on the node's transfer function output and governs the output range of the node, $y = \Pi(x)$.

In Tables 12.1 and 12.2 we show some of the most popular input operators and discriminatory functions. The transfer functions that are popular are the unit operator (i.e., the output of the input operator goes directly into the discriminatory function) and the integral operator (where the output of the input operator is integrated before it enters the discriminatory function). In

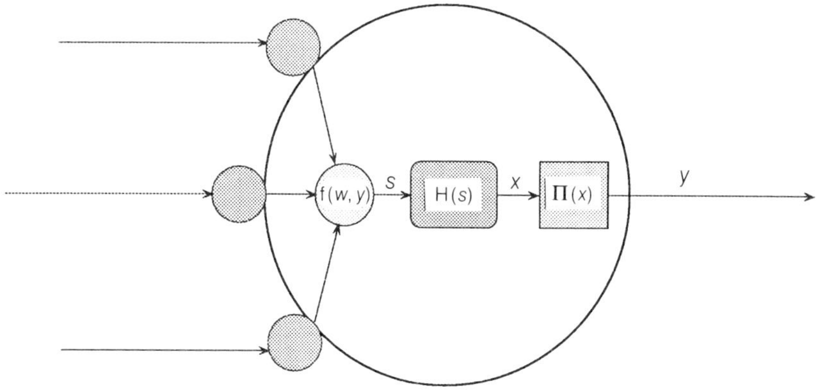

Fig. 12.2 Genetic ANN node.

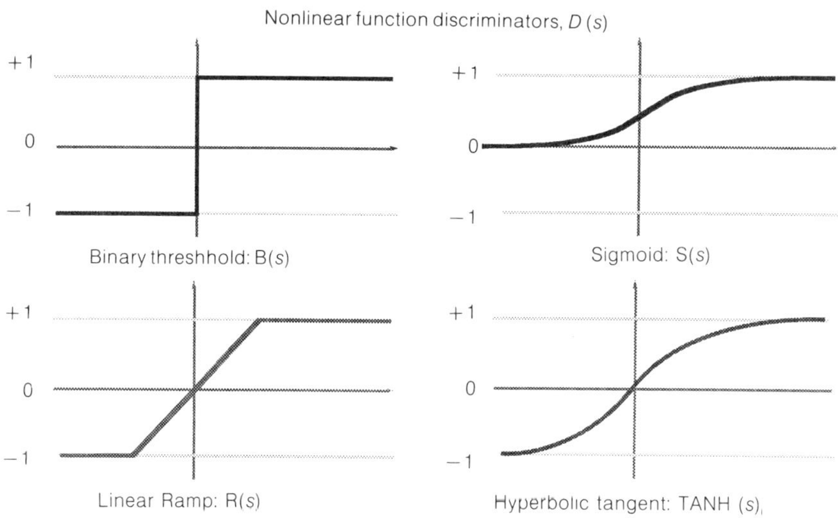

Fig. 12.3 Discriminatory functions.

most cases the integral action is performed using Euler approximation through a delayed recurrent link of the node's output. As you can see, there is a lot of flexibility and several degrees of freedom allowed in our choice of ANNs.

12.3.4 Learning rules

The learning rule's function is to correlate the input and output values of the PEs (nodes) by adjusting the interconnection weights. The neurodesigner has to consider the close match between the design and the application.

Table 12.1 Common discriminatory functions

Discriminator	*Advantages*	*Disadvantages*
Binary threshold, $B(s)$	Fast and easy use, especially in hardware Makes hard decisions	Not invertible No linear zone Cannot smoothly imitate functions
Linear ramp, $R(s)$	Has a linear zone Can smoothly imitate functions Easy to implement.	Not invertible
Sigmoid, $S(s)$	Invertible and C^∞ Semilinear zone to imitate Can make soft and fuzzy decisions Similar firing that biol. neuron.	More difficult to implement than piecewise linear functions Output limited to positive values
Hyperbolic tangent, $\mathrm{Tanh}(s)$	Invertible and C^∞ Semilinear zone to imitate Can make soft and fuzzy decisions Can produce pos. and neg. outputs.	More difficult to implement than piecewise linear functions

- What are the 'optimal' variables to measure?
- What is a good performance measure to evaluate the training?
- Do we have a purely dynamical system?
- Will the learning phase be performed on-line or off-line?
- Will the learning phase rely on local or global information?
- Is convergence speed important?

Again, there any many types of learning rules, and Table 12.3 presents the most widely used today.

In reading Table 12.3 the following notes apply:

1. from Parker (1986);
2. where $Y_j = \mathscr{B}(\sum_k W_{jk} X_k)$;
3. also known as steepest descent or least mean square (LMS), this is the most commonly used algorithm

$$Y_j = \mathscr{R}\left(\sum_k W_{jk} X_k\right)$$

4. the LMS algorithm is derived from this one, $e_j = d_j - Y_j$, $\eta = \alpha \cdot \Delta t$, where Δt is the time step, and

$$Y_j = \mathscr{R}\left(\sum_k W_{jk} X_k\right)$$

Table 12.2 Common input operators and transfer functions

Type	*Input Operator*	*Discriminator*	*Comments*
Simple linear	$x_j = \sum_i W_{ij} y_i$	none, $y_i = x^j$	Pure linear function used for associative recall
Weighted sum	$x_j = \sum_i W_{ij} y_i + \theta$	$y_j = \Pi(x_j)$	Commonly used; some case $\theta = 0$
Feedback	$x_j = \alpha x_{\text{old}} + \beta \sum_i W_{ij} y_i$	$y_{j,\text{new}} = R(x_j)$	Gives persistence to output state
Sigma pi	$x_j = \sum_i W_{ij} \pi y_j$	$y_j = \pi(x_j)$	One input can gate another or act as a gain control
Thermodynamic	$x_j = \{\sum_i W_{ij} y_i + \theta\} T^{-1} S$	$p(y_j = 1) = S(x_j)$ $p(y_j = 0) = 1 - S(x_j)$	p is the probability that a specific state is realized T is the temperature that is systematically lowered

Note: θ is the transfer function threshold, α and β are feedback weighting constants.

Table 12.3 Common learning rules

Algorithm	*Continuous/discrete formulas*	*Precondition for convergence*[1]
Hebb's Rule[2] (1949)	$\Delta W_{jk} = \eta Y_j X_k$	none
Widrow-Hoff Rule[3] (1960)	$\Delta W_{jk} = \eta(d_j - Y_j) X_k$	$0 < \eta < 1/tr(A)$
Gradient Descent[4]	$\Delta W_{jk} = \eta(e_j - A_{n+1} W_{jk})$	$0 < \eta < 2/tr(A)$
Generalized Delta Rule (1986)	$\Delta W_{jk} = \eta Y_j \delta_k$	none
(Back error propagation, BEP)	$Y = S(\sum_k W_{jk} X_k + \theta_k)$	none
	$\delta_k = \begin{cases} Y_k(1 - Y_k)(d_k - Y_k) & \text{(output layer)} \\ Y_k(1 - Y_k)\sum_k \delta_j W_{kj} & \text{(hidden layer)} \end{cases}$	
Greedy Delta Rule[5] (1960)	$\Delta W_{jk} = (d_j - Y_j) X_k / N_p$	none
2nd and LMS[6]	$\Delta W_{jk} = \eta e_j X_k + (I - \alpha A_{n+1} \Delta W_{jk}^{n-1})$	$(1 + \eta/2)\alpha < \eta < 2/tr(A)$
Exact Least Squares[7]	$W_{jk} = A_{n+1}^{-1} X_k$	none
Updating Exact Least Squares[8]	$W_{jk} = \eta A_{n+1}^{-1} e_j X_k$	$0 < \eta < 2$

5. N_p = number-of-input-ones;
6. ΔW_{jk}^{n-1} is the previous weight increment, momentum term;
7. all other learning algorithms are basically variations (approximations) of this one;
8. this is the updating form of the LMS, and is the algorithm from which 2nd order LMS is derived;
9. $\mathrm{tr}(A)$ is the trace of the matrix A; and
10. the martix A is the average input correlation matrix, so

$$\mathbf{A} = \mathrm{avg}(\mathbf{x}^{\mathrm{T}}\mathbf{x}) = \begin{bmatrix} \mathrm{avg}(x_1^2) & \cdots & \mathrm{avg}(x_1 x_p) \\ \vdots & \ddots & \vdots \\ \mathrm{avg}(x_p x_1) & \cdots & \mathrm{avg}(x_p^2) \end{bmatrix}$$

12.4 ARTIFICIAL NEURAL NETWORK PARADIGMS IN MANUFACTURING

The application of artificial neural systems (ANS) in manufacturing is quite broad, but there are three major tasks that ANS perform: classification (C), prediction (P), and function approximation (FA). Although all these task could be grouped into a single underlying paradigm, i.e., association, the three tasks will be used here for performance measures purposes. **Function approximation** is the association of a set of independent variables $\mathbf{x} = [x_1, x_2, \ldots, x_m]^{\mathrm{T}}$ with an 'output' signal $y = f(x_1, x_2, \ldots, x_m)$ into pairs $\{y, f(\mathbf{x})\}$. Prediction is the association of previous values of 'states', $\mathbf{X} = [x(k),\ x(k-1), \ldots, x(k-n)]^{\mathrm{T}}$ and current and previous values of the 'inputs', $\mathbf{U} = [u(k), u(k-1), \ldots, u(k-m)]^{\mathrm{T}}$ with future values of the states, $x(k+1) = f(\mathbf{X}, \mathbf{U})$ into pairs $\{x(k+1), \{\mathbf{X}, \mathbf{U}\}\}$. Classification is the association of a set of inputs $\{x_i^{c_j},\ i = 1, \ldots, N_j\}$ with a class C_j into pairs $\{C_j, \mathbf{x}^{c_j}\}$.

In what follows, we briefly describe the most popular networks that are in use within manufacturing systems. The following is a list of widely used paradigms.

- Multilayer perceptrons (MLP)
- Associative memories (LAM, BAM, OLAM).
- Hopfield networks
- Adaptive resonance theory (ART)
- Learning vector quantizer (LVQ)
- Sparse distributed memory (SDM)
- Fuzzy associative memory (FAM)
- Self-organizing map (SOM)

12.4.1 Multilayer perceptrons

Also known as backpropagation networks (BPN), multilayer perceptrons are the most popular and most used in practice today. Being the extension of the pioneer single-layer perceptron, of McCulloch and Pitts (1943), they are feedforward networks with one input layer, one output layer, and may have one or more hidden layers. MLPs are universal approximators of model-free function mappings, hence its popularity. They have been studied extensively in the literature and applied in industry and academia (Wasserman and Octzel, 1990).

12.4.2. Associative memory paradigms

Associative memory (Kohonen, 1977) is an emulation of the biological memory – the capability of association. Associative memory uses distributed structure for memory function. Memory retrieval is done by association, unlike computer memory, which uses indexing or addressing. The associative memory is a data-to-data mapping device. There are two primary mapping mechanisms: autoassociative and heteroassociative (see Fig. 12.4). There are also two primary operation modes of associative memories: learning (encoding) and recall (decoding). Encoding is a process that the weights of the connections between neurons are set. The recall is a process where a pattern is retrieved given a 'cue' (or 'key') pattern.

Linear associative memory (LAM)

LAM (Anderson, 1968) is a two-layer heteroassociative analog pattern encoder. It uses Hebbian learning (correlation equation) to encode patterns $W=\sum_k A_k \cdot B_k$, where the network is trying to associate patterns A_k with patterns B_k into pairs $\{A_k, B_k\}$. The recall is done by applying matrix multiplication $\mathbf{B} = \mathbf{A} \cdot \mathbf{W}$, which is an interpolative type instead of the more popular nearest-neighbor recall. LAM has the advantage of fault tolerance. When the weights are corrupted in a slight degree, the recall operation still can retrieve the desired output pattern. But it can store only linear mappings and the storage capacity is limited to the number of input nodes.

Bidirectional associative memory (BAM)

BAM (Kosko, 1987a) (see Fig. 12.4(a)) is a two-layer heteroassociative content-addressable memory (CAM). The encoding is Hebbian (Kosko, 1988) and uses outer product of the first and the second pattern vectors, $\mathbf{W} = \mathbf{A} \cdot \mathbf{B}$. The BAM recall uses resonation like ART. When the input and output vectors stop changing, the heteroassociative pattern (A, B) is obtained. BAM has unconditional stability (Kosko, 1988). The capacity, C, of BAM is estimated to be $C = \sqrt{\min(m, n)}$ (Wang *et al.*, 1990), where m and n are the number of nodes of the two layers of the BAM network. An adaptive version of BAM called ABAM also exists – also a continuous implementation exists called CABAM for encoding of analog signals.

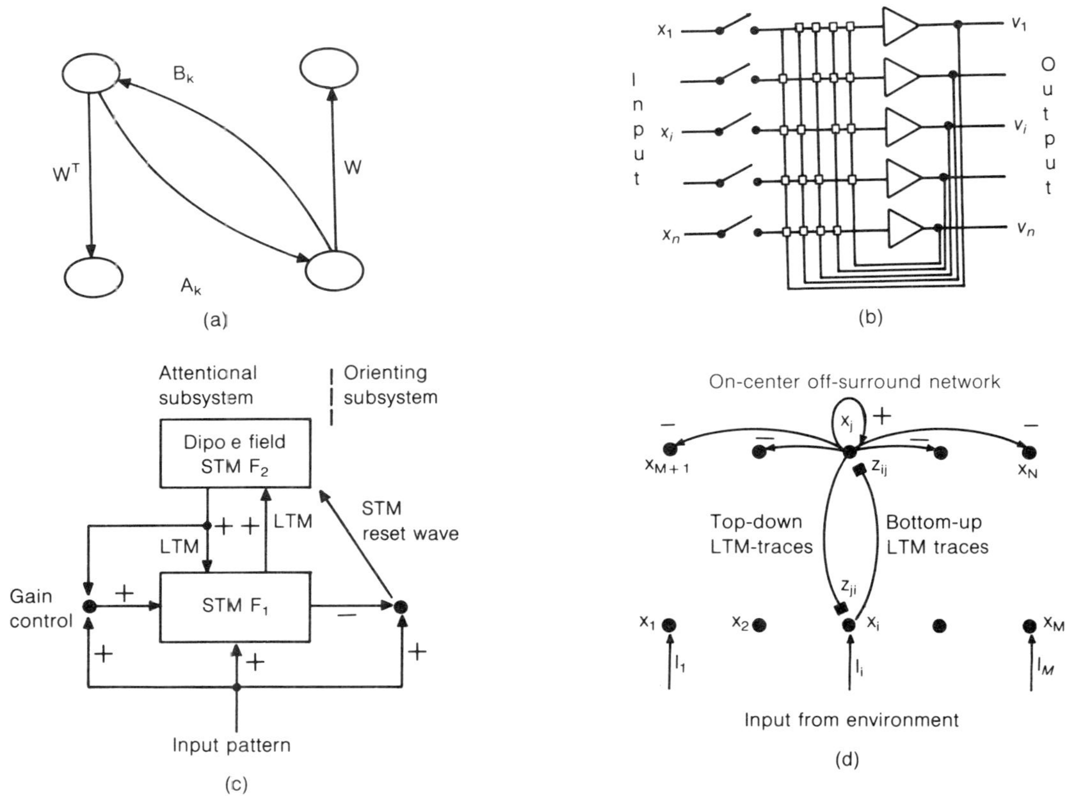

Fig. 12.4 Associative memories: (a) BAM; (b) Hopfield; (c) ART; and (d) LVQ.

Optimal linear associative memory (OLAM)

OLAM (Wee, 1968; Kohonen and Ruohonen, 1973) employs an encoding algorithm that gives optimal correlation (in the sense of least square) using the pseudo-inverse technique. OLAM can be used as heteroassociative encoder, autoassociative encoder, or new pattern detector. The storage capacity of OLAM exceeds the number of input nodes but it needs more computational time compared to LAM. It is still limited to linear mappings.

Hopfield networks

The Hopfield (1982, 1984) model is a single layer, symmetric, non-linear neural network. It learns off-line and is an autoassociative encoder. It is capable of retrieving the entire stored pattern by given a partial key. It has global stability for memory recall. Hopfield model has two versions: discrete and continuous time. The convergence rate to stored patterns is inversely proportional to the percentage of corrupted bits and number of stored patterns in the network, but is proportional to the Hamming distances between stored patterns (Zurada, 1992). The capacity, C, is the maximum number of patterns that can be stored by the network. C is estimated (McEliece *et al.*, 1987) to be bounded by $n/(4\ln(n)) < C < n/(a\cdot\ln(n))$, where n is the number of nodes. The Hopfield model has been applied to combinatorial optimization problems successfully.

Adaptive resonance theory (ART)

The adaptive resonance theory (Carpenter and Grossberg, 1986; 1987) first proposed by Grossberg (1968, 1988) explains why feedback connections are required to solve instability of feedforward-only networks. It also explains how the system makes expectations, pays attention, and learns patterns. ART serves the purpose of cluster discovery. Similar to Kohonen networks, ART learns patterns unsupervised. The network produces clusters by itself, storing clustering information about patterns or features without a priori information regarding number or type of clusters. When an input pattern sufficiently different from previously stored patterns, a new cluster is formed. 'Sufficiently different' is determined by a global internal parameter called the **vigilance parameter** which must be adjusted externally to provide the desired sensitivity to differences in the input patterns. The network has two layers, an attentional subsystem, and an orienting subsystem. (see Fig. 12.4(c)).

Learning vector quantizer (LVQ)

The learning vector quantizer (Pratt, 1978) is a heteroassociative pattern classifier. It uses competitive learning for encoding algorithm. Each weight vector represents the centroid of one decision region, i.e. the network learns prototype patterns as synaptic points, m_i, in some pattern space R^n (see Fig. 12.4(d)). If an input vector is close to the centroid, it converges to it. The prototypes 'move around' (adapt) as learning proceeds. It is useful for data compression and combinatorial optimization problems.

LVQ is optimal according to some criteria. The prototypes may spread themselves to minimize the mean-squared error of vector quantization, or to minimize some other performance criteria. More generally, the prototypes should estimate the underlying (unknown) probability distribution of the training patterns.

Temporal associative memory (TAM)

TAM is a binary-valued spatio-temporal pattern sequence encoder (Amari, 1972). The learning rule is Hebbian learning; the recall uses threshold operation. It is basically a modified version of BAM with identical weights in the first and the second layers. The capacity has the bound as $C \leqslant \min(m, n)$.

Sparse distributed memory (SDM)

SDM (Kanerva, 1984) is a three-layer heteroassociative pattern classifier. Its main attractiveness is that the storage capacity is larger, but needs many more nodes. SDM can be applied to temporal sequence encoding like speech processing.

Fuzzy associative memory (FAM)

FAM (Kosko, 1987b) is a heteroassociative pattern classifier, which can be used for storing fuzzy rules. The encoding is done by using a fuzzy version of Hebbian learning, which means the weight between two nodes is the minimum of the membership function values at the input node and the output node. The recall can be two-way, but it is still a feedforward operation. Fuzzy composition is used in pattern (rule) recall. That is, $B = A \cdot W$. FAM can be applied to fuzzy control systems.

Self-organizing map (SOM)

SOM (Kohonen, 1984) is an extension of LVQ and is mainly used for systems that need dimensionality reduction of data without loss of information. Its applications include speech recognition, data compression, and combinatorial optimization.

The user should choose a suitable type of associative memory for his work and design the proper size (number of nodes in each layer) to optimize the performance under the constraints/requirements of memory capacity, fault tolerance, and execution time.

Boltzmann machine (BM)

Combinatorial optimization networks developed by Hopfield and Tank (1985, 1986) do not always find the best solution in constrained optimization problems. They can be trapped in local minima wells of energy. This problem can be alleviated by adding a stochastic part to the search algorithm called **simulated annealing** (Kirkpatrick *et al.*, 1983). A popular network, the BM (Ackely *et al.*, 1985; 1986), uses this strategy solving the credit assignment problem for a special case of recurrent networks with symmetric connections. Although it is theoretically interesting, and one of the ways to guarantee global minima convergence, simulated annealing causes BM training to be too long for practical applications on current hardware.

12.5 PERFORMANCE ANALYSIS

There are two, sometimes conflicting, measures of the network capabilities: the accuracy and the generalization of the solution. The **accuracy** of the match depends on the complexity of the non-linearities, the size of the network, the training time, and the learning algorithm used for training. There is a current issue in research regarding the existence of at least one (there might be multiple solutions depending on the network basis functions) minimal network size (hence optimal) for the representation of the system non-linearities within a prescribed accuracy. Another issue is that given the fact that there exists a network size for which the network could reduce the matching error to zero (for a finite training set) the achievability of that solution from a network of the same dimension with arbitrary initial conditions is not guaranteed. The responsibility lies with the learning/training algorithm to tune the network parameters to achieve at least one such set of parameters. The other issue is the fact that although the network could achieve a small error for the data points in the training set, generalization might be unsatisfactory. A trivial example (see Fig. 12.5) is when you try to fit six points that almost lie in a straight line. Accuracy is achieved by a polynomial, but the resulting solution varies wildly between points.

There is no clear criteria for evaluating and comparing the performances of different paradigms. But the important factors to be considered for performance analysis include (Hassoun and Youssef, 1989)

- stability/convergence
- execution time (synchronous or asynchronous operation)
- memory capacity (number of patterns that can be stored)
- noise tolerance
- avoidance of spurious memory states
- spatial or temporal patterns

Each of these criteria needs an appropriate measure, i.e., we need to parameterize the criteria and then weight each measure with respect to the

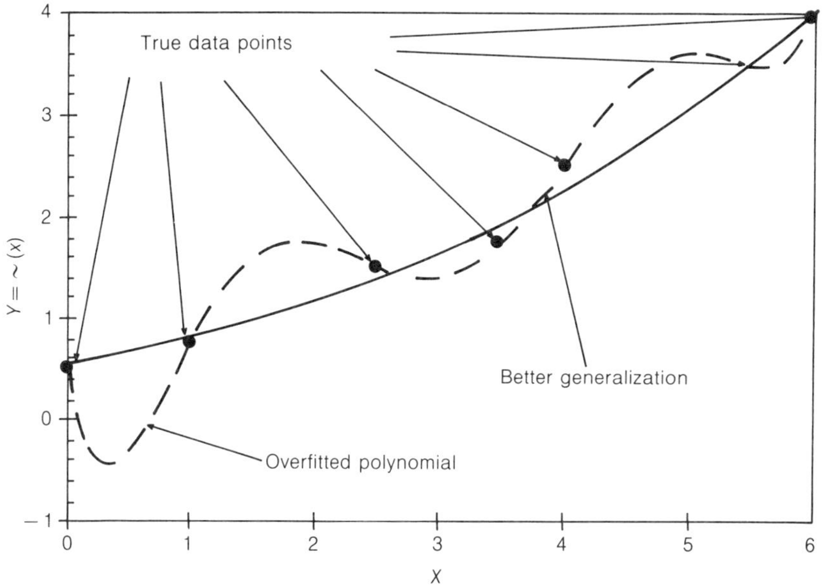

Fig. 12.5 Generalization versus overtraining.

others into a decision matrix. The following is a list of guidelines that can be used for the performance analysis of ANNs.

DARPA's Neural Network Study (1988) suggested tasks where neural networks research should focus.

- **Classification** Classifiers are trained with supervised data, labeled to group it in different classes.
- **Self-organization/category formation** Self-organizing networks partitions the input data into clusters according to certain criteria. The clusters are formed by the network itself and reduces the information needed for processing at higher levels of decision.
- **Associative memory** Associative or content-addressable memory provides a complete memory item from a 'key' consisting of a partial or corrupted version of the memory.
- **Sensory data processing** An enormous amount of sensory data processing is required for real-time application of vision and hearing problems.
- **Computational problems** Specific computational applications require custom ANN architectures.
- **Multi-sensor automata** A number of problems require the fusion of multiple sensors data.

There are some fundamental theoretical issues to be addressed when evaluating ANN performance.

- **Capability** What task can the network perform? Is there enough (rich) data available to present a solution to the problem? Does the network have the computational power to solve the problem?

- **Capacity** How much information can be stored in the network?
- **Learnability** What types of problem are learnable? Can the network be redesigned or the data representation altered to make the problem learnable?
- **Data representation** How are the data represented in the network? How does data representation affect structure, capacity, capability, and learnability?
- **Data extraction** How are the data used to generate a model of the world? Is the information stored 'understandable'? How does internal data representation affect knowledge acquisition?
- **Network design** How should the network be designed to perform specific functions? What topology, node functionality is required?
- **Credit assignment** Given an existing set of variable weights and training data, how is credit assigned to each weight during training for a correct response?
- **Training issues**
  - *Effectiveness* How much training data is required for a given level of performance?
  - *Learning efficiency* How long does it take for training to the desired weights values?
  - *Complexity* How does training time change as the problem is scaled up?
  - *Generalization* How well does a classifier generalize?
  - *Plasticity* Can new items, classes, or behaviors be added without disrupting current performance?
  - *Sensitivity* Is the network response invariant to irrelevant variability in the input? How is this invariance obtained?
  - *Adaptability* How is training time affected by internal data representations?
- **Performance/operation issues**
  - *Band width* How long does it take until the output settles and reaches the desired solution?
  - *Accuracy* How accurate is the solution?
  - *Size* How does network performance and capacity vary with network size?
  - *Robustness* How sensitive is the solution to component inaccuracies?
  - *Fault tolerance* How fault tolerant is the network to component failure?
  - *Portability* How easy is the network to implement with different types of hardware?
  - *Stability* Is the network operation (globally) stable?
  - *Separability* How is performance affected by internal data representations?

Experience with many different types of neural network models and data representations is beginning to provide design guidelines. The designer has several constraints apply upon him/her.

- **Resource requirements:** number of nodes, complexity of nodes functionality, fidelity of nodes, robustness of nodes, number of connections, weights adaptability, etc.
- **Performance requirements:** retrieval time (bandwidth), learnability, scaling, error recovery, fault tolerance, training protocols, generalization, accuracy, repeatability, etc.

To support a particular application, a neural network could be stochastic, deterministic, or hybrid; it could use static, dynamic or cyclic processes; the signals processed could be binary, analog, continuous or discrete; it could learn off-line or adapt in real-time; etc.

12.5.1 Qualifying performance

The DARPA study (1988) requested in a 'broad area announcement' inputs from researchers and industry on a set of application areas selected by the 'system applications panel'. The study compared several networks applied to 25 applications (selected from a pool of 268). The majority of the applications-oriented responses fell into eight categories: vision, speech, sonar, radar, signal processing, robotics, dynamical systems, and cognitive functions. The main characterization of the problems' features were those where ANN solutions seemed advantageous: data-intensive problems, and problems dealing with difficult and changing environments. The results were entered in a matrix table relating application to 'benefit'. The categories for performance evaluation were: speed, parallelism, ease of implementation, supervised learning, hierarchical control, self-organization, data fusion, unsupervised learning, generalization, fault-tolerance, spatial-temporal, large data, noise immunity. The table was filled with primary and secondary requirements for the application. Some of these categories are hard to parameterize and are very subjective, e.g., ease of use. The weighting of each category and the parameterization procedure need to be 'weighted' accordingly to each application.

One of the most difficult problems faced by neural networks today is evaluating the performance of proposed neural network paradigms (architecture/functionality/learning). Rating the quality of the networks is very much problem dependent. As mentioned before, it is often quite easy to define a 'toy' problem where one paradigm outperforms the rest of the field, yet application to real data results are hard to evaluate and not conclusive.

Another factor is the cost of evaluating performance, measured in resource allocation: personnel (with expertise in the field) to select the data, personnel, time and cost to gather data, personnel (with expertise in neural networks) to design the network, personnel, time and cost for training the network, time and cost of implementing solution.

Too often representative data are not readily available and simulations might be required which could seriously bias the results. The amount of data required to adequately train the network may also prove to be difficult to predict, as well as suitable test for the quality of the network progress.

Uncle Bernie's rules

To alleviate some of the problems confronted with the application of neural networks, a set of rules that Uncle Bernie (Bernard Widrow) suggests follows:

- **Data:** 10 pieces of data for every free parameter
- **Noise:** Add noise to data to maximize generalization
- **Scale:** Scale the training set whenever possible

This and the other suggestions in this chapter are general in nature. There is no implied guarantee that they will work in every case. There are obviously limitations for each of the techniques presented here. The intent is to share some of the 'tricks' we have used over the years in combating the slow and sometimes painful learning in neural networks.

Network capacity

An important measure for comparing networks is the amount of information they can store, although the quality and availability of that information should also be considered when comparing different network topologies and functionalities.

Network capacity is of little interest if the network is not capable of generalization. For pure encoding and recall a table look-up is far more efficient. A fundamental trade-off exists in neural networks between the network capacity and its generalization capabilities.

Computational requirements

The application spectrum of neural networks is wide and abundant. It ranges from signal processing applications like radar pulse identification where signal traces are measured every 50 sec, to robotics and speech, with sampling times of about 10 ms. Emulation, analog, digital, and hybrid implementations exist. To measure the computational requirements, a different measure is needed. The requirements are in

- **Memory requirements** (storage, number of nodes and layers),
- **processing speed** (interconnects/sec.)

Table 12.4 Neural networks computation units

	Storage (memory)	*Speed (operations/sec)*
Digital computer	words	IPS, flops
Neural network	interconnects	interconnects/sec

Most of the computational requirements for available and developing technology are usually compared in a speed vs. storage figure (DARPA, 1988); see Table 12.4.

ANNs using supervised learning

- **Computation time of each epoch:** The estimation of the average execution time of the neural network for each epoch is important, since it gives an indication of how much time it will take to complete learning if the total number of training cycles can be estimated or tested. For small problems, this may not be critical, but it has great influence on the overall training time if network is huge (Becker and le Cun, 1988).
- **Number of epochs for convergence:** When the maximum allowable learning rates are used, the computation time for each epoch is a good index of speed, but if a network needs a tremendously large number of iterations to reach the minimum point, it might not be useful unless slow learning is not a concern to the users. The standard backpropagation algorithm, in some cases, needs a lot of training time, therefore almost every research group has its own variation of BP algorithm. The total execution time is the number of epochs multiplied by the execution time for each epoch. A multi-layer perceptron type of network uses random weights as initial weights, therefore a batch of different initial conditions has to be tried and the average time can be found. The total execution time is the total CPU time for neural networks using digital computer simulations. For VLSI circuit implementations, it is the time for the circuit to reach the stable states.
- **Computer memory storage for variables:** Resource allocation/consumption is also an important factor for determining which neural network to use. If the computer storage is abundant, then a more complex architecture and sophisticated learning algorithm can be utilized. This depends on the application.
- **Network size:** The number of hidden layers, the total number of nodes (states and biases), and the total number of connections (weights) is the index for 'size' of a network. To test the performance of an NN completely, one has to use small, medium, as well as large problems. This is termed **problem (network) scalability**. One learning method may perform better for smaller problems, but takes too much time for larger problems. This depends on the order of magnitude to computational effort needed for each epoch. For a specific application, there is an optimal size of the NN. There are two approaches for **tuning** the network size: start with a small network and add nodes when necessary; start with a large network and prune unimportant nodes/links.
- **Data representation:** The ranges of input and output data have some influence on the speed of convergence of ANNs. If proper scaling is done, significant time could be saved in training. Scaling of training data changes the eigenvalues of the system, if the eigenvalues are of the same order of

magnitude, the gradient descent algorithm is likely to reach to the minimum faster for random initial conditions.

- **Static and dynamic problems:** A network that learns static mapping problems may not learn a dynamic problem at all. For engineering applications, it is desirable to use the same NN to learn static and dynamic problems. For dynamic problems, 'dynamic' links (feedback, crosstalk, and recurrent) are usually needed.
- **Generalization/test:** One NN may learn a mapping quickly enough but fail to generalize given a different set of inputs. Some NNs need a large training set to have good generalization capability. Some NNs generalize well even when the training set is small. There is always some trade-off between the training time and the generalization capability. It is possible that **overtraining** could happen and an NN fails to generalize well but the cost function for the training set is very small.
- **Benchmark problems:** There are some problems used by researchers as benchmark problems for various aspects of NNs. For example, XOR or parity problems are considered as hard static mapping/classification problem because the network output is totally different if one bit of data is changed in the input pattern. An N–M–N encoder problem is also used for testing. A good test problem for dynamic problems would be chaotic time series prediction (Lapedes and Farber, 1987). Character recognition problems are also popular benchmarks (Le Cun *et al.*, 1989).

ANNs using unsupervised learning

- **Stability/convergence:** ANNs using unsupervised learning are usually used as associative memories or non-parametric encoding mechanisms. The stability is the most important consideration for these networks. Given an input (cue), the NN can retrieve the associated pattern – no limit cycle or spurious state occurs. A **limit cycle** is oscillation between two patterns forever. A **spurious state** is a stable yet wrong pattern associated with the specific input.
- **Fault tolerance:** If the inputs are corrupted slightly, it is desirable that an NN can retrieve the stored memory correctly. An input within the attractor basin of a stored pattern can result in the correct recall of that pattern.
- **Capability of creating new categories:** A good NN is able to create a new category if it cannot find a proper classification for the input. This feature is specially useful for pattern clustering problems.
- **Memory capacity vs. network size:** Memory capacity is the number of maximum stable patterns can be stored by a neural network. If the encoding algorithm of an NN is properly designed, large memory capacity is feasible. What we want is the high ratio of memory capacity to network size.

12.6 BENCHMARKS

In what follows we present some 'problem types' that have been used throughout the development of artificial neural systems to benchmark the effectiveness of these systems solving complex problems.

12.6.1 The XOR problem

Rumelhart, Hinton, and Williams (Rumelhart *et al.*, 1986) study the question of learning representation for multilayered networks. Minsky and Papert (1969) provided a well carried out analysis of the necessary conditions for one-layer networks to effectively reproduce input–output mappings. As an example they show the XOR problem.

The eXclusive-OR (XOR) problem is the classic problem requiring hidden units. Also many problems include the XOR functionality as subproblems. The XOR as compared with other logic operations (AND, $\wedge$, OR, $\vee$, and their negates NAND, $\bar{\wedge}$, NOR, $\bar{\vee}$) is not linearly separable. The XOR, $\not\equiv$ problem has the training set:

$$\{(0,0,1),(0,1,1),(1,0,1),(1,1,0)\}$$

The training set is finite and precise representing a set where no measurement errors and/or noise occurs. For this benchmark, the probability density function is discrete valued only at the training points.

Rumelhart, *et al.* (1986) report that on several occasions the system encountered local minima depending on the initial conditions of the network rendering the problem unsolvable.

12.6.2 The *N*-parity problem

The N-input parity problem has been a popular benchmark among researchers in NN such as Minsky and Papert (1969), Rumelhart *et al.* (1986), Tesauro and Janssens (1988), Fahlman (1988) to name a few. The problem consists in mapping an N-bit wide binary number into its parity, i.e. if the input pattern contains an odd number of 1s then the parity is 1, else is 0. This is considered a difficult problem because patterns that are closer (using euclidean distance) in the measure (sample) space, i.e., numbers that differ in only one bit require

Table 12.5 XOR problem truth table

$\not\equiv$	0	1
0	0	1
1	1	0

Table 12.5 4-bit parity training set

b_3	0	0	0	0	0	0	0	0	1	1	1	1	1	1	1	1
b_2	0	0	0	0	1	1	1	1	0	0	0	0	1	1	1	1
b_1	0	0	1	1	0	0	1	1	0	0	1	1	0	0	1	1
b_0	0	1	0	1	0	1	0	1	0	1	0	1	0	1	0	1
Parity	0	1	1	0	1	0	0	1	1	0	0	1	0	1	1	0

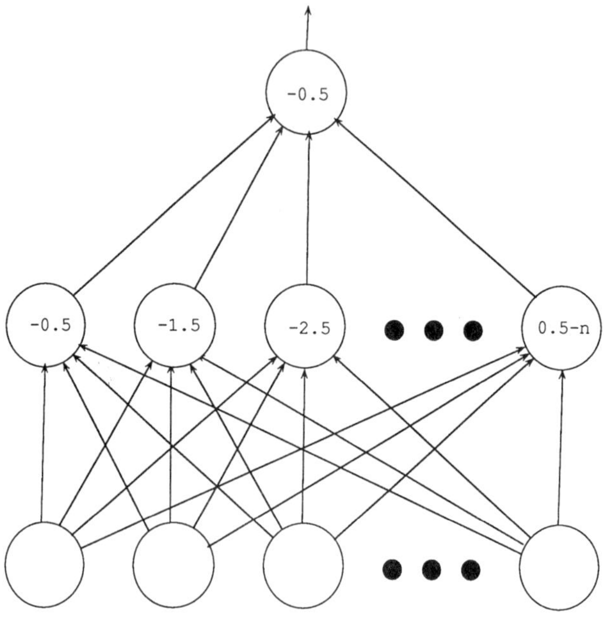

Fig. 12.6 Parity bit logic for n-bit address bus.

their answer to be different. This benchmark represents a perfect training set, e.g. for an $N = 4$, the training set is as given in Table 12.6.

The N-parity problem benchmark has been used to compare different learning algorithms and comparing the epochs necessary to produce a perfect result (no misclassification). The XOR problem is a 2-input parity problem, and the general solution for the parity problem is a group of XORs circuits. Rumelhart, *et al.* showed a general solution network as shown in Fig. 12.7.

12.6.3 The two-spirals problem

The two spirals problem, first proposed by Alexis Wieland of MITRE Corp., was used by Fahlman and Lebiere (Fahlman, 1988) to test cascade correlation

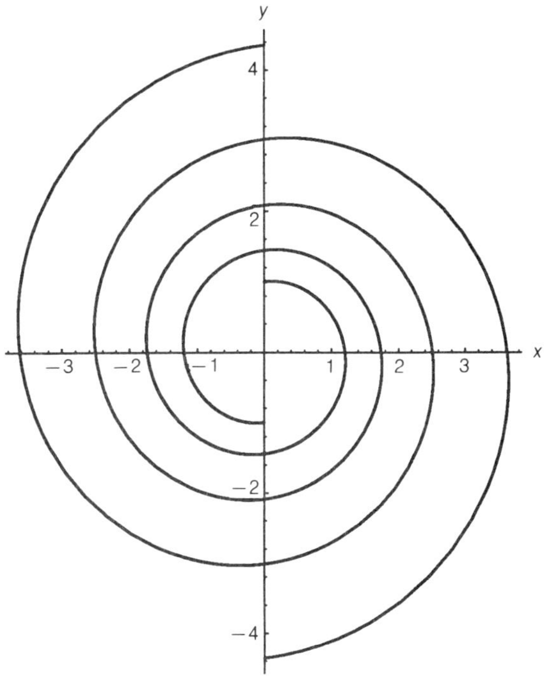

Fig. 12.7 Parity bit logic for y-bit address bus.

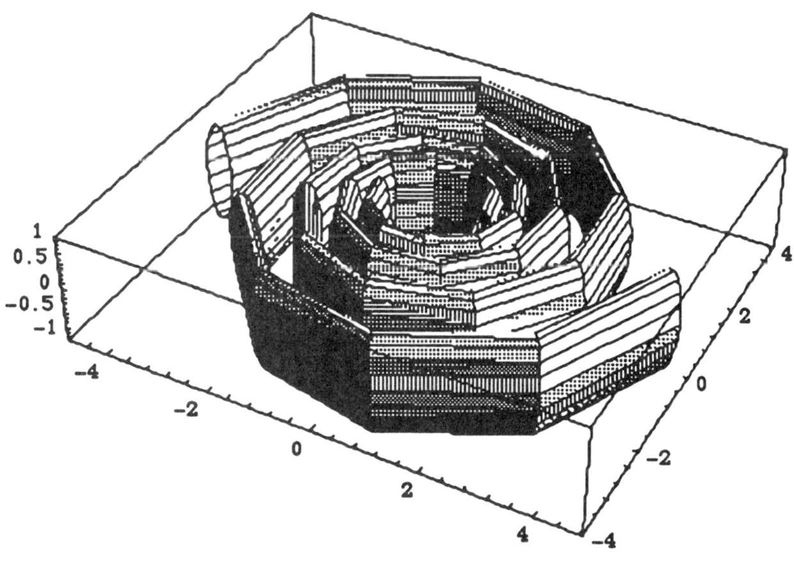

Fig. 12.8 Two-spirals problem.

learning. It was selected as the primary benchmark because 'it is an extremely hard problem for algorithms of the backpropagation family to solve'. The problem consist of $N_p = 194$ (x, y) pairs of points that lie in interlocking spirals ($N_p/2$ on each) that go around a common origin $C = 3$ times (see Figs 12.7 and 12.8). Each spiral belongs to a different class and the network has to associate points in the sample space with the correct class $\{+1, -1\}$. Lang and Witbrock (1988) also tried the problem using a $2 \times 5 \times 5 \times 5 \times 1$ 'umbrella' network where the nodes of any layer has inputs from all the nodes of *all* previous layers, not just from the previous layers as in the MLP network.

12.6.4 The encoding/decoding problem

Ackely, Hinton, and Sejnowski (1986) posed a problem for internal representation testing in which a set of N_I orthogonal input patterns are mapped to a set of N_O orthogonal output patterns through a small set of hidden units, N_H (see Fig. 12.9). This benchmark forces the hidden layer's nodes to be efficient. The expectation is that the hidden units will form a binary encoding of the N_I-bit input pattern onto a N_H-bit pattern and then form a binary decoding of the N_H-bit pattern onto a N_O-bit output pattern. This benchmark problem is important to evaluate the network storage capacity and it has application to data compression and transmission. In general (Zurada, 1992), given the capacity of linearly separable patterns, $N_H = \log_2(N)$ for $N_I = N_O = N$.

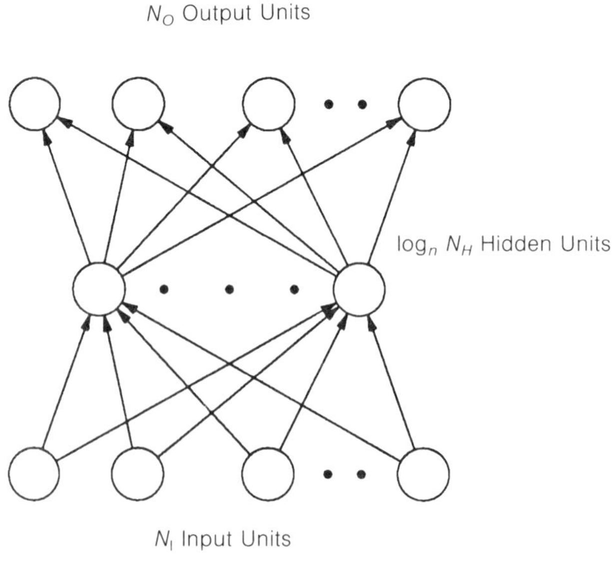

Fig. 12.9 Encoding/decoding problem.

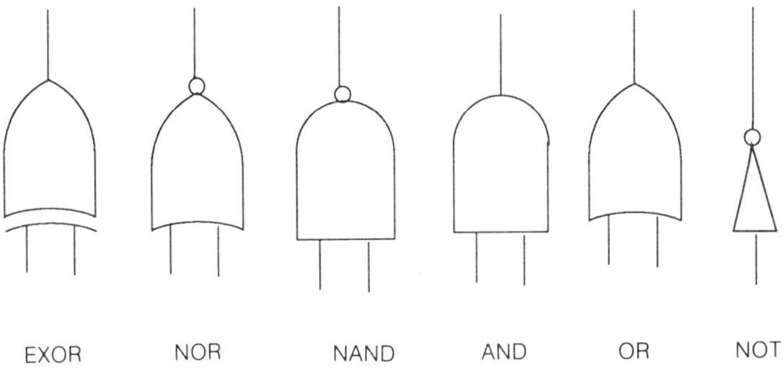

Fig. 12.10 Symbols for logic gates.

12.6.5 Logic/arithmetic operations

The Logic/Arithmetic Operations problems have been studied since the conception of artificial neural networks. Minsky and Papert (1969) used the XOR problem to show the limitations of single layer perceptrons. Figure 12.10 show the basic symbol for combinatorial logic gates. By combining these gates any logical statement can be implemented. They can also be combined to generate temporal components useful for sequencing. Artificial neural networks can also be formed (see Fig. 12.13 later) that imitate the same logic elements.

Rumelhart, Hinton, and Williams, (in Rumelhart *et al.*, 1986), studied two more logic/arithmetic problems: the binary addition and the binary negation. The binary addition has the feature of being able to find local minima. Hence it can be studied for networks and learning algorithms to check for the way they react to these conditions, to discover how learning algorithms find and avoid local minima. The binary addition of two-bit binary numbers gives at most a 3-bit binary number. This problem has a digital logic solution as shown in Fig. 12.11 where a half-adder and a full-adder are shown, see also Tables 12.7 and 12.8.

$$S = \text{Sum} = (A \vee B) \wedge \overline{(A \wedge B)} - A \not\equiv B$$
$$C = \text{Carry} = A \wedge B$$

Digital logic can be translated into neural networks directly (see Fig. 12.12). For example the NAND, $\overline{\wedge}$, and the NOR, $\overline{\vee}$, functions can be described by the networks shown in Fig. 12.13 as described by Zurada (1992) when simulating McCulloch–Pitts Neuron models.

$$o^{(k+1)} = \begin{cases} 1 & \text{if } \sum_{i=1}^{n} w_i x_i^{(k)} \geqslant T \\ 0 & \text{if } \sum_{i=1}^{n} w_i x_i^{(k)} < T \end{cases}$$

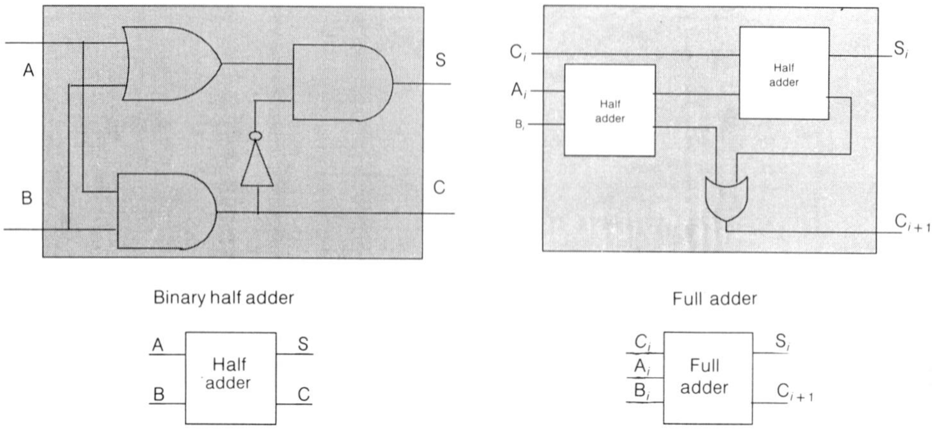

Fig. 12.11 Binary addition logic implementation.

Table 12.7 Half Adder Truth Table

INPUT		*OUTPUT*	
A	*B*	*S*	*C*
0	0	0	0
0	1	1	0
1	0	1	0
1	1	0	1

Table 12.8 Full Adder Truth Table

INPUT			*OUTPUT*	
A	B	C_i	S	C_{i+1}
0	0	0	0	0
0	0	1	1	0
0	1	0	1	0
0	1	1	0	1
1	0	0	1	0
1	0	1	0	1
1	1	0	0	1
1	1	1	1	1

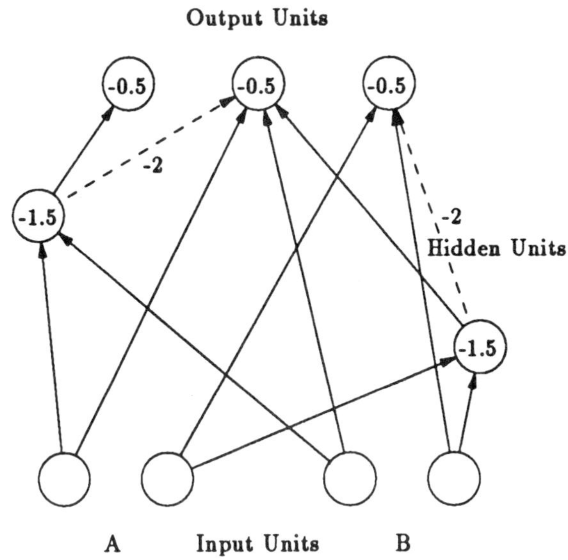

Fig. 12.12 ANN minimum size adder of 2-bit numbers.

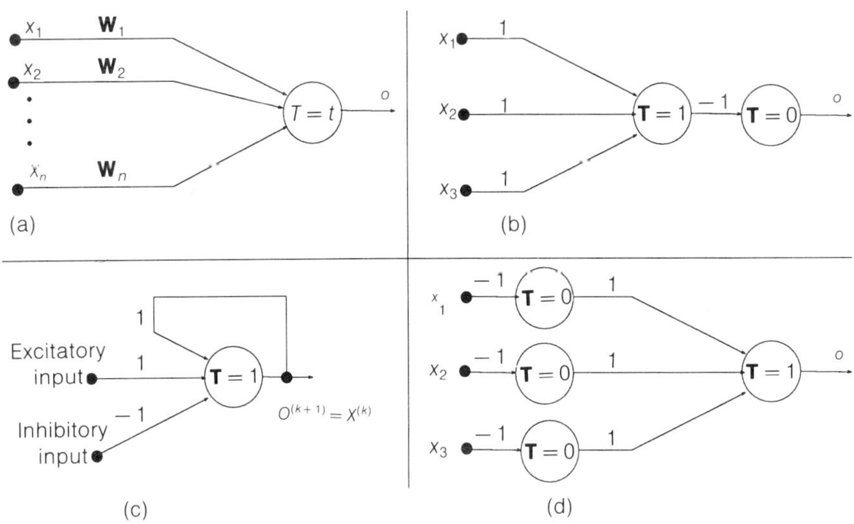

Fig. 12.13 McCulloch-Pitts neuron model (network) digital logic: (a) general neuron model; (b) NOR gate; (c) memory cell; (d) NAND gate.

The negation problem consists of an $(n+1)$-bit number, $(b_0, b_1, \ldots, b_n, b_{n+1})_i$, as input ($n$-bit data and an extra negation-bit, $b_N = b_{n+1}$), the output is an n-bit number, $(b_0, b_1, \ldots, b_n)_o$, that is equal to the input pattern if b_N is 0, and it outputs the input's pattern complement when b_N is set to 1, i.e.

$$(b_0, b_1, \ldots, b_n)_o = \begin{cases} (b_0, b_1, \ldots, b_n)_i & \text{if } b_N = 0 \\ (\bar{b}_0, \bar{b}_1, \ldots, \bar{b}_n)_i & \text{if } b_N = 1 \end{cases}$$

where $\bar{b}_k$ is the complement of b_k. The problem reduces Rumelhart *et al.* (1986) to a set of n XOR problems between the negation bit b_N and each of the inputs $(b_k)_o$, such that

$$(b_k)_o = (b_k)_i \not\equiv b_N$$

For n-tuple bipolar binary vectors, the error measure is the Hamming distance,

$$\mathrm{HD}((b_k)_o, (b_k)_d) \stackrel{\text{def}}{=} \frac{1}{2} \sum_{j=1}^{n} |(b_{kj})_o - (b_{kj})_d|$$

where $(b_k)_d$ is the desired output or target vector.

12.6.6 The traveling salesperson problem (TSP)

The Traveling Salesperson Problem (TSP) is a classic of difficult optimization (Hopfield and Tank, 1985). It is simple to describe, mathematically well characterized, and makes a good vehicle for describing the use of neural networks to solve difficult optimization problems. Different groups have continued to use this idea of using analog networks to solve mathematical problems (Sanchez-Sinencio and Lau, 1992) and it dates back to the days of analog computation.

The problem can be stated as follows. Given a set of n cities, $C = \{C_1, C_2, \ldots, C_n\}$, having pairwise distances d_{ij}, the problem is to find a closed tour which visits every city once, returning to the starting city, and has a minimum total path length. A tour is defined by the sequence $T = \{m, p, q, \ldots, s, m\}$ that tells the sequence of cities to be visited, i.e. starting at C_m go to C_p then C_q then $\cdots$ then to C_s, then to C_m. The total path length, d is given by

$$d = d_{mp} + d_{pq} + \cdots + d_{sm}$$

The actual 'best' solution is computationally very hard. The problem is NP-complete (Garey and Johnson, 1970), and the time required to solve this problem grows exponentially with n. There are $n!/2n$ distinct paths for closed routes and $2n$ paths of equal distance.

To solve the problem, Hopfield and Tank (1985) propose an energy function where the lowest level of energy state (the most stable state of the network) corresponded to the best path.

12.6.7 The accuracy problem

Another measure of performance is the network accuracy. A way of testing this is to define a classification problem where there is a boundary layer between classes where the error is acceptable, i.e. not penalized or less penalized. A general geometry can be specified for the classification. Makram-Ebeid *et al.* (1989) proposed the half plane, a corner, and a rectangle; another geometries include the spirals, circle is square, concentric circles, etc. The basic idea is shown in Fig. 12.14 where the decision boundary is fuzzy, i.e. a 'boundary layer' (BL) is defined where no training data is present, and perfect recall is expected. The BL thickness, Δ_C (euclidean distance between the classes) and is the relative accuracy for separation for a given network. The smallest the Δ_C for a network, the most accurate it can be. Obviously, the discriminatory functions play an important role in determining what network is more suitable for the problem. For example Π–Σ networks are more suitable for the concentric circles problem. Knowledge of the system helps to determine the best set of discriminatory functions.

This problem can be used to determine the effectiveness of learning algorithms to converge to the desired accuracy of the problem. Conventional forms of backpropagation yields a learning time that diverges like $1/\varepsilon^2$, where $\varepsilon = \Delta_C/C_{\max}$.

Equivalently for function approximation, an accuracy measure can be established. Cybenko (1989) using sigmoidal discriminants, $\sigma(\cdot)$, proves that

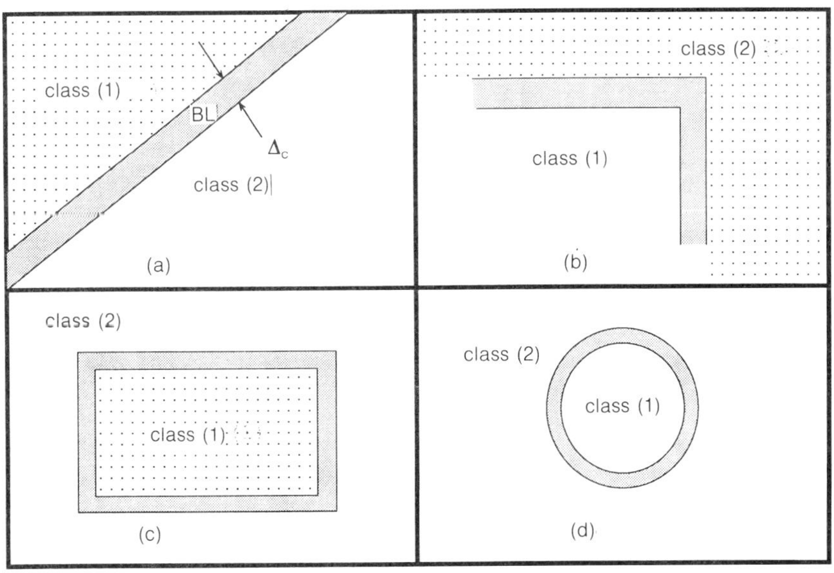

Fig. 12.14 The accuracy problem, geometric interpretation; (a) line; (b) corner; (c) rectangle; (d) circle.

the finite sums of the form

$$G(x) = \sum_{j=1}^{N} \alpha_j \sigma(y_j^{\mathrm{T}} x + \theta_j)$$

are dense in $C(I_n)$, i.e. given any function $f(x) \in C(I_n)$ and $\varepsilon > 0$, there is a sum, $G(x)$, for which

$$|G(x) - f(x)| \leqslant \varepsilon$$

Here the measure is given by the boundary layer of size ε. An equivalent results is shown for $\mathscr{L}^1$, i.e.,

$$\| G(x) - f(x) \|_{\mathscr{L}^1} = \int_{I_n} |G(x) - f(x)| \, \mathrm{d}x \leqslant \varepsilon$$

Hetch-Nielsen (1987) shows that a three-later-perceptron network can approximate any $\mathscr{L}_2$ function within ε squared error accuracy,

$$\int_{I_n} |G(x) - f(x)|^2 \, \mathrm{d}x \leqslant \varepsilon.$$

12.6.8 The sin (x) sin (y) problem

General function mapping problems have been used by different researchers to test their networks' capabilities, learning algorithms efficiency, etc. A popular mapping, shown in Fig. 12.15 is $z(x, y) = \sin(x) \sin(y)$. This function gets more complex as the norm of the input vector (x, y) grows.

12.6.9 The predicate calculus problems

Minsky and Papert (1969) studied a general class of problems related to pattern recognition, learning machines, and threshold logic. They presented their study in an elegant math.

Let $\mathscr{R}$ be an euclidean space, and let $\mathscr{X}$ be a geometrical object in this space, for example a figure in a 2D plane. In general $\mathscr{X}$ is simply a subset of points of $\mathscr{R}$. Let $\Phi(\mathscr{X})$ be a function (of objects $\mathscr{X}$ on $\mathscr{R}$) that have two values $\{0, 1\}$. By taking these values as TRUE and FALSE, we can think of $\Phi(\mathscr{X})$ as a predicate, i.e., a variable statement whose truth or falsity depends on the choice of $\mathscr{X}$. One of the simplest predicates 'recognizes' when a particular single point is in $\mathscr{X}$: let $p \in \mathscr{R}$,

$$\phi_p(\mathscr{X}) = \begin{cases} 1 & \text{if } p \text{ is in } \mathscr{X} \\ 0 & \text{otherwise} \end{cases}$$

The next step in complexity is the predicate that tells when a subset $\mathscr{S}$ is a subset of $\mathscr{X}$:

$$\phi_{\mathscr{S}}(\mathscr{X}) = \begin{cases} 1 & \text{if } \mathscr{S} \in \mathscr{X} \\ 0 & \text{otherwise} \end{cases}$$

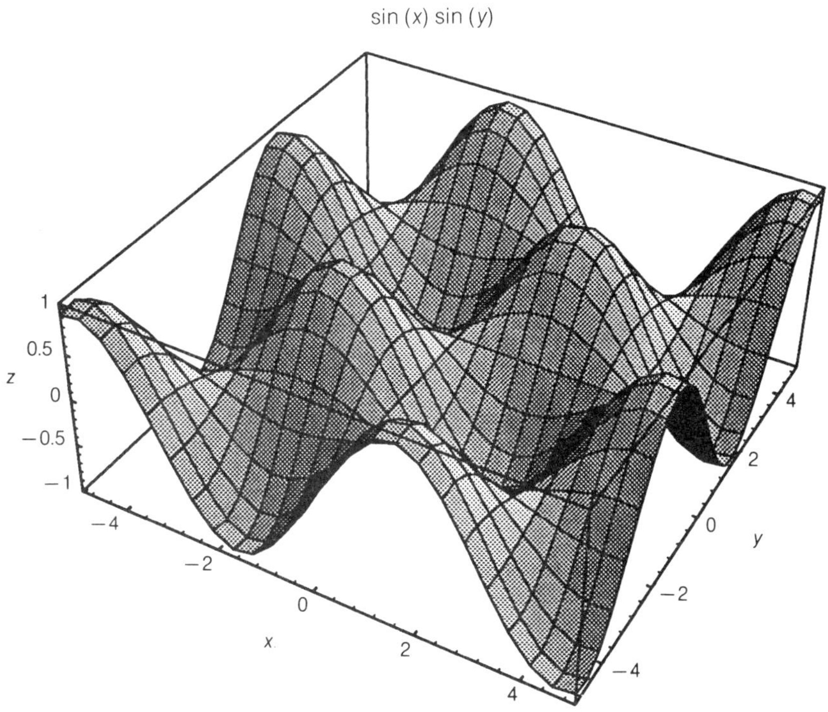

Fig. 12.15 The $\sin(x)\sin(y)$ problem.

Other predicates of interest are

$$\phi_{\text{CIRCLE}}(\mathscr{X}) = \begin{cases} 1 & \text{if } \mathscr{X} \text{ is a circle} \\ 0 & \text{otherwise} \end{cases}$$

$$\phi_{\text{CONVEX}}(\mathscr{X}) = \begin{cases} 1 & \text{if } \mathscr{X} \text{ is a convex figure} \\ 0 & \text{otherwise} \end{cases}$$

$$\phi_{\text{CONNECTED}}(\mathscr{X}) = \begin{cases} 1 & \text{if } \mathscr{X} \text{ is a connected figure} \\ 0 & \text{otherwise} \end{cases}$$

$$\phi_{\mathscr{A}}(\mathscr{X}) = \begin{cases} 1 & \text{if } \mathscr{A} \subset \mathscr{X} \\ 0 & \text{otherwise} \end{cases}$$

The order of the predicate is the number points on which the predicate depends, so a predicate Φ of order k, its value depends upon no more than k points in $\mathscr{R}$. Using this, Minsky and Papert defined a perceptron as a device capable of computing all predicates which are linear in some given set Ψ of partial predicates. A particular test proposed by Minsky and Papert is the recognition of geometrical figures invariant to translation and rotation. This

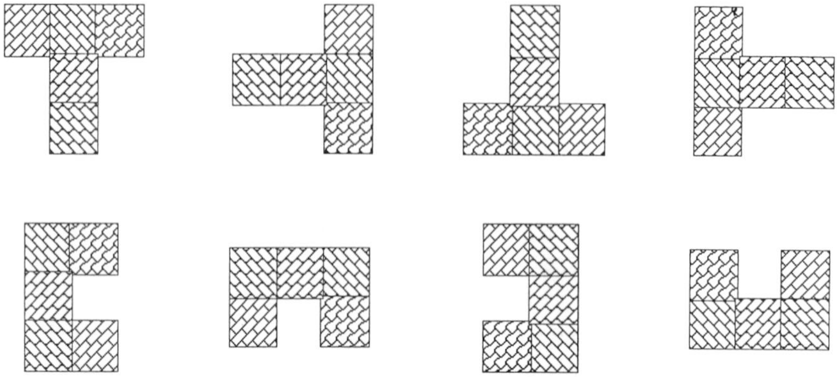

Fig. 12.16 TC problem-training exemplars.

problem (of order three) was also presented by Rumelhart *et al.* (1986) as the T–C problem. This problems tries to discriminate between a 'T' and a 'C' independent of translation and rotation. Figure 12.16 shows some of the possible 'exemplars' of the training set. In order to solve this problem, one must look, at least, at configurations of triplets of square (hence the order three). The concept of receptive field is used, where the activation of a group of local nodes is passed onto the next layers nodes in a pyramidal fashion. This approach is also used in other character recognition networks like neocognitron (Fukusima, 1980).

Minsky and Papert also defined an interesting set of families of perceptrons. Among these families are

- **diameter-limited perceptrons** – for each ϕ in Ψ, the set of points upon which ϕ depends is restricted not to exceed a certain fixed diameter in the plane;
- **order-restricted perceptrons** – a perceptron has order $\leqslant n$ if no member of Ψ depends on more than n points;
- **gamba perceptrons** – each member of Ψ may depend on all the points but must be a 'linear threshold function';
- **random perceptrons** – the perceptrons are order restricted and Ψ is generated by a stochastic process according to the assigned distribution function; and
- **bounded perceptrons** – Ψ contains an infinite number of ϕs, but the α_ϕ lie in a finite set of numbers.

12.6.10 The static/dynamic mapping problem

System identification is probably the most crucial step in several engineering disciplines such as fault detection, diagnosis, prediction, and control. In general, it is the backbone of most of the available design techniques, and its importance has been noted since the beginning of science. The accurate extraction of dynamics from system observations (sensor readings) is the topic of system identification. It is clear that this subset of system and control theory is of great importance in many practical applications. This is especially true, when the dynamics of the system to be identified are time-varying and/or non-linear. In any case, obtaining an accurate description of the system under consideration is of utmost importance for fault diagnosis and control purposes.

Dynamic mapping, i.e., the function approximation of deterministic dynamic systems has been a goal of many branches of engineering and science since humans tried to identify, reproduce, and predict natural phenomena. System identification and prediction has been studied by several authors (Lapedes and Farber, 1987; Narendra and Parthasarathy, 1990; Fernández *et al.*, 1992). During system identification, the neural network adjust its weights such that the dynamics of the ANS reproduces the sampled-data system from which the training data has been drawn. Two basic approaches have been used: the time delayed feedforward networks (based on the concept of transfer function) that emulate linear auto-regressive moving-average (ARMA) processes; and the recurrent neural networks based on state-space description of the system.

In general, the class of non-linear systems considered for identification are assumed to have dynamics that are well represented by a set of non-linear, finite-dimensional, time-invariant ordinary differential equations (ODEs), which are linear in the controls, as follows (Isidori, 1989):

$$\begin{aligned} \dot{\mathbf{x}}(t) &= \mathbf{f}(\mathbf{x}(t)) + \mathbf{g}(\mathbf{x}(t))\mathbf{u}(t) \qquad && \mathbf{x}(t)\in\mathscr{R}^n,\ u(t)\in\mathscr{R}^m \\ \mathbf{y}(t) &= \mathbf{h}(\mathbf{x}(t)) && \mathbf{y}(t)\in\mathscr{R}^p \end{aligned}$$

where $\mathbf{f}(\mathbf{x}(t))$, $\mathbf{g}(\mathbf{x}(t))$ and $\mathbf{h}(\mathbf{x}(t))$ are in general smooth functions of $\mathbf{x}(t)$, of appropriate dimensions. These equations represent a multi-input/multi-output (MIMO) system, with m inputs and p outputs.

There are two strong conditions that a system should satisfy in order to be identifiable: **observability** is required so that the reduced order manifold dynamics do not corrupt the system measurements which are fed to the identification algorithm; and **controllability** is necessary so that any output response can be associated to a particular input. Failure to comply with either one or both of these conditions does not necessarily result in catastrophic behavior but only in erroneous predictions. If on-line identification is contemplated, then these conditions could be relaxed to detectability and stabilizability, as long as the learning algorithm is 'faster' than the reduced order manifold dynamics.

For linear systems, the dynamics are well represented by a set of linear, finite-dimensional, time-invariant ordinary differential equations (ODEs), as follows

$$\dot{\mathbf{x}}(t) = \mathbf{A}\mathbf{x}(t) + \mathbf{B}\mathbf{u}(t)$$

$$\mathbf{y}(t) = \mathbf{C}\mathbf{x}(t) + \mathbf{D}\mathbf{u}(t)$$

where $\mathbf{A}$, $\mathbf{B}$ and $\mathbf{C}$ are matrices of appropriate dimensions.

This system can be expressed in the discrete-time domain as follows:

$$\mathbf{x}(k+1) = \mathbf{\Phi x}(k) + \mathbf{\Gamma u}(k)$$
$$\mathbf{y}(k+1) = \mathbf{Cx}(k+1) = \mathbf{C\Phi x}(k) + \mathbf{C\Gamma u}(k)$$

where the matrices $\mathbf{\Phi}$ and $\mathbf{\Gamma}$ are the discretize version of the continuous **A** and **B** matrices, generated through the exponential matrix.

$$\mathbf{\Phi} = \mathbf{\Phi}(\mathscr{T}) = \exp\{-\mathbf{A}\mathscr{T}\}$$
$$\mathbf{\Gamma} = \mathbf{\Gamma}(\mathscr{T}) = \int_0^{\mathscr{T}} \exp\{-\mathbf{A}\mathscr{T}\}\mathbf{B}\,\mathrm{d}t$$

and $\mathscr{T}$ is the discretization/sampling time. Continuing the above representation the following form is obtained:

$$\mathbf{x}(k+2) = \mathbf{\Phi x}(k+1) + \mathbf{\Gamma u}(k+1)$$
$$\mathbf{y}(k+2) = \mathbf{Cx}(k+2) = \mathbf{C\Phi}^2\mathbf{x}(k) + \mathbf{C\Phi\Gamma u}(k) + \mathbf{C\Phi u}(k+1)$$

and, generally,

$$\mathbf{x}(k+j) = \mathbf{\Phi x}(k+j-1) + \mathbf{\Gamma u}(k+j-1)$$
$$\mathbf{y}(k+j) = \mathbf{Cx}(k+j) = \mathbf{C\Phi}^j\mathbf{x}(k) + \mathbf{C\Phi}^{j-1}\mathbf{\Gamma u}(k) + \mathbf{C\Phi}^{j-2}\mathbf{\Gamma u}(k+1)$$
$$+ \cdots + \mathbf{C\Phi\Gamma u}(k+j-2) + \mathbf{C\Phi u}(k+j-1)$$

The state $\mathbf{x}(k)$ can be eliminated from the above equation, resulting in the following autoregressive-moving average (ARMA) model:

$$\mathbf{y}(k+1) + \sum_{j=1}^{K_1} \mathscr{D}_j(\mathscr{T})\mathbf{y}(k-j+1) = \sum_{j=1}^{K_2} \mathscr{E}_j(\mathscr{T})\mathbf{u}(k-j) + \mathscr{B}\mathbf{u}(\mathscr{T})$$

The ARMA form given is widely used in the system identification community. However, it should be noted that ARMA models are 'linear', and therefore suffer from all the short-comings of linear model structures.

For non-linear systems, the following formulation is obtained in the discrete time domain.

$$\mathbf{x}(k+1) = \mathbf{f}_1(\mathbf{x}(k)) + \mathbf{g}_1(\mathbf{x}(k))\mathbf{u}(k)$$
$$\mathbf{y}(k+1) = \mathbf{h}_1(\mathbf{x}(k+1)) = \mathbf{H}_1(\mathbf{x}(k), \mathbf{u}(k))$$

Continuing the above representation the following form is obtained.

$$\mathbf{x}(k+2) = \mathbf{f}_2(\mathbf{x}(k)) + \mathbf{g}_2(\mathbf{x}(k))\mathbf{u}(k) + \mathbf{g}_{21}(\mathbf{x}(k))\mathbf{u}(k+1)$$
$$\mathbf{y}(k+2) = \mathbf{h}_1(\mathbf{x}(k+2)) = \mathbf{H}_2(\mathbf{x}(k), \mathbf{u}(k), \mathbf{u}(k+1))$$

and, generally,

$$\mathbf{x}(k+j) = \mathbf{f}_j(\mathbf{x}(k)) + \mathbf{g}_j(\mathbf{x}(k))u(k) + \mathbf{g}_{j1}(\mathbf{x}(k))\mathbf{u}(k+1)$$
$$+ \cdots + \mathbf{g}_{jj-1}(\mathbf{x}(k))\mathbf{u}(k+j-1)$$
$$\mathbf{y}(k+j) = \mathbf{h}_1(\mathbf{x}(k+j)) = \mathbf{H}_j(\mathbf{x}(k), \mathbf{u}(k), \mathbf{u}(k+1), \ldots, \mathbf{u}(k+j-1))$$

So general non-linear dynamic systems can be represented using a recurrent architecture such as DyNet (Fernández *et al.*, 1989) (as shown in Fig. 12.17).

A good problem used in the literature is that of chaotic systems. Lapedes and Farber (1987) selected for their study the Glass–Mackey equation,

$$\dot{x} = \frac{ax(t-\tau)}{1+x^{10}(t-\tau)} - bx(t)$$

a non-linear equation that has an infinite dimensional phase space, similar to other infinite dimensional system such as partial differential equations. For $a = 0.2$, $b = 0.1$, and $\tau = 17$, $x(t)$ is a chaotic series with a fractal attractor of dimension 2.1 (Lapedes and Farber, 1987). For $\tau = 30$, the fractal attractor is of order 3.5.

Fernández *et al.* (1992) proposed a structured recurrent network based in bond graph theory that generates a sparse network keeping physical resemblance to the actual system. This approach gives the neural network and identification community (Ljung, 1991) a specific architecture to work with so when the training is done, the 'knowledge' acquired can be correlated to

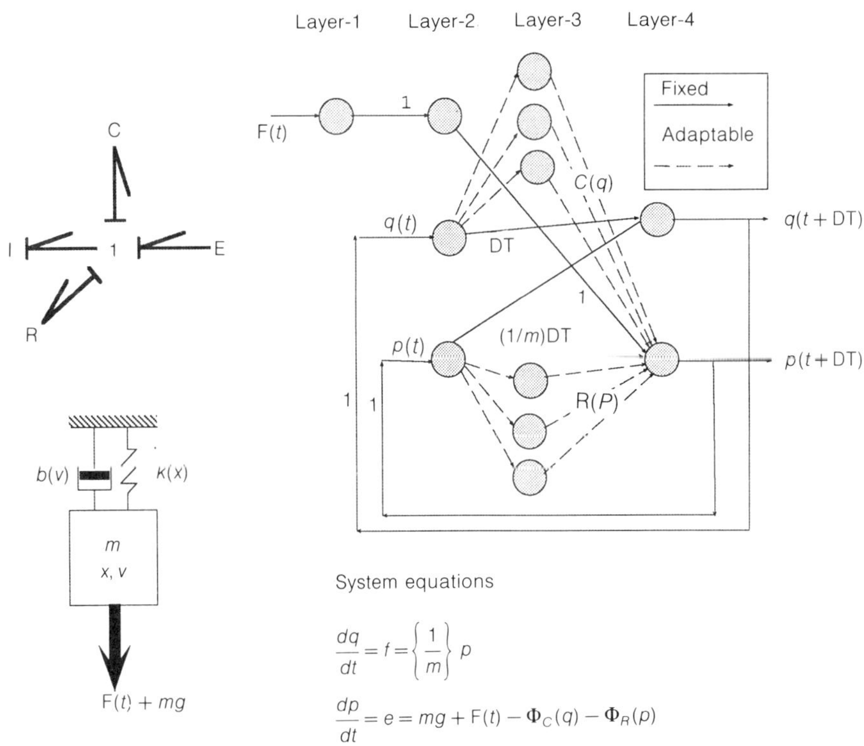

Fig. 12.17 Neuro bond graph.

physical parameters. In addition, the bondgraph community acquires a tool to represent and identify generalized non-linear systems.

Bond graphs, an energy-based modeling technique developed at M.I.T. by Paynter (1961), provides a structured approach to modeling of physical systems (Rosenberg and Karnopp, 1983). Bond graphs use bipartite graphs to describe the power interactions among the different elements of a system. Three basic elements (storage, transmission, and dissipation) each having specific constitutive properties that can be classified into (nine) generic element types: effort source (force) **E**, flow source (velocity) **F**, inductor **I**, capacitor **C**, resistor **R**, transformer **TF**, gyrator **GY**, common-effort junction 0, and common-flow junction 1. Since bond graphs are based on power flows through a system, a representative bond graph of a complicated system can be constructed from physical insight. Assigning proper causality to each element leads directly to the system state equations. Bondgraphs of distributed systems result in field elements and reduce to finite elements when discretized through reticulation.

Bond graphs are used to give a structured interaction among the different components in machining processes. The identification of the functionality of the different interactions is obtained through artificial neural networks paradigms (Fernández *et al.*, 1992).

12.6.11 The sequencing problem

Sequencing is another form of prediction where the network acts as a solid state machine. Different vertices points are directed towards specific points in the state hypercube. The sequence is in general finite, although cyclic operation is possible. Gating between different paths can be controlled through external inputs. Examples are sequences to start/stop processing units, grasping subtasks for a robot, etc.

A popular problem is that of controlling a system in finite time (finite time horizon), such as the 'truck backer-upper' (Nguyen and Widrow, 1989; Kosko, 1992) where a finite number of maneuvers is required to move a truck from its initial condition, to a desired final condition. The goal is to make the truck (originally at a general initial condition within a box, (x_0, y_0)), (Fig. 12.18) arrive (or get as close as possible) at a right angle $\phi(t_f) = 90$ and align its position, (x, y), to the loading dock (x_f, y_f) within a specified time. The problem is a classical problem in optimal control. A performance index can be specified and the time optimal control is a bang-bang action. An interesting feature of optimal control is that the next control action does not depend on the previous actions, i.e., if the trajectory is optimal, one can redefine the problem at each time step. Techniques such as backpropagation through time (BTT) can be employed. In general the problem is solved suboptimally, in the sense that only a finite horizon (a subset of available data) is used to calculate the performance index, and this data used for network weights adaptation.

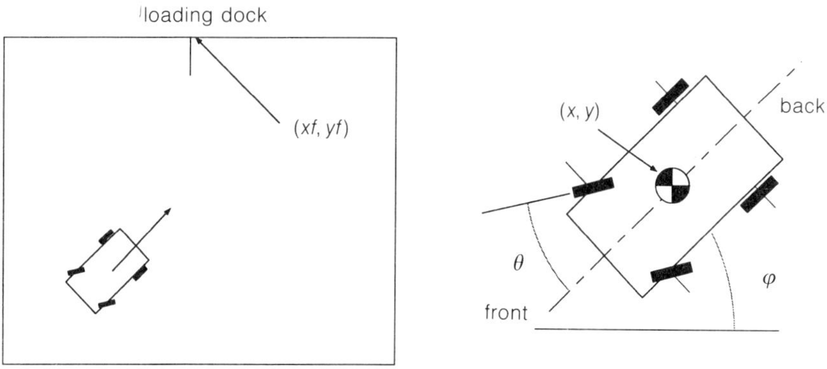

Fig. 12.18 Truck backer-upper problem (backer-up?).

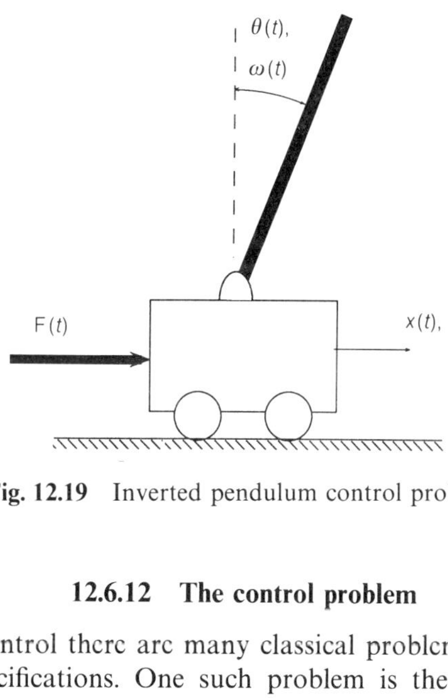

Fig. 12.19 Inverted pendulum control problem.

12.6.12 The control problem

In the area of control there are many classical problems that test different performance specifications. One such problem is the inverted pendulum problem, also known as the broom balancer problem.* The problem, as depicted in Fig. 12.19 has as input the horizontal force $F(t)$ and requires four state variables to describe the system: the cart position, $x(t)$, and velocity, $v(t)$, and the pendulum angle, $\theta(t)$, and angular velocity, $\omega(t)$. The problem is open-loop unstable, which is a challenge, and external loads (wind), unmodeled dynamics (flexible modes, friction), parameter uncertainty (unknown mass),

*Attitude control of the space shuttle reduces to the same problem.

sensor noise, saturation in the actuators ($|F(t)| \leqslant F_{max}$) etc. can be added to the problem to simulate realistic problems.

The use of an adaptive network to stabilize a mechanical inverted pendulum was first studied by Widrow (1987) and Smith (Widrow and Smith, 1963) where they controlled the pendulum with a simple adaline network. Adaptive critic networks, fuzzy neural networks, ART, and other networks have been tested in this benchmark. In a recent study Hung and Fernandez (1992) showed a comparative analysis among different controllers in an experimental setup.

12.7 SIMULATION PARADOX

Given that we need to simulate neural networks to test their performance, the way the simulation is carried out has a direct influence upon the decision, hence its performance evaluation. We need to specify different scenarios and try to develop a suitable set of experiments that test the network capabilities to its fullest.

A network is defined by

- its **topology**, i.e., the arrangement of the layers, specify feedforward, recurrent, crosstalk, and/or feedback paths for the links;
- the network **size**, i.e., the number of layers, the number of nodes on each layer;
- the **functionality**, i.e., the nodes input operator, transfer function, and discriminatory function;
- the **training set**, i.e., the type of data used for training, e.g. analog, binary, discrete, fuzzy, etc.;
- the **learning algorithm**, e.g. backpropagation, clustering techniques, resonant theory, static vs. dynamic learning, vector quantization, delta rule, Hebbian learning, etc.;
- the **learning parameters**, e.g. learning rate, momentum term, vigilance parameter, etc.;
- the **weights** initial conditions, e.g. random weights, normalized, etc.;
- **constraints** on the network during learning, and while in operation; and
- the **testing set**, i.e. the type of data used to validate the network.

Some times for specific learning rates and training patterns the learning enters a limit cycle (**attractor**). For example, assume that the network is at the global minimum for a line approximation as shown in Fig. 12.20 with three points in the training set $\{(0,0), (3,4), (3,2)\}$. The line is $x_1 - x_2 = 0$.

Makram-Ebeid *et al.* (1989) made sequential-computer simulations to study the validity of the network results. Each simulation was made by averaging over 100 different randomly chosen initial settings of an MLP's synaptic weights and thresholds. For each initial set, a learning 'trial' was made. The corresponding percentage $\mathscr{P}(s)$, (s success) of those trials leading

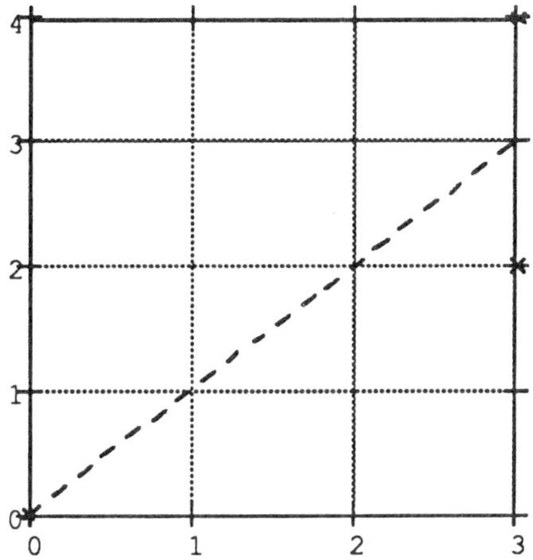

Fig. 12.20 Learning attractor for single-step updating.

to a complete learning of the examples in the training set was computed. Then the effective number of learning epochs, K_{eff} is defined as the average total number of epochs for successful and unsuccessful trials needed to obtain one satisfactory training.

12.7.1 Basic principles and derivation of the loss function

Early work in this area (Werbos, 1974), common to pattern recognition, was based on classical Bayesian approach to parameter estimation. That approach starts with the formulation of a static model. Then the goal is to find the parameters, **W**, which have the highest probability of being the true set of parameters, based on knowledge gained after inspecting the data. Basically, we try to maximize the conditional probability $\Pr(\mathbf{W}|\mathscr{D})$, where $\mathscr{D}$ represents the training set data. This term is not well defined, in an objective manner. We can shed some light using Bayes:

$$\Pr(\mathbf{W}|\mathscr{D}) = \frac{\Pr(\mathscr{D}|\mathbf{W})\Pr(\mathbf{W})}{\Pr(\mathscr{D})}$$

Here the 'likelihood term' $\Pr(\mathscr{D}|\mathbf{W})$ is well specified for a well-specified stochastic model. $\Pr(\mathscr{D})$ can be ignored, since it does not affect the optimization of **W**. $\Pr(\mathbf{W})$ represents a priori knowledge of the probability distribution of the weights, before any data become available. In the maximum likelihood approach, the term $\Pr(\mathbf{W})$ is discarded assuming that all values of **W** are equally likely (a priori). Solomonoff (1964) proposed a priori probability for any model should be e^{-kS} where k is some constant and S is the number of symbols required to describe the model, i.e., model parameters.

To cope with this problem, two major techniques are in use (Werbos *et al.*, 1992): **ridge regression** (Dempster, 1977), which effectively penalizes large estimates of parameters; and **stepwise regression**, in which parameters are deleted when they fail a basic test of statistical significance.

In ridge regression, the parameters, **W**, come from a gaussian distribution, instead of infinite variance distribution (maximum likelihood theory). The regression tries to minimize an error/cost function of the form:

$$L = \sum_i e_i^2 + k \sum_{i,j} W_{ij}^2$$

This approach puts excessive penalty in large weights. A less constrictive distribution is the lognormal probability distribution which leads to a loss function:

$$L = \sum_i e_i^2 + k \sum_{i,j} \log \frac{1}{|W_{ij}|}$$

12.7.2 Alternative cost functions

The conventional loss function, L, (standard back error propagation, BEP) is given by the $\mathscr{L}_2$-norm of the error:

$$L(y, y_d) = \sum_i ((y_i)_d - y_i)^2$$

Alternative loss functions includes those by Werbos and Titus (1978)

$$L(y, y_d, W) = \sum_i \left(\frac{((y_i)_d - y_i)^2}{\sigma_{y_i}^2 (1 - |w_i|)^2} \right)$$

and

$$L(y, y_d, W) = \sum_i \frac{|(y_i)_d - y_i|}{\sigma_{y_i}(1 - |w_i|)}$$

and Fernández (1991):

$$L = \sum_{k=1}^{N_p} \frac{1}{2} \frac{(x - x_d)^{\mathrm{T}} Q (x - x_d)}{1 + x_d^{\mathrm{T}} x_d} \tag{12.1}$$

which for 'large' values of error becomes 'relative error' (Gill and Wright, 1981) and for 'small' values of the error it becomes 'absolute error', i.e.

$$\frac{|x - \hat{x}|}{1 + |\hat{x}|} = \begin{cases} \dfrac{|x - \hat{x}|}{|\hat{x}|} & \text{when } |x| \ll 1 \\ |x - \hat{x}| & \text{when } |x| \gg 1 \end{cases} \tag{12.2}$$

This doesn't change the BEP technique except that it weights the errors differently according to their (absolute/relative) 'magnitude'.

This technique works better when the data are sparse, in the sense that the errors are 'equally' distributed among the data points. When the points are 'well balanced', i.e. within the same scale (order of magnitude), the technique doesn't show great improvement. Having some additional computation is not recommended for these cases. Note that other norms could be used in the functional.

12.8 PERFORMANCE MEASURES

Performance measures are ways for decision theory to evaluate the goodness (performance) of the desired 'cognitive' task or goal of the network. In that sense, ANS basic operation is pattern association, and an evaluation will result with a correct or incorrect association of two patterns in the 'probable' input space. Given that, most of the measurements are themselves a statistical value of the exact but unknown physical variable, bounds on acceptable levels of error need to be furnished to classify the network association as correct or incorrect. Most of the notation in this section is after Chen (1973).

Elementary Bayes decision theory (Chen, 1973; Amari, 1990; Schalkoff, 1992) can be used to describe the performance of the network (Fig. 12.21). Given two classes: correct association, ω_1, and incorrect association, ω_2, with a priori probabilities of occurrence P_1 and P_2 ($P_1 + P_2 = 1$). Let the observation (measurement) x be an N-dimensional vector which is uniquely classified into one of the two classes by the decision criterion, i.e. the decision criterion partitions the sample space of all observations $\mathscr{R}$ into $\mathscr{R}_1$ (class ω_1) and $\mathscr{R}_2$ (class ω_2). Let $p(x|\omega_1)$ and $p(x|\omega_2)$ represent the conditional probability densities associated with the measurement x, given that the unknown pattern came from class ω_1 or class ω_2 respectively. For all regions of the space $\mathscr{R}$ in which both probability densities are greater than zero, a finite probability of misclassification exists. The overall probability of correctly classifying patterns from class ω_1 is then given by

$$\mathrm{P}(d_1|\omega_1) = \int_{\mathscr{R}_1} p(x|\omega_1)\,\mathrm{d}x$$

while the probability of *mis*classifying patterns from class ω_1 is given by

$$\mathrm{P}(d_2|\omega_1) = \int_{\mathscr{R}_2} p(x|\omega_1)\,\mathrm{d}x$$

where d_j, $j = 1, 2$, denotes the decision that the jth class is true and $\mathrm{d}x = [\mathrm{d}x_1, \mathrm{d}x_2, \ldots, \mathrm{d}x_N]^{\mathrm{T}}$.

Let $L(\omega_i, d_j)$, $i, j = 1, 2$, be the loss incurred with the decision d_j when the ith class is true. Assuming that the cost of taking the observation is zero, $L(\omega_1, d_1) = L(\omega_2, d_2) = 0$. Also $L(\omega_i, d_j) > 0$ for $i \neq j$. Then the average (expected) loss from misclassification is

$$\mathscr{L} = P_1 L(\omega_1, d_2)\mathrm{P}(d_2|\omega_1) + P_2 L(\omega_2, d_1)\mathrm{P}(d_1|\omega_2)$$

It is this average loss that we would like to minimize. That is, we want the network to divide the sample space into regions $\mathscr{R}_1$ and $\mathscr{R}_2$ such that the average loss $\mathscr{L}$ is as small as possible. The procedure to minimize $\mathscr{L}$ is called a Bayes procedure. When the a priori probabilities are unknown, the conditional expected loss if the observation is from class 1 is

$$r(\omega_1, \mathscr{R}) = L(\omega_1, d_2)\mathrm{P}(\omega_2|\omega_1)$$

and the expected loss if the observation is from class 2 is

$$r(\omega_2, \mathscr{R}) = L(\omega_2, d_1)\mathrm{P}(\omega_1|\omega_2)$$

A procedure is minimax if the maximum expected loss $r(\omega_1, \mathscr{R})$ is a minimum.

When the a priori probabilities are known, the decision rule is to minimize the average loss $\mathscr{L}$. The Bayes rule is to choose $\mathscr{R}_1$ and $\mathscr{R}_2$ according to

$$\mathscr{R}_1 := P_1 p(x|\omega_1)L(\omega_1, d_2) \geqslant P_2 p(x|\omega_2)L(\omega_2, d_1)$$
$$\mathscr{R}_2 := P_1 p(x|\omega_1)L(\omega_1, d_2) \leqslant P_2 p(x|\omega_2)L(\omega_2, d_1)$$

When the equality holds, the observation can be classified as either class 1 or class 2 (the point belongs to the boundary between both classes).

Let

$$\mathscr{K} = \frac{P_2 L(\omega_2, d_1)}{P_1 L(\omega_1, d_2)}$$

Another way to choose $\mathscr{R}_1$ and $\mathscr{R}_2$ is

$$\mathscr{R}_1 : \frac{p(x|\omega_1)}{P_2 p(x|\omega_2)} \geqslant \mathscr{K}$$

$$\mathscr{R}_2 : \frac{p(x|\omega_1)}{P_2 p(x|\omega_2)} \leqslant \mathscr{K}$$

Consider two multivariable normal distributions with means M_1 and M_2 and covariable matrix V.

$$p(x|\omega_i) = \frac{1}{(2\pi)^{N/2}|V|^{1/2}} \exp\left[-\frac{1}{2}(x - M_i)^{\mathrm{T}} V^{-1}(x - M_i)\right] \quad i = 1, 2$$

The logarithm of the ratio of densities is

$$\mathscr{U} = \log\frac{p(x|\omega_1)}{p(x|\omega_2)} = -\frac{1}{2}[(x - M_1)^{\mathrm{T}} V^{-1}(x - M_1) - (x - M_2)^{\mathrm{T}} V^{-1}(x - M_2)]$$
$$= x^{\mathrm{T}} V^{-1}(M_1 - M_2) - \frac{1}{2}(M_1 + M_2)^{\mathrm{T}} V^{-1}(M_1 - M_2)$$

The first term is known as the (Fisher's) linear discriminant function. It is a linear function of the observation vector x. $\mathscr{U}$, the test statistic, can be compared

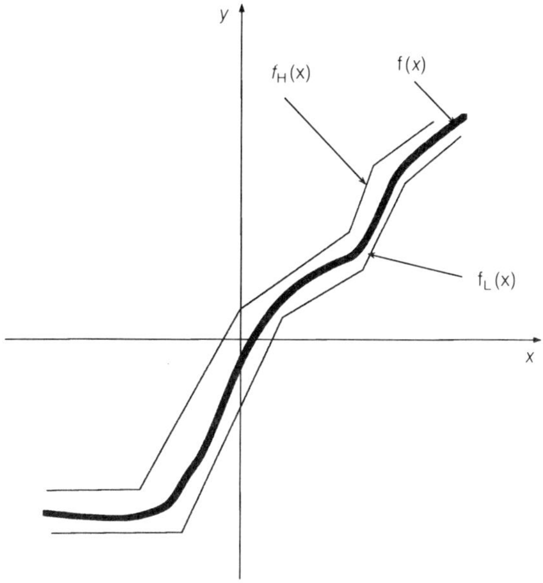

Fig. 12.21 Bayes decision theory for function mapping.

with the threshold $\log \mathscr{K}$. In particular when $P_1 = P_2$ and $L(\omega_2, d_1) = L(\omega_1, d_2)$, $\log \mathscr{K} = 0$.

Let α

$$\alpha = (M_1 + M_2)^{\mathrm{T}} V^{-1}(M_1 - M_2)$$

denote the distance between normal distributions which is the square of the euclidean distance between M_1 and M_2 where $V = I$, the identity matrix.

The test statistic $\mathscr{U}$ has also a normal distribution with means

$$E_i\mathscr{U} = \int p(x|\omega_i) \log \frac{p(x|\omega_1)}{p(x|\omega_2)} \mathrm{d}x$$

$$= (-1)^{i+1} \frac{1}{2} \alpha \quad i = 1, 2$$

$$\sigma_i^2(\mathscr{U}) = (M_1 - M_2)^{\mathrm{T}} V^{-1}(M_1 - M_2) = \alpha$$

where $E_i(\mathscr{U})$, $i = 1, 2$ are called the directed divergences representing the mean of $\mathscr{U}$ when class i is true. $\sigma_i^2(\mathscr{U})$ is the variance of $\mathscr{U}$, and the divergence between the two distributions is

$$J = E_1(\mathscr{U}) - E_2(\mathscr{U}) = \alpha$$

The probability of misclassification if the observation is from class i and classified as belonging to class j is given by $P(d_j|\omega_i)$

$$P(d_2|\omega_1) = \int_{-\infty}^{c} \frac{1}{\sqrt{2\pi\alpha}} e^{-(z-\alpha/2)^2/2\alpha} dz$$

$$= \frac{1}{\sqrt{2\pi}} \int_{-\infty}^{(c-\alpha/2)/\sqrt{\alpha}} e^{-y^2/2} dz$$

$$P(d_1|\omega_2) = \int_{-\infty}^{c} \frac{1}{\sqrt{2\pi\alpha}} e^{-(z+\alpha/2)^2/2\alpha} dz$$

$$= \frac{1}{\sqrt{2\pi}} \int_{(c+\alpha/2)/\sqrt{\alpha}}^{\infty} e^{-y^2/2} dz$$

where $c = \log \mathscr{K}$. The random variable z is equal to the rest statistic $\mathscr{U}$ given that class i is true. Figure 12.22 shows the two probabilities of misclassification as the shaded portions of the tails of each probability density function.

If the a priori probabilities are not available, an option is to select the threshold c to make expected losses equal to the misclassification,

$$L(\omega_1, d_2)P(d_2|\omega_1) = L(\omega_2, d_1)P(d_1|\omega_2)$$

This is called the **minmax solution**. In practice the statistics are estimated from the available samples. For a set of samples of size N_p, let N_1 and N_2 denote the number of samples belonging to classes 1 and 2 respectively. The set of samples mean and convariance matrices are

$$\hat{M}_i = \frac{1}{N_i} \sum_{j=1}^{N_i} x_{ji}$$

$$\hat{V}_i = \frac{1}{N_i - 1} \sum_{j=1}^{N_i} (x_{ji} - \hat{M}_i)(x_{ji} - \hat{M}_i)$$

where x_{ij} is the jth sample of the ith class, $i = 1, 2$.

The bayesian discriminant function $f_{\mathscr{B}}(x)$ that defines the decision boundary

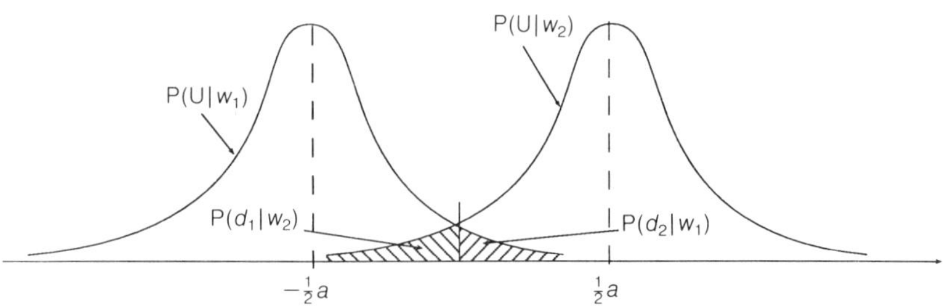

Fig. 12.22 Probability of misclassification.

between classes 1 and 2 is obtained by

$$f_{\mathscr{B}}(x) = \frac{1}{p(x)}[P_1 p(x|\omega_1)L(\omega_1, d_2) - P_2 p(x|\omega_2)L(\omega_2, d_1)]$$

which is normalized by

$$p(x) = \sum_{i=1}^{2} P_i p(x|\omega_i)$$

This bayesian decision rule is optimum in the sense that minimizes the expected risk with respect to the a priori probability and loss functions. There are other criteria for optimality that result in different decision/discriminatory functions. Two other popular techniques are:

- the **maximum distance** which tries to find a boundary as far as possible from the class clusters – this method, postulated by Fisher (1963), regards $\mathscr{U}_i$ as random variables and dictates as measure of the effectiveness of $\mathscr{U}$

$$L = \frac{(\bar{\mathscr{U}}_1 - \bar{\mathscr{U}}_2)^2}{\sigma_1^2(\mathscr{U}) + \sigma_2^2(\mathscr{U})}$$

  where $\bar{\mathscr{U}}_i = E_i \mathscr{U}$ and $\sigma_i^2(\mathscr{U}), i = 1, 2$ are the mean and variance respectively – to maximize L, the distance between the means, i.e., the interclass distance, must be large and the samples must be clustered around its mean, i.e., the intraclass distance must be small, and
- the **least mean square (LMS)** that measures the mean square distance $\bar{e}^2$ between M_i and its desired value (when ω_i is true) denoted by δ_i.

$$\bar{e}^2 = \sum_{i=1}^{2} E_i[(\mathscr{U} - \delta_i)^2]$$

  The closer $\bar{e}^2$ is to zero, the better $\mathscr{U}$ is in a LMS sense.

12.9 DECISION FUNCTIONS

For every pair of pattern classes, one hyperplane can be constructed to separate them (Highleyman, 1962). For m pattern classes, there are $\frac{1}{2}m(m-1)$ possible linear boundaries, which is the maximum number. In general, there is no guarantee that linear decision boundaries (as shown in Fig. 12.23) will meet all the classification requirements. Nonlinear decision hyperplanes might be required. In Fig. 12.23 the point A is classified as belonging to class 1, 3, or 2 by using the decision lines B_{12}, B_{13}, and B_{23} respectively. These boundaries classify a point in sector B as belonging to classes 2, 1, and 3 respectively. Thus sector B is rejected and does not belong to any class.

Linear discriminant functions are given $f_j(\mathbf{x}) = \mathbf{w}_j \cdot \mathbf{x} + \mathbf{w}_{j0}$, where $\mathbf{w} \cdot \mathbf{x}$ is the dot product of the measurement vector $\mathbf{x}$ and the weighting vector $\mathbf{w}$. The hyperplane is then

$$f_j(\mathbf{x}) = \mathbf{w}_j \cdot \mathbf{x} + w_{j0} = \sum_{i=1}^{N} w_{ij} \cdot x_i + w_{j0}$$

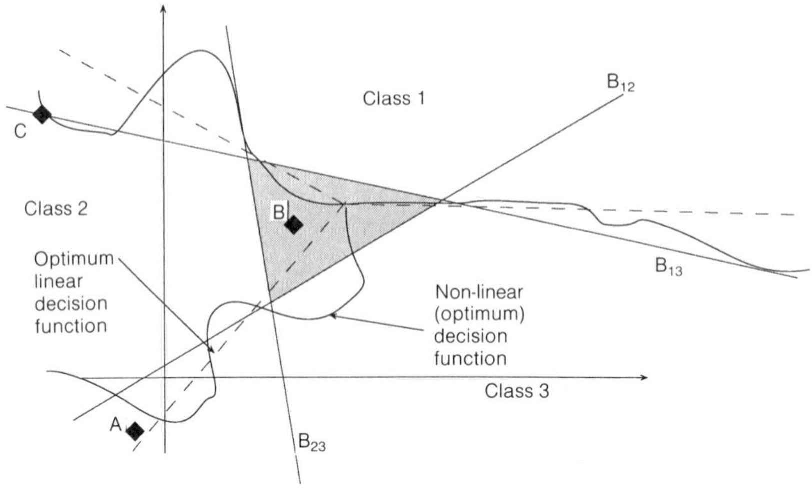

Fig. 12.23 Decision functions.

where $w_{ij}, i = 1, \ldots, N$ are the coefficient (weights) of the jth hyperplane and w_{j0} is the jth hyperplane threshold. An unknown pattern $\mathbf{x}$ is assigned to class j if $f_j(\mathbf{x}) \geqslant 0$.

The weights and threshold can be adjusted iteratively by changing the network parameters to reach optimum. This is done by using a sequence of training samples whose correct classification is known. ADALINE (*Ada*ptive *Lin*ear *E*lement) (Widrow and Hoff, 1960; Widrow, 1962) implements this hyperplane concept (Fig. 12.24). Defining the error of the presentation of the kth pattern $\varepsilon(k) = d(k) - f(\mathbf{x})$, where $d(k)$ is the desired output of the kth training sample. Then ADALINE minimizes the mean square error averaged over k.

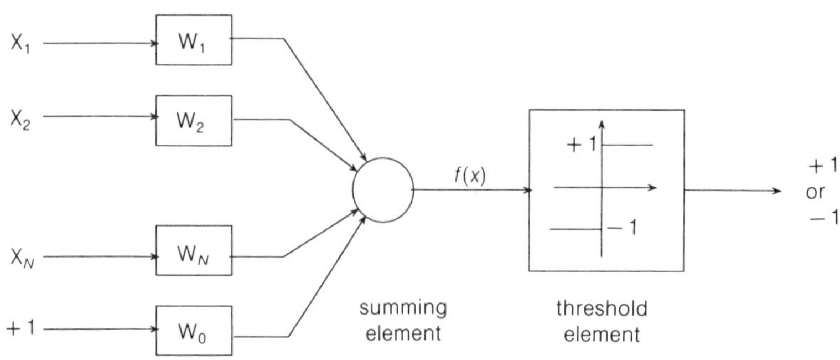

Fig. 12.24 Adaptive linear element network (ADALINE).

12.9.1 Minimum distance classifier

Let each pattern be associated with an exemplar, $\mathscr{P}_i$, $i = 1, \ldots, m$ (a reference point which may be, for example, the mean of the number of samples belonging to the ith class). The euclidean distance, $d(x, \mathscr{P}_i)$, from an arbitrary point x to the exemplar $\mathscr{P}_i$ is given by

$$d(x, \mathscr{P}_i) = |x - \mathscr{P}_i|_2 = \sqrt{(x - \mathscr{P}_i)(x - \mathscr{P}_i)}$$

The minimum distance classifier compares $d(x, \mathscr{P}_i)$, $i = 1, \ldots, m$ and the point x is assigned to the class which has the smallest distance $|x - \mathscr{P}_i|_2$.

When multiple (disjoint) clusters belong to the same class, a non-euclidean distance may be used (Sebestyen, 1962)

$$d(x, \mathscr{P}_i) = \min |x - \mathscr{P}_i^{(j)}|_2, \quad j = 1, \ldots, n_i$$

where n_i denotes the number of points (clusters), $\mathscr{P}_i^{(j)}$ associated with the ith class. The linear separation of the classes requires piecewise linear boundaries.

12.9.2 Non-linear classifiers

Non-linear classification theory provides a more general method of partitioning the sample space. Due to the complexity of the sample space, non-linear decision boundaries are often required. The non-linear discriminant function can be written in the general form

$$f(\mathbf{x}) = \sum_{i=1}^{r} w_i f_i(\mathbf{x}) + w_0$$

where $f_i(\mathbf{x})$, $i = 1, \ldots, r$ are functions of $\mathbf{x}$ representing the basis functions for the definition of the boundaries. An important class of the decision boundaries take the form of polynomials (also known as $\Pi - \Sigma$ networks (Sebestyen and Chu (1962); Specht (1962))

$$f_i(\mathbf{x}) = \underbrace{\prod_{i=0}^{N-1} \cdots \prod_{j=0}^{N-1}}_{N-1 \text{ terms}} x_p^i \cdots x_q^j \quad \text{where} \quad \begin{cases} p, \ldots, q-1, \ldots, N \\ i + \cdots + j \leqslant N-1 \end{cases}$$

Exemplars of these polynomials are $f_1(\mathbf{x}) = x_1$, $f_2(\mathbf{x}) = x_2$, $f_{32}(\mathbf{x}) = x_1 \cdot x_2$, $f_{45}(\mathbf{x}) = x_1^2 \cdot x_3, \ldots$, etc.

12.10 METRICS FROM MEASURE

For patterns with numerical components, a simple measure is the unweighted euclidean distance

$$d_{ij}^2 = (x_i - x_j)^{\mathrm{T}} \cdot (x_i - x_j) = \sum_{k=1}^{N} (x_{ik} - x_{jk})^2$$

where x_i and x_j are vectors of dimension N. The euclidean distance is considerably affected by the scale factor associated with each pattern component. For binary patterns the most common measure (Chen, 1973) is the correlation $x_i^T x_j$ which is often normalized by the geometric mean of the number of ones in both vectors, i.e., the normalized correlation

$$\rho_{ij} = \frac{x_i^T x_j}{(x_i^T x_j)\cdot(x_i^T x_j)}$$

which can also be used for numerical measurements (Okajima *et al.*, 1963).

Another similarity measure for binary variables which takes pairwise correlation (Bonner, 1964) into account is given by

$$S_{\alpha\beta} = \sum_{i=1}^{N}\sum_{j=1}^{N} \rho_{ij}[1 - |x_{\alpha i} - x_{\beta i}|][1 - |x_{\alpha j} - x_{\beta j}|][1 - 2|x_{\alpha i} - x_{\beta j}|]$$

where $S_{\alpha\beta}$ is defined as the **similarity** between patterns α and β and ρ_{ij} is the correlation coefficients for features i and j.

Another weighted euclidean distance may take into account the dispersion of the samples within a cluster (Sebestyen, 1962). Let $\mathscr{C}_k$ be the weighting coefficient, the weighted distance is given by

$$\delta_{ij}^2 = \sum_{k=1}^{N} \mathscr{C}_k (x_{ik} - x_{jk})^2$$

A more complicated measure with some desirable information theoretic properties but without all the properties of a metric is given by (Rogers and Tanimoto, 1960)

$$d_{ij} = -\log[S_{ij}]$$

where S_{ij} is a similarity measure given by

$$S_{ij} = \frac{x_i^T x_j}{x_i^T x_i + x_j^T x_j - x_i^T x_j}$$

12.11 CLUSTER ANALYSIS

Cluster analysis almost always involves the processing of large amounts of data. The applicability of any given cluster-seeking technique is determined by five factors.

- Computational complexity
- Memory requirement
- Sample size requirement
- Nature of the data
- Availability of classification information

A measure of the interaction between clusters pairs x and y used by Bonner (1964) is given by

$$I_{xy} = \frac{1}{N_x N_y} \sum_{\alpha=1}^{N_x} \sum_{\beta=1}^{N_j} S_{\alpha\beta}$$

where N_x and N_y are the number of members in cluster x and y respectively, $S_{\alpha\beta}$ is 1 if member α of cluster x is similar to member β of cluster y; 0 if they are not similar. I_{xy} is the percentage of possible similarity 'links' which are actually present between the members of cluster x and the members of cluster y.

For clustering of fuzzy sets, Ruspini (1970) considered the numerical classification as the process of assigning to each data sample a certain degree of 'belongingness' to each cluster S_i, $i = 1, \ldots, m$. However the membership functions are considered probabilistics. Being N the total number of samples in the set,

$$\mathscr{P}(S_j) = \sum_{i=1}^{N} p(x_i)\mathscr{P}(S_j | x_i)$$

is a measure of the relative size of cluster

$$S_j \quad \mathscr{P}(S_j) \geqslant 0 \quad \sum_{j=1}^{m} \mathscr{P}(S_j) = 1 \quad \text{and} \quad p(x) = \sum_{i=1}^{m} \mathscr{P}(S_j) p(x | S_j)$$

This way, the clustering process can be viewed as the decomposition of the probability density function $p(x)$ into the weighted sum of the component cluster densities $p(x | S_j)$ with weights $\mathscr{P}(S_j)$. This concept is similar to that of unsupervised learning.

A common drawback to all clustering algorithms is that their performance depends highly upon the values of several parameters by the user. As a matter of fact, the setting procedure is in general a trial and error method.

Nearest-neighbor decision procedure

The nearest-neighbor (NN) the decision rule is a nonparametric method for decision theory. The NN rule, first formulated by Fix and Hodges (1951), assigns to an unclassified sample point the class most heavily represented among its k_n nearest neighbors in some appropriate metric space.

Let $\mathscr{X} = \{x_1, \ldots, x_n\}$ be a group of n independent measures of known classes, and x' a measurement to be classified. Among $\mathscr{X}$, let the measurement with the smallest (euclidean) distance from x' be denoted as x_k, i.e.,

$$d(x', x_k) = \min_{x_k \in \mathscr{R}^{\mathscr{X}}} |x_i - x'| \quad i = 1, \ldots, n$$

Then the nearest-neighbor assigns x' to the class of x_k. Figure 12.25 illustrates the NN rule for three classes in a two-dimensional measurement space.

Cover and Hart (1967) showed that as the number of samples, $n \Rightarrow \infty$, the 'risk' incurred when the classification based on NN rule is incorrect, $\Re_n$, is

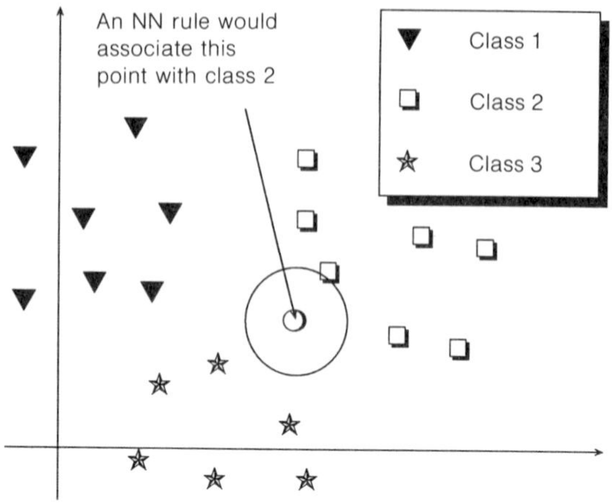

Fig. 12.25 Nearest-neighbor rule for three classes in a two-dimensional space.

upper and lower bounded by a function of the Bayes risk, $\mathfrak{R}^*$ when the statistics are known exactly, as follows

$$\mathfrak{R}^* \leqslant \mathfrak{R}_n \leqslant \left(2 - \frac{m}{m-1} \cdot \mathfrak{R}^*\right) \cdot \mathfrak{R}^*$$

where m is the number of pattern classes. For $m = 2$, then $\mathfrak{R}^* \leqslant R_n \leqslant 2\mathfrak{R}^*$. Figure 12.26 sketches the risk bounds.

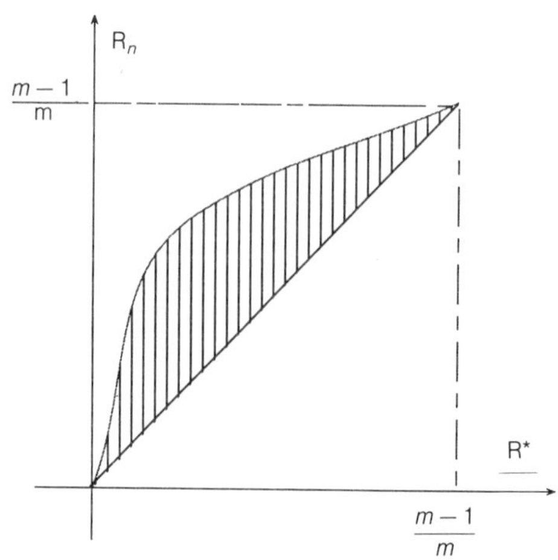

Fig. 12.26 Risk bounds for nearest-neighbor rule.

12.11.1 Measures for network selection

The goal of learning systems is to extract decision rules from sample data. To determine what variables are to be measured is the first and most important step toward the designing of a successful and efficient neural network application. This task of selecting the variables is called **feature selection or extraction**, and is usually preceded by pre-processing of the data available. If not, engineering judgment about the relevant information contained in the different variables is required. The pre-processing includes sorting, normalization, centralization, filtering, etc. of the input patterns (Chen, 1973). Important in the selection is the information available in the different variables that describe the feature of the pattern. Also important are the easiness for processing, the invariance to scaling, translation, and rotation of the patterns, the dimensionality of the problem, and the sample size required to guarantee certain degrees of confidence to the results. The selection of features that satisfy all these requirements is almost impossible. In reality, the designer reduces the dimension of the physical problem based on experience. Also due to resources constraints certain algorithms and network topologies may be discarded. Feature selection criteria using information theory and statistics have been proposed (Chen, 1973), where the features selected minimize indirectly the probability of misrecognition or the average risk.

Error estimation

The underlying objective of learning classifications from sample data is to classify and predict successfully on new data, i.e., to generalize. The most common measure of success/failure is the classifier's error rate. The **true error rate**, $\mathscr{E}$, is defined statistically as the limit of the error rate on a large number of cases, which asymptotically converge to the population distribution.

$$\mathscr{E} = \frac{N_e}{N_c}$$

where N_e and N_c represents the number of errors and cases respectively.

An error is a misclassification. Errors in general do have an importance ranking according to their type. To distinguish among error types, a **confusion matrix** can be used to lay out the different errors. The confusion matrix lists the correct classification against the predicted classification for each class.

While the primary measure of performance is the error rate, there are a number of alternatives, extensions, and variations possible. A natural alternative to an error rate is a **misclassification cost**. In this case, instead of designing a classifier to minimize the error rates, the goal would be to minimize the classification costs.

12.11.2 Measures for recall

The categorization problem of a 'good' or 'bad' recall can be considered as a statistical hypothesis testing. The 'distance' between two hypothesis or pattern classes has been defined in several ways in the statistical literature. All distance measures, however, are qualitatively related to the probability of misrecognition in the same manner: the largest the distance between the two pattern classes as provided by the variables selected, the smaller the probability of misrecognition.

Consider for example two pattern classes $\{\omega_1, \omega_2\}$. Many distance measures can be written in the form

$$f\{E[\phi(\mathscr{L})/\omega_1]\}$$

where $\mathscr{L} = p(x|\omega_1)/p(x|\omega_2)$, $\phi(\cdot)$ is a continuous convex function, $f(\cdot)$ is an increasing real-valued function of a real variable. Typical distances in pattern recognition algorithms are given below:

- The directed divergence (also called the **Kullback–Leibler number**) (Das, 1971)

$$I(1,2) = \int p(x|\omega_1) \log \frac{p(x|\omega_1)}{p(x|\omega_2)} \mathrm{d}x = E[\log(\mathscr{L})|\omega_1]$$

$$I(2,1) = \int p(x|\omega_2) \log \frac{p(x|\omega_1)}{p(x|\omega_2)} \mathrm{d}x = E[\mathscr{L}^{-1} \log(\mathscr{L})|\omega_1]$$

- The **divergence distance** (J-distance) (Estes, 1965)

$$J = I(1,2) + I(2,1) = E[(1 - \mathscr{L}^{-1}) \log(\mathscr{L})|\omega_1]$$

- The **Bhattacharyya distance** (B-distance) (Fu, 1971)

$$B = -\log(\rho)$$

  where

$$\rho = \int \sqrt{p(x|\omega_1) p(x|\omega_2)}\, \mathrm{d}x = E[\mathscr{L}^{-\frac{1}{2}}|\omega_1]$$

- The **Matusita's measure** of distance (Gonzalez and Tou, 1968)

$$d = \left[\int (\sqrt{p(x|\omega_1)} - \sqrt{p(x|\omega_2)})^2 \, \mathrm{d}x \right]^{\frac{1}{2}} = E[(\mathscr{L}^{-\frac{1}{2}} - 1)^2|\omega_1]$$

- The **Kolmogorov variational distance** (Henderson and Lanistis, 1970)

$$d = \frac{1}{2} \int |P_1 p(x|\omega_1) - P_2 p(x|\omega_2)| \mathrm{d}x = \tfrac{1}{2} P_1 E[|P_2 \mathscr{L}^{-1}/P_1 - 1||\omega_1]$$

12.12 ANN PARADIGM SELECTION IN MANUFACTURING

Among the measures used in ANS applied to manufacturing are the $\mathscr{L}_2$ norm of the error, also used are the TSS (total sum of the square errors) (Huang *et al.*, 1991), the MSE (mean square error), the RMS (root mean squared error), etc. For evaluating the learning algorithm, the amount of time required to perform the evaluation of the network (i.e., forward propagation, error evaluation, and backpropagation of the error gradients) and updates the network weights for an epoch is relevant, as well as the total number of epochs to bring the MSE down to a prescribed level. In essence the 'effective' CPU time for training. Note that the effectiveness will depend on the coding of the algorithm, so the actual number of operations (i.e., multiplications, additions, etc.) are required. The number of epochs by itself is not an accurate measure for the complexity of the learning algorithm. Some learning algorithms, like those that use second order methods, use global information, e.g. the pseudo-inverse of the error gradient. Given a training algorithm, for a specific problem (i.e. training set), the convergence rate (number of epochs for convergence) depends heavily on the initial conditions and learning parameters, such as learning rate, hence epochs may not be a good measure of the learning algorithm effectiveness, since there might be an 'optimal' set of learning parameters for each specific problem. To avoid the influence of initial conditions, statistics of the measures are calculated a set of training sessions each with different random weights. The effectiveness of this approach on the other hand depends on the scale of the problem and the range of the randomization of the weights. When multiple algorithms are being compared, sometimes some algorithms get stuck in local minima or never converge. In these cases, we propose to assign a factor of the largest number of epochs among the cases that converged. This way, these cases will not be thrown out, penalizing unsuccessful attempts.

Breaking down the different costs, we may propose:

$$[\text{pattern}]_{\text{cost}} = \sum_i [\#\text{ of } i\text{-operations}]\cdot[\text{time cost of operation}]_i$$

$$[\text{epoch}]_{\text{cost}} = [\#\text{ of patterns}]\cdot[\text{pattern}]_{\text{cost}}$$

$$[\text{session}]_{\text{cost}} = [\text{epoch}]_{\text{cost}}\cdot[\#\text{ of epochs-for-training}]$$

$$[\text{average-session-init}]_{\text{cost}} = [\text{average}]_{k\text{-init-trials}}([\text{epoch}]_{\text{cost}}(k\text{-trial}))$$

$$[\text{average-session-parameter}]_{\text{cost}} = [\text{average}]_{j\text{-parameter-trials}}([\text{average-session-init}]_{\text{cost}}(j\text{-trial}))$$

12.13 TOOLS THAT INCREASE PERFORMANCE

This section presents tools that aide the learning algorithms in artificial neural networks (ANNs) that use back error propagation (BEP/BP) as the adaptation mechanism for learning and have 'smooth' discriminatory functions (Fernández *et al.*, 1989a, b, c; Fernández and Murty, 1991). Two fundamental ideas and a possible implementation are presented. First, a family of methods to improve the initialization of artificial neural networks, called neural space initialization (NSI) are presented in the context of feedforward neural networks. Second, an alternative (to standard backpropagation) learning algorithms is developed that results in a dynamic learning algorithm denominated adaptive back-propagation (ABP). These proposed algorithms are founded on non-linear control theory, in particular in sliding mode control (Fernández, 1988) and traditional (gradient-descent) back error propagation (BEP). The end result is a learning algorithm that can be viewed as one that inherently 'adapts' its learning to the network loss function and gradient. Simple examples are shown for the benefit of the reader. More elaborate applications can be found in the aforementioned publications.

12.13.1 NSI–neural space initialization

The objective of this subsection is to show aides to initialization procedures for neural networks that improve the converge time compared with common procedures such as random weights initialization. The proposed procedures use pseudo-inverses of data pair sets to initialize the loss function in a least squares minimum sense, normalization of the weights to keep the loop $\mathscr{H}_\infty$ norm of the mapping, and averaging methods to maintain the network in the linear region. These procedures have been shown to reduce the learning by orders of up to 100% (i.e. no learning). For a specific pattern association problem with delta rule that takes 100 iterations to converge to within a 2% error, it can be adjusted to learn in one iteration. The basic idea is to lump groups of data that are 'similar' into a reduced order training set. This will in turn change the problem into a completely determined system, such as reducing a set of data points to two in linear curve-fitting. Effectively, this procedure can be viewed as reducing the 'kinkiness' of the loss function, just like throwing a 'blanket' over a rough terrain (see Fig. 12.27), creating a global fictitious minimum.

The network operates using a learning algorithm (such as backpropagation) along with the set of input and target output pairs $\{(x, y)_k\}$, and by letting the system learn the weights $\{W_i^L, b_i, W_i^R\}$. For a three hidden nodes network (Fig. 12.28) we might have the non-linear curve fitting shown in Fig. 12.29. Note

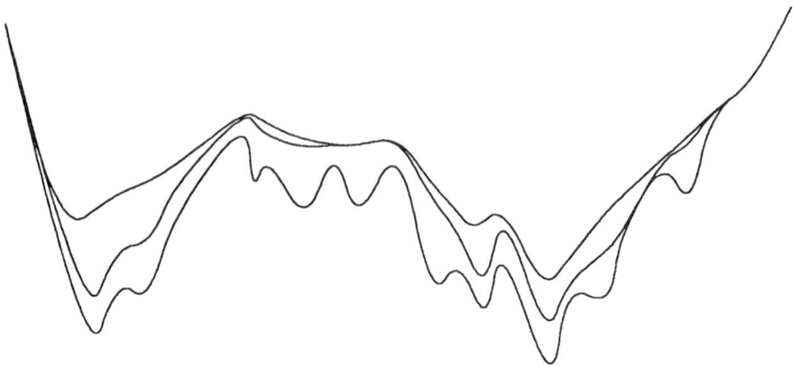

Fig. 12.27 Neural space initialization ('blanket') approach.

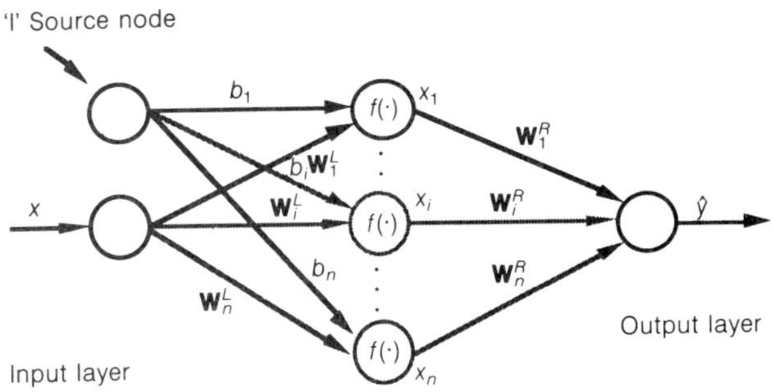

Fig. 12.28 Three-layer perceptron (3LP) function approximation.

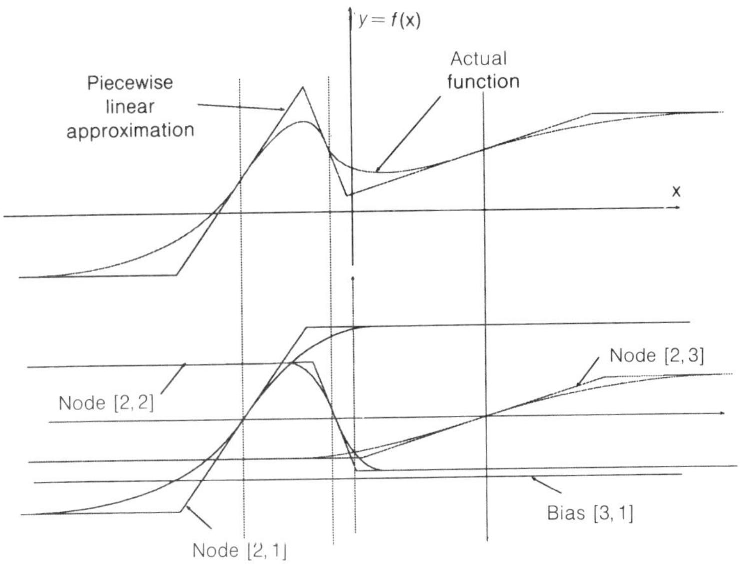

Fig. 12.29 Three-layer perceptron (3LP) function approximation.

that more layers will have complex effect on the output, and more nodes will asymptotically fit $f(x)$ (Lapedes and Farber, 1987; Cybenko, 1989).

$\mathscr{H}_\infty$ normalization

The basic principle here is to keep the randomization of the weights but to scale them so as to keep the mapping 'norm', i.e. use

$$W_{\text{Loop}} = \sqrt[L-1]{\frac{1}{N_P}\sum_{k=1}^{N_P}\left|\frac{\mathbf{y}_k}{\mathbf{u}_k}\right|} \tag{12.3}$$

where L represents the number of layers and N_P represents the size of the training set. An alternative addition to this technique is to use a modified version of Hopfield's initialization, that makes the network symmetric (for stability reasons), i.e.

$$\begin{aligned}
W_{ij} &= \text{random}(0, W_{\text{Loop}}) && \forall i,j \quad \text{(first pass)} \\
W_{ij} = W_{ji} &= \frac{1}{2}\left(W_{ij} + W_{ji} + \frac{W_{ii}}{N(i)}\right) && \forall i,j \mid i \neq j \\
W_{ii} &= W_{ii}\left(-\frac{1}{N(i)}\right) && \forall i
\end{aligned} \tag{12.4}$$

Linear interpolation

The technique suggested here clusters the training set into related groups. The number of clusters is a function of the network dimension, i.e. the overall number of nodes. The technique that makes piece-wise approximations between clusters. The simpler way of visualizing this is with a 3LP example.

Let us assume a $1 \times 3 \times 1$ three-layer-perceptron (3LP) as shown earlier in Fig. 12.28. If we take a closer look a hidden node, we can see the node's output as a function of the input (see Fig. 12.30). The contribution of that node's output, z_i to the network's output y is given by

$$z_i = \Pi(x_i) \quad \text{where} \quad x_i = b_i + W_i^{\text{L}} x$$

Here $\Pi(\cdot)$ is the discriminatory function, for example a sigmoid, a hyperbolic tangent, a saturation, etc. The '1' source node shown in Fig. 12.28, helps the network to provide the bias term b_i for each discriminatory node. The output node can simply be a summing function, without a discriminatory function, i.e. the discriminatory function is 'linear'. This is shown in the following

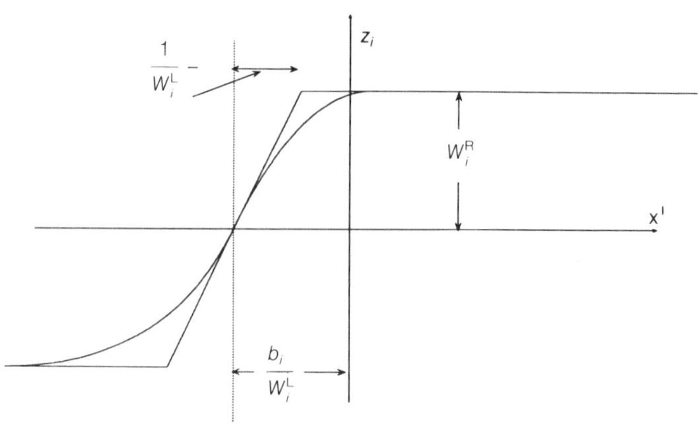

Fig. 12.30 Three-layer perceptron hidden node response.

expression:

$$\hat{y} = \sum_{i=1}^{N} W_i^{\mathrm{R}} \Pi(b_i + W_i^{\mathrm{L}} x) = \hat{f}(x) = \hat{f}(x, W) \tag{12.5}$$

It would also have been possible to use a discriminatory function in the output layer, however, in that case the function in the output pairs would have to be scaled. Note that locally only a couple of the z_i layer nodes are active, i.e. change locally with x, therefore a locally linear surface, 'tangent' to the actual surface, is obtained.

We then can find the values of the weights explicitly:

$$W_i^{\mathrm{R}} = \frac{y_i^{\mathrm{R}} - y_i^{\mathrm{L}}}{2}$$

$$W_i^{\mathrm{L}} = \frac{2}{x_i^{\mathrm{R}} - x_i^{\mathrm{L}}}$$

$$b_i = \frac{x_i^{\mathrm{R}} + x_i^{\mathrm{L}}}{2 \cdot W_i^{\mathrm{L}}}$$

$$b = \sum_{i=1}^{N=3} W_i^{\mathrm{R}} + y_0^{\mathrm{L}}$$

where $\{(x_i^{\mathrm{L}}, y_i^{\mathrm{L}}), (x_i^{\mathrm{R}}, y_i^{\mathrm{R}})\}$ are the points associated with the ith node of the hidden layer, which weights are $\{W_i^{\mathrm{L}}, W_i^{\mathrm{R}}, b_i\}$, and b is the output node bias.

Scaling

An alternate approach is scaling, by which we mean to select the network weights so the network operates in the 'linear' region of the discriminatory functions. The objective then is to minimize the norm of each x_i so the output z_i is an 'almost' linear operator on x_i. The method results in a recurrent test that averages the signals over the training set. These equations can be summarized by the following algorithm.

Step 1. $(b_i)_0 = -\overline{W_i^{\mathrm{L}} x(k)}$

Step 2. $W_i^{\mathrm{L}} \leftarrow (\overline{W_i^{\mathrm{L}} x(k)}) x^{\dagger}$
where $x^{\dagger}$ is the input pseudo-inverse. Go to Step 1 until convergence.

Step 3. $z_i(k) = \mathscr{D}(b_i + W_i^{\mathrm{L}} x(k))$

Step 4. $(b)_0 = \bar{y}(k) - W_i^{\mathrm{R}} \bar{z}_i$

Step 5. $W_i^{\mathrm{R}} \leftarrow \sum_{k=1}^{N_P} (y(k) - \bar{y} + W_i^{\mathrm{R}} \bar{z}) z^{\dagger}$
where $z^{\dagger}$ is the pseudo-inverse of the z_i. Go to Step 4 until convergence.

Step 6. $y = b + \sum_{i=1}^{N} W_i^{\mathrm{L}} z_i$

This algorithm has been successful, in most cases effectively reducing the training time on average by 10%.

Boundary decision regions

For design/control purposes, we might be more interested in bounding the function to be identified first, rather than accurately estimating it. This could be achieved more rapidly using two concurrent layers, as shown in Fig. 12.31. The two bounding functions can be expressed as follows:

$$\mathbf{y}_{\mathrm{H}} = f_{\mathrm{H}}(\mathbf{W}_{\mathrm{H}}, \mathbf{z})$$
$$\mathbf{y}_{\mathrm{L}} = f_{\mathrm{L}}(\mathbf{W}_{\mathrm{L}}, \mathbf{z})$$

It is now proposed to select $\mathbf{W}_{\mathrm{L}}$ and $\mathbf{W}_{\mathrm{H}}$, such that:

$$f_{\mathrm{L}}(\mathbf{W}_{\mathrm{L}}, \mathbf{z}) \leqslant f(\mathbf{W}, \mathbf{z}) \leqslant f_{\mathrm{H}}(\mathbf{W}_{\mathrm{H}}, \mathbf{z}) \tag{12.6}$$

A similar approach is also useful when acceptable bounds in errors are included in the decision process. Around the desired mapping, a 'boundary layer' is prescribed where the weights of the errors that fall within this layer (small errors) are weighted less that if they are outside (i.e., larger errors). A suitable implementation would be

$$J = \frac{1}{2} \sum_{k=1}^{N_P} \Upsilon(\mathbf{e}^{\mathrm{T}}\mathbf{e})$$

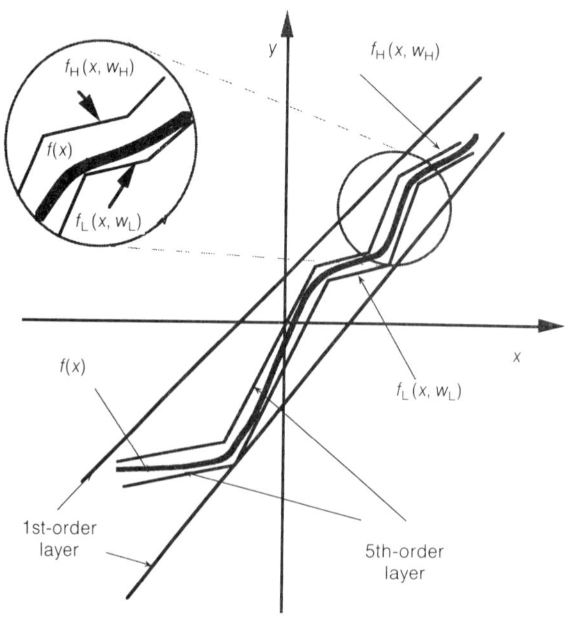

Fig. 12.31 Popov-like bounding functions.

where $\Upsilon(\cdot)$ is a discriminant of the error function, e.g.,

$$\Upsilon(\xi) = \begin{cases} \rho \cdot \xi & \text{when } |\xi| \leqslant \xi_0 \\ \dfrac{\xi}{1+\rho} & \text{otherwise} \end{cases}$$

here, ρ is the weighting of the 'misclassification', and ξ_0 is the width of the acceptable error band. If $\rho = 0$, then there is no penalty for points that map within the error band.

Another supporting factor to use this approach is when the user does not have sufficient confidence in the training set, or if there is a sizeable experimental error that enters in a random way and needs to be smoothed out. As a result, the network could add an extra dimension to the quality of the data it generates. The boundary layer concept could accomplish the equivalent to a 'confidence' layer as shown in Fig. 12.32. This confidence band could be both a measure or a specification for training.

Other forms of discriminants $\Upsilon(\cdot)$ can be used to provide special features according to the problem at hand. For example different weights according to the sign of the error, e.g.,

$$\Upsilon(\xi) = \begin{cases} \rho_{\text{pos}} \cdot \xi & \text{when } \xi \geqslant 0 \\ \rho_{\text{neg}} \cdot \xi & \text{otherwise} \end{cases}$$

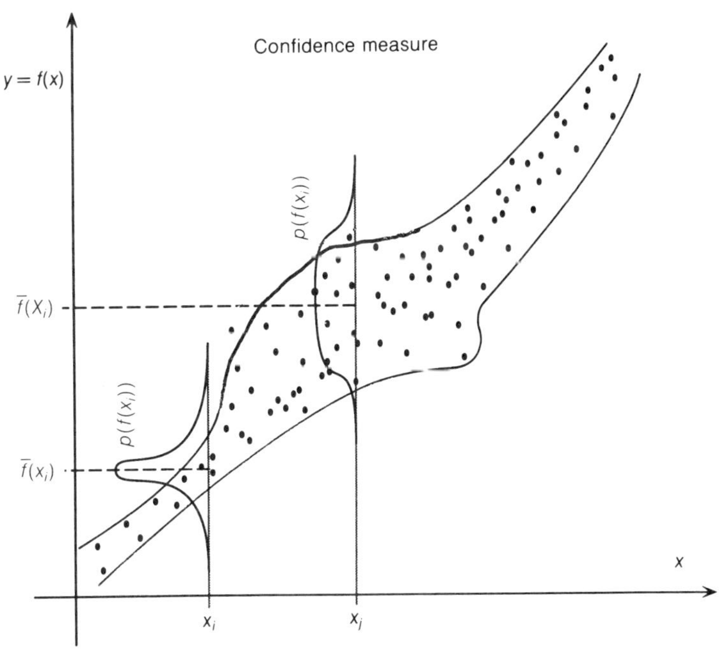

Fig. 12.32 Network confidence measure.

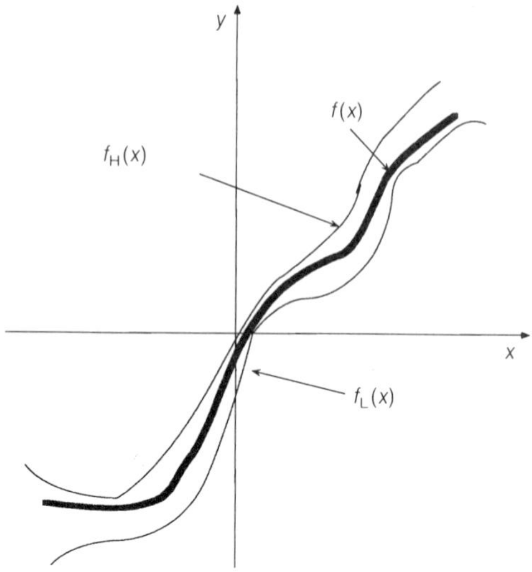

Fig. 12.33 Boundary layer as a function of region and sign of error.

In its more general form, $Y(\cdot)$ may be also a function of the input pattern. For example, certain mappings might be more critical in certain regions of the input space. By the same token, the boundary layer, ξ_0, need not be constant, e.g., the training data may have a subset of points where the data collection was noisy and the confidence of the values for that set is low, so a larger ξ_0 is justified. Figure 12.33 depicts these cases.

12.13.2 ABP–adaptive back-propagation

A modified back error propagation (BEP) algorithm, based on sliding mode control, is proposed that provides flexibility to specify how the learning behaves. The motivation is to define a 'sliding surface' that specifies the dynamics of the loss function, J, that measures the accuracy of the network and is used as a Lyapunov function. Then we find the dynamic characteristic of the learning to make this surface 'attractive' hence forcing the system to behave close to the desired dynamics.

The learning mechanism

A commonly used generalization of the gradient descent procedure is the backpropagation method. One of the steps evaluating the gradient requires the calculation of the gradient with respect to the output. This quantity is difficult to compute in networks with hidden units. The basic idea of back-propagation is the recognition that these calculations are easier if one starts the evaluations at the output layer and work backwards.

Adaptive backpropagation on the other hand, tries to derive the behavior or change in the network weights, **W** such that the network measure of accuracy, the loss function, $J = J(\mathbf{W}, \mathbf{u}, \mathbf{y}_d)$ is specified. J is in general the $\mathscr{L}_2$-norm of the output error,

$$J = \tfrac{1}{2}\mathbf{e}^{\mathrm{T}} \cdot \mathbf{e}$$

ABP-1 energy/sliding-based method

The first obvious approach is to try to force the rate of change of J with the learning time, τ:

$$\frac{\mathrm{d}J}{\mathrm{d}\tau} = -\eta(J)$$

Then, given that the derivative of J can be chained to **W**, we have

$$\frac{\mathrm{d}J}{\mathrm{d}\tau} = \frac{\partial I}{\partial \mathbf{W}} \cdot \frac{\mathrm{d}\mathbf{W}}{\mathrm{d}\tau} = -\eta(J) \tag{12.7}$$

$$\frac{\mathrm{d}\mathbf{W}}{\mathrm{d}\tau} = -\lambda^{\dagger}\eta(J)$$

where, λ represents the gradient of J with respect to **W**

$$\lambda \overset{\Delta}{=} \frac{\partial J}{\partial \mathbf{W}} \tag{12.8}$$

$$\lambda^{\dagger} = \lambda^{\mathrm{T}}[\lambda \cdot \lambda^{\mathrm{T}}]^{-1}$$

In summary, the ABP-1 approach proposes as a learning algorithm

$$\frac{\mathrm{d}\mathbf{W}}{\mathrm{d}\tau} = -\frac{\lambda^{\mathrm{T}}}{|\lambda|^2}\eta(J) \tag{12.9}$$

ABP-2 energy/smoothing-based method

The second approach, ABP-2 results from trying to minimize the 'jumping' characteristics of ABP-1 has about local minima due to the fact that the gradient norm, $|\lambda|$ goes to zero. To smooth the jumping, we add to the desired behavior of J, the norm of λ

$$\frac{\mathrm{d}J}{\mathrm{d}\tau} = -\eta(J)|\lambda| \tag{12.10}$$

which results

$$\frac{\mathrm{d}\mathbf{W}}{\mathrm{d}\tau} = -\frac{\lambda^{\mathrm{T}}}{|\lambda|}\eta(J)$$

This form will produce a continuous convergence toward $J = 0$ but it will have excessive switching. To reduce the chattering, several forms are appropriate, the most recommended is adding a 'boundary layer', λ_0 and using a 'squashing' function as tanh (·),

$$\frac{d\mathbf{W}}{d\tau} = -\tanh\left(\frac{\lambda^T}{\lambda_0}\right)\eta(J)$$

ABP-3 energy/entropy-based method

Another alternative is to augment the 'loss functional' usually referred to as a measure of the 'energy' of the system with a measure of the 'entropy'. We define this measure as the norm of the gradient, i.e. the tendency of the system to change.

$$J = \sum_{k=1}^{N_P} \frac{1}{2}\{\mathbf{e}^T\mathbf{Q}\mathbf{e} + T\lambda^T\mathbf{R}\lambda\} \qquad (12.11)$$

where T represents the 'artificial' temperature of the system. The resulting learning is given by

$$\text{[ABP-3]} \qquad \frac{d\mathbf{W}}{d\tau} = -\tanh\left(\frac{\lambda^T}{\lambda_0}\right)\left\{\eta(J) + \tfrac{1}{2}|\lambda|^2\frac{dT}{d\tau}\right\}\{\mathbf{I} + T\mu\}^{-1} \qquad (12.12)$$

where μ represents the second gradient of J

$$\mu \stackrel{\Delta}{=} \frac{\partial\lambda}{\partial\mathbf{W}} \qquad (12.13)$$

Other learning mechanisms

The Greedy Delta rule (Whitcomb and Augusteijn, 1991) adjusts the weights for each training for each training pair so that the output will be immediately correct for each pattern. For binary MLPs, the weight update rule is

$$\Delta W_{ij} = \frac{((y_j)_d - y_j)\cdot x_i}{1_\#(x)}$$

where $1_\#(x)$ is the number of ones of the input pattern.

12.13.3 Network size

Huang *et al.* (1991) use the **total sum of squared errors (TSS)** to determine the optimal number of hidden nodes in an MLP network. They suggest that BPN low training efficiency can theoretically increase with larger learning rates but the risk of falling in a local minimal also increases. For a dynamic learning (adaptation of the learning rate based on the effectiveness of training) the speed of convergence increases tenfold.

12.14 SUMMARY

This chapter probed at the question of evaluation and performance assessment of artificial neural systems applications. We presented some of the common performance and analysis criteria for artificial neural networks. We also presented measures to quantify those criteria. From pattern recognition and statistical tools, methods to quantify network and learning performance were presented.

A set of tools that can considerably improve the learning of artificial neural networks were also presented. Suitable implementation techniques for the methods were presented. As a disclaimer, it has to be said that the techniques presented should, but need not, work for most problems. In order to be effective while using artificial neural systems (and any other technology for that matter) requires from the user to wear two 'hats'. First, the user needs to know the technology to be used, its potential (the know-how) and limitations, in our case neural networks. Another important hat is the application's hat, i.e., the user needs to be familiar with the intended application (system, technology, science, e.g. vision, control, etc.). If not, the application user will not be able to ask the right questions to be answered by the neuro-engineer.

REFERENCES

Ackely, D.H., Hinton, G.E. and Sejnowski, T.J. (1985) A learning algorithm for Boltzman machines. *Cognitive Science*, **9**, 147–60.

Ackely, D.H., Hinton, G.E. and Sejnowski, T.J. (1986) Learning and relearning in Boltzman machines, in *Parallel Distributed Processing: Explorations in Microstructure of Cognition* (eds D.E. Rumelhart, J.L. McLelland and the PDP Research Group), Vol. I: Foundations, MIT Press/Bradford Books, Cambridge, MA.

Amari, S.-I. (1972) Learning patterns and pattern sequences by self-organizing nets of threshold elements. *IEEE Transactions on Systems, Man and Cybernetics*, **SMC-2**, 643–57.

Amari, S.I. (1990) Mathematical Foundations of Neurocomputing. *Proceedings of IEEE*, **78**(9), 1443–63.

Anderson, J. (1968) A memory storage model utilizing spatial correlation functions. *Kybernetik*, **5**, 113–19.

Becker, S. and le Cun, Y. (1988) Improving the convergence of back-propagation learning with second order methods, in *Proceedings of the 1988 Connectionist Models Summer School*, Morgan Kaufmann Publishers, San Mateo, CA, pp. 29–37.

Bonner, R.E. (1964) On some clustering techniques. *IBM Journal of Res. and Def.*, 23–32, January.

Carpenter, G.A. and Grossberg, S. (1986) Adaptive resonant theory: Stable self-organization of neural recognition codes in response to arbitrary lists of input patterns, in *Eighth Annual Conference of the Cognitive Society*, **45**, 45–62, Lawrence Erlbaum Assocs, Hillsdale, NJ.

Carpenter, G.A. and Grossberg, S. (1987) ART2: Self-organization of stable category recognition codes for analog input patterns. *Applied Optics*, **26**, 4919–30.

Chen, C. (1973) *Statistical Pattern Recognition*, Spartan Books, Hayden Book Company, Inc. Rochelle Park, New Jersey.

Cover, T.M. and Hart, P.E. (1967) Nearest neighbor pattern classification. *IEEE Trans. on Information Theory*, **IT-13**, 21–7, January.

Cruz, C.A. (1988) *Understanding Neural Networks: A Primer*, Graeme Publishing Corporation, Amherst, NH.

Cybenko, G. (1989) Approximation by superposition of a sigmoidal function. *Mathematics of Control, Signals, and Systems*, **2**, 303–14.

Das, S.K. (1971) Feature selection with linear dependence measure. *IEEE Trans. on Computers*, **C-20**(9), 1106–9, September.

DARPA Neural Networks Study (1988) Fairfax, Va: AFCEA International Press.

Dempster, A.J. (1977) A Simulation Study of Alternatives to Least Squares. *J. American Statistical Association*, **72**, 77, March.

Din, W.V. (1990) Neural nets software that learns by example. *Computer Engineering*, **9**, pp. 62–6.

Dornfeld, D.A. (1990) Neural network sensor fusion for tool condition monitoring. *Annals of the CIRP*, **39**(1), 101–5.

Estes, S.E. (1965) Measurement selection for linear discriminants used in pattern classification, IBM Research report RJ-331, April.

Fahlman, Scott E. (1988) An empirical study of learning speed in back-propagation networks, CMU Report CMU-CS-88-162, June.

Fernández, B., Parlos, A.G. and Tsai, W.K. (1989a) A neural network architecture for enhanced diagnostics and control of nuclear power plant component, Technical report 89-NN-01, Decision and Information Systems Laboratory, Department of Nuclear Engineering, Texas A&M University, College Station, TX, February.

Fernández, B., Parlos, A.G. and Tsai, W.K. (1989b) Discrete time learing of recurrent dynamic networks, Technical report 89-NN-02, Decision and Information Systems Laboratory, Department of Nuclear Engineering, Texas A&M University, College Station, TX, October.

Fernández, B., Parlos, A.G. and Tsai, W.K. (1989c) CNET: An interactive environment for the development of recurrent dynamic networks, Volumes I, II & III, Technical report 89-NN-03, Decision and Information Systems Laboratory, Department of Nuclear Engineering, Texas A&M University, College Station, TX, December.

Fernández, B. (1988) Generalized sliding mode control for MIMO nonlinear systems, Ph.D. dissertation, Mechanical Engineering Department, MIT, Cambridge, MA, May.

Fernández, B. and Murty, V. (1991) DyNet – Dynamic Network: A neural network for nonlinear system identification, Technical report NERDLab-TR-91-002, Neuro-Engineering Research and Development Laboratory, Department of Mechanical Engineering, The University of Texas at Austin, Austin, TX.

Fernández, B. (1991) Tools for artificial neural networks learning, in *Intelligent Engineering Systems Through Artificial Neural Networks* (eds C.H. Dagli, S.R.T. Kumara and Y.C. Shin), 69–76, ASME Press, New York.

Fernández, B., Murty, V.V. and Chang, W.R. (1989) *DyNet: Dynamic Network, User Manual*, NeuroEngineering Research and Development Laboratory, Department of Mechanical Engineering, The University of Texas at Austin, Austin, TX, Report 91-NERDLab-01, December.

Fernández, B., Prabhudesai, A.V., Murty, V.V., Gupta, R. and Chang, W.R. (1992) Neurobondgraphs: Development environment for modeling of nonlinear dynamics systems using neural networks and bondgraphs, 1992 ASME Winter

Annual Meeting, Anaheim, CA., Nov. 9–13 (also in *Neural Networks in Manufacturing and Robotics*, PE-Vol. 57 (eds Y.C. Shin, A.H. Abodelmonem and S. Kumora), ASME Press, New York, NY, 1992, pp. 75–90.)

Fisher, R.A. (1963) *Statistical Methods for Research Workers*, Hafner, N.Y.

Fix, E. and Hodges, Jr., J.L. (1951) Discriminatory analysis, nonparametric discrimination: Consistency properties, USAF School of Aviation Medicine, Randolph Field, Texas, Project 21-49-004, Report 4, Contract AF 41 (128)-31, Feb.

Fu, K.S. (ed.) (1971) Special issue on feature selection and extraction in pattern recognition. *IEEE Trans. on Computers*, **C-20**, September.

Fukusima, K. (1980) Neocognitron: A self-organizing neural network model for a mechanism of pattern recognition unaffected by shift position. *Biological Cybernetics*, **36**, 193–202.

Garey, M.R. and Johnson, D.S. (1970) *Computers and Intractability*, W.H. Freeman, New York.

Gill, P.E. and Wright, M.H. (1981) *Practical Optimization*, Academic Press.

Gonzalez, R.C. and Tou, J.T. (1968) Some results in minimum-entropy feature extraction, IEEE Region III Convention, Cocoa Beach, Florida.

Grossberg, S. (1988) Nonlinear neural networks: Principles, mechanisms, and architectures. *Neural Networks*, **1**(1), 17–61.

Grossberg, S. (1968) Some nonlinear networks capable of learning a spatial pattern of arbitrary complexity. *Proceedings of the National Academy of Sciences*, **59**, 368–72.

Hassoun, M.H. and Youssef, A.M. (1989) High performance recording algorithm for Hopfield Model Associative Memories. *Opt. Eng.*, **28**(1), 46–54.

Hecht-Nielsen, R. (1987) Theory of backpropagation neural network. *Proceedings of the International Joint Conference on Neural Networks*, **I**, 593–605.

Henderson, T.L. and Laniotis, D.G. (1970) Application of state-variable techniques to optimal feature extraction-multichannel analog data. *IEEE Trans. on Information Theory*, **IT-16** (4), July, 396–406.

Highleyman, W.H. (1962) Linear decision functions with applications to pattern recognition. *Proc. of IRE*, **50**, 1501–14, June.

Hopfield, J.J. (1982) Neural networks and physical systems with emergent collective computational abilities. *Proceedings of the National Academy of Sciences*, **79**, 2554–8.

Hopfield, J.J. (1984) Neurons with graded response have collective computational properties like those of two-state neurons. *Proceedings of the National Academy of Sciences*, **81**, 3088–92.

Hopfield, J.J. and Tank, D.W. (1985) Neural computation in optimization problems. *Biological Cybernetics*, **52**, 141–52.

Hopfield, J.J. and Tank, D.W. (1986) Computating with neural circuits. *Science*, **233**, 625–33.

Huang, H.H., Zhang, C., Lee, S. and Wang, H.P. (1991) Implementation and comparison of neural network learning paradigms: Back propagation, simulated annealing, and tabu search, in *Intelligent Engineering Systems Through Artificial Neural Networks* (eds C.H. Dagli, S.R.T. Kumara and Y.C. Shin), 95–100, ASME Press, New York.

Hung, C.C. and Fernández, B. (1992) Comparative analysis of control systems, 1992 Workshop on Fuzzy Applications in Industry, College Station, TX, December 2–4.

Isidori, A. (1989) *Nonlinear Control Systems*, Springer-Verlag, Communications and Control Engineering Series, Berlin.

Kanerva, P. (1984) Self-propagating search: A unified theory of memory, Stanford Center for Study of Language and Information, Technical report CSLI-84-7.

Kirkpatrick, S., Gelatt, C.D. and Vechi, M.P. (1983) Optimization by simulated annealing. *Science*, **229**, 671–9.

Kohonen, T., Ruohonen, M. (1973) Representation of associated data by matrix operators, *IEEE Transactions on Computers*, **C-22**, 701–3.

Kohonen, T. (1977) *Associative Memory: A System-Theoretical Approach*, Springer-Verlag, Berlin.

Kohonen, T. (1984) *Self-Organization and Associative Memory*, Springer-Verlag, Berlin.

Kosko, B. (1987a) Bidirectional associative memories, *BYTE*, September.

Kosko, B. (1987b) Fuzzy associative memories, in *Fuzzy Expert Systems* (ed. A. Kandel), Reading, MA, Addison-Wesley.

Kosko, B. (1988) Bidirectional associative memories. *IEEE Transactions on Systems, Man, and Cybernetics*, **SMC-18**, 42–60.

Kosko, B. (1992) *Neural Networks and Fuzzy Systems*, Prentice Hall.

Lang, K.J. and Witbrock, M.J. (1988) Learning to tell two spirals apart, in *Proceedings of the 1988 Connectionist Models Summer School*, Morgan Kaufman.

Lapedes, A. and Farber, R. (1987) How neural nets work, in *Proceedings of the International Joint Conference on Neural Networks*, Denver, CO, July. Also Technical Report LA-UR-87-2662, Los Alamos National Laboratory, Los Alamos, NM.

Le Cun, Y., Boser, B., Denker, J.S., Henderson, D., Howard, R.E., Hubbard, W. and Jackel, L.D. (1989) Backpropagation applied to handwritten zip code recognition. *Neural Computation*, **1**, 541–51.

Liu, L. and Liu, G.H. (1990) Neural network computing applied in design for manufacturability, in *Proc. of Int. Semiconductor Manufacturing Science Symposium*, pp. 107–9.

Ljung, L. (1991) Issues in system identification. *IEEE Control Systems*, **11**(1), 25–9, January.

Makram-Ebeid, S., Sirat, J.-A. and Viala, J.-R. (1989) A rationalized error back-propagation learning algorithm, in *Proc. of the International Joint Conference on Neural Networks*, Vol. II, Washington, D.C., pp. 373–80, June.

Manufacturing Studies Board (1986) *Towards a New Era in US Manufacturing: The Need for a National Vision*, National Academy Press, Washington DC.

McAulay, A.D., Danset, P. and Wicker, D. (1990) Assembly line inspection using neural networks. *Hybrid Image and Signal Processing*, **1217**, 339–49.

McCulloch, W.S. and Pitts, W.H. (1943) A logical calculus of the ideas imminent in nervous systems. *Bull. Math. Biophy.*, **5**, 115–33.

McEliece, R.J., Posner, E.C., Rodemich, E.R. and Venkatesh, S.V. (1987) The capacity of the Hopfield associative memory. *IEEE Transactions on Information Theory*, **IT-33**(4), 461–82.

McLelland, J.L., Rumelhart, D.E. and the PDP Research Group (eds) (1986) *Parallel Distributed Processing: Explorations in Microstructure of Cognition, Vol. II: Psychological and Biological Models*, MIT Press/Bradford Books, Cambridge, MA.

Minsky, M. and Papert, S. (1969) *Perceptrons: An Introduction to Computational Geometry*, MIT Press, Cambridge, MA.

Nguyen, D. and Widrow, B. (1989) The truck backer-upper: An example of self-learning in neural networks, in *Proceedings of the IEEE International Joint Conference on Neural Networks*, Vol. II, IEEE Press, New York, pp. 357–63, June.

Narendra, K.S. and Parthasarathy, K. (1990) Identification and control of dynamical systems using neural networks. *IEEE Transactions on Neural Networks*, **1**(1), March.

Okajima, M., Stark, L., Whipple, G. and Yasui, S. (1963) Computer pattern recognition

PART THREE

Neurocomputing for intelligent manufacturing: Execution level applications

CHAPTER 13

Process monitoring and control

Michel Guillot, Riadh Azouzi and Marie-Claude Cote
Mechanical Engineering Department, Laval University, Quebec, Canada, G1K 7P4.

13.1 INTRODUCTION TO PROCESS MONITORING AND CONTROL

In everyday life, several mechanisms and processes surrounding us are monitored or controlled in view of achieving automated, efficient and more reliable operations. Process monitoring generally accounts for the *continuous surveillance* of events like process malfunctions, that require a specific and often discrete action. As opposed to monitoring, a controller *acts continuously* in order to regulate or control certain process outputs, or to optimize a given performance index. Thus process control usually involves the dynamics of the process, sensors and actuators in a closed-loop arrangement.

One can find in the literature, and in real life, numerous examples of process monitoring and control problems that have only partly been solved using traditional techniques or even worse, that remain without solutions often because of the complexity of their interacting variables or in reason of highly non-linear behavior. The neural network monitoring and control techniques represent an appealing and efficient solution not only to such difficult problems, but also to those, more traditional, that could be solved in less time and effort, and offering more intelligent capabilities.

Consequently, this chapter will first review typical process monitoring and control problems, and present traditional techniques of solution. Thereafter, the neural-based monitoring and control techniques will be presented and exemplified with two implementation cases: monitoring grasping and assembly operations on an industrial robot; and neurocontrol and optimization in machining.

Artificial Neural Networks for Intelligent Manufacturing. Edited by Cihan H. Dagli.
Published in 1994 by Chapman & Hall, London. ISBN 0 412 48050 6

13.1.1 Process monitoring problems and traditional techniques of solution

Monitoring is used in many, if not, most manufacturing processes. It is often applied to the simple detection of malfunctions like an overtravel of a machine tool axis or a lack of lubricant in the slideways. In most cases, programmable logic controllers are employed to monitor such simple events. However, there exist several other more complicated cases that require intelligent monitoring. In machining for instance, tool condition monitoring allow tool breakage detection from pattern identification of ultrasonic acoustic emissions (Hayashi *et al.*, 1988), sounds (Sata *et al.*, 1985), vibrations and forces (Braun *et al.*, 1982), or power signals. Progressive tool wear can also be estimated from a change in the signal magnitude (Sadat and Shivakimar, 1987; Koren *et al.*, 1986), thus allowing automatic tool replacement as the tools wear out. In robotics, the success of an assembly operation can be evaluated by monitoring any slipping between the gripper and the handled part using an ultrasonic acoustic emission sensor (Rangwala *et al.*, 1988). This last application will be described in section 13.5.1.

As depicted in Fig. 13.1, a process monitoring system comprises at least one sensor conditioned and pre-processed such that to emphasize certain important features and reduce the noise level. Signal pre-processing refers generally to techniques like filtering and fast Fourier transform (FFT) for frequency-domain feature identification. The next, and probably the most difficult step, consists in interpreting correctly the signal characteristics which can be viewed without restrictions as signal amplitudes, peaks, as well as complex patterns. Finally, the selection of a control action based on the characteristics identified is usually reduced to a simple classification task.

The most common process monitoring techniques fall into four categories.

1. As depicted in Figs 13.2(a) and 13.2(b), the simplest technique consists in monitoring the signal for an excessive magnitude. As the signal crosses a preset threshold level, an action is undertaken. For example, tool breakages in machining are sometimes monitored commercially using an acoustic emission (AE) sensor and such a monitoring technique. A tool breakage generates a peak in the AE RMS signal, which can be detected as it crosses the preset threshold, thus triggering an emergency stop of the machine

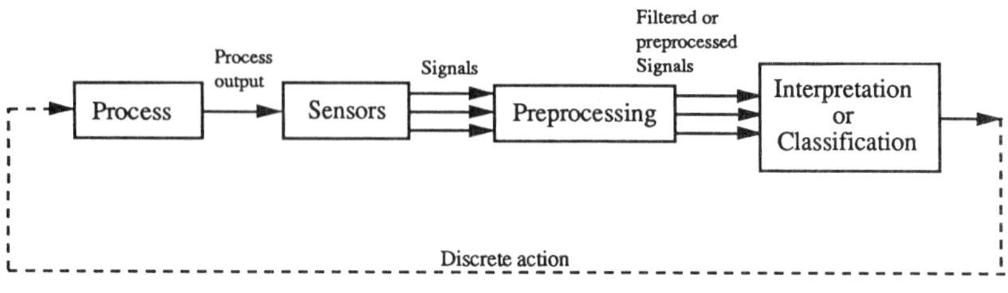

Fig. 13.1 Process monitoring functions.

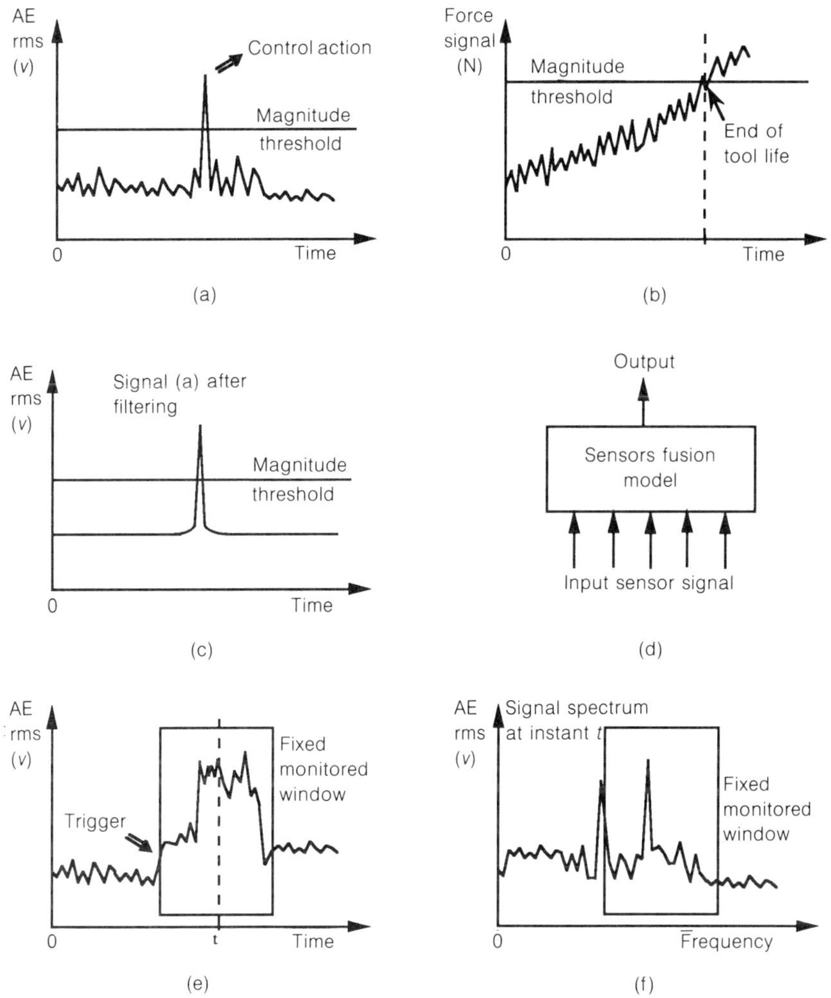

Fig. 13.2 Process monitoring techniques: (a) peak magnitude monitoring; (b) magnitude level monitoring; (c) addition of filtering; (d) sensors fusion: (e) stationary time-domain pattern recognition; (f) stationary frequency-domain pattern recognition.

tools. The drawback of this technique is that it can be very unreliable under noisy signal conditions or when peaks appear for reasons other than tool breakages for the machining example (Braun *et al.*, 1982).

2. In order to reduce the noise level and improve the readability of certain signal features, traditional or adaptive filters (Goodwin and Sin, 1984) can be added (see Fig. 13.2(c)) prior to the use of a signal magnitude threshold of the first technique. An application example from Takata *et al.* (1985) demonstrated tool breakage monitoring using Kalman filtering. However,

very often, the required selectivity in terms of both complexity (e.g. frequency selections, behaviors, etc.) and quality (signal-to-noise ratio) is so important that traditional techniques become ineffective.

3. It is also possible to combine several sensor signals in order to improve the reliability of the monitoring system, particularly when a single sensor cannot provide a good correlation with all the events that need to be detected. The so-called sensor fusion technique sometimes consists in a simple weighted addition of the sensor signals, although in most cases, more sophisticated and intelligent fusion models are necessary (Fig. 13.2(d)).
4. In the fourth category of monitoring techniques, more complicated patterns must be identified on-line. Until now, most signal pattern identification techniques have dealt with stationary patterns relative to the monitoring window, using two different methods: first, as a function of time (Fig. 13.2(e)) by assuming a constant starting position often triggered by signal magnitude, and second, as a function of frequency (Fig. 13.2(f)), which is rendered possible by pre-processing the signal with a FFT algorithm. Obviously, pattern matching of a frequency spectrum can be applied only on sensors showing a good dynamic response. In reason of its complexity, the identification of moving patterns as a function of time, in a way somewhat similar to the action of a specialized filter, has not been developed very extensively so far. An example of such specialized filtering technique implemented for tool breakage identification from a cutting force signal and using neural techniques can be found in Guillot and El Ouafi (1991).

Obviously, there exist numerous variations to these monitoring techniques as well as new ones emerging from several fields of research. Among those, artificial neural-based techniques appear very promising for several reasons:

- the neural network models can learn complex, non-linear relationships that will be used to integrate several sensors, to act as specialized filters, or as pattern recognizers,
- their response is fast and can be noise resistant when properly trained, and
- they can evolve as new information or patterns become available.

13.1.2 Process control problems and classical techniques of solution

The theory behind classical and adaptive process control is well documented and proposes a rich variety of process control schemes intended for different purposes and objectives (e.g. Goodwin and Sin, 1984). Accordingly, only four typical feedback control schemes will be presented next and compared later in section 13.4, to neural equivalents. As depicted in Fig. 13.3(a), the first control scheme represents a classical regulator that maintains the output y of the process close to a desired set point. To simplify the representation of the scheme, the actuators, process and sensors fit in a single bloc. In the second scheme, in Fig. 13.3(b), a process model help the optimizer finding the next

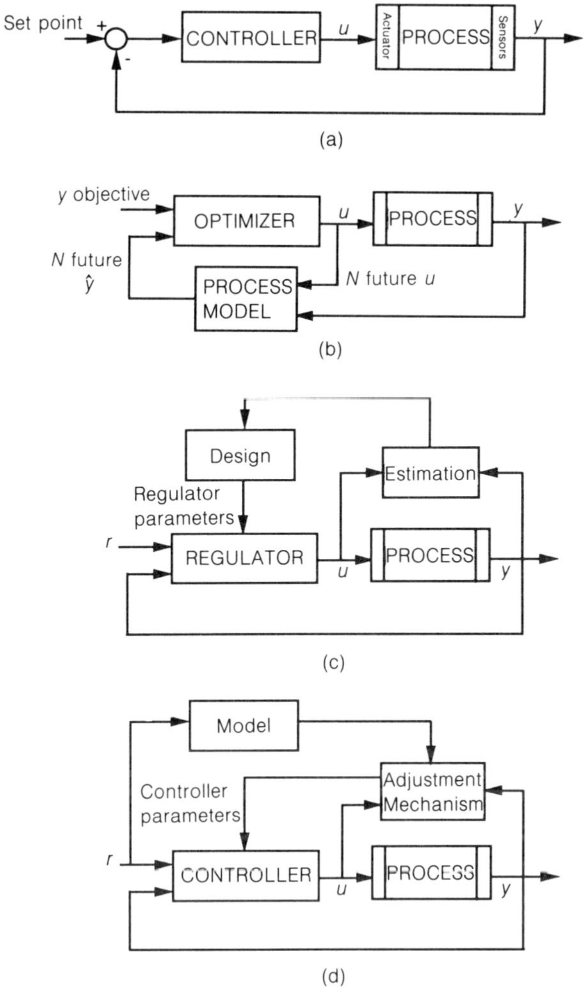

Fig. 13.3 Process control techniques: (a) simple register; (b) optimizing controller; (c) self-tuning regulator (STR); (d) model reference adaptive controller (MRAC).

control input vector *u*. The self-tuning regulator (STR) can be viewed as a simple classical feedback controller with on-line adjustment of the regulator parameters (Fig. 13.3(c)). These adjustments can allow the regulator to operate in a broader range of process inputs particularly with nonlinear processes, or with changing dynamic characteristics. However, the design adaptation is still fixed in the sense that it is intended for a certain range of conditions and for a given set point. The performance of the regulator may deteriorate drastically as the set point is modified. Finally (Fig. 13.3(d)), the model reference adaptive controller (MRAC) is designed so that the output of the process eventually follows the output of the model. During the learning phase, the controller parameters are adjusted in a way to reduce or even minimize the error. Consequently, the performance of the MRAC scheme depends strongly on the quality of its process model.

The problem with most classical and adaptive control schemes lies first in the need to identify separately or as a whole, the transfer functions and models of the actuators, process and sensors, and second, in the lack of adaptability and accuracy of these models, particularly faced to non-linear behavior. For example, in the course of developing a classical chemical process controller, one may have to identify the transfer function of the valves that control the inflow and outflow of chemical solutions, the process dynamics, and the level, pressure, and temperature sensors. Because of the usual non-linear behavior of these components and process, such models will generally be difficult to realize and often simplified to linear functions valid around a given operating point.

Further explanations will be provided in section 13.4 as we will discuss respectively about neurocontrol approaches.

13.2 NEURAL NETWORK MODELS FOR PROCESS MONITORING AND CONTROL

Because of the usually fast response required both for process monitoring and control, the multilayer feedforward network shown in Fig. 13.4, is commonly

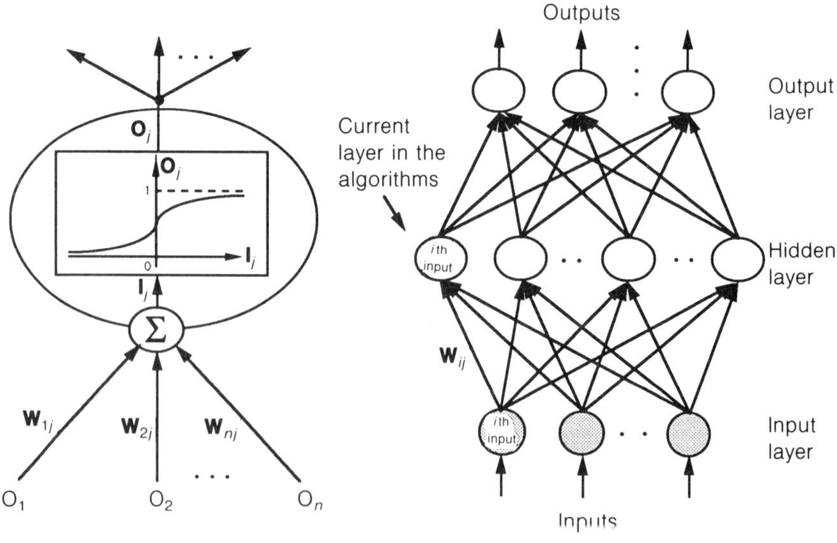

Fig. 13.4 Multi-layer feedforward network.

used, sometimes as specialized filters or feature identifiers (Guillot and El Ouafi, 1991), or as process models (Rangwala and Dornfeld, 1989) and neuro-controllers (Thibault and Grandjean, 1991). For convenience, this well-known neural network model and the backpropagation training technique (e.g. Rumelhart and McClelland, 1986) will be described briefly next.

Each neuron of the network is represented by a sigmoid function of the form:

$$O_j = \frac{1}{[1 + \exp(-(I_j + \Theta_j))]} \tag{13.1}$$

where

$$I_j = \sum_i W_{ij} O_i \tag{13.1a}$$

The outputs of the current neuron and of the neurons of the preceding layer are represented respectively by O_j and O_i. Usually, all the neurons of a given layer are fully connected to those of the adjacent layers. The input values are fed forward through the network and provide a model estimate at the output.

This neural network learns from examples usually obtained from experimentations carried out on the process. Each input–output vector is called an **exemplar**. Accordingly, model training will require several exemplars that may be obtained using a structured and efficient design of experiments (Box *et al.*, 1978) providing a smaller but more representative set of exemplars. After having gathered the exemplars, training is achieved using one of the algorithms presented or mentioned below. It is recommended to start with smaller models (e.g. one hidden layer of 5 neurons) so that to avoid overfit of the experimental data. The size of the hidden layer can be increased until the improvements in accuracy become negligible. If the error is still important, a second hidden layer may be appropriate.

One of the best-known neural net training algorithms was presented by Rumelhart and McClelland (1986). In this technique, the model is trained by propagating backward through the network the error ε between the model estimates of the exemplars and the exemplar outputs themselves. The error is backpropagated according to the generalized delta rule given by (13.2) (Rumelhart and McClelland, 1986), which set the amount of correction for the neuron of hidden and output layers.

$$\delta_j = \varepsilon O_j(1 - O_j) \quad \text{for the node of the output layer} \tag{13.2a}$$

$$\delta_j = O_j(1 - O_j) \sum_k \delta_k W_{kj} \quad \text{for all other nodes} \tag{13.2b}$$

where, the indices j and k are respectively referring to the current layer and the next layer of the network. Thresholds Θ_j and weights W_{ij} are evaluated according respectively (13.3) and (13.4).

$$W_{ij}(n + 1) = W_{ij}(n) + \eta \delta_j O_i + \alpha(W_{ij}(n) - W_{ij}(n - 1)) \tag{13.3}$$

$$\Theta_j(n + 1) = \Theta_j(n) + \eta \delta_j + \alpha(\Theta_j(n) - \Theta_j(n - 1)) \tag{13.4}$$

The subscript i corresponds to the preceding layer, and n, to the current number of iterations. The gain, η, and momentum, α, may vary between 0 and 1 providing settings for the learning rate and filtering against high-frequency variations of the error surface.

The exemplar values input in the neural network are linearly mapped to a 0 to 1 range. Similarly, the neural net outputs that range between 0 and 1 can be mapped back to full range.

Several improved algorithms (e.g. adaptive gain adjustment, quasi-newton based on a second order gradient descent) that learn generally much faster than the basic backpropagation technique presented above, can be found in the literature.

In addition to the multilayer feedforward network, several other neural network models have been attempted recently. In process monitoring for example, probabilistic neural networks have been applied to spike detection from a vibration time-domain analysis (Branagan and Wasserman, 1992), and to the fusion of informations provided by vibration signature analyses (Loskiewicz-Buczak and Uhrig, 1992). This latter implementation allows faulty operating conditions such as misalignment, bearing damage or wear of machine components to be identified. In Tansel *et al.* (1992), severe tool damages during microdrilling operations are monitored and classified using the ART2 neural network. Finally, Kohonen self-associators seem also promising in chemical process control (Tryba and Croser, 1991) or for controlling an arbitrary plant (Grassi and Grunewald, 1992).

13.3 NEURAL NETWORK APPROACHES TO PROCESS MONITORING

Multilayer feedforward networks can be applied very successfully in categories 2, 3 and 4 of process monitoring techniques described in section 13.1.1.

13.3.1 Fusion of sensors

As depicted in Fig. 13.2(d), the fusion of sensors can be viewed as a simple input–output function. With the proper number of layers and neurons, and having carried out adequate training of a representative set of input–output exemplars, a neural network should easily model this function more accurately than many other techniques (Chryssolouris and Guillot, 1990).

The monitoring system selects a control action according to the output of the fusion model. Using a normalized binary output, the zero and one usually represent respectively adequate and faulty conditions. Several outputs may be required in order to identify one characteristic from a number of characteristics. An example of sensor fusion applied to process monitoring is presented in Loskiewicz-Buczak and Uhrig (1992).

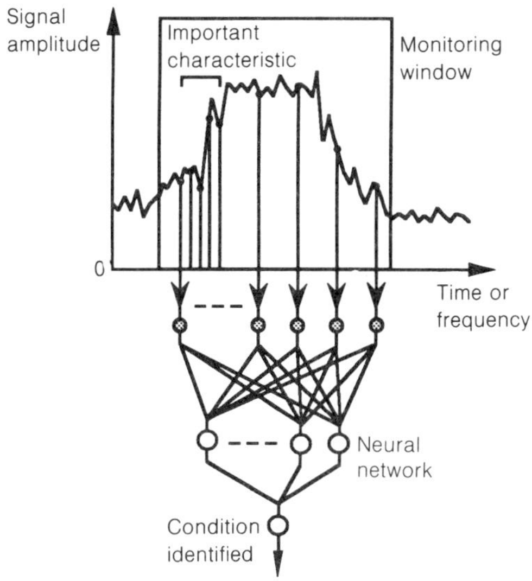

Fig. 13.5 Neural network for stationary time or frequency-domain pattern recognition.

13.3.2 Stationary time-domain and frequency-domain pattern recognition

Process monitoring involving the recognition of a stationary pattern is illustrated in Fig. 13.5. The neural net inputs are selected at constant positions within the monitoring window. Sections of the pattern that are more important for the recognition process may be more densely represented. In the time domain, the starting location of this window is usually triggered by signal magnitudes. In the frequency domain, a FFT algorithm transforms periodically the time-domain signal into its frequency domain. The signal magnitude at specific frequencies is input in the neural network. The network outputs classify the pattern families usually in two states: present (1) or absent (0). Each output will represent a family of patterns associated with the same problem. For example, in on-line machine diagnosis, one output may represent a bearing failure, another output, a gear failure, and so on.

The set of exemplars must contain all typical patterns that may be observed in the signal, those representing the characteristics we are looking for, as well as the others. For instance, in on-line machine diagnosis, the network will be trained with patterns typically representing part failures and with others corresponding to good operating conditions.

13.3.3 Specialized signal filtering

At first, specialized signal filtering is similar to time-domain pattern recognition, except for its moving window (Fig. 13.6(a)). Indeed, because the window

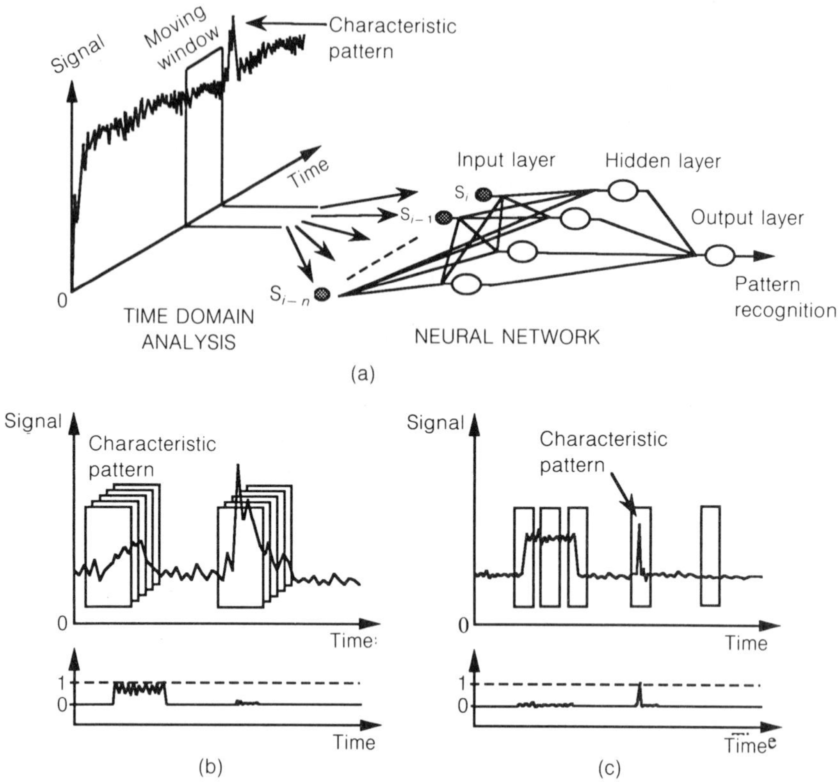

Fig. 13.6 Process-monitoring using specialized filtering (b) specialized signal filtering of complex patterns; (c) specialized signal filtering of simple patterns.

contains the last N inputs of the sensor signal, incremented every monitoring cycle, the patterns appear to the network as moving.

To realize such a specialized signal filter, two approaches may be available. First, for complex signal patterns, it becomes necessary to learn not only the pattern, but also several subsets located at different time increments before and after the central location of the pattern (see Fig. 13.6(b)). This technique is exemplified in section 13.5.1. When these patterns are relatively simple or when they can be subdivided into simple ones, only a few may be utilized as shown in Fig. 13.6(c). An example of this last technique is presented in Guillot and El Ouafi (1991) for tool breakage monitoring of a machining process.

13.4 NEURAL NETWORK APPROACHES TO PROCESS CONTROL

The neurocontrol area has evolved over the last ten years toward generalized and powerful techniques extremely promising for the control of processes involving several parameters and hard to define process models and controllers.

Werbos (1989) classified the neurocontrol techniques into five broad categories: supervised control, inverse control, neural adaptive control, backpropagation of utililty and adaptive critic methods. We will borrow part of this classification and adapt it to the following description of process control.

13.4.1 Supervised control

In the supervised control scheme depicted in Fig. 13.7(a), the neural network represents the mapping that exists between sensor signals and adequate control action. To learn this mapping, input–output exemplars must be provided. This set of exemplars could be obtained simply by replacing the neural controller with a human operator. In this case, the operator must be careful to assess his control action solely on sensor information and performance estimation derived from these sensors, not on his own senses. Indeed, the human senses have different and much richer information content than simple sensors, and this input would no longer be available to the neurocontroller later during process control which would otherwise lead to large control error and instabilities. By learning a priori from these examples, the dynamics of the process and time response of the human operator would both be embedded in the neurocontroller shown in Fig. 13.7(b), and could then be applied to process control. It should be noted that the desired process output y_{des} can be implicitly defined in the operator's mind as an objective to optimize, and consequently, may not figure explicitly. This kind of objectives

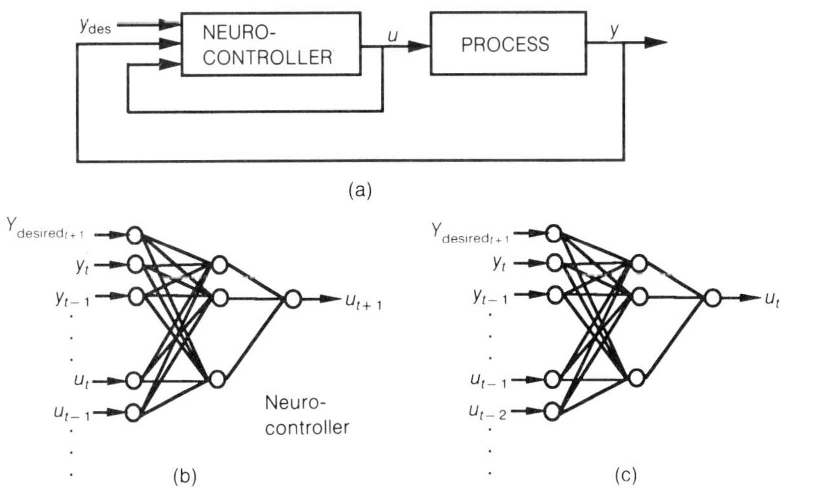

Fig. 13.7 Neurocontrollers: (a) feedforward scheme; (b) supervised neurocontroller; (c) inverse process neurocontroller.

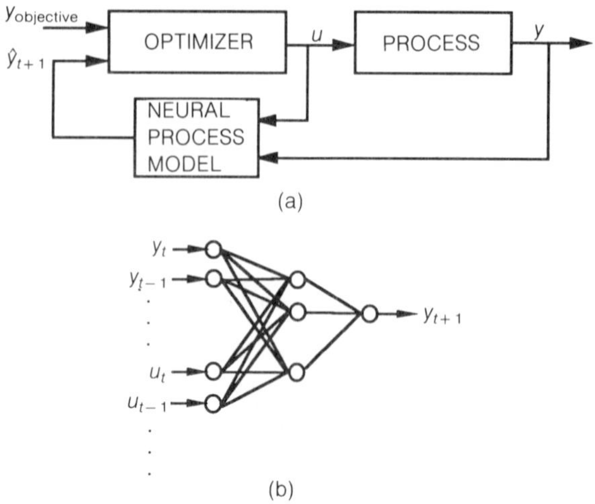

Fig. 13.8 (a) Explicit scheme for process optimization; (b) neural process model.

would nevertheless be present in the exemplar and in the knowledge acquired by the neural network.

Like in many optimizing process control schemes, it is also possible to express explicitly the objective and the optimization strategy as shown in Fig. 13.8. In this scheme the neural model of the process provides an estimate of the next process output y. The optimizer compares the estimated output with the objective output, and proposes, based on a certain strategy, the best control action. An example of such a strategy consists in minimizing the following cost function J without violating some inequality constraints (Thibault and Grandjean, 1991; Rangwala and Dornfeld, 1989):

$$J = \sum_{i=0}^{N-1} (y_{\text{objective}, t+i} - \hat{y}_{t+i})^2 + \lambda \sum_{i=0}^{N-1} (u_{t+i} - u_{t+i-1})^2 \qquad (13.5)$$

where N is the number of sampling periods of the prediction horizon, and λ, the weighting factor representing the relative importance of each member.

13.4.2 Inverse process neurocontrol

One of the most striking characteristics of the neural networks is their ability to learn almost any relationship from input–output exemplars. Let us assume a set of input–output exemplars representing both the physical and dynamic behavior of the process. The neural network can learn the inverse of this relationship simply by switching the input and the output in the exemplar set prior to learning, thus providing a model of the inverse process dynamics.

Interestingly, this model can be used as a predictive neurocontroller to estimate explicitly the next control move. Indeed, by studying the process, one can obtain the output y of the process at time $t+1$ as a function of control inputs u and outputs y at time $t-i$, where $i=0$ to n. The value for n depends especially on the order of the system dynamics. After learning the inverse of this relationship (Fig. 13.7(c)), the neurocontroller can be integrated in different control schemes like the feedforward scheme of Fig. 13.7(a). The inverse control approach can be utilized in order to follow a reference model or to maintain the process to a set point as with the self-tuning regulator. A very comprehensive process control example of a wood chip refiner is available in Kooi and Khorasani (1991), making interesting comparisons between the self-tuning regulator and an inverse process neurocontroller trained a priori. As expected, the results are the same because the parameter adaptation of the STR is embedded in the neurocontroller.

13.4.3 Adaptive neurocontrol

The adaptive controllers as defined by the control community, can adjust to slowly-varying unobserved system parameters. In neurocontrol, this adaptation will be dealt by learning not only a priori, but also dynamically as the neurocontroller is operating. For example, the feedforward inverse model scheme of Fig. 13.7(a) can become neuro-adaptive with on-line learning capabilities as illustrated in Fig. 13.9. In this specific case, the neural net controller will adapt to process variations according to two possible schemes: based on a direct model of the process, or based on an inverse model of variations between process inputs and outputs.

In the direct model-based scheme shown in Fig. 13.9, a neural model of the process, trained a priori, is used as a reference. The variation of the output $(y_{\text{desired}}-y)_t$ is propagated backward in this process model, so that to obtain the corresponding variation of inputs $(u_{\text{desired}}-u)_{t-1}$. Thereafter, this input variation is used to adjust by learning, the inverse process neurocontroller. Meanwhile, the direct process model remains in its original state. In the second scheme, a model of the transfer function between output and input variations is trained a priori. During process control, the output variations $(y_{\text{des}}-y)_t$ is processed forward in this model which gives an estimate of input variations $(u_{\text{des}}-u)_{t-1}$. Likewise, the inverse process neurocontroller learns this variation on-line. Interestingly enough, the preliminary investigations (e.g. Kooi and

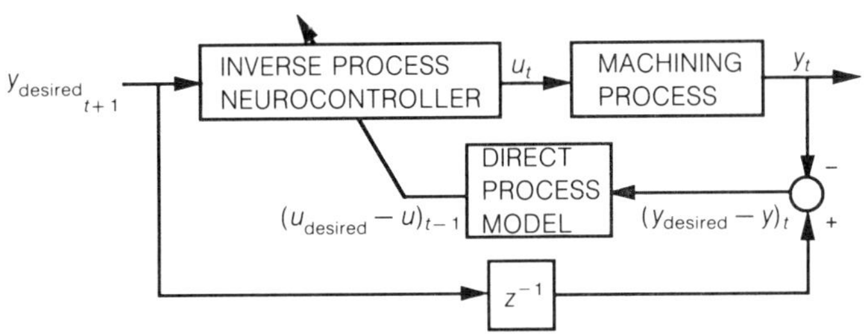

Fig. 13.9 Dynamically learning neurocontrol scheme.

Khorasani (1991)) suggest that these adaptive neurocontrol schemes may sustain process variations much more important than classical adaptive control schemes, especially because the neural network can reconfigure entirely its knowledge as it learns.

Similarly, the optimizing scheme shown in Fig. 13.8, could be modified such that its neural process model would be adapted on-line using the last input–output exemplar collected. Chapter 14 gives an in-depth information on these adaptive neurocontrollers.

13.5 IMPLEMENTATION CASES

The following examples of implementation will attempt to describe not only the type of monitoring and control problems that can be solved using neural techniques, but also the methodology of implementation, as well as the type of results that can be experienced.

The first example involves on-line robotic monitoring of grasping and assembly operations. Acoustic emissions are used to characterize the contact state between the part and the gripper fingers during part grasping, handling and assembly. We propose a four-step process monitoring approach allowing a multilayer feedforward network to identify and interpret time-domain acoustic emission signals.

In the second example, a scheme that uses an inverse model of the turning process to synthesize optimal process outputs is presented. The inverse model is implemented in a feedforward neurocontroller that adapts on-line its outputs based on indirect measurements of the process state. The synthesized process inputs should optimize process performance by automatically achieving simultaneously the best compromise between maximum productivity and part quality. The simulation and experimental results, herein presented, demonstrate that neural networks can control complex manufacturing processes using currently available technology.

13.5.1 Process monitoring implementation: robot grasping and assembly monitoring

Introduction to the problem

In industrial robotic applications such as part loading and unloading, material handling and assembly, untended operation depends strongly on the reliability of the manufacturing system, particularly if it involves several robot manipulators and many operations carried out sequentially. In such cases, an undetected problem at a certain step can have disastrous consequences on the following operations. It can also delay the entire system and may cause high part defect rates. Thus, a robotic grasping and assembly condition monitoring system should detect these problems, allowing the robot to recover adequately from faulty operations. For example, during an assembly operation, a robot may grasp a component in the wrong orientation. If detected, the recovery routine could allow the robot to put this component away and resume with a new one.

Monitoring of faulty grasping and assembly operations requires the use of a sensor responding to surface friction and contacts particularly between the part handled and the gripper. As explained extensively by Rangwala *et al.* (1988), the acoustic emission (AE) phenomenon can be observed during any plastic deformation in a material. Even microscopic surface deformations that originate from surface friction or contacts between a gripper and an object, can generate very high frequency waves transmitted through the material. When monitoring AE signals as a function of time, we can observe characteristics and patterns associated with various robotic operation conditions as shown later in Fig. 13.12.

Interpretation of the AE signal is a relatively trivial task for the human intelligence which can easily discriminate signal patterns representing faulty conditions like when a part slips in the gripper or when a collision occurs during an assembly task, from good ones. However, this is quite different when it comes to computerize such a flexible and reliable pattern recognition technique for robotic condition monitoring.

The process monitoring approach

As depicted in Fig. 13.10, on-line robotic condition monitoring requires four major steps. During robotic operations (1) such as grasping, handling and assembly, a high frequency acoustic emission sensor (2) can provide as a function of time, signal patterns containing information about the contact state between the object manipulated and the gripper. However as mentioned previously, these contact or operation states still need to be extracted from the signal patterns, which requires a certain form of intelligent interpretation (3) that will be accomplished using the multilayer feedforward network. Once the state of a robotic operation has been identified, a control action (4) can be undertaken accordingly.

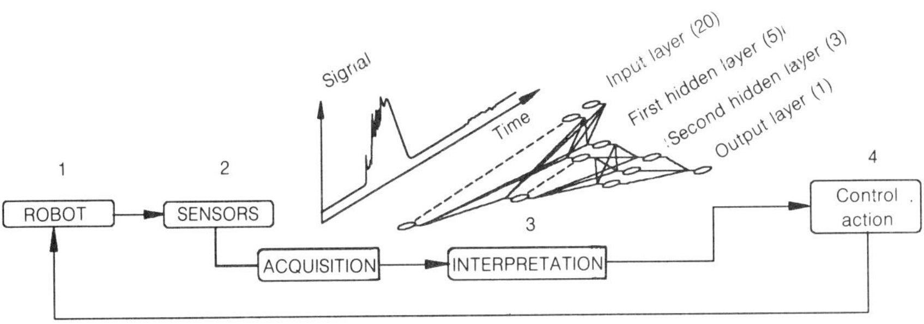

Fig. 13.10 Time-domain monitoring of robotic operations using a neural network.

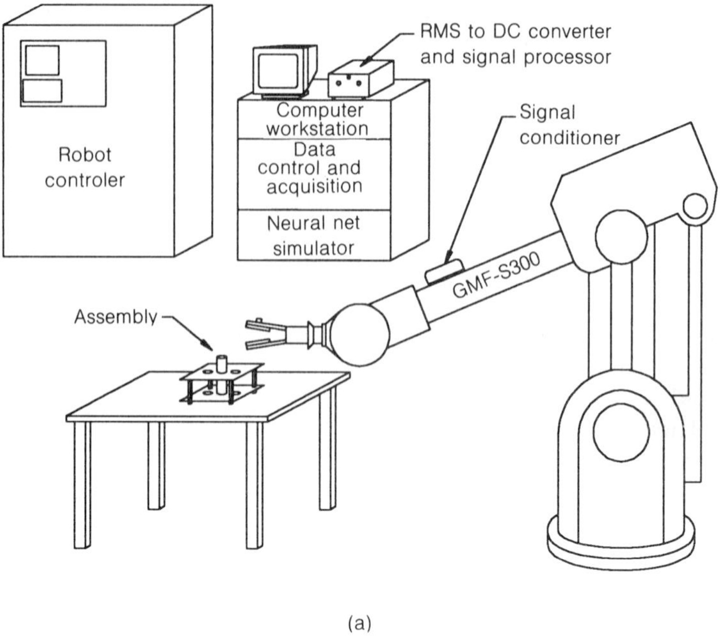

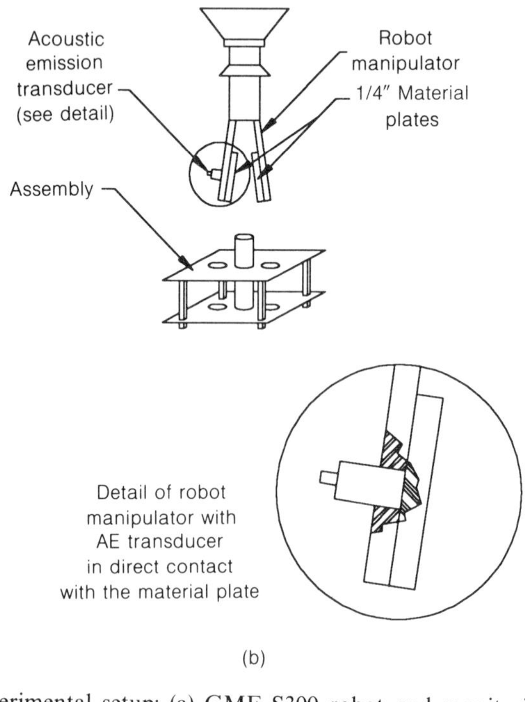

Fig. 13.11 Experimental setup: (a) GMF S300 robot and monitoring system; (b) details of the gripper and acoustic emission transducer.

Typically, this neural network consists of 20 inputs reading time-domain signal, one or two hidden layers whose number of neurons is selected according to the complexity of the signal patterns, and a single output set to 1 on faulty conditions and 0 otherwise. An AE signal magnitude ranging between 0 and 500 mV RMS is linearly scaled to a 0 to 1 range before neural input. In this approach, the neural network is trained with time-domain patterns representing both normal and faulty conditions during grasping, handling and assembly operations. This last example can be categorized as a specialized filter with complex patterns as shown earlier in Fig. 13.6(b).

Experimentation and data acquisition

The application of the proposed condition monitoring approach requires first, neural model training using experimental data. In the experimental set-up shown in Fig. 13.11, a two-finger gripper mounted on a FANUC GMF-S300 robot has been utilized to achieve part grasping, handling and assembly operations. For the purpose of this demonstration, these operations have been carried out using ANSI 1020 steel and 6061-T6 aluminum finger plate materials. The manipulated parts consisted of a one inch ANSI 1020 steel cylinder.

The high frequency acoustic emission signal has been collected using a model AC175L acoustic emission transducer from Acoustic Emission Technology Co. The transducer face was mounted directly on the back of one of the gripper plates as shown in Fig. 13.11(b). The transducer output was first conditioned by a 125 kHz to 2 MHz bandwidth filter model FL12X and then converted to RMS values using a model 201 processor also from AET. RMS values, ranging from 0 and 50 mV, were then amplified with a gain of 10 and recorded by a computer. The acquisition frequency was set to 1000 samples per second. For every sampling cycle, a window comprising the last 400 readings was monitored (400 msec duration) and 20 of these (1 every 20 readings) have been input in the neural net.

In order to train the neural model, acoustic emission signals have been collected for six typical robotic operations as indicated in Table 13.1 and in Fig. 13.12 for the aluminum finger plate material.

Table 13.1 Description of the patterns utilized for training

	Operation	*Status*
#1	Grasping of cylindrical object	normal
#2	Handling of cylindrical object	normal
#3	Handling of cylindrical object	fault (slipping and dropping)
#4	Displacement of an empty gripper	normal
#5	Assembly of a cylinder in a hole	normal
#6	Assembly of a cylinder in a hole	fault (slipping)

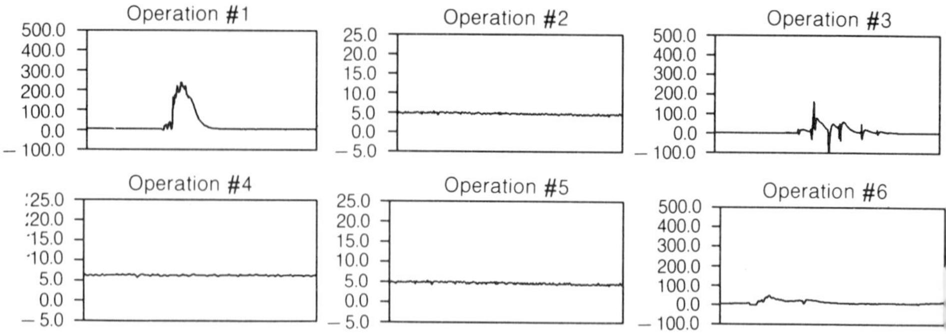

Fig. 13.12 Typical rms acoustic signal-time patterns for model training.

Since the signal is monitored every sampling cycle, the characteristic pattern that has to be identified may not figure entirely in the sample or at a unique location relative to the monitoring window of 20 neural net inputs. Accordingly, the neural net has been trained with a series of input–output subsets representing all six patterns of Fig. 13.12. Based on the complexity and number of patterns to learn (patterns #1, 3 and 6 were more complicated), a total number of 50 subsets have been found adequate for the steel fingers and 27 subsets for the aluminum fingers.

Neural network training

Based on these subsets, different network configurations have been trained for steel and aluminum finger plates under various gains and momenta. The best model found is presented in Table 13.2 which also indicates the number of iterations needed to obtain the estimated answer with an error down to 0.5% on training data.

Testing the monitoring approach

In practice, signal patterns are never perfectly alike as shown in Fig. 13.13 for grasping operations. Under such variations, the neural network monitor behaved relatively well. Indeed, the neural net output shown in the lower part of those graphics, indicated as expected, that the grasping operation was successfully achieved.

Table 13.2 Model training data

Plate Matl	*Gain*	*Mom.*	*Neural net*	*% Error max.*	*% Error avg.*	*Std deviation*	*Number of passes*
Steel	0.9	0.9	20 × 5 × 5 × 1	0.0049	0.0009	0.000003	5127
Aluminum	0.5	0.5	20 × 5 × 5 × 1	0.0049	0.0015	0.000006	57 592

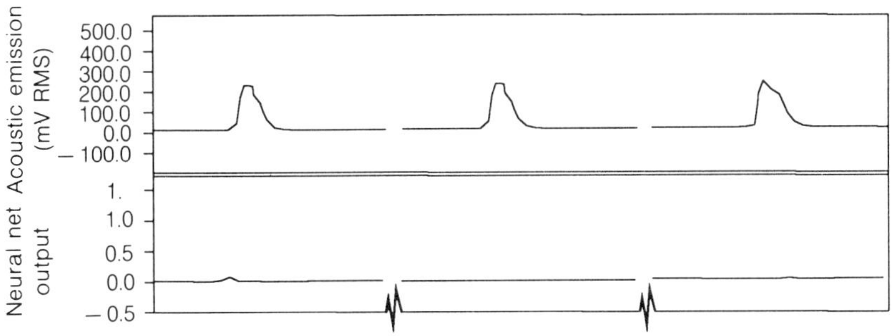

Fig. 13.13 Neuromonitoring responses of a repeated grasping operation using aluminum plates.

The following results describe how the neuromonitor reacts to different practical situations. Typically, we programmed the robot to approach the cylindrical part, to take it, to carry and insert it in a hole. We voluntarily provoked grasping and assembly problems.

The results shown in Fig. 13.14 illustrate typical signals and neural net responses during normal and faulty conditions. The neuromonitor interpreted correctly and entirely these new AE signals. Indeed, the neural net output was

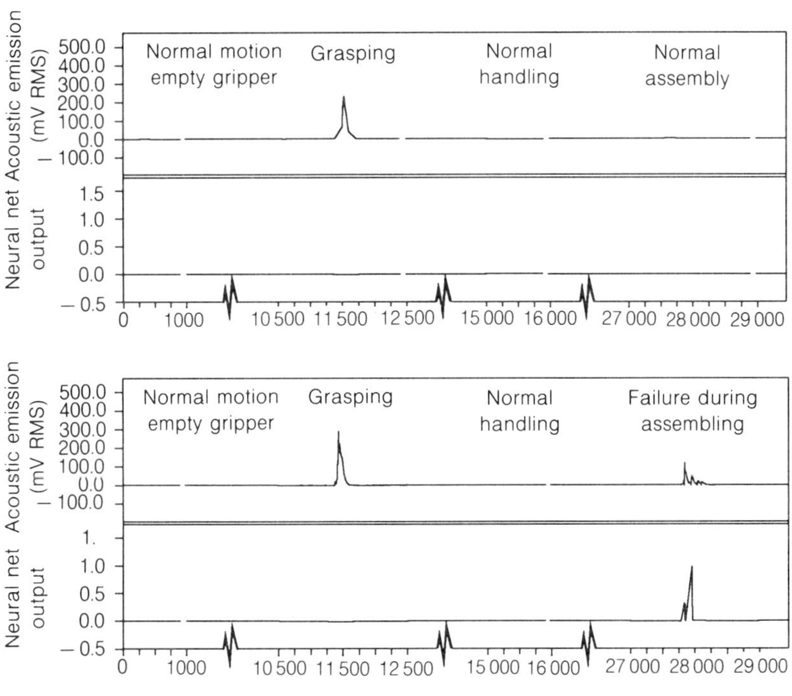

Fig. 13.14 Complete robotic grasping, handling and assembly test using steel plates (unknown pattern).

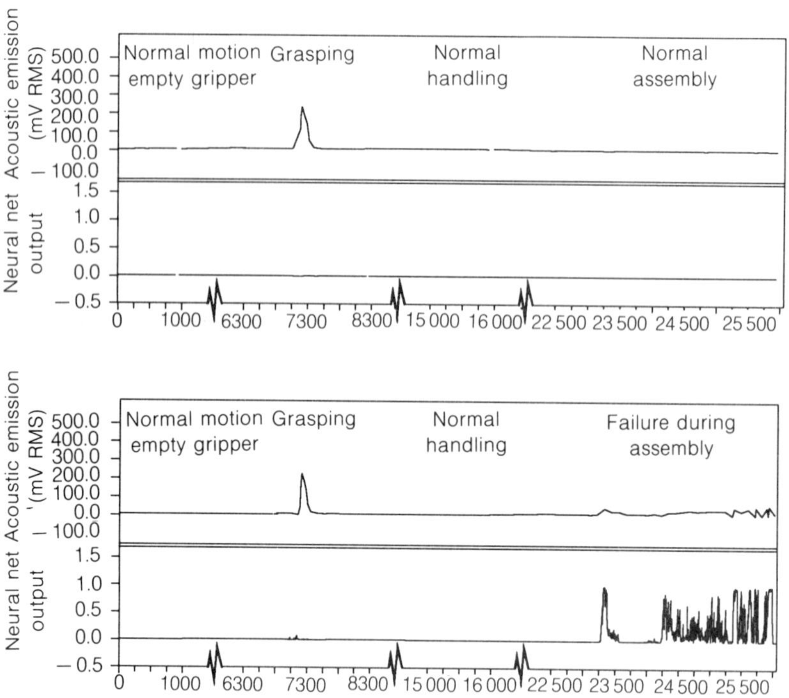

Fig. 13.15 Complete robot grasping, handling and assembly test using aluminum plates.

near zero everywhere but during the assembly failure where it reached a near one value. Equally good results were found using the aluminum finger plates as depicted in Fig. 13.15.

The neural networks demonstrated good capabilities in classifying between 27 to 50 different patterns in two classes, the good and faulty operations. This technique has also been tested for several other monitoring applications ranging from tool breakage detection to on-line predictive monitoring and diagnosis of quality control charts with a great deal of success.

13.5.2 Process control: Performance optimization of the machining process

Introduction to the problem

In the past decades the number of computer numerical control (CNC) systems has grown tremendously in almost every field of manufacturing. A common problem with most CNC machining systems is the need for a part programmer to select proper operating parameters (e.g. feed rates, cutting speeds and depths of cut) based on experience and data found in machining handbooks. In most cases, the metal cutting conditions vary significantly as the part is

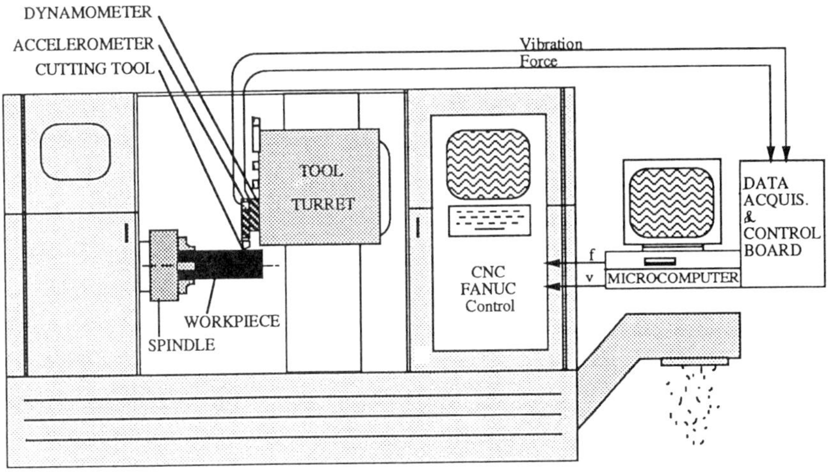

Fig. 13.16 CNC turning center and control set-up.

machined, thus letting no choice to the part programmer, but to select extremely conservative parameter values that prevent any potential tool breakage or machine vibration. This obviously results in less than optimum conditions.

The solution we propose in this example, consists in developing a neurocontroller that will optimize single-point turning operations (see Fig. 13.16). On-line adjustments of feed rate and cutting speed parameters are thus carried out based on a cost/performance index, estimated from force and vibration sensor measurements. Systems which possess such capabilities are referred to as adaptive control (AC) systems in the machining literature.

The process control and optimization approach

The proposed process control scheme is shown in Fig. 13.17. It consists of an inverse process neurocontroller, partly similar to the one presented in Fig. 13.7(c). The control input vector u comprises the tool feed rate, f, and the cutting speed, v. The output y corresponds to an estimated cost function L. Process optimization is assured by minimizing on-line, the cost function L and particularly L^* which is the new value proposed by the minimization strategy. Given the desired minimum loss for the next iteration, L^*_{t+1}, the current loss, L_t, and the last inputs of the process, f_{t-1} and v_{t-1}, the inverse neurocontroller predicts the optimal process input parameters to be used every sampling period. Since the machine tools responds very quickly to new control actions, a delay n of only one sampling period is considered.

Ideally, a true adaptive controller would be one capable of designing its own logic for optimal process performances in response to sensors, thereby automatically achieving simultaneously the best compromise between

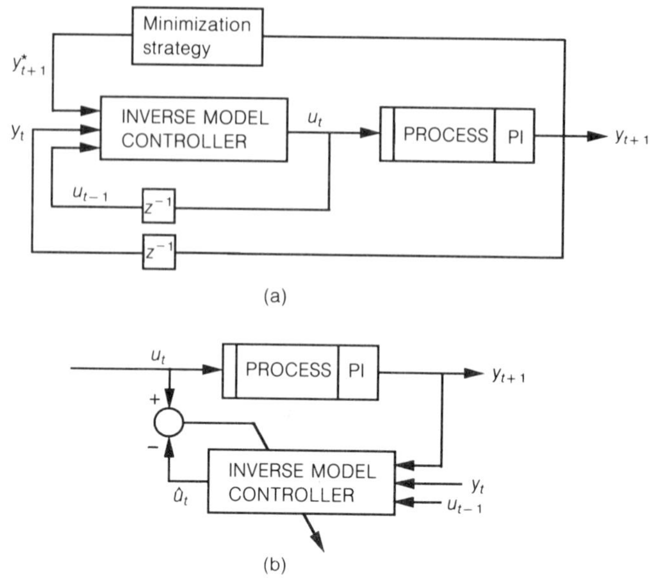

Fig. 13.17 Scheme of the optimizing inverse process neurocontroller.

maximum productivity, surface quality and dimensional tolerance. Our cost function thus combines metal removal rates, dimensional tolerances and surface finishes of the part, estimated from force and vibration sensor signals.

The metal removal rate is given by

$$\mathrm{MRR} = fvd \tag{13.6}$$

where d is the depth of cut. The cost function L (given in dollars) becomes at instant t:

$$L(t) = L_1(t) + L_2(t) + L_3(t) \tag{13.7}$$

where

$$L_1(t) = k_1(\phi(t) - \phi)^{-2} \tag{13.8a}$$

$$L_2(t) = k_2(\mathrm{Ra}(t))^2 \tag{13.8b}$$

$$L_3(t) = k_3 \frac{1}{\mathrm{MRR}(t)} \tag{13.8c}$$

where ϕ is the error due to the deflection of the tool, Ra, the surface finish, and t, the time. Parameters k_1, k_2 and k_3 were set respectively to 111 000 \$/mm^2, 39.06 \$/μm^2, and 20 000 \$ mm^3/s for single point turning of titanium parts.

To evaluate the cost function L, we need in addition to the metal removal rate, on-line estimates of surface finishes and tolerances. However, in-process surface finish and tolerance characterization is very difficult to achieve analytically because of the environmental conditions (lubricant, chips,...). To overcome these problems, we have trained another neural model that predicts surface finishes and dimensional tolerances from on-line force and vibration measurements. For the sake of conciseness, this particular model will not be presented.

Experimentation, data acquisition and off-line model training

Off-line training of the inverse process neurocontroller required a set of exemplars obtained by conducting some experiments on the machine tools.

As indicated above, the Mori Seiki CNC turning center has been interfaced with a data acquisition and control computer system. The instrumentation consists of a Bruel and Kjear accelerometer for vibration measurements and a Kistler 3-component dynamometer for cutting force measurements. On this turning center, the input variables are constrained by the Fanuc 15TF CNC controller to the following ranges.

$$0\% f_i \leqslant f \leqslant 255\% f_i$$
$$50\% v_i \leqslant v \leqslant 120\% v_i$$

where f_i and v_i are respectively, the initially programmed tool feed rate and cutting speed.

A total of 24 exemplars have been found by experimenting various situations on the turning center. It took about 50 000 training iterations to drive the sum of squared errors down to about 0.1%.

On-line training strategy

The metal cutting process is subject to significant changes caused for example, by tool wear, hardness variations and by different tool and workpiece geometries. Even the best inverse process neurocontroller may not suffice to maintain optimal operating conditions under such variations. Accordingly, a neuro-adaptive version of the scheme proposed in Fig. 13.17(a) will also be tested in the next section. The dynamic adaptation of both weights and thresholds is carried out every sampling period of two seconds. As described in Fig. 13.17(b), the last process input u_t is compared with $\hat{u}_t$ which is estimated by the inverse process neurocontroller that becomes for the circumstance an inverse model. This estimated $\hat{u}_t$ is provided by the model in response to the actual process output y at time $t + 1$ and t, and process input u at time $t - 1$.

Experimental performance evaluation of the inverse neurocontroller

In this section, both standard and adaptive versions of the inverse process neurocontroller are tested. Their results are presented respectively in Figs 13.18 and 13.19. In regard with these tests, a constant 2 mm depth was machined off a 4 inch diameter titanium part. The initially programmed feed and speed were respectively 0.35 mm per part revolution, and 37 m/min. As shown in

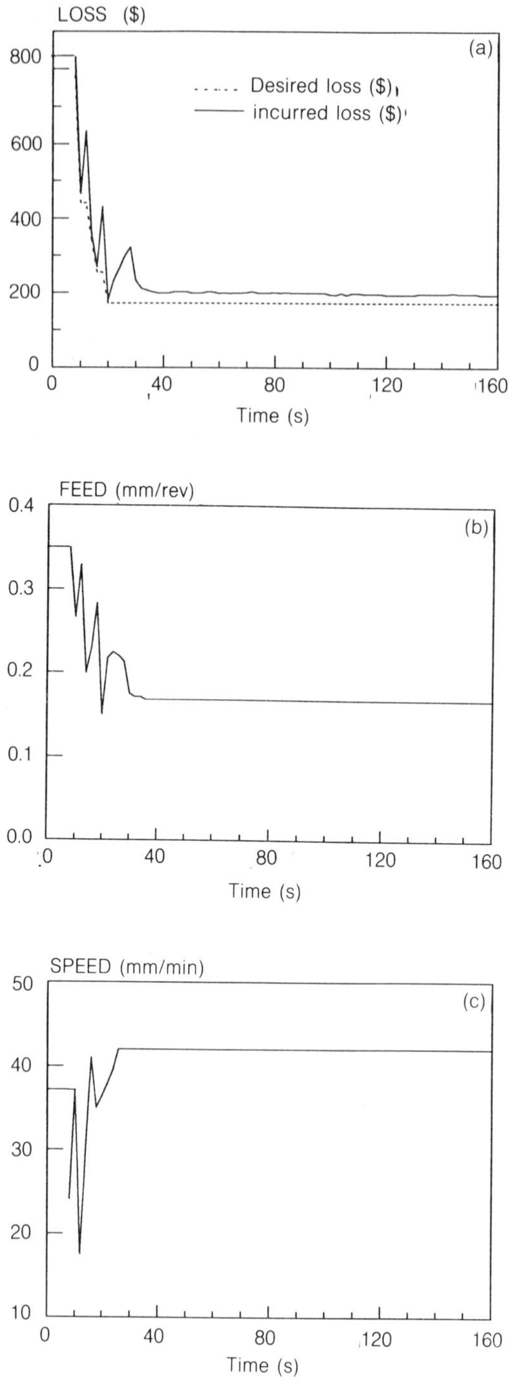

Fig. 13.18 Experimental results using the standard version of the inverse process neurocontroller.

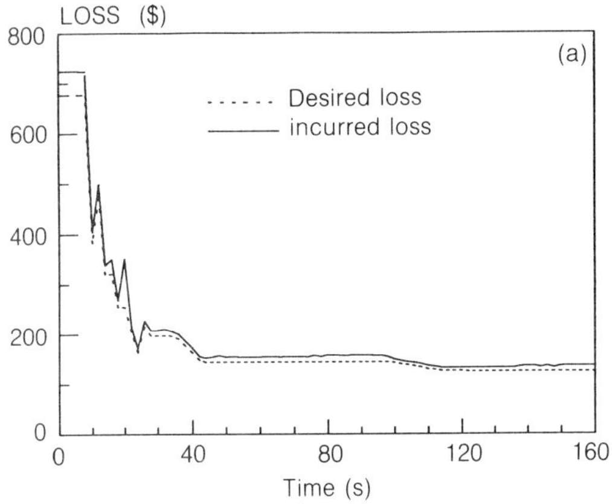

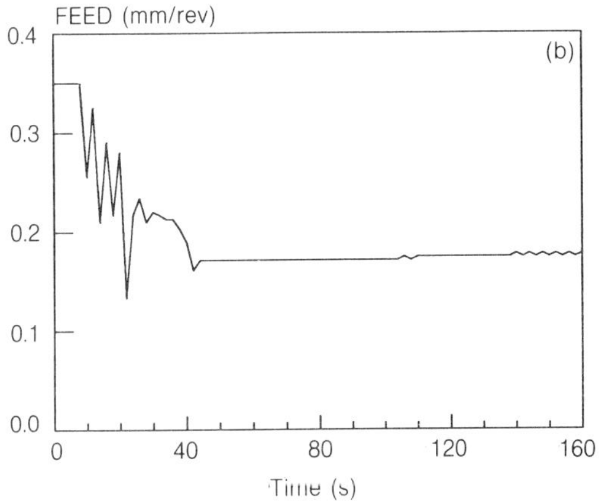

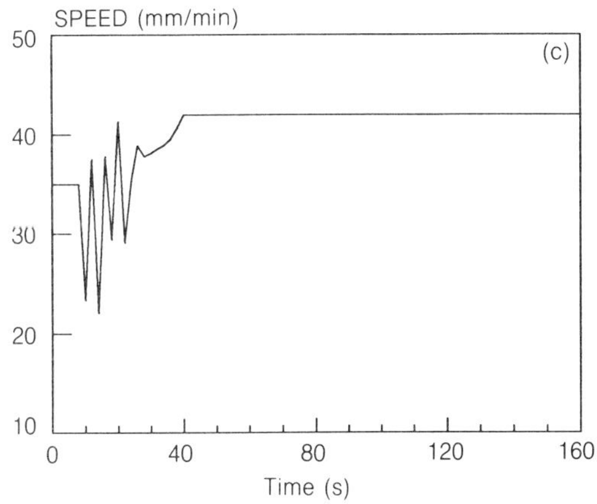

Fig. 13.19 Experimental results using the adaptive version of the inverse process neurocontroller.

these graphics, the lower cost (loss) has been achieved rapidly after about 22 seconds or 16 iterations. After a few oscillations, the parameters settled to a final feed of 0.171 mm/rev and speed of 41.5 m/s providing a cost reduction of 78%. By comparing the two figures, we can see that the adaptive version performs better, especially after 100 seconds, where the cost is reduced to 133$ as opposed to 200$ for the standard non-adaptive neurocontroller.

13.6 SUMMARY

This chapter has reviewed some typical process monitoring and control problems, and presented traditional techniques of solutions. Because these techniques are often limited when it comes to solving complex non-linear problems, several neural-based alternatives have been introduced. In process monitoring, sensor fusion, pattern recognition and specialized filtering appear as very promising techniques. Neurocontrollers and neuromodels have been applied very successfully in process control and optimization. Different neurocontrol schemes such as the direct and the inverse process controllers, or the optimizers have been presented. Finally, the neural-based monitoring and control techniques have been exemplified with two implementation cases, the first on monitoring grasping and assembly operations using an industrial robot, and the second, applying neurocontrol and optimization in machining.

The neural networks have found in process monitoring and control a very technically and economically important ground of development and industrial implementation. The very promising results provided by the increasing number of researchers interested in this area is certainly a significant factor and motivation for maintaining such a pace. Finally, as pertinently stated by Werbos (1989), 'the human brain is living proof that it is possible to build an analog, distributed controller capable of learning and effective planning (long-term optimization) under conditions of noise, qualitative uncertainty, non-linearity, and millions of variables to be controlled at once, all with a very low incidence of falling down'.

REFERENCES

Box, G.E.P., Hunter, W.G. and Hunter, J.S. (1978) *Statistics for Experimenters*, John Wiley & Sons, New York.

Branagan, L.A. and Wasserman, P.D. (1992) Introductory use of probabilistic neural networks for spike detection from an on-line vibration diagnostic system, in *Intelligent Engineering Systems Through Artificial Neural Networks, Vol. 2* (eds C.H. Dagli, L.I. Burke and Y.C. Shin), ASME Press, New York, pp. 719–24.

Braun, S., Lenz, E. and Wu, C.L. (1982) Signature analysis applied to drilling. *Journal of Mechanical Design*, **104**, April, 268–76.

Chryssolouris, G. and Guillot, M. (1990) A comparison of statistical and AI approaches to the selection of process parameters in intelligent machining, ASME Trans. *Journal of Engineering for Industry*, **112**(2), 122–31.

Goodwin, G.C. and Sin, K.S. (1984) *Adaptive Filtering Prediction and Control*, Prentice-Hall, Englewood Cliffs, N.J.

Grassi, E. and Grunewald, A. (1992) Controlling an arbitrary plant using a modified Kohonen map, in *Intelligent Engineering Systems Through Artificial Neural Networks, Vol. 2* (eds C.H. Dagli, L.I. Burke and Y.C. Shin), ASME Press, New York, pp. 637–42.

Guillot, M. and El Ouafi, A. (1991) On-line identification of tool breakage in metal cutting processes by use of neural network, in *Intelligent Engineering Systems Through Artificial Neural Networks* (eds C.H. Dagli, S.R.T. Kumara and Y.C. Shin), ASME Press, New York, pp. 701–9.

Hayashi, S.R., Thomas, C.E. and Wildes, D.G. (1988) Tool break detection by monitoring ultrasonic vibrations. *Annals of the CIRP*, **37**(1), 61–4.

Kooi, S.B.L. and Khorasani, K. (1991) Control of wood chip refiner using neural networks, in *Intelligent Engineering Systems Through Artificial Neural Networks* (eds C.H. Dagli, S.R.T. Kumata and Y.C. Shin), ASME Press, New York, pp. 567–72.

Koren, Y., Ulsoy, A.G. and Danai, K. (1986) Tool wear and breakage detection using a process model. *Annals of the CIRP*, **35**(1), 283–8.

Loskiewicz-Buczak, A. and Uhrig, R.E. (1992), Probabilistic neural network for vibration data analysis, in *Intelligent Engineering Systems Through Artificial Neural Networks, Vol. 2* (eds C.H. Dagli, L.I. Burke and Y.C. Shin), ASME Press, New York, pp. 713–18.

Rangwala, S.S. and Dornfeld, D.A. (1989) Learning and optimization of machining operations using computing abilities of neural networks. *IEEE Trans. on Systems, Man, and Cybernetics*, **19**(2), March/April, 299–314.

Rangwala, S., Forouhar, F. and Dornfeld, D. (1988), Application of acoustic emission sensing to slip detection in robot grippers. *Int. J. Mach. Tools Manuf.*, **28**(3), 207–15.

Rumelhart, D.E. and McClelland, J.L. (eds) (1986) *Parallel Distributed Processing: Explorations in the Microstructures of Cognition, Vol. 1: Foundations*, MIT Press, Cambridge, MA.

Sadat, A.B. and Shivakimar, R. (1987) Detection of tool wear using acoustic signature analysis. *Wear*, **115**, 265–72.

Sata, T., Kimura, F., Matsushima, K., Takata, S. and Otsuka, J. (1985) Identification of machine and machining states by use of pattern recognition technique. *Manufacturing Systems*, **14**(3), 273–86.

Takata, S., Ogawa, M., Bertok, P., Outsuka, J., Matushima, K., and Satat, A. (1985) Real-time monitoring system of tool breakage using Kalman filtering, *Robotics & Computer Integrated Manufacturing*, **2**(1), 33–40.

Tansel, I., Mekdeci, C., Rodriguez, O., and Vragun, B. (1992) Monitoring microdrilling operations with wavelets and neural networks, in *Intelligent Engineering Systems Through Artificial Neural Networks, Vol. 2* (eds C.H. Dagli, L.I. Burke and Y.C. Shin), ASME Press, New York, pp. 681–6.

Thibault, J. and Grandjean, B.P.A. (1991) Neural networks in process control – A survey, in *IFAC Inter. Symp. ADCHEM-91 Proceedings*, Toulouse, France, 14–16 Oct., pp. 295–304.

Tryba, V. and Goser, K. (1991) Self-organizing feature maps for process control in chemistry, in *Artificial Neural Networks* (eds T. Kohonen, K. Makisara, *et al.*), Elsevier Science Publishers B.V., North-Holland, pp. 847–52.

Werbos, P. (1989) Backpropagation and neurocontrol: A review and prospectus, in *Proceedings of the Int. Joint Conf. on Neural Neworks* (IJCNN), IEEE, June.

CHAPTER 14

Adaptive control in manufacturing

Yung C. Shin

School of Mechanical Engineering,

Purdue University, West Lafayette, IN 47907.

14.1 CHARACTERISTICS OF ADAPTIVE CONTROL SYSTEMS

14.1.1 Introduction

Increasing requirements for high accuracy and productivity have forced many manufacturing firms to pursue automation. In order to meet the stringent requirements imposed on products, various forms of controllers have been used to regulate manufacturing processes and systems. Due to the extensive research on control theories, a wide variety of control schemes are now available. The controllers being used in practice, however, are mostly the types with fixed parameters designed by an engineer based on the physical system and specific control objective, which can address a single objective. This type of controller is acceptable for many simple mechanical systems. However, many manufacturing systems are inherently complex, non-linear, multivariable and subject to unknown external disturbance. When an autonomous operation of such a system is pursued, controllers must be able to adapt to the dynamic change of the system.

Adaptive control is conceived as an attractive means to achieve such a task. Recent years have seen some explosive research and some major break-throughs in adaptive control theories. The objective of adaptive control is to adjust controller parameters in order to achieve a desired response of an unknown or time variant system with or without an estimated model and/or a reference model. If properly implemented, it can reduce the variability of parts being produced and increase productivity. Therefore, substantial

Artificial Neural Networks for Intelligent Manufacturing. Edited by Cihan H. Dagli. Published in 1994 by Chapman & Hall, London. ISBN 0 412 48050 6

economic benefits are easily justifiable. Currently, several types of adaptive control applications to various processes are available. Specific examples of these will be described in section 14.2.

14.1.2 Traditional adaptive control methods

Adaptive control has been a major topic of technical inquiry in the control community since the mid 1950s (Landau, 1979; Astrom, 1980). Research in adaptive control was motivated by a desire to perform real-time control of partially or completely unknown plants. This extremely attractive idea led to some early ingenious but intuitive schemes. One of such initial approaches was simultaneously to estimate unknown parameters of the plant and generate control signals (Kalman, 1958). However, these earlier schemes did not guarantee the system stability. As system dynamics change, such a system can become unstable or lead to an unsatisfactory performance. Lately, parametric adaptive control strategies based on rigorous stability theorems have been developed to overcome the limitations of earlier schemes (Landau, 1979; Astrom, 1980).

Current adaptive control strategies can be divided into two categories: indirect and direct. The **indirect approach** is based on on-line system identification of the plant and simultaneous adjustment of control parameters. Self-tuning adaptive control strategies, which have become very popular schemes, supported by a tremendous amount of work on system identification, fall into this category. Self-tuning regulators seek to generate control laws for unknown systems, which would converge to the optimal strategies that could be chosen if the system were known. The basic structure of self-tuning control is shown in Fig. 14.1.

In general, these adaptive controllers consist of three parts: a recursive parameter estimator, a design calculator and a feedback controller with adjustable parameters. Many different versions of self-tuning control are now available, mainly differing in controller design principles and on-line parameter estimation methods. Some of the widely used design principles in self-tuning control entail pole-zero placement, minimum variance control, and linear quadratic gaussian control (LQGC). One important requirement of this

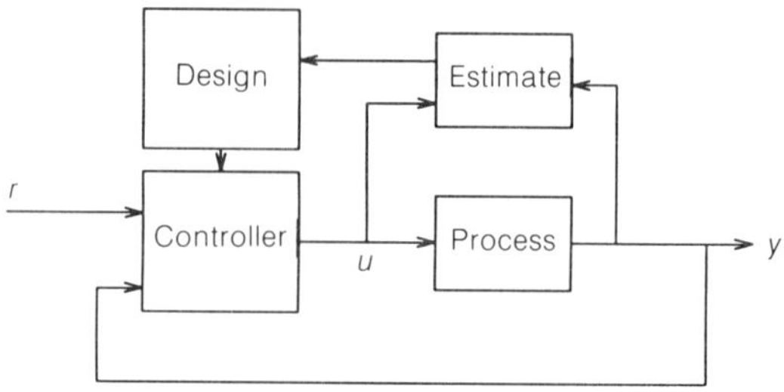

Fig. 14.1 Self-tuning control.

method is that on-line identification must yield an accurate estimate of the 'locally linearized' process model, when the system is non-linear, so that the control design strategy can generate accurate control parameters. However this requirement often limits the success of this method since most parameter identification techniques are open-loop while the actual system works in a closed loop with the feedback controller. Lack of richness in an input signal may not perturb full complexity of the system, thus hampering parameter estimation. To circumvent this problem, persistent excitation has been used. However, the use of these dual inputs certainly degrades the performance of the overall adaptive control system.

Direct approach does *not* require explicit model identification. Instead, a desired process model is specified and the controller parameters are adjusted so as to minimize the error between the model and actual system outputs. Model reference adaptive control (MRAC), derived from linear model following control, has gained great popularity lately and falls into this category. Many different variations of this approach are now found in the literature (Landau, 1974). The basic configuration of model reference adaptive control is shown in Fig. 14.2.

The reference model specifies the desired response of the system. The controller parameters are adjusted to minimize the difference between the outputs of the adjustable system and those from the reference model. The objective in this case is to minimize the error between outputs from the reference model and the physical plant. Since the parameters of the plant are unknown, the adaptation mechanism modifies the parameters of the controllers so that the outputs from the actual plant can follow those from the reference model. Based on the adaptation principle, structure and other criteria, MRAC systems can be largely divided into four groups (Landau, 1974). While it is difficult to conclude which method is most acceptable, the parallel structure is currently most popularly used.

There has been a substantial amount of progress in the research of this area.

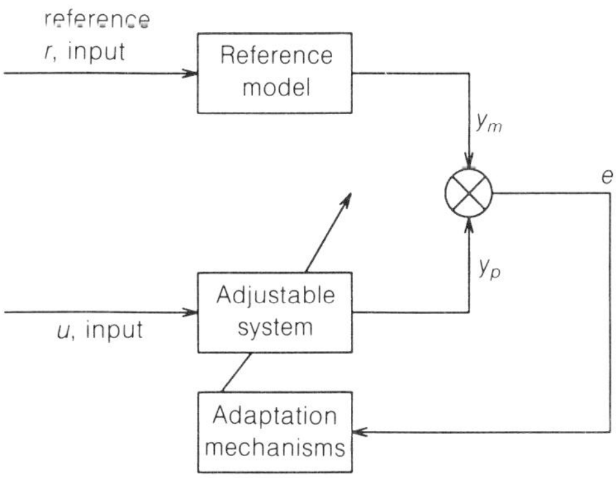

Fig. 14.2 Basic of model reference adaptive control concept.

Recent work was concerned with explicit design procedures of such systems based on stability theorems. Some globally stabilizing approaches have been proposed. However, the performance issue has been largely ignored despite its importance in engineering applications. Due to the time required for calculation and slow convergence of parameters, the response of the controlled system becomes often unacceptable especially when the system is highly non-linear or fast time-variant. This may not be necessarily a problem when the plant dynamics is slowly time varying. Therefore, many successful implementations are found in chemical process controls.

14.1.3 Contributions of artificial neural networks

Feasibility of adapting artificial neural networks to adaptive control of non-linear systems has been actively explored recently. As discussed in the previous section, linear adaptive control theories do not yield satisfactory performance when the plant to be controlled is highly non-linear. Many manufacturing systems are highly complex, non-linear and multi-variate. Therefore, there is a need to develop high performance controllers overcoming deficiencies of present linear adaptive controllers. Undoubtedly, if a model can be constructed to represent a non-linear plant and to generate control actions, the performance of the overall system can be improved substantially. More desirably, if a scheme can be developed to learn to model non-linear systems it would be possible to design an autonomous system.

Neural networks have demonstrated the capabilities of modeling a large class of non-linear systems, and hence representing input–output relationships robustly. The neural network model can be used to generate proper control signals and hence neural networks offer a great potential for adaptive control of non-linear systems due to their generalization capabilities. Particularly, the parallel processing architecture of neural networks provides the capability of processing a large amount of information in real-time once a network is trained. Considering the characteristics of manufacturing processes these attributes are very attractive in the context of adaptively controlling such systems. While developing stable adaptive control methods with neural networks seems to be of primary interest to researchers, certain neural networks show stabilizing characteristics if certain conditions are met.

Almost all sorts of neural adaptive controllers equivalent to existing linear adaptive controllers are found in the literature: feedforward, feedback, model reference adaptive, self-tuning controllers, etc. Recently, Narendra and Parthasarathy (1990) presented several schemes for neural network control. Different types of neural networks have also been considered. The major underlying issue is how to train the network robustly and reliably when the system is rapidly time varying. The popular backpropagation has been the primary choice of training when multilayered feedforward networks are used. Although there is no theoretical proof on the convergence of training, such an approach has been found acceptable through many case studies.

14.2 SAMPLE PROBLEM DEFINITION

14.2.1 Machining processes

With the rapid advent of microprocessor-based computer numerical controllers (CNC), adaptive control of machining processes has been actively pursued recently. The major motivation for adaptive control is to account for the process variability due to the change of operating conditions, workpiece geometry, hardness and temperature. The most common control objectives are maintaining constant cutting forces, chatter suppression, and improving geometric accuracy. Regardless of specific control objectives, all of these control systems must take the cutting process into consideration. The cutting process is usually non-linear with respect to feed rate and is represented in a simplified form by:

$$\frac{\text{cutting force}}{\text{feed rate}} = \frac{F_c(s)}{f(s)} = \frac{K_c f^{\rho-1} \cdot b}{\tau_c s + 1} \tag{14.1}$$

where the gain and time constant, τ_c, vary as cutting conditions (f = feed, b = width of cut) change.

Figure 14.3 shows the overall schematic diagram for achieving constant force control. In this case, the controller parameters are adjusted so that the actual cutting force approaches the command force. If chatter suppression is pursued, vibration amplitudes induced by cutting forces will be used as feedback signals, while cutting forces will act as disturbance input in geometric adaptive control. While adaptive control is to account for the non-linear process dynamics, linear models are usually used for feed drives and servomotors to reduce the computational load, thus neglecting the effects of friction, backlash and non-linear spindle dynamics. While this assumption is acceptable, accurate identification of such a model is still a difficult task (Kulkani and Srinivasan, 1984). Furthermore, the linear models are valid only near the tested operating conditions due to the non-linearity of the amplifier used for servo-control. It is fairly obvious from this point that higher performance adaptive control can be achieved if the non-linear process is controlled by using a non-linear controller. Despite this potential, application of neuro-adaptive control is yet to be attempted.

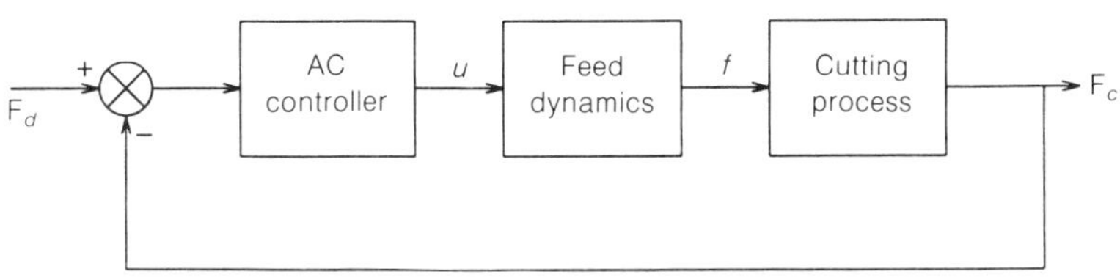

Fig. 14.3 Schematic of force adaptive control.

14.2.2 Injection molding

Injection molding processes play an important role in producing desired shapes for products made of advanced materials such as plastics and composites. In order to improve product quality and reduce failure, various control schemes have been adopted (Argawal *et al.*, 1987). Presently, closed-loop controls are used to regulate machine and process variables during an injection process. As feedback, variables such as ram velocities, chamber pressure or temperatures which can be measured through imbedded sensors are used. Obviously, the success of this type of control hinges upon the ability to relate machine variables to process conditions which are not directly observable. The relationships between process conditions and measurable sensor signals are non-linear or may not be available. In practice, machine settings are empirically determined by trial and error (Srinivasan and Srinivasan, 1991). When process conditions change due to the environmental changes such as ambient temperature change, large variability in part quality may result.

Although product attributes are not directly accessible during operation, they can be related to the processing variables reasonably. If these relationships are captured by neural networks, it would be possible to control the process in order to achieve a desired objective. A schematic diagram of a possible control system is shown in Fig. 14.4.

A typical injection molding process is shown in Fig. 14.5. Granules of raw material are fed into a chamber from the hopper and then pushed into a heated chamber by a plunger. In the heated chamber, sometimes with a torpedo, the material is melted. Then it leaves the chamber and enters the mold through a nozzle and fills the cavity. While the die remains cool, the molten material solidifies. In order to attain proper filling without premature solidification, the material must be injected into the cavity rapidly under considerable pressure.

As can be seen from this brief description, there are several process variables

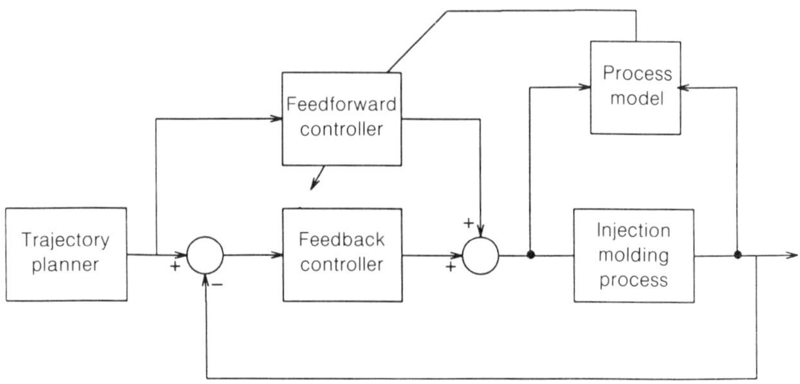

Fig. 14.4 Adaptive control scheme for injection molding processes.

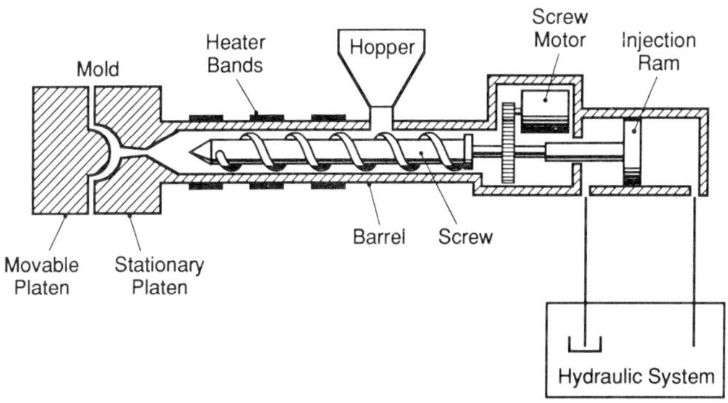

Fig. 14.5 A schematic diagram of an injection molding process.

which need to be regulated to ensure proper processing. Some of these include injection pressure, stock temperature, injection speed, nozzle size, mold temperature, gate size, back pressure, etc. The most commonly controlled variables found in the literature (Fara *et al.*, 1985; Kamal *et al.*, 1987; Fara *et al.*, 1990) are the melt pressures at the nozzle and cavity gate. Unfortunately, the complexity of the process does not allow for explicit analytical relationships between process variables and product attributes. Such process characteristics and difficulties in instrumentation impose a severe challenge when conventional controllers are used. If the relationship between the profiles of process variables and product attributes can be modeled by a suitable means, e.g., by neural networks, based on knowledge and experience, a substantial improvement in the performance of the control system can be attained.

14.2.3 Robotics

Robotic systems have been a target of extensive control theory applications. Motivated by the thrust to replace human work forces, particularly in hazardous and repetitive jobs, the number of robots being used in industry is increasing rapidly. A major challenge in the early stage was generating control input for each joint by solving inverse kinematics to achieve a prescribed trajectory of the end effector. While such a problem still remains a technical difficulty for complex robotic systems with rapid motions, the recent focus of research has migrated into robotic systems performing tasks in an uncertain dynamic environment.

Typical examples would be an autonomous vehicle moving on a shop floor or a material handling robot with different payloads. Sensors are used to provide a feedback on position or force. However, when a robot is working in an uncertain environment, real-time control based on the feedback sensors is not easy. For example, if a vision system is used, the time delay due to image processing will cause undesirable feedback delay in addition to the time required to solve inverse kinematics. Other signals such as force signals may not provide the feedback that can be directly related to the input commands.

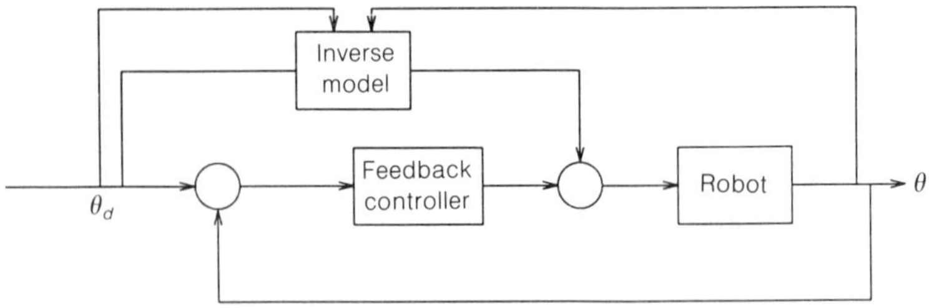

Fig. 14.6 An inverse model control of a robot.

Various adaptive control schemes have been applied to robotic systems. An attractive feature of such schemes is that no prior knowledge is required about the robot dynamics which may constantly change with different joint coordinates and environmental change. However, the heavy computational requirements for real-time identification and control have limited application of AC to simple manipulators.

In order to circumvent these difficulties, recently, neural net based controllers have been proposed. The basic idea is to employ a neural network to learn repeatedly the inverse kinematics or dynamics of a robot and then use it to generate control inputs. The major advantage of this approach is that it can serve as a learning controller for a robot which must operate in an uncertain environment. A basic feedforward control scheme of a robot can be defined as shown in Fig. 14.6.

Once the robot kinematic or dynamic relationships are learned, real-time control can be easily achieved. A possible limitation is that the neural network can learn only in the neighborhood of the specified trajectory and hence will require several iterations if input trajectory is changed.

14.3 ADAPTIVE NEURO-CONTROL ARCHITECTURE

14.3.1 Neuro-control architecture

In general, neuro adaptive controller architectures can be categorized based on learning schemes, control law generation, open-loop (feedforward) or closed-loop (feedback) configuration. Psaltis *et al.* (1988) proposed three different learning architectures: indirect learning, general learning and specialized learning.

The **indirect learning** scheme provides a means of generating a feedforward law with the inverse model of the plant. A schematic of this learning architecture is shown in Fig. 14.7. If a correct inverse model is established, the output of the overall system can track the input commands closely. The training is carried out to minimize the error between u and v. However, this scheme alone cannot lead to satisfactory controller design since the neural network model may not necessarily represent the true plant. This is because

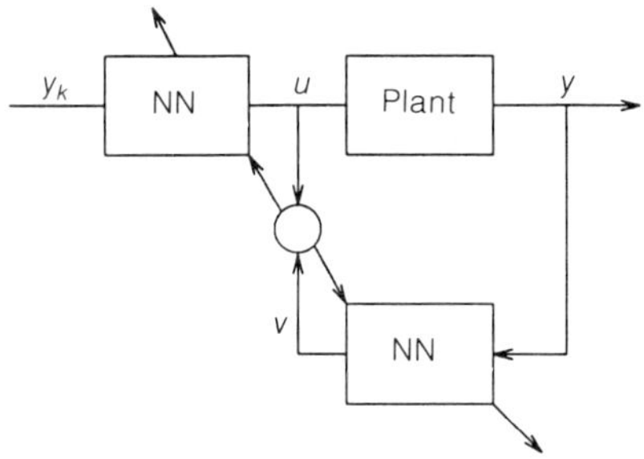

Fig. 14.7 Indirect learning scheme.

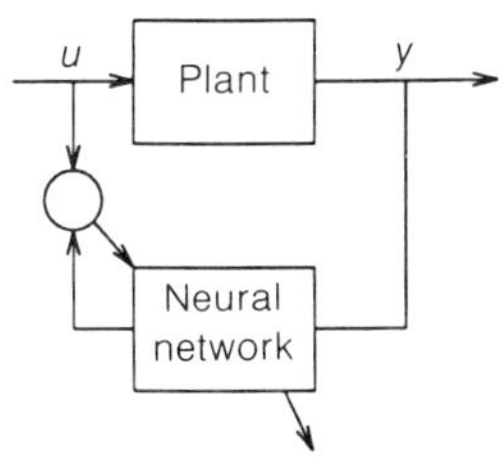

Fig. 14.8 General learning algorithm.

the network has the tendency of converging to a single value of u when input and output values of the plant are used for network training.

The **general learning** method attempts to generate a neural model for the plant by adopting a specified input space and generating corresponding output responses. This learning architecture is schematically shown in Fig. 14.8. A major difficulty of this learning architecture is that one must choose an input set u during training to generate outputs y which are sufficiently close to the desired input r. Since no detailed knowledge about the plant is known a priori, this scheme, if used for adaptive control, can lead to an improperly trained network and may require persistent excitation which usually degrades the performance of a system.

One other scheme is to connect the neural network and plant serially and adjust the parameters of the neural net so that output of the plant follows the desired net so that output of the plant follows the desired input. The major requirement of this procedure is for the Jacobian of the plant to be known. Such a requirement may impose a severe limitation on the usefulness of this technique for controlling unknown or time variant systems. Nonetheless, this method has been used with some success by calculating the Jacobian matrix numerically near the operating point (Chen and Pao, 1989).

All the above learning schemes are primarily concerned with multilayer neural networks which provide static mapping and require continuous parameter updating. On the other hand, recurrent networks are directly suitable for dynamic feedback systems. In this case, a single layer network

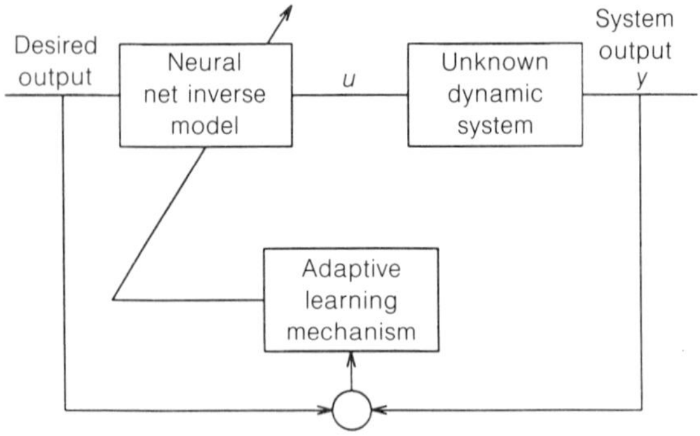

Fig. 14.9 Open-loop feedforward neuro adaptive control.

includes a feedback. When a proper function is used to represent a dynamical system, the states (or could be outputs) will reach an equilibrium.

In the following, existing neuro adaptive control architectures will be described based on control law generation mechanism and learning methods. The need to add delay elements or feedback as in recurrent networks will depend on the specific control objective.

The most intuitive designs are to use neural networks as feedforward controllers with or without a linear feedback controller in order to achieve good tracking performance. In the simplest scheme, a neural network can be used as a feedforward controller to represent an inverse model of an unknown plant in an open-loop manner as shown in Fig. 14.9.

This type of controller, however, will have no capability of rejecting disturbances. If a feedback controller is used along with a feedforward controller, the controller architecture can be modified as shown in Fig. 14.10. In this case, the feedback controller can be a conventional PID controller or a non-linear neuro-controller while a neural network is used as a feedforward

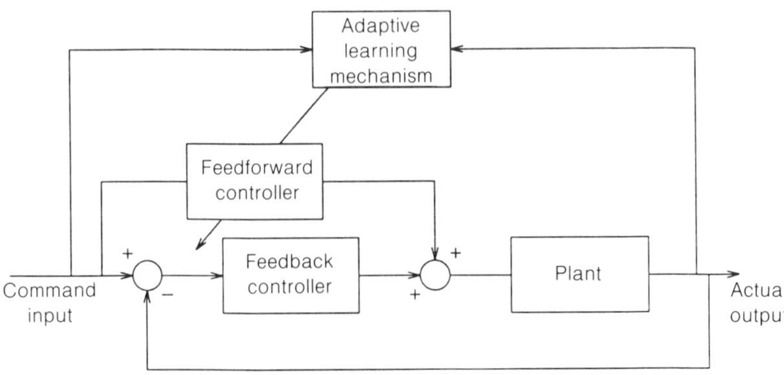

Fig. 14.10 Feedforward adaptive control with neural network.

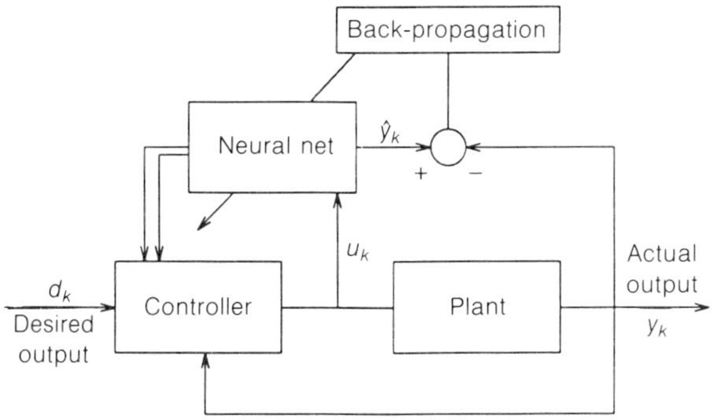

Fig. 14.11 Self-tuning control with a neural network.

controller. As the learning process continues, the error signal will gradually become small, thus switching from feedback control to feedforward control.

The non-linear version of self-tuning control has been proposed by adopting a neural network for system identification (Chen, 1990). In this case, the parameters of a neural net are gradually adjusted based on the input and output signals of the plant and the estimated plant model by the neural net is used to generate control signals directly (Fig. 14.11).

This architecture is suitable for plants which can be represented by

$$y_{k+1} = f(y_k) + g(y_k)u_k$$

The control law is generated to make actual output track the desired output:

$$u_k = -\frac{f(\cdot)}{g(\cdot)} + \frac{d_{k+1}}{g(\cdot)}$$

The neural network will provide estimated plant dynamics $\hat{f}(\cdot)$ and $\hat{g}(\cdot)$, since the plant model is unknown.

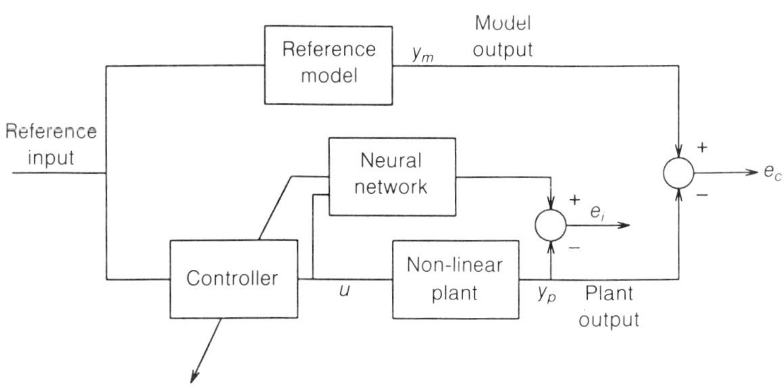

Fig. 14.12 Indirect model reference adaptive control with neural networks.

Similarly, a non-linear model reference adaptive control architecture has been proposed (Narendra and Parthasarathy, 1990). In this case, non-linear neural networks replace the linear controllers. While both direct and indirect approaches have been used for the linear case, only the indirect method with neural networks has been implemented so far. The non-linear plant is first identified by a neural network and then the parameters of the controller are adjusted as shown in Fig. 14.12.

There are other types of adaptive schemes which have been proposed. One such scheme is to use a neural network to perform the tuning of PID parameters. However, these schemes cannot be considered as truly adaptive.

14.3.2 Neural networks paradigm

In the previous section, we reviewed various existing architectures for adaptive control with neural networks. Various neural networks paradigms have been used to construct such control schemes. Usually, the specific neural network which must be adopted will depend on the dynamic characteristics of the plant. If the plant is time dependent, the neural network must possess the capability to model the temporal input–output relationships. Recurrent networks have inherent feedback and hence possess a suitable structure for temporal data. On the other hand, multilayer neural networks provide static non-linear mappings and hence have been used mostly for modeling time-invariant non-linear systems. The most popularly used paradigm is the multilayer feedforward network with back propagation training algorithm. Although its original structure is for static representation, it can be easily adapted to dynamic modeling by introducing delayed elements and feedback from the output layer to the input layer.

Another noticeable neural model is the CMAC based (cerebellar model arithmetic computer) paradigm. Due to the relatively simple structure and training rule such as least mean square, CMAC networks have been frequently used for control of robots, welding processes and hydroforming processes.

14.4 PERFORMANCE OF ADAPTIVE NEURO-CONTROL SYSTEMS

On several occasions, attempts have been made to compare the performances of adaptive neuro-controllers with those of conventional controllers. In general, neural network based controllers are slower in learning, compared with other conventional adaptive controllers. However, once fully trained, they are much more robust and immune to noisy signals (Kraft and Campagna, 1989). Particularly, neuro controllers can provide global control of non-linear plants while linear conventional controllers show poor performance or instability when the plant is highly non-linear.

14.5 CONCLUSIONS

In this chapter, we have introduced general adaptive control theories, the role of neural network for adaptive control, problem definitions for manufacturing applications, and various controller architectures. Presently, this field is still in an embryonic stage and most studies are limited to theoretical investigations only. However, implementation results with real systems, although limited, show real promising future. Since many manufacturing systems are highly non-linear and time-variant, the applications of neuro adaptive control are expected to increase substantially in the next few years.

REFERENCES

Agarwal, A.R., Pandelides, I.O. and Pecht, M. (1987) Injection molding process control – A review. *Polymer Engineering and Science*, **27**, 1345–57.

Astrom, K.J. (1980) *Application of Adaptive Control*, Academic Press, pp. 1–68.

Chen, F.C. (1990) Back-propagation neural networks for nonlinear self-tuning adaptive control. *IEEE Control Systems Magazine*, **10**, 44–8.

Chen, V.C. and Pao, Y. (1989) Learning control with neural networks, *IEEE Int. Control on Robotics and Automation*, **3**, 1448–53.

Fara, D.A., Kamal, M.R. and Patterson, W.F. (1990) Comprehensive strategies for sequential closed loop pressure control throughout the injection molding cycle, in *SPE ANTEC Conf. Proceedings*, pp. 239–42, May.

Fara D.A., Kamal, M.R. and Patterson, W.F. (1985) Evaluation of simple dynamic models and controllers for hydraulic and nozzle pressure in injection molding. *Polymer Engineering and Science*, **25**, 714–23.

Kalman, R.E. (1958) Design of a self-optimizing control system. *Trans. ASME*, **80**(2), 123–62.

Kamal, M.R., Patterson, W.I., Conley, N., Fara, D.A. and Lohfink, G. (1987) Dynamics and control of pressure in the injection molding of thermoplastics. *Polymer Engineering and Science*, **27**, 1403–10.

Kraft, L.G. and Campagna, D.P. (1989) A comparison of CMAC neural network and traditional adaptive control systems, *Proc. of American Control Conference*, pp. 884–9.

Kulkani, P.K. and Srinivasan, K. (1984) Identification of discrete time dynamic models for machine tool feed drives, *ASME Symp. on Sensors and Controls for Automated Manufacturing and Robotics*, pp. 1–20.

Landau, I.D. (1974) A survey of model reference adaptive techniques – Theory and applications. *Automatica*, **10**, 353–79.

Landau, I.D. (1979) *Adaptive Control: A Model Reference Approach*, Marcel Dekker, New York, NY.

Narendra, K.S. and Parthasarathy, K. (1990) Identification and control of dynamical systems using neural networks. *IEEE Trans. on Neural Networks*, **1**, 4–27.

Psaltis, D., Sideris, A. and Yamamura, A.A. (1988) A multilayered neural network controller. *IEEE Control Systems Magazine*, **9**, 17–25.

Srinivasan, K. and Srinivasan, T. (1991) Learning control of melt pressure in injection molding processes. *ASME DSC Control of Manufacturing Processes*, **28**, 151–9.

CHAPTER 15

Fuzzy neural control

L. H. Tsoukalas, A. Ikonomopoulos and R. E. Uhrig
Department of Nuclear Engineering and Center for Neural Engineering, The University of Tennessee and Oak Ridge National Laboratory, Knoxville, TN 37996-2300.

15.1 PROBLEM OF FUZZY CONTROL

Fuzzy neural control refers to the use of fuzzy logic and neural networks to control motors, actuators, and in general, the behavior of processes. The rapidly increasing number of fuzzy logic applications in process control and consumer electronics, as well as theoretical and hardware advances in neuro-control, provide significant incentives for studying fuzzy neural control. Of particular interest to control designers may prove the maturity of novel hardware systems, such as fuzzy logic and neural network boards, customized chips, and even fuzzy computers, where the data will be stored and arithmetically processed as fuzzy numbers.

Fuzzy logic and neural networks are thought of as complementary technologies. They are both well suited for non-linear and time-varying systems. The main difference between fuzzy logic approaches and neural network techniques (such as backpropagation) is that the former use qualitative information articulated in a set of IF–THEN rules, whereas the latter require considerable input–output data. In this chapter, we survey the basic terms and concepts of fuzzy neural control through examples and present a methodology for measuring fuzzy numbers through neural networks. We offer more attention to the fuzzy logic side of fuzzy neural control since earlier chapters provide considerable depths on neural networks, learning, and applications.

Neural networks consist of many simple processing units, called **neurons**, that can be globally programmed to perform a computation (Khanna, 1990).

Artificial Neural Networks for Intelligent Manufacturing. Edited by Cihan H. Dagli. Published in 1994 by Chapman & Hall, London. ISBN 0 412 48050 6

They may be programmed to store, recognize and retrieve patterns, to filter noise from measurement data, and to control ill-defined processes. Neural networks are similar to fuzzy logic in that they both are model-free estimators, i.e., they can estimate a function without an explicit analytical model of how outputs depend on inputs (Kosko, 1992).

Fuzzy logic is based on the mathematics of fuzzy sets introduced in the mid-sixties by Professor Lotfi Zadeh (1965, 1968). Conventional sets, also called **Cantorian** or **crisp sets**, allow only full membership or no membership into a set whereas fuzzy sets allow partial membership. Given a universal set U, also called **universe of discourse**, the degree of membership of an element u of U to a fuzzy set A is described by the *membership function* of A, $\mu_A(u)$, a function taking values ranging from 0 to 1, and mathematically denoted as:

$$\mu_A(u): U \rightarrow [0, 1]$$

A fuzzy set is defined by its membership function and thus fuzzy classes, categories, numbers, and clusters, and any other mathematical object that rests on the notion of a fuzzy set, need to be defined through an appropriate membership function. Different fuzzy objects (numbers, linguistic terms, hedges, etc.) impose different mathematical demands, on their corresponding membership functions. **Fuzzy set operations** such as complementation, union and intersection, and their logical counterparts of negation (NOT), disjunction (OR), and conjunction (AND), are defined in terms of operations that involve the membership functions of their constituent fuzzy sets (Keller and Tahani, 1992). Given, for example, two fuzzy sets A and B, and their respective membership functions, μ_A and μ_B, the union of A and B is a new fuzzy set with membership function:

$$\mu_{A \cup B} = \max\{\mu_A, \mu_B\}$$

Similarly, the intersection of A and B is defined by the membership function:

$$\mu_{A \cap B} = \min\{\mu_A, \mu_B\}$$

and the complement of A by:

$$\mu_{\bar{A}} = 1 - \mu_A$$

A **fuzzy variable**, also called a **linguistic variable**, has arguments that are fuzzy sets describing a specific linguistic categorization of the crisp values of the variable. For example, if we consider **temperature** to be a fuzzy variable (we indicate fuzzy variables by bold-face letters and fuzzy values by italics) then its values may be {*low*, *medium*, *high*} as shown in Fig. 15.1. The values of a fuzzy variable are collectively referred to as the **term set**. Each fuzzy value is described analytically through a membership function $\mu(T)$. In Fig. 15.1. we have three membership functions, $\mu_{low}(T)$, $\mu_{medium}(T)$ and $\mu_{high}(T)$ denoted for convenience as *low*, *medium* and *high*, respectively. Notice that the term set of **temperature** is defined over a universe of discourse, e.g., the range of real numbers between 0 and 100°C. Any crisp value of temperature, e.g., 60°C, has

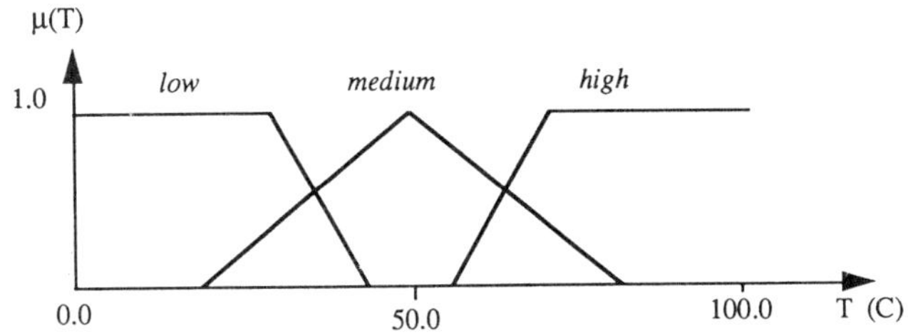

Fig. 15.1 Membership functions describing the temperature categories low, medium and high.

a unique degree of membership to the three fuzzy values of **temperature**; for example, 60°C may be a *medium* temperature to a degree 0.7 and also a *large* temperature to a degree 0.5. We can even think of it as being a (trivial) member of the category *small* to a degree of 0. Of course, a different categorization of the values of a fuzzy variable would entail a different structure of the term set, involving a different number of terms and different membership functions (Zadeh, 1978; 1979; 1983).

In fuzzy control the algorithm for controlling a process is described by a set of IF–THEN rules such as:

IF **X** is *small* THEN **Y** is *positive-large*

Where, **X**, **Y** are fuzzy variables, and *small* and *positive-large* are values from their corresponding term sets. A collection of such rules may represent the control law, i.e., a function $y = f(x)$ that relates the state of the system, x, to a prescribed control action, y. The geometric interpretation of fuzzyfying the control law is shown in Fig. 15.2, as indicated, each rule covers a region of the situation–action (or state–control) space, not merely a point. The development of a fuzzy control algorithm requires the definition of fuzzy variables and term sets at the appropriate level of granularity. For example, variable **Y** may initially have a term set with 3 fuzzy values, {*small*, *medium*, *large*}, and later the control designer may decide that a term set of 5 is most appropriate,

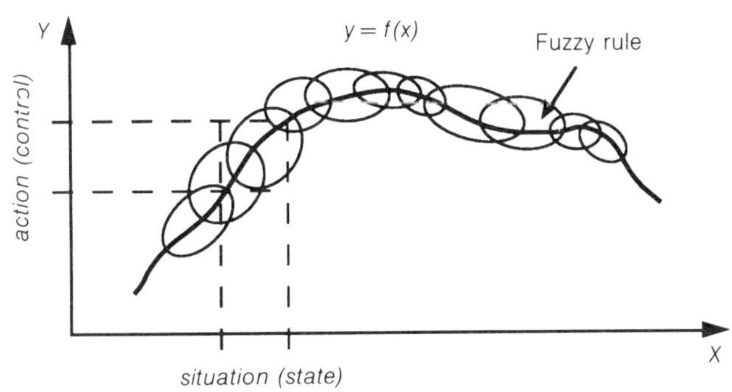

Fig. 15.2 In fuzzy control, a control law is represented by a collection of fuzzy rules.

e.g., {*very-small, small, medium, large, very-large*}. Generally, the process of deciding the appropriate term set is complex and several iterations may be required. Note that if **X** is thought of as the situation (or state) variable and **Y** as the action (or control) variable the set of IF–THEN rules performs a mapping analogous to the mapping that an artificial neural network may perform from an input vector, **X**, to a target vector, **Y**.

Generally, the control policy is a set of IF–THEN rules associating situation or antecedent clauses, s^j, and action or consequent clauses, a^j, through an inferencing procedure referred to as fuzzy or generalized *modus ponens** where, each clause is of the form '**X** is A'.

Typically, in process control algorithms, at time t_i the actual situation, s_i, calls for inferring the value of an action, a_i. Classified or Aristotelian *modus ponens* guarantees the logical truth of a consequent given that the antecedent of a syllogism (rule) is true. In generalized *modus ponens*, however, the antecedent is only true to some degree, and hence it is desired to compute the degree to which the consequent is true. Fuzzy logic is eminently suited for this kind of computation. Linguistic variables (fuzzy variables) are used assuming as values a set of membership functions (fuzzy values) each uniquely and unambiguously quantifying the degree of membership to a class. Thus, for a rule with one condition and one action variable we write:

$$\begin{array}{lcc} s^j \rightarrow a^j: & \text{IF } \mathbf{X} \text{ is } A \text{ THEN } \mathbf{Y} \text{ is } B \\ s_i \quad : & \mathbf{X} \text{ is } A' \\ \hline a_i \quad : & \mathbf{Y} \text{ is } B' \end{array} \tag{15.1}$$

This procedure is applied successively for each rule. Consider a rule base R^N composed of n such rules:

$$\begin{aligned} R &= \{R^1, R^2, R^3, \ldots, R^n\} \\ &= \{s^1 \rightarrow a^1, s^2 \rightarrow a^2, s^3 \rightarrow a^3, \ldots, s^n \rightarrow a^n\} \end{aligned} \tag{15.2}$$

Generally, the condition s^j can be thought of as the logical product of m variables, $s^{j1} \times s^{j2} \times \cdots \times s^{jm}$, and the action part of a rule can be thought of as the union of r variables, $a^{j1} + a^{j2} + \cdots + a^{jr}$. Thus, the set of rules that describe the system may be written as:

$$R^N = \underset{j=1}{\overset{n}{f_\alpha}} (s^{j1} \times s^{j2} \times \cdots \times s^{jm} \rightarrow a^{j1} + a^{j2} + \cdots + a^{jr}) \tag{15.3}$$

where, f_α is an operator that depends on the particular inferencing procedure used, such as max-product of max-min (Ragheb and Tsoukalas, 1986, 1988; Tsoukalas and Ragheb, 1988; Lin and Takefuji, 1990; Kosko, 1992). Using

*In classical logic *modus ponens* or rule of detachment refers to the principle or rule of inference that whenever a conditional statement and its antecedent are both given to be true, then its consequent may be validly inferred. Generalized *modus ponens* is the fuzzy logic generalization of classical *modus ponens*.

distributivity we rewrite (15.3) as:

$$R = \left\{ \begin{array}{l} \overset{n}{\underset{j=1}{f_\alpha}} (s^{j1} \times s^{j2} \times \cdots \times s^{jm} \rightarrow a^{j1}), \\ \overset{n}{\underset{j=1}{f_\alpha}} (s^{j1} \times s^{j2} \times \cdots \times s^{jm} \rightarrow a^{j2}), \\ \cdots \\ \overset{n}{\underset{j=1}{f_\alpha}} (s^{j1} \times s^{j2} \times \cdots \times s^{jm} \rightarrow a^{jr}) \end{array} \right\}$$

$$= \bigcup_{p=1}^{r} \overset{n}{\underset{j=1}{f_\alpha}} (s^{j1} \times s^{j2} \times \cdots \times s^{jm} \rightarrow a^{jp})$$

or,

$$R = \bigcup_{p=1}^{r} \mathrm{RC}^p \tag{15.4}$$

Thus, $R = \mathrm{RC}^1 + \mathrm{RC}^2 + \cdots + \mathrm{RC}^r$, i.e., the rule base, R, can be decomposed into r rule clusters (RC) of the form shown above. In each rule cluster there are: n rules, m conditions, and one action. Thus, totally we have r actions (taking all RCs into account). This offers a simple, yet powerful way for describing the behavior of a multiple-input multiple-output (MIMO) system and designing related controllers. The algorithm of a fuzzy expert system infers the action, a_i, at time $t = t_i$, on the basis of the actual state of the system, s_i, and sends the crisp signal, a_i, to appropriate transducers.

In fuzzy control, the fuzziness of the antecedents eliminates the need for a precise match with the input. Each rule is fired to a degree that it is a function of the degree of match between the antecedents and the input. The mechanism of imprecise matching provides a basis for interpolation, which, in turn, serves to minimize the number of fuzzy IF–THEN rules needed to describe a fuzzy input–output relation. There exist, however, two major issues to be addressed: determining the membership functions; and, adapting the rules to changes in the system and/or its environment. Several approaches involving neural networks have been suggested (Takagi and Hayashi, 1991; Barenji, 1992; Hayashi *et al.*, 1992; Kosko, 1992) to ameliorate both of these problems, and considerable research has been performed on this subject. In the following sections we will examine some of them.

15.2 FUZZY NEURAL ARCHITECTURES

There are two approaches to fuzzy neural architecture which we refer to as fuzzy-centric and neuro-centric. The **fuzzy-centric approach** emphasizes a fuzzy logic description, i.e., the control algorithm is formulated through fuzzy IF–THEN rules. Neural networks are used as ancillary tools either to determine membership functions, or to change the set of rules adaptively. The **neuro-centric approach** gives primacy to the neural side of the description. The control algorithm is basically neural (Werbos, 1992). Fuzzy arithmetic is used as a way of improving neuronal and/or entire network behavior. Both approaches exploit heavily the learning and interpolative abilities of neural computing paradigms.

In this chapter, although we focus on fuzzy-centric approaches, a few words on neuro-centric approaches may be appropriate. Neuro-centric approaches endeavor to improve the performance of neural networks using fuzzy mathematics and fuzzy reasoning. The fundamental difference with conventional approaches is that the control parameters of the neuron and the dendritic connection weights are replaced by fuzzy regions. To illustrate how this works we look at a typical neuron used in a backpropagation network (McClelland and Rumelhart, 1986). Consider the multi-input, single-output neuron shown in Fig. 15.3. The relationship between the output y_i and the input vector $\mathbf{x}$ is given by:

$$y_i = f\left(\sum_{j=1}^{N} \omega_j x_j + \theta_0\right) \tag{15.5}$$

where $f(z) = 1/(1 + e^{-z})$ is the sigmoidal transfer function, $\omega_{\varphi l}$ are connection weights and θ_0 is a bias term.

A fuzzified neuron uses fuzzy sets as connection weights, $\omega_{\varphi l}$, and the transfer function is a fuzzy set representing the compatibility between the signal input strength and the firing strength of the neuron. Thus the output becomes a fuzzy set with various degrees of strength. When the fuzzy output is received by another neuron it is summed along with other active input signals to generate the cumulative signal strength and the resultant signal is mapped to a fuzzy region corresponding to the possible activation levels associated with the number of active signals or the attenuated strength of the cumulative signal.

As already mentioned in the fuzzy-centric approaches, determining the term sets of fuzzy variables and adaptively changing the rules are the two major challenges of fuzzy control where neural networks hold considerable promise.

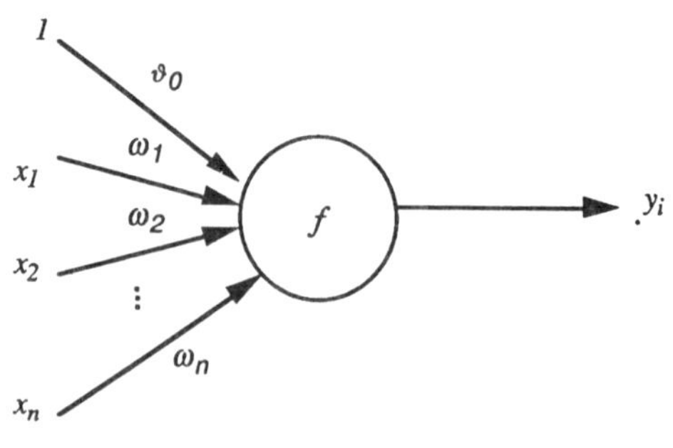

Fig. 15.3 A typical processing unit of a neural network.

In the first problem, that is the problem of deciding the term set, it is particularly difficult to design membership functions in multidimensional spaces by using experience and intuition. Since the situation/action variables are supposed to be independent in fuzzy control, the membership function is designed on a variable by variable case (Takagi and Hayashi, 1991). If a membership function is found to be unfit, then heuristic tools are used to tune the membership function to the appropriate position/shape. Consider, however, a rule that relates temperature and humidity to power, for example, a rule that can be utilized in an air-conditioning control problem.

IF **room temperature** is *higher* AND **humidity** is *lower*
THEN **power** is *negative small*

This rule says that if the room temperature is increasing and the humidity remains low, then the controller has to reduce power slightly. Power and humidity, however, are not completely independent. Thus, determining the membership function requires a curved surface in three-dimensional space consisting of the membership value axis and the temperature–humidity plane. In general, the higher the dimensionality of the operating space, the more difficult it becomes to design and improve the term set of fuzzy variables. Another difficulty in fuzzy control is the fact that it is fairly difficult to use a learning function. This makes the optimization of the fuzzy algorithm for a time-varying problem relatively difficult. Both of these difficulties may be addressed by utilizing the mapping and learning functions of neural networks.

In fuzzy-centric approaches, neural networks are used either to produce the membership functions describing the values of fuzzy variables, or to help a system adapt to the environment, i.e., to change the rule set. Consider the classical benchmark problem for non-linear algorithms, the inverted pendulum (Kosko, 1992). We use two state variables and one control variable for the fuzzy algorithm. The two state fuzzy variables are: the angle θ the pendulum shaft makes with the vertical, and, the angular velocity $\Delta\theta/\Delta t$ which is the difference between the two angular positions over a sampling period, i.e., $\Delta\theta = \theta_t - \theta_{t-1}$ ($\Delta t = 1$). The fuzzy control variable is the angular velocity υ of the motor which relates to the current a motor should have in order to keep the inverted pendulum in the upright position ($\theta = 0$). All three fuzzy variables are defined over the universe of real numbers. We categorize each universe of discourse into a number of fuzzy values, typically five are chosen for the inverted pendulum case. For each variable we build the term set, i.e., a lexicon of fuzzy values. Figure 15.4 shows a typical set of values for the fuzzy variable θ, i.e., the term set of θ, where PM stands for *positive medium*, PS stands for *positive small*, ZE stands for *zero*, NS stands for *negative small*, and NM stands for *negative medium*. Figure 15.5 shows a set of values for the fuzzy variable $\Delta\theta$ and Fig. 15.6 a typical term set for the control fuzzy variable υ. Notice that although the labels of the values of the term set are the same, the membership functions for the three variables are actually different (since they are defined over different universes of discourse).

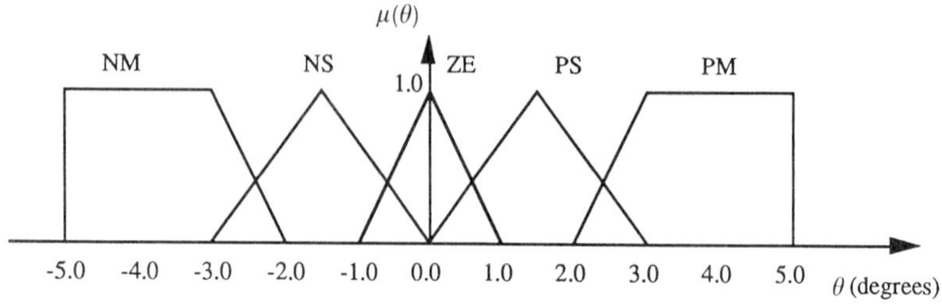

Fig. 15.4 Values of the fuzzy variable θ.

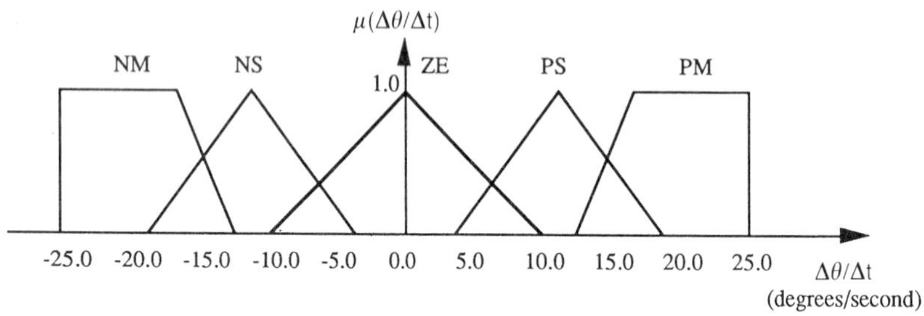

Fig. 15.5 Values of the fuzzy variable $\Delta\theta$.

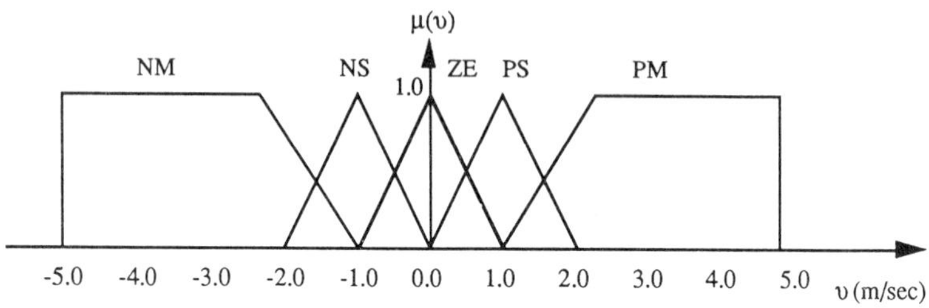

Fig. 15.6 Values of the fuzzy variable v.

The control algorithm is a collection of rules like:

$$\text{IF } \theta \text{ is } NM \text{ AND } \Delta\theta/\Delta t \text{ is } ZE \text{ THEN } \upsilon \text{ is } PM$$

Each rule has two situation variables in the left-hand side (LHS), or antecedent, and one action variable in the right-hand side (RHS), or consequent. Thus, a specific control algorithm is a 5×5 matrix with fuzzy values as its entries, as shown in Fig. 15.7. The empty cells in the table indicate that there are no rules for the particular combination of situation variables, e.g., there is no action to be taken if the angle, θ, is NM and the angular velocity, $\Delta\theta/\Delta t$, is NS.

At any given time, physical sensors provide crisp values to the controller.

$\Delta\theta/\Delta t$ \ θ	NM	NS	ZE	PS	PM
NM			PM		
NS			PS	NS	
ZE	PM	PS	ZE	NS	NM
PS		PS	NS		
PM		PM	NM		

Fig. 15.7 Fuzzy control algorithm.

These values are matched against each rule of the control algorithm. The matching is called **fuzzification** since the values of sensor measurements are matched against the fuzzy values of the LHS variables. Suppose, for example, that in the control of the inverted pendulum we have a rule cluster of two rules.

Rule 1: IF θ is NM AND $\Delta\theta/\Delta t$ is ZE THEN υ is PM

Rule 2: IF θ is NS AND $\Delta\theta/\Delta t$ is ZE THEN υ is PS

Let θ_0 and $\Delta\theta_0$ the crisp sensor readings at some specific time. The degree to which the readings are members of the values NM and ZE is given by the membership functions evaluated at these points, i.e., $\mu_{NM}(\theta_0)$ and $\mu_{ZE}(\Delta\theta_0)$ respectively, for Rule 1. The strength d_1, of Rule 1 is the minimum of these two values, since we use the conjunction operator. (This is called the **max–min method** of inference, other methods exist, a more appropriate one for certain applications calls for taking the product of these two numbers, and thus it is referred to as the **max-product method** of inferencing.) Therefore,

$$d_1 = \min\{\mu_{NM}(\theta_0), \mu_{ZE}(\Delta\theta_0)\} \tag{15.6}$$

Similarly for Rule 2 we have

$$d_2 = \min\{\mu_{NS}(\theta_0), \mu_{ZE}(\Delta\theta_0)\} \tag{15.7}$$

The control contribution of Rule 1 is calculated by applying the strength of Rule 1 to its RHS.

$$a_1 = \mu_{c_1}^{-1}(d_1) \tag{15.8}$$

Similarly for Rule 2 we have

$$a_2 = \mu_{c_2}^{-1}(d_2) \tag{15.9}$$

The contribution of both Rules 1 and 2 is calculated in the defuzzification step, typically using Tsukamoto's method (Bernard, 1988; Barenji, 1992).

$$a^* = \frac{\sum_{i=1}^{n} d_i a_i}{\sum_{i=1}^{n} d_i} \qquad (15.10)$$

where n is the number of rules with firing strength d_i. The result of each rule reading sensor values θ_0 and $\Delta\theta_0$ is to recommend a control action. The defuzzified combination of all rules produced through equation (15.10), gives a crisp value to be sent to the motor transducer at a specific time.

Each element in the matrix of Fig. 15.7 may equal any one of the five values of the action variable υ. The entries of the matrix can be adaptively generated by a neural network (Kosko, 1992) from process sample data. Adaptive clustering algorithms are used to estimate an unknown rule, e.g., a triplet such as (NM, ZE; PM), from training samples of the form ($\theta, \Delta\theta$: υ). In other words a network is trained with the membership functions for θ and $\Delta\theta$ as inputs, and the values of υ as outputs. Various combinations of inputs and outputs have appeared in the literature, ranging from all fuzzy inputs to some fuzzy and some crisp. Competitive adaptive vector quantization algorithms can adaptively estimate both the unknown rules ($\theta_\kappa, \Delta\theta_\kappa; \upsilon_\kappa$) and the unknown number of rules necessary to produce a fuzzy system description on the basis of a set of input–output data.

15.3 DEVELOPMENT OF SYSTEM ARCHITECTURE

As alluded to in the previous section, a major problem in fuzzy control is the difficulty in generating membership functions, particularly, when there are more than a few variables in the LHS of rules. For variables that are highly interdependent, heuristic correlation amongst these variables may be hard to establish. In addition, common sense and engineering judgment are difficult to apply in fuzzy algorithms involving a large number of variables.

In this section, we present an approach to membership function generation that employs neural networks. The basic idea is to teach the network to associate a set of time-dependent input data with a membership function that represents a fuzzy value, e.g., *medium*, *positive open*, *acceptable*, *inadequate*, etc. Each network learns one value and thus a collection of networks is used to represent the term set of a fuzzy variable dynamically (Tsoukalas and Ragheb, 1988; Uhrig, 1989; Tsoukalas *et al.*, 1991). The salient point in such an approach is that the term set may change dynamically as the system changes in the sense that membership functions are allowed to shift while maintaining their relative order over the universe of discourse. The overall architecture of such a system is shown in Fig. 15.8. The methodology presented here was applied to the control of the high flux isotope reactor (HFIR) of the Oak

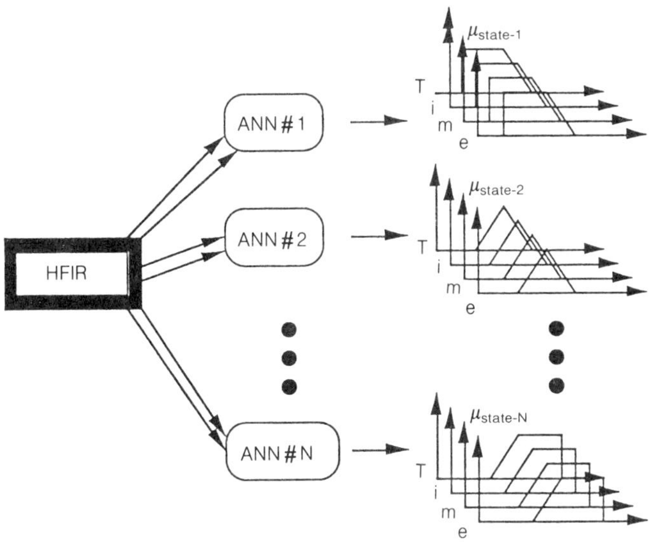

Fig. 15.8 Schematic representation of a fuzzy neural system for determining a term set.

Ridge National Laboratory. An important control element in the reactor is the secondary flow control valve. Its function is to regulate secondary side flow in order to maintain the neutronic power in the reactor core at a desirable level. This constitutes a delicate and relatively difficult control task.

In the fuzzy neural approach we present in this chapter, a collection of neural networks ANN#1 through ANN#*N* is used to map time-dependent signals representing the state of a nuclear system to a set of membership functions that describe the term set of a fuzzy variable with control significance. Once trained, the neural networks are the receivers of several on-line time-series corresponding to vital nuclear system parameters. They filter the noise of the time-series and calculate other system parameters not in the form of time-series but in the form of a membership function. Each membership function corresponds to a different value of the monitored variable and has a unique shape. The particular fuzzy variable examined here is called **VALVE_POSITION**, and it provides an indication of the valve position of an important control valve called the secondary flow control valve.

Neural networks are used to design, in effect, what is referred to as a *virtual measuring device*, i.e., a software-based instrument for the 'measurement' of user-specified fuzzy variables. Usually, these variables cannot be measured directly, or the failure of a sensor requires that a variable be inferred from other measurements (Uhrig, 1989). A promising feature of virtual instruments is that their function may be modified by changing their software, not the hardware. Generally, measurement involves a mapping of complex input patterns to simple output patterns. Neural networks can map a complex input pattern of variables to a simplified set of membership functions representing the values of a fuzzy variable. They produce membership functions that uniquely and unambiguously represent the values of variables that are fuzzy, such as performance, risk, operability, and availability. The fundamental requirements for useful measurements are the *precision* and *reproducibility* of the input–output relation and the functional value of the entire operation of the system doing the measuring. The requirement of reproducibility means that the measuring device must be **isolatable** from the system being measured and **resettable** so that the measurements can be repeated an arbitrary number of times to give the same output for the same input pattern. Both requirements are met by the artificial neural networks contemplated here.

15.3.1 Fuzzy neural network

The neural networks used for determining the fuzzy values of the fuzzy variable **VALVE_POSITION** are three-layer networks, as illustrated in Fig. 15.9, (input, hidden, output layers). At the input layer there are five input nodes each receiving a time series form a physically measurable variable, 10 nodes in the middle (or hidden) layer and four nodes at the output layer representing the corners of a trapezoidal membership function.

A number of input–output pairs, called examples, are presented to the network and the connection weights are adjusted until the network has

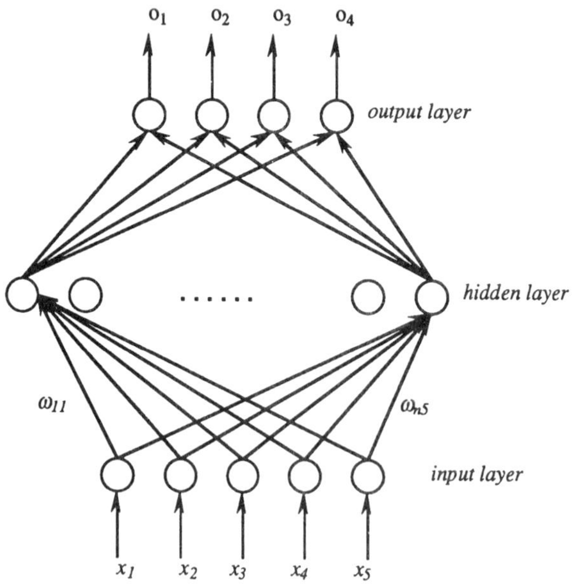

Fig. 15.9 Each neural network maps a number of time-series onto a set of numbers describing a trapezoidal fuzzy number.

'learned' the underlying relationship that the examples represent. This is called **supervised learning** and the process of weight adjustment is referred to as **training**. The algorithm for training in the methodology presented is back-propagation with generalized delta rule and momentum term (see Chapter 3 for details).

The collection of neural networks essentially performs a mapping, $f: X \rightarrow Y$, where, the domain X is the hyperspace of physically measured variables, such as temperatures and pressures, and the output range Y is a set of fuzzy numbers that constitute the term set of the fuzzy variable **VALVE_POSITION.** A fuzzy number is the value of a fuzzy variable at any given time and is uniquely represented by a membership function as shown in Fig. 15.10. The fuzzy number **C** shown is called a trapezoidal fuzzy number over the universe of discourse, U, which is the set of real numbers between 0 and 1, (individual elements of U are denoted by lower case u) (Fig. 15.10). Trapezoidal membership functions are very common in fuzzy logic applications since they are uniquely and conveniently described by a set of four numbers, e.g., $\mathbf{C} = \{o_1, o_2, o_3, o_4\}$, and therefore offer the advantage of speed in computations performed by a computer.

After the networks comprising the mapping mechanism of the virtual instrument are trained, (a process similar to 'calibration'), they receive on-line time signals as inputs and produce a set of membership functions as outputs. Generally, the outputs will be somewhat different than the membership functions the networks were trained for, and moreover one or at most two (if we allow overlap of membership functions) will represent correct values while the rest need to be ignored. It is thus important to identify the correct output. Since we consider each neural network output to be a fuzzy number, we use a measure of similarity of fuzzy numbers, called the **dissemblance index** (Kaufmann and Gupta, 1991) to estimate the outputs that are closest to the set of prototype membership functions on which we trained the networks, and select one as the actual fuzzy value.

Suppose, for example, that the network recognizing the disk position virtual measurement value *closed* has been trained on the output membership function μ_{closed}, and after training it produces an actual output membership function μ^*_{closed}. We consider the two membership functions as two fuzzy

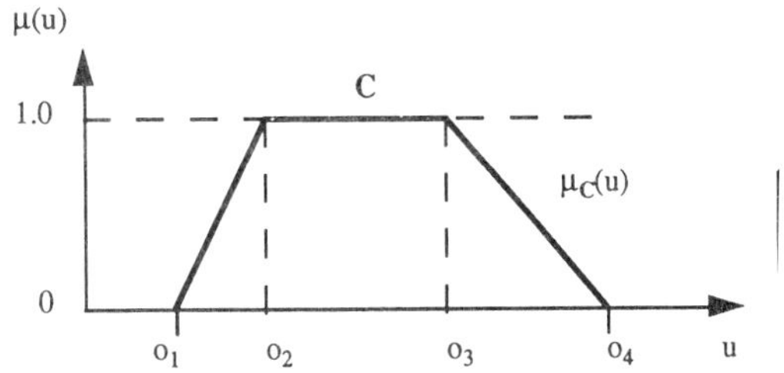

Fig. 15.10 The output of each network is a trapezoidal fuzzy number **C**.

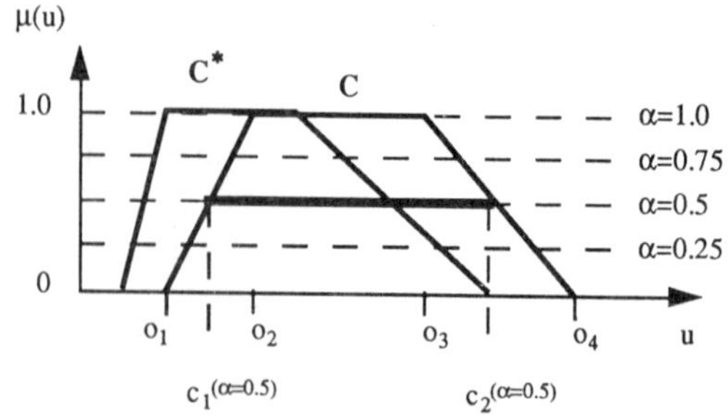

Fig. 15.11 The index of dissemblance is a measure of the difference between fuzzy numbers **C** and **C***.

numbers, call them **C** and **C***, each with a trapezoidal shape and support on the universe of discourse [0, 1]. Figure 15.11 illustrates the two fuzzy values.

The support of each function is an interval, i.e.,

$$\mathbf{C} = [c_1, c_2]$$
$$\mathbf{C}^* = [c_1^*, c_2^*]$$

We can compute a numerical function, i.e., the dissemblance index, $\delta(\mathbf{C}, \mathbf{C}^*) \in [0, 1]$, which is the distance between **C** and **C***. The α-cuts** for the two membership functions are denoted by, $\mathbf{C}_\alpha, \mathbf{C}^*_\alpha$, respectively. The dissemblance index of the prototype output, **C**, and the actual output, **C***, is defined as:

$$\delta(\mathbf{C}, \mathbf{C}^*) \equiv \int_{\alpha=0}^{1} \delta(\mathbf{C}_\alpha, \mathbf{C}^*_\alpha)\, d\alpha \tag{15.11}$$

When the universe of discourse is the interval of real numbers between 0 and 1, i.e., [0, 1], we can easily derive from (15.11) a value for the dissemblance index in terms of the support of each fuzzy value, and the α-cuts, (with α in the interval 0 to 1 also, i.e., $\alpha \in [0, 1]$):

$$\delta(\mathbf{C}, \mathbf{C}^*) = \frac{1}{2} \int_{\alpha=0}^{1} (|c_1^{(\alpha)} - c_1^{*(\alpha)}| + |c_2^{(\alpha)} - c_2^{*(\alpha)}|)\, d\alpha \tag{15.12}$$

The dissemblance index is a number ranging from 0 to 1, representing the distance between two fuzzy numbers. If $\delta(\mathbf{C}, \mathbf{C}^*) \approx 0$ then **C** and **C*** are almost identical, on the other hand, if $\delta(\mathbf{C}, \mathbf{C}^*) \approx 1$, then **C** and **C*** are totally different. For a virtual measuring device with three outputs, i.e., three trained networks,

The α-cut of a fuzzy set A is defined as the ordinary (non-fuzzy) set of the elements of A which belong to the fuzzy set at least to a degree α. For example, the 0.5-cut of the fuzzy number **C in Fig. 15.11 is the interval of real numbers between $c_1^{(\alpha=0.5)}$ and $c_2^{(\alpha=0.5)}$. This is an important notion in the theory of fuzzy sets useful for representing an arbitrary fuzzy set in terms of constituents crisp sets and for performing fuzzy arithmetic operations; Kaufmann and Gupta (1991) for more details.

the output with the lowest dissemblance index is chosen as the actual output at any given time, with two outputs (overlapping membership functions) being given by the pair of values with the two lowest dissemblance indices.

15.3.2 Input representation

In order to demonstrate the proposed methodology we utilized actual data obtained during a start-up of the high flux isotope reactor (HFIR). Five parameters were chosen for describing the secondary flow control valve position: neutron flux, primary flow pressure variation (DP), core inlet temperature, core outlet temperature and secondary flow (Figs 15.12, 15.13, 15.14, 15.15 and 15.16). All but the last one of the above mentioned time series, contain average values of the corresponding parameters of the three loop system. These parameters are selected in order to provide sufficient description of both the primary and secondary sides of HFIR during start-up. The time

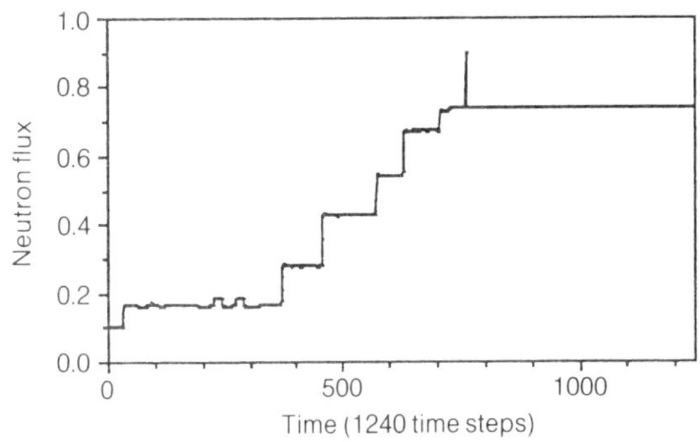

Fig. 15.12 Neutron flux as a function of time (average three loop values).

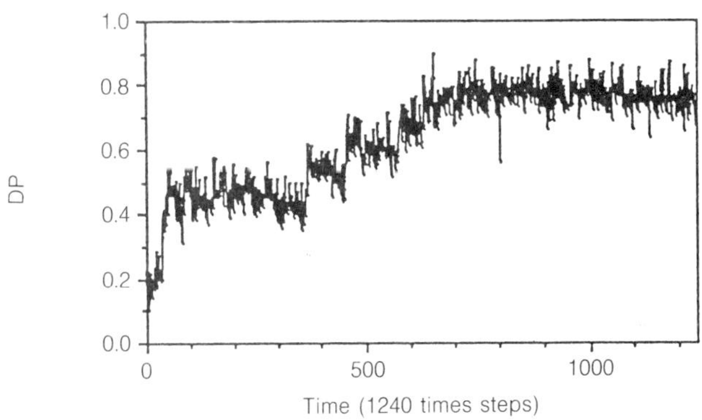

Fig. 15.13 Primary flow pressure variation as a function of time (average three loop values).

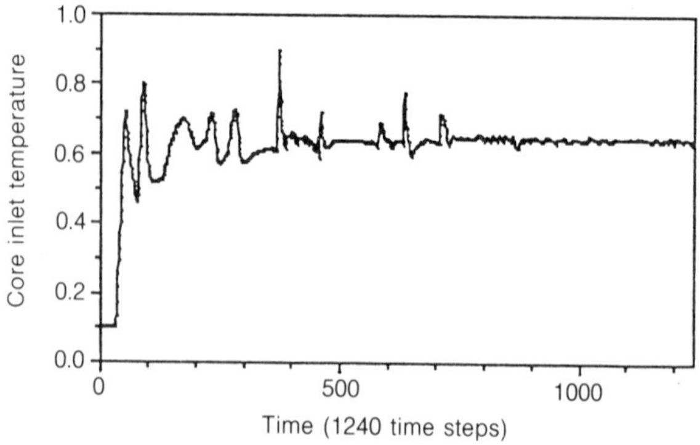

Fig. 15.14 Core inlet temperature as a function of time (average three loop values).

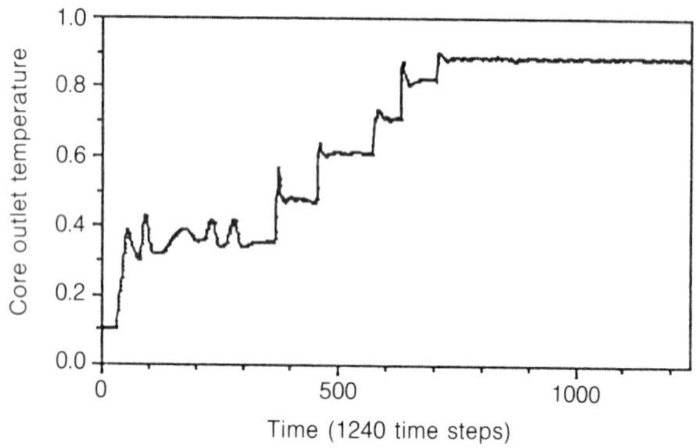

Fig. 15.15 Core outlet temperature as a function of time (average three loop values).

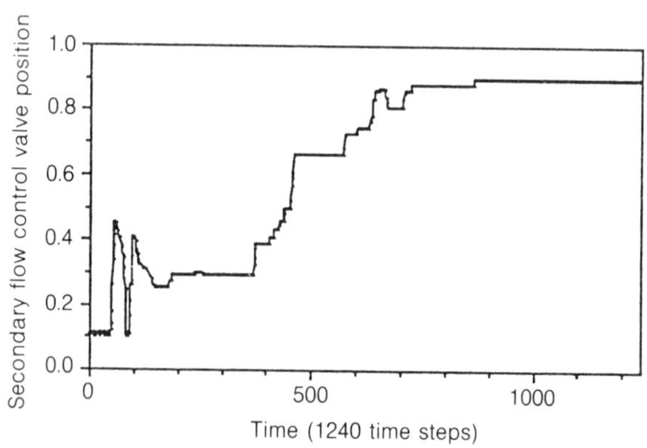

Fig. 15.16 Secondary flow as a function of time.

series of these five parameters are used to train five neural networks where each one of them has five nodes at the input layer and four nodes at the output layer (Fig. 15.9). The output of each of the five neural networks is a membership function uniquely labeling a particular position of the secondary flow control valve.

15.3.3 Output representation

The bahavior of the secondary flow control valve is represented in the space of alternatives (universe of discourse) with the fuzzy variable **VALVE_POSITION**, which may take five fuzzy values: *closed, partially_closed, medium, partially_open*, and *open*. Each one of these fuzzy values is represented with a membership function: μ_{closed}, $\mu_{\text{partially-closed}}$, μ_{medium}, $\mu_{\text{partially-open}}$, and μ_{open}. These three membership functions describe the position of the valve at every instant during the start-up period. A schematic of the membership functions representing the values of the fuzzy variable **VALVE_POSITION**, is shown in Fig. 15.17.

The membership functions have trapezoidal shapes or the degenerated (triangular) form of it, which is very useful for computations in the fuzzy control area (Tsoukalas and Ragheb, 1988; Ikonomopoulus *et al.*, 1991). The membership function μ_{closed} (Fig. 15.17) is defined by a trapezoid with peak coordinates $\{(0,0), (0,1), (0.1,1), (0.2,0)\}$.

15.3.4 Training set

In order to test the ability of each trained neural network to predict the valve position by calculating the right membership function at any particular time step, different levels of noise were introduced in the input signals. Initially up to 10% noise was introduced to all five input signals and the set of networks was tested with the 'noisy' vectors. The appropriate networks fired at the corresponding time steps calculating the coordinates of the peaks of the corresponding membership functions with 98% accuracy. Henceforth, there was an excellent prediction of the position of the disc valve during the whole time interval under consideration. Going one step further, 20% noise was

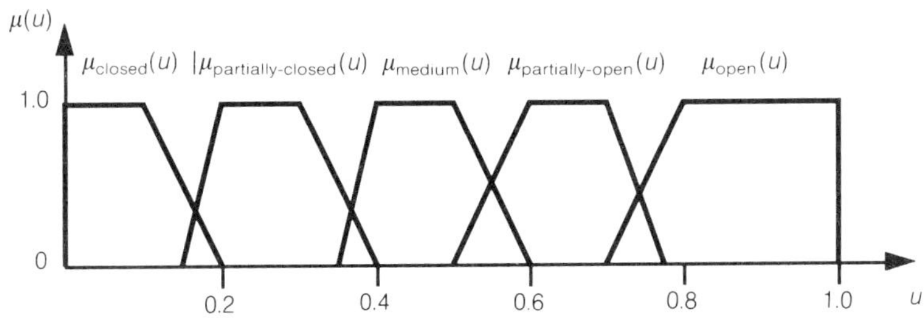

Fig. 15.17 Output membership functions the networks were trained on.

introduced to all five input signals and the networks were tested again. The response of the system was the same as in the previous case, but this time the accuracy of the coordinate prediction dropped to 95%. Once again, the accuracy of the network response was adequate to define the boundaries of the membership function responsible for firing.

15.4 FUZZY NEURAL NETWORK IMPLEMENTATION AND PERFORMANCE

The fuzzy neural methodology outlined in the previous section is able to determine, dynamically, the fuzzy values of linguistic variables describing the behavior of a complex system in a process analogous to measurement. It was implemented on a SPARC work station and extensively tested with real data from the high flux isotope reactor. Various levels of noise were used to check the filtering abilities of the networks. One of the most important tests was the following: we eliminated one of the inputs and substituted random noise in its place. Later, each input was eliminated one by one in order to check the robustness of the methodology and to decide which input variable was actually the most important. In order to perform this task, the fuzzy value measuring device depends on information supplied from other system parameters and infers a linguistic description of the state of the system under investigation. Thus, it seems important to explore the abilities of the virtual measuring device to execute its task when it is supplied with partially correct information. Failure of one or more measuring devices is a rather common phenomenon in industrial applications. The problem during this situation is that the measuring device records totally random values without a particular pattern, and thus it becomes very difficult to detect and isolate the fault signal. In order to simulate this situation as realistically as possible, we substituted each one of the five time series with a signal corresponding to 100% random noise, i.e., one of the original signals is completely dropped and a random, created by a random number generator, time series replaces it. We use this random signal along with the other four original signals in order to test that the methodology can successfully perform its task with partially missing information.

The results presented here depict the substitution of the average DP signal by a totally random one, generated from the random number generator. We have supplied the virtual measuring device with vectors containing the other four signals in their original form and the random noise in the place of the average DP. The results are plotted in Figs 15.18 and 15.19. In Fig. 15.18 the responses of the virtual measuring device are plotted with a frequency of appearance equal to one every ten time steps for a time span covering the first 200 steps of the start-up duration. The reason we have chosen this particular time span lies in the high activity in the alteration of the valve-position taking place during this time span (operator actions followed by compensatory adjustments), and thus it is worthwhile to examine the virtual measuring device responses during the same time interval. In Fig. 15.19 we see that the

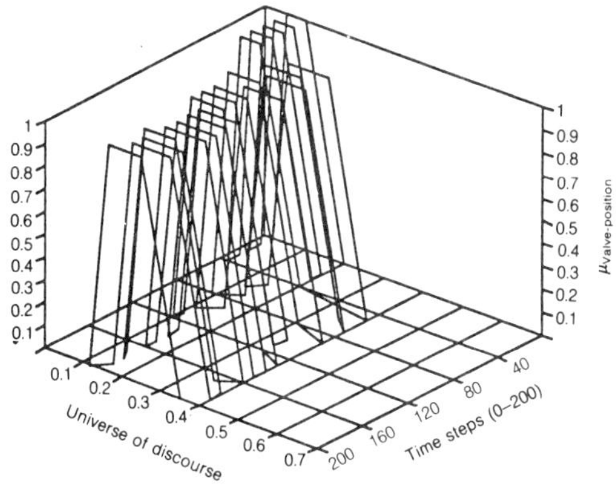

Fig. 15.18 Membership functions for time steps 0–200 when the average DP signal was substituted with 100% noise.

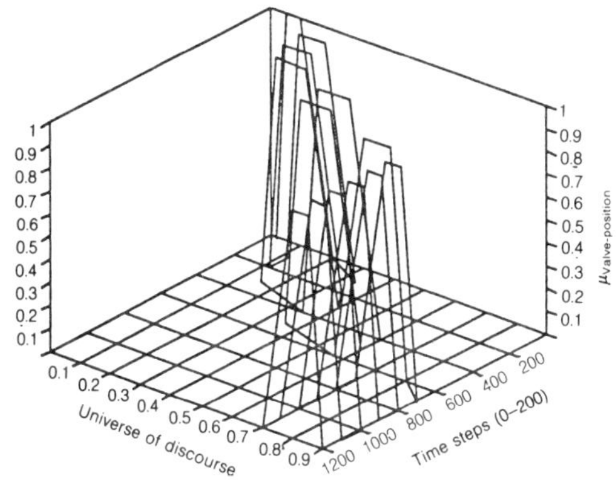

Fig. 15.19 Membership functions for time-steps 0–1200 when the average DP signal was substituted with 100% noise.

predictions about the valve-position coincide with the crisp measurements obtained from HFIR. The two abrupt openings of the valve at around 80 and 100 time steps after the start-up initiation have been accurately recorded and the valve-position has been identified as *medium.* Considering that the crisp measurements of the position of the valve attain a maximum value close to 0.5 we deduced that the characterization is correct. As expected, the resultant membership functions do not concur exactly to those existing in the original term set.

Figure 15.19 shows the accuracy of the predictions throughout a reactor start-up period, depicted here with one membership function every 100 time steps. We see that around 700 time steps after the transient initiation (time $t = 0$ is found the inner corner of the three-dimensional plot), the valve-position is characterized as *open*, and the same characterization is repeated thereafter. By taking into account that at the same moment the crisp value of the valve-position becomes 0.8 and later increases to 0.9, we see that the identification of the state of the valve is quite accurate. Furthermore, during the time interval between time steps 200 and 700, the state of the valve is initially defined as *medium* and later as *partially_open*, following the transition of the valve-position from one state to another.

In order to check the accuracy of the response when one of the input signals has been replaced by a random time series, we have calculated the centroids of the resultant membership functions at every time step. The results are compared to those obtained when all five input signals are supplied to the measuring tool and the euclidean distance is computed between the coordinates of the centroids of the membership functions. The results of the comparison are shown in Fig. 15.20, where we see that more than half of the membership functions have been predicted with 100% accuracy. In addition, about one-third of them has been calculated with a norm, between expected and predicted values, equal to or less than 0.05. For the remaining time steps the membership functions have been estimated with a euclidean distance from the expected values, equal to or less than 0.40. The results presented exhibit the system robustness to fault signals. The substitution of the DP signal by a random time series had a negligible effect on the system response as it can be seen from the average euclidean norm between the centroids of the noise-free and 'noisy' membership functions. These results suggest that even with 20% of its information missing the error is statistically insignificant.

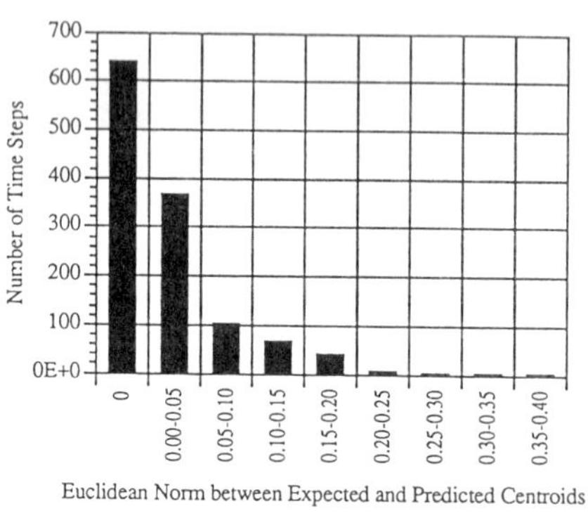

Fig. 15.20 Distribution of the Euclidean norms between expected and predicted centroids when the average DP signal was eliminated

15.5 SUMMARY

In this chapter we surveyed the fundamentals of fuzzy neural control and the related area of measurement. We discussed some of the relations between fuzzy logic and neural networks and how best to combine them. Fuzzy logic and artificial neural networks in combination present unique opportunities for enhancing the performance of control methodologies that utilize either one of them separately. They are essentially complementary technologies and when exploited synergistically may offer considerable benefit for the development of flexible, user-adaptive and system-specific control applications.

Fuzzy control benefits a great deal by using the mapping and learning abilities of neural networks to address the problems of determining membership functions and adaptively changing a rule base. In this chapter we looked at how a set of complicated time series may be mapped to the universe of discourse of human linguistics through a collection of neural networks acting as an interpreter of vital information supplied from a process system. The information encoded in a time series is in the form of rate of increase/decrease, and maximum/minimum values attained over a period of time. The networks are trained to represent this kind of 'hidden' information in the form of membership functions which can be used by a control algorithm. The membership functions provide sufficient information to describe a flow control valve position, for example, at a particular time but also to predict the actual valve position in the near future. This feature of the networks in conjunction with recent advancements in fuzzy hardware (Lim and Takefuji, 1990) may make it possible to implement very robust, highly distributed, anticipatory control algorithms.

The method presented to dynamically determine fuzzy values may be useful for designing virtual measuring devices, i.e., software-based (programmable) instruments for the 'measurement' of user-specified variables with operational significance, e.g., performance, risk, availability, cost, benefit, etc. The function of such devices may be modified by changing their software, not hardware, a promising feature for application-specific monitoring, diagnostic and control tasks. The procedure to determine membership functions can be summarized as follows.

1. We decide on the number of partitions necessary to adequately cover the range of the variable.
2. We determine the physically measurable variables that will be the input to the neural network.
3. We train a neural network per fuzzy value; this amounts to 'calibration'.
4. Finally, we use the index of dissemblance to select which output is the actual fuzzy value for our variable at any given time.

REFERENCES

Barenji, H.R. (1992) A reinforcement learning-based architecture for fuzzy logic control. *International Journal of Approximate Reasoning*, **6**, 267–92.

Bernard, J.A. (1988) Use of rule-based system for process control. *IEEE Control Systems Magazine*, **17**, 3–13, October.

Hayashi, I., Nomura, H., Yamasaki, H. and Noboru, W. (1992) Construction of fuzzy inference rules by NDF and DDFL. *International Journal of Approximate Reasoning*, **6**, 241–66.

Ikonomopoulos, A., Tsoukalas, L. and Uhrig, R. (1991) A hybrid neural network – fuzzy logic approach to nuclear power plant transient identification, in *Proceedings of AI91 Frontiers in Innovative Computing for the Nuclear Industry*, Jackson, WY, September, pp. 217–26.

Kaufmann, A. and Gupta, M.M. (1991) *Introduction to Fuzzy Arithmetic*, Van Nostrand Reinhold, New York.

Keller, J.M. and Tahani, M. (1992) Implementation of conjunctive and disjunctive fuzzy logic rules with neural networks. *International Journal of Approximate Reasoning*, **6**, 221–40.

Khanna, T. (1990) *Foundations of Neural Networks*, Addison-Wesley Publishing Co., Boston.

Kosko, B. (1992) *Neural Networks and Fuzzy Systems*, Prentice Hall, Englewood Cliffs, NJ.

Lim, M.-H. and Takefuji, Y. (1990) Implementing fuzzy rule-based systems on silicon chips. *IEEE Expert*, February, 31–45.

McClelland, J. and Rumelhart, D. (1986) *Explorations in Parallel Distributed Processing*, MIT Press, Cambridge.

Ragheb, M. and Tsoukalas, L. (1986) A coupled probability–possibility method for decision-making in knowledge-based systems, in *Knowledge-Based Expert Systems or Manufacturing* (eds S. C-Y. Lu and R. Commanduri), ASME, New York.

Ragheb, M. and Tsoukalas, L. (1988) Monitoring performance of devices using a coupled probability–possibility method. *International Journal of Expert Systems*, **1**, 111–30.

Takagi, H. and Hayashi, I. (1991) NN-driven fuzzy reasoning. *International Journal of Approximate Reasoning*, **5**, 191–212.

Tsoukalas, L., Ikonomopoulos, A. and Uhrig, R. (1991) Hybrid expert system – Neural network methodology for transient identification, in *Proceedings of the American Power Conference*, Chicago, IL, April, pp. 1206–11.

Tsoukalas, L. and Ragheb, M. (1988) Performance monitoring and diagnosis in a process environment using a probability–possibility approach, in *Manufacturing International '88: Symposium on Manufacturing Systems – Design, Integration, and Control*, G. Chryssolouris, B. Von Turkovich and P. Francis, American Society of Mechanical Engineers, pp. 231–8.

Uhrig, R.E. (1989) Opportunities for automation and control of the next generation of nuclear power plants. *Nuclear Technology*, **88**, Nov., 157–65.

Werbos, P.J. (1992) Neurocontrol and fuzzy logic. *International Journal of Approximate Reasoning*, **6**, 185–219.

Zadeh, L.A. (1965) Fuzzy sets. *Information and Control*, **8**, 338–53.

Zadeh, L.A. (1968) Probability measure of fuzzy events. *Journal of Mathematical Analysis and Applications*, **23**, 421–7.

Zadeh, L.A. (1978) Fuzzy sets as a basis for theory of possibility. *Fuzzy Sets and Systems*, **1**, 3–28.

Zadeh, L.A. (1979) Fuzzy sets and information granularity, in *Advances in Fuzzy Set Theory and Applications* (eds M.M. Gupta, R.K. Ragade and R.R. Yager), North-Holland Publishing Company, pp. 3–18.

Zadeh, L.A. (1983) A computational approach to fuzzy quantifiers in natural languages. *Computer and Mathematics*, **9**, 149–84.

CHAPTER 16

Neural networks in continuous process diagnostics

Najwa S. Merchawi and Soundar R. T. Kumara
Intelligent Design and Diagnostics Research Laboratory, Department of Industrial and Management Systems Engineering, The Pennsylvania State University, University Park, PA 16802.

In this chapter, the use of neural networks in diagnostics is investigated. Diagnostics is critical to any continuous process set up, to detect the occurrence of failures with accuracy and speed, particularly when failures can be catastrophic. Previous approaches to diagnostics problems have been rule-based. They suffered from slow response time and failure under missing information. Neural networks are fast because they use parallel processing concepts. They are tolerant of missing information because information is not stored locally but distributed over several nodes. In this research, a two-layer feedforward neural network is proposed as a modeling paradigm for continuous process diagnostics. Specific diagnostic problems in the domain of nuclear reactor control are studied. The diagnostic knowledge gathered by experts is represented in binary patterns. Different input and output representations are derived and the performance under each is analyzed. The results indicate that input representation is particularly important in determining the neural network's recognition capability and the amount of tolerable noise in parameter readings.

Artificial Neural Networks for Intelligent Manufacturing. Edited by Cihan H. Dagli. Published in 1994 by Chapman & Hall, London. ISBN 0 412 48050 6

16.1 INTRODUCTION

Recent developments in neural network theory, models, and learning algorithms have sparked interest of the research community of various disciplines in possible application of neural network based approaches to their problems. Extensive research efforts have been devoted to the use of neural networks in vision (Feldman, 1985; Ballard, 1986, 1987), speech recognition tasks (Burr, 1986; Elman and Zipser, 1987; Bourlard and Wellekens, 1987; Lippmann and Gold, 1987), robotics, and signal processing. Most of these tasks involve pattern recognition or classification: neural networks match an arbitrary input pattern to a corresponding output pattern, with the fundamental property that 'small' variations in the input pattern (due to noise, inaccuracy, failure of a subsystem, or any other reason) do not result in misclassification.

Given this fundamental property, we suggest that neural networks are a good approach for solving diagnostic problems. Here, using the appropriate knowledge representation of the problem description is critical. A neural network architecture can then be developed based on the representation scheme.

In this chapter, we develop a neural network approach to continuous process diagnostics. In the next section, we argue the case for using neural networks in diagnostics. The characteristics of a typical continuous process diagnostics problem are outlined. We discuss knowledge representation issues and then we consider a specific diagnostic problem: the TMI-2 nuclear reactor, for which we developed neural network models to perform real-time diagnostics.

16.2 NEURAL NETWORKS FOR DIAGNOSTICS

Reasons for using neural networks for diagnostics.

1. Previous approaches have been too slow to satisfy the real-time requirements of some systems (Nann, 1988). The symbolic processing involved in processing rules in rule-based systems is time consuming. In particular, the sequential nature of search tasks slows down performance in large knowledge bases drastically. The parallelism inherent in neural network processing is appealing from the perspective of speed.
2. Neural networks are often described as universal pattern matchers; they are thus good candidates for diagnostics at the fourth level of Milne's (1987) hierarchy (pattern matching level).
3. Neural networks are noise and fault tolerant. If the input pattern representing the symptoms contains errors, it can still be classified. Often in diagnostics, an incoming case does not fall exactly into any of the predefined known cases. In this situation, a diagnosis should be made based on how close the current case is to the predefined ones. Neural networks

are well suited to this task because they extract the closest pattern to the patterns stored in its connection matrices. On the other hand, rule-based systems often fail completely in the presence of input errors.

4. Rule-based systems development requires the gathering of explicit knowledge to write the rules. However, much of human expert knowledge is implicit. Neural networks have the potential to represent implicit knowledge since knowledge is distributed throughout the network in the form of weight matrices.

Yoon *et al.* (1989) suggested using a neural network approach in the diagnosis of skin diseases. The input to their network is a pattern representing the observed symptoms, such as location, distribution, number of lesions, etc. The output is the corresponding disease, represented by a node which will have a high activation level when the disease is diagnosed. The network is trained with 250 hypothetical cases with typical symptoms of diseases, and tested with 99 real cases. The average success rate of this network was 70%, ranging from 30% to 80% for different groups of diseases.

Dietz *et al.* (1989) have used neural networks for diagnosis of jet and rocket engine failures. For the jet engine diagnostic neural network, the input comes from sensors that measure engine parameters such as combustion temperature and turbine rotation speeds. The network is trained to recognize two types of failures (bearing failures and fuel interruptions) based on the shapes of the curves of the measured parameters – not their magnitude. A different neural network is designed for each sensor, to perform diagnosis based on that particular sensor's reading. The overall diagnosis is the average of the outputs of all these neural networks.

16.3 PROBLEM DESCRIPTION

A typical continuous process is characterized by on-line readings of a number of parameters. Each parameter reading can be considered either normal (if it falls into a specified normal range), too high, or too low (both of which indicate failure). At any instant, the process status description is the set of readings of all the parameters. For the class of problems we are dealing with, it is not necessary to know the exact continuous value of a parameter. For diagnostic purposes, it is sufficient to characterize each parameter as being in one of the following seven discrete levels:

1. high
2. very high
3. extremely high
4. low
5. very low
6. extremely low
7. normal

For simplicity, we use the symbols H, HH, HHH, L, LL, LLL, and N, respectively to represent these levels. All levels except for N are called out-of-bounds (oob) levels. Levels L, LL, and LLL are called low-type oob levels. Similarly, H, HH, and HHH are high-type oob levels. Experts consider this classification adequate to draw conclusions about the failures of the process. In other words, reasoning in diagnostic systems is qualitative, not quantitative. The advantage of qualitative reasoning is that it reduces the range of data we have to deal with. Instead of analyzing absolute numbers (which fall into a wide range), qualitative reasoning allows us to focus on the interpretation and meaning of values which have a smaller range.

If all input parameters are in the normal range, then the process is functioning normally and no fault is to be detected. When a fault does occur, a number of parameters could be out-of-bounds. Experts in the process can identify all possible non-trivial failure scenarios, which parameters each one of them affects, and how these parameters are affected (lowered or increased in value) by the failure scenario. Non-trivial failure scenarios are failure scenarios that affect more than just one parameter. Any parameter can be affected by more than just one scenario. Under these assumptions, the diagnostic process is not straightforward, but rather involves analysis of the entire knowledge base to select the scenario whose description is closest to the current description of the process.

The first and most important issue involved here is appropriate knowledge representation of the problem. Let $S1, S2, \ldots, Sn$ represent the n failure scenarios, and $P1, P2, \ldots, Pm$, the m parameters relevant to the diagnostics of the n scenarios. Each scenario affects the reading of certain parameters by pushing their values into lower or higher levels. The more extreme the value becomes, the more likely it is that the corresponding scenario has occurred. These facts can be described as in Table 16.1: reading the first line – scenario 1 causes parameters $P1$ and $P2$ to read lower and parameter $P3$ to read higher, but does not affect the rest of the parameters. The rest of the table is interpreted in a similar manner. We must next convert this tabular knowledge representation into a form suitable for neural network processing.

Table 16.1 A typical continuous process diagnostics table

	Parameters		
Scenarios	*P*1	*P*2	*P*3
*S*1	Lower	Lower	Higher
*S*2		Lower	
*S*3		Higher	Higher
⋮			
Sn	Higher		Lower

16.4 KNOWLEDGE REPRESENTATION FOR CONTINUOUS PROCESS DIAGNOSTICS BY A NEURAL NETWORK

Since a scenario is characterized by a pattern of qualitative parameter values, the diagnosis problem falls into the category of pattern recognition, or pattern matching/classification. An input pattern of parameter values (that we can also call a **symptom pattern**) is to be matched with or classified into a failure scenario. Different input patterns could result in the same diagnosis. For example, if scenario *S*1 is described by the following pattern

*P*1	*P*2	*P*3	*P*4	⋯	*Pm*
Lower	Lower	Higher	—	⋯	—

then all of the following patterns

	*P*1	*P*2	*P*3	*P*4	⋯	*Pm*
Pattern 1	LLL	LLL	HHH	N		N
Pattern 2	LLL	L	L	N		N
Pattern 3	N	L	HHH	N		N

are classified into scenario *S*1. Pattern 1 fits into the *S*1 category best because its symptom pattern satisfies the description exactly. Pattern 2 should also be diagnosed as *S*1 even though *P*3 is in level L instead of being high. This can be interpreted as incomplete or inaccurate information. Thus, it is clear that the diagnostic problem is a classification/pattern matching problem.

Since there are seven discrete levels that determine the diagnosis, it is appropriate to use discrete binary patterns to represent input patterns. The approach taken here is analogous to that in visual pattern recognition: a neural network can be trained to recognize digitized visual patterns. An image is converted into a binary string where a '1' represents a dark pixel (picture element) and '0' represents a light pixel. The picture is usually scanned from left to right, top to bottom. A neural network's tolerance to small differences in input patterns is illustrated in terms of visual pattern recognition in Fig. 16.1.

A neural network can be trained to recognize the pattern in Fig. 16.1(a) as

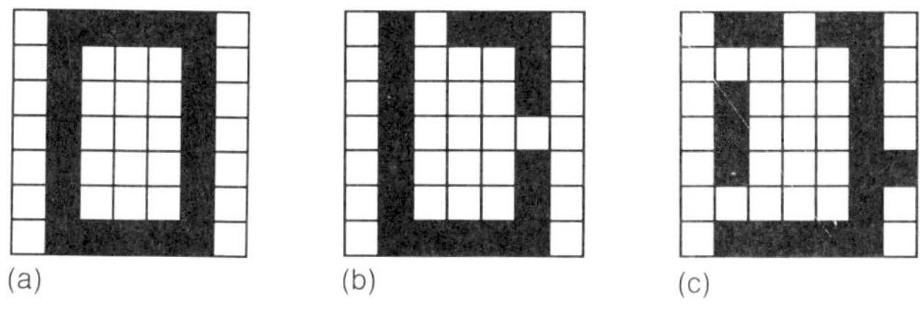

Fig. 16.1 Typical binary patterns recognized by a neural network. (a) Complete rectangular pattern; (b) Disturbed pattern 1; (c) Disturbed pattern 2.

being of class 'rectangle'. After successful training, this network will also classify the patterns in Fig. 16.1(b) and 16.1(c) in the class 'rectangle'. This is because the binary patterns of these pictures do not differ in many bits from the original noiseless rectangular pattern. In fact these patterns are:

original pattern:

0111110 0100010 0100010 0100010 0100010 0100010 0111110

disturbed 1:

0101110 0100010 0100010 0100000 0100010 0100010 0111110

disturbed 2:

0110110 0000010 0100010 0100010 0100011 0000010 0111110

where disturbed 1 differs from the original by 2 bits and disturbed 2 by 4 bits (difference bits are underlined). The Hamming distance between two binary patterns is defined as the number of bits in which the two patterns differ.

The inherent behavior of neural networks makes them classify patterns that look similar (differ in only a few bits, regardless of the position of the differing bits) into the same class. Furthermore, the more alike a pattern looks to the original pattern used in training, the more confident the classification is. Similarly, for the diagnostics problem, binary representations of the failure scenarios can be derived satisfying the criterion: patterns that look similar have close interpretations. The more different the interpretations are, the more different the corresponding patterns will be.

The overall patterns representing the continuous process and its state of operation will be the collection of the representations of each parameter. To derive the overall input patterns we need a binary representation for each of the seven levels. Let us refer to the binary representation of a level as a subpattern.

The application of the criterion established above can be applied to the subpattern level. If the subpatterns satisfy the criterion, then the overall pattern (made up of the concatenation of these patterns) will also satisfy the criterion. Levels that have a close interpretation will have representations that do not differ in many bits (i.e. look close). Therefore, the further the interpretations are, the more difference bits there should be between them. For example, levels LL and LLL have similar interpretations. If a failure scenario *S* that lowers the value of a parameter *P* occurs, whether *P* is in level LL or LLL is only an indication of the severity of the failure, but the interpretation is the same: other things being equal, scenario *S* has occurred.

Since levels LLL and HHH have completely different and opposite interpretations, their binary representations should be completely different: LLL will be represented by a string of zeros, HHH by a string of ones. As the level ranges from LLL to HHH, the representations should start looking increasingly similar to HHH and increasingly different from LLL. A minimum of six bits is required to satisfy this condition. The minimum length of the representation for the subpatterns is thus 6 bits.

LLL	is represented by	000000
LL	is represented by	100000
L	is represented by	101000
N	is represented by	101010
H	is represented by	101011
HH	is represented by	101111
HHH	is represented by	111111

Let us refer to this representation scheme as In1. Overall patterns are formed by concatenating *m* of these subpatterns; recall that *m* is the number of parameters. For example, the complete pattern for a process described by readings of nine parameters, *P*1 through *P*9, in levels L, N, N, N, H, LL, N, N, and N, respectively is:

101000 101010 101010 101010 101011 100000 101010 101010 101010

(The spaces between subpatterns are included for clarity only.)

Representation scheme In1 is clearly not the only scheme satisfying our requirements. It uses the minimum number of bits per subpattern; representations using more bits can be derived. This might be necessary if other relationships between the levels are to be reflected in the subpatterns. In In1, all levels differ from their neighboring levels in exactly one bit. If, in a certain problem, one level is more distinct than others (for example, it has more importance and/or a bigger role in determining the output), then more bits are needed to make its representation more distinct from others. In our TMI-2 diagnostic problem, it may be argued that the normal level N is distinct from the oob levels – the importance of the N level in making diagnostic decisions is greater. A representation scheme that reflects this fact, together with the previous criterion, uses 8 bits for each subpattern and is shown below.

LLL	is represented by	00000000
LL	is represented by	10000000
L	is represented by	10100000
N	is represented by	10101010
H	is represented by	10101111
HH	is represented by	10111111
HHH	is represented by	11111111

We refer to this scheme as In2. In In2, any two consecutive levels differ in one bit (just like In1), but the N level differs in 2 bits from each of its neighboring levels. It is then easier for the neural network to distinguish between the N level and its neighboring oob levels. Furthermore, the low type oob levels become even more distinct than the high type oob levels. For example, level L differs in 4 bits from level H (instead of just 2 bits as in In1). Note that these subpatterns are 8 bits long. The overall pattern for the example process description mentioned above, using In2 would be:

10100000 10101010 10101010 10101010 10101111 10000000 10101010 10101010 10101010

A diagnostic input pattern then has an equivalent visual pattern that can be formed using an $m \times 6$ matrix (for In1) or an $m \times 8$ matrix (for In2).

Output representation

The output of the diagnostic system is a message telling us what failure, if any, has occurred, or an 'unidentified situation' message, if the system does not recognize the fault. Since there are n failure scenarios and one normal operation scenario, there are $n+1$ different outputs to be encoded in the diagnostic system. There are two ways to accomplish this using neural networks.

1. Associate an arbitrary binary pattern with each of the scenarios and store this information in a look-up table. In general, to represent n scenarios requires a minimum of $|\log_2(n)|$ bits. More bits can be used to satisfy other requirements, such as network training and convergence. After the neural network processes an input pattern and produces the output pattern, the look-up table is searched to find the scenario name and the corresponding corrective action. If the output pattern is not in the look-up table, the situation is unidentified. We refer to this method of output representation as Out1.
2. Associate each node at the output level with a distinct failure scenario. Only one node will be active high at a time, indicating the occurrence of that failure scenario. Thus, the number of nodes at the output level is equal to the number of failure scenarios plus one (normal operation).

For more accurate diagnostics, the continuous valued activation of the output nodes (before conversion to binary) can be used to output the message. The node with the highest output (activation level) is selected to output the diagnostic message. If the highest continuous value is not very close to 1, the match is less than perfect. Therefore, a confidence level can be associated with each message, equal to the activation of the output node. When the confidence level is low, an alternative diagnosis should be suggested. The second highest scoring node is a good candidate for this alternative diagnosis. This method of output representation and messaging is referred to in this chapter as Out2.

10.5 EXAMPLE PROBLEM: THE TMI-2 NUCLEAR REACTOR

The TMI-2 nuclear reactor is a continuous process that requires a fast reliable diagnostic system. We use it as the basis for our diagnostic neural network model.

It has been established (Nann, 1988) that nine parameters play a crucial role in diagnosing the failures of the TMI-2 nuclear reactor.

1.	Pressurizer pressure	(in psig)
2.	Pressurizer level	(in inches)
3.	Hot leg temperature	(in deg F)
4.	Cold leg temperature	(in deg F)
5.	Steam generator #1 pressure	(in psig)
6.	Steam generator #1 level	(in inches)
7.	Steam generator #2 pressure	(in psig)
8.	Steam generator #2 level	(in inches)
9.	Drain tank pressure	(in psig)

For simplicity we refer to these parameters as $P1, P2, \ldots, P9$. There are eleven possible failure scenarios:

1. Pressurizer leak
2. Block valve leak
3. Pipe rupture (drain)
4. Drain tank faulty
5. Pipe rupture type 1
6. Pipe rupture type 2
7. Reactor pump
8. Steam generator type 1
9. Steam generator type 2
10. Pipe rupture type 3
11. Feedwater pump

We refer to these scenarios as $S1, S2, \ldots, S11$. The normal operation scenario is referred to as N. Tables 16.2 and 16.3 are derived from the experience of personnel on the TMI reactor diagnostics. Table 16.2 gives the operating ranges of the nine parameters and the level associated with each range. Table 16.3 gives the scenario descriptions.

Table 16.2 Setpoint database from TMI-II accident

Parameter	LLL	LL	L	N	H	HH	HHH
*P*1	1200	1900	2055	2150	2250	2355	2400
*P*2	45	150	200	222	240	260	280
*P*3	300	400	500	606	610	619	630
*P*4	800	850	900	940	1050	1070	1105
*P*5	10	30	45	160	170	180	190
*P*6	1	2	2.5	3	35	80	122
*P*7	800	850	900	940	1050	1070	1150
*P*8	10	30	45	160	170	180	190
*P*9	300	400	500	558	610	619	630

Table 16.3 Scenario database

Scenario	*P*1	*P*2	*P*3	*P*4	*P*5	*P*6	*P*7	*P*8	*P*9
*S*1	Lower	Lower		Lower	Lower		Lower	Lower	
*S*2	Lower	Lower		Lower	Lower	Higher	Lower	Lower	
*S*3	Lower	Lower				Lower			
*S*4	Lower	Higher				Higher			
*S*5				Lower	Lower		Lower	Lower	Lower
*S*6			Higher	Higher	Lower	Higher	Higher	Lower	
*S*7			Higher	Higher			Higher		Higher
*S*8			Higher	Lower	Lower		Lower	Lower	Higher
*S*9				Higher	Higher		Higher	Higher	Higher
*S*10				Higher	Lower		Higher	Lower	Higher
*S*11				Lower	Higher		Lower	Higher	Higher

16.5.1 The training set

Using the representation schemes described in the previous section, we first derive a training set for the TMI-2 nuclear reactor problem. A training set is a set of pairs of patterns; each pair is made of an input pattern and the corresponding output pattern. A training set is used to train a neural network to match an input pattern with the corresponding output pattern. The key issue here is deciding what the training set should contain: what data should the network be trained with in order to guarantee accurate and reliable diagnostics once the network is in operational mode? Since there are 12 classes of scenarios to recognize, the training set will consist of 12 pairs. A pair is an input pattern of a failure scenario and an output representation of the scenario. But what should the input pattern of a failure scenario be? From Table 16.3, a typical scenario, say scenario 4, is described by

*P*1:	lower
*P*2:	higher
*P*6:	higher
Rest of the parameters:	unaffected.

When a parameter value is lower, it could be in any of the levels L, LL, or LLL. Similarly, if a parameter value is above normal, it could be in level H, HH, or HHH. Hence, many combinations of parameter levels correspond to the same scenario. Which one should be used for training? This question can be answered by noting that the more extreme the values are, the more likely the corresponding failure has occurred. Thus, the pattern used for training should be formed by taking the lowered parameter values close to extremely low (level LL) and the increased parameter values to be close to extremely high (level HH). Unaffected parameters will be considered to be at normal level. We assume that this is equivalent to forming a pattern that best describes the failure scenario. Note that a similar idea is used in visual pattern recognition. The fact that the perfect square shape (Fig. 16.1(a)) is the best description of a square shape is why it is used for training.

The output patterns using representation scheme Out1 are 9-bit patterns arbitrarily assigned to represent each scenario. Four-bit patterns are adequate but we use 9 bits for a more reasonable reduction in size from input pattern length (54) to output pattern length (9 instead of 4). This will help in training the neural network.

The output patterns using representation scheme Out2 are 12-bits long since there are 12 scenarios to be detected in the TMI-2 example.

*S*1	is represented by	100000000000
*S*2	is represented by	010000000000
*S*3	is represented by	001000000000
*S*4	is represented by	000100000000
*S*5	is represented by	000010000000
*S*6	is represented by	000001000000

*S*7	is represented by	000000100000
*S*8	is represented by	000000010000
*S*9	is represented by	000000001000
*S*10	is represented by	000000000100
*S*11	is represented by	000000000010
N	is represented by	000000000001

16.5.2 Neural network model and architecture

The neural network model used in this research is the two-layer perceptron network with backpropagation (Rumelhart *et al.*, 1986). The reasons behind this choice are primarily:

1. this model is best suited for pattern matching/classification tasks and diagnostics is such a task;
2. this model shows potential for learning arbitrary patterns, such as those derived for diagnostics; and
3. the backpropagation algorithm has shown powerful convergence behavior in previous applications (Rumelhart *et al.*, 1986).

Figure 16.2 shows the structure of a two-layer perceptron network. There are three layers of neuron nodes (arranged vertically in Fig. 16.2); this network is also referred to as a three-layer network. However, it is also considered a two-layer network since the input layer does not contribute in the processing. The input nodes are merely sensors that transmit the information they hold to the next layer. They are conceptual nodes whose use makes the model and its learning paradigm clearer. The middle layer is often called a hidden layer since it is hidden to the user, who interacts only with inputs and outputs. The right-most layer is the output layer. This network is fully connected: there is a connection from every node in one layer to every node in the following layer.

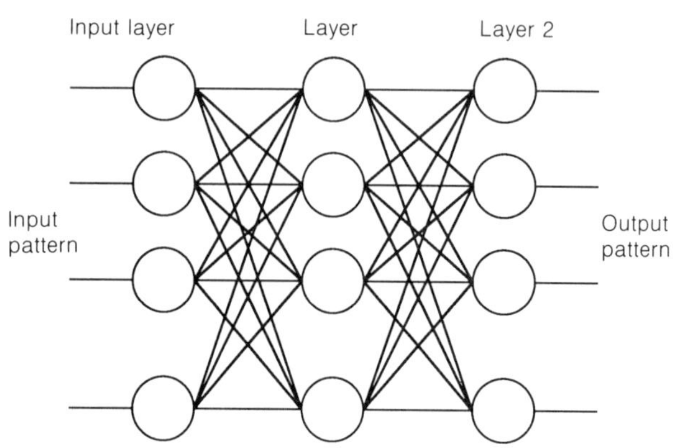

Fig. 16.2 A two-layer neural network.

Note, however, that during training, some of the connection weights may reduce to 0, which is equivalent to no connection.

This model is designed to learn associating an input pattern to an output pattern, where a pattern is a vector of binary or real values. Each input node inputs one value of the input vector and each output node holds one value of the ouput vector. Thus the number of input and output nodes is application dependent. No theoretical scheme has yet been developed to determine the number of nodes required in the hidden layer.

The network is trained with a training set derived for the particular application. Each element of the training set is a pair of patterns: an input pattern and its corresponding output pattern, which is the target output. During an iteration of training, all input–output pairs of the training set are successively presented and the connection weights are modified using the back-propagation rule (Rumelhart *et al.*, 1986). Training is complete when the difference between the calculated output pattern and the target output pattern is smaller than a predefined value referred to as the Z-value. Using such a training procedure a 'neural network learns from past experience'. When training is complete, the network can enter operational mode. In this mode, an input pattern, which may be one of the patterns used in the training set or a partially distorted version of it, is presented to the network. The nodes in the hidden layer perform their characteristic calculations (sigmoid function of a sum of products) and transmit their output to the output layer. The output layer performs the same calculations, and the output pattern appears at the output nodes. No weight changes occur in operational mode.

In the current model, the number of input nodes is equal to the length of the input pattern; all nodes simultaneously hold one bit of the pattern. Similarly, the number of output nodes is equal to the length of the output pattern. Thus, if In1 is used, there are 54 input nodes; if In2 is used, there are 72 input nodes. At the output level, if Out1 is used, there are 9 output nodes; if Out2 is used, there are 12 output nodes. The number of hidden layer nodes was arbitrarily selected as 27 for all network models except In2/Out2 where the number of hidden layer nodes selected was 40.

16.6 IMPLEMENTATION AND SIMULATION RESULTS

The diagnostic neural network is simulated and its behavior is studied. The simulation programs are written in the C programming language running on a Vax 11/785. The simulation is done in two phases: training phase and operational phase.

16.6.1 Training phase

The training program reads input from a training set file. Initially, the weight values of all connections are randomly assigned, uniformly distributed between -1 and 1. The program then cycles through all 12 training pairs and for each pair, the input pattern is processed by the hidden and output nodes. Then at each output node i, an error delta1(i) is calculated using the following rule (from backpropagation algorithm).

$$\text{delta1}(i) = \text{out}(i) \times (1 - \text{out}(i)) \times (\text{target}(p, i) - \text{out}(i))$$

Here, out(i) is the output or activation level of node i, target(p, i) $-$ out(i) is the difference between the target output for pattern p at node i and the actual output at the same node.

The weights coming into that node are then modified using the rule

$$W2(j, i) = W2(j, i) + \text{delta1}(i) \times \text{hid}(j) \tag{16.1}$$

where $W2$ is the weight matrix between output layer and hidden layer, and hid(j) is the activation level of hidden node j.

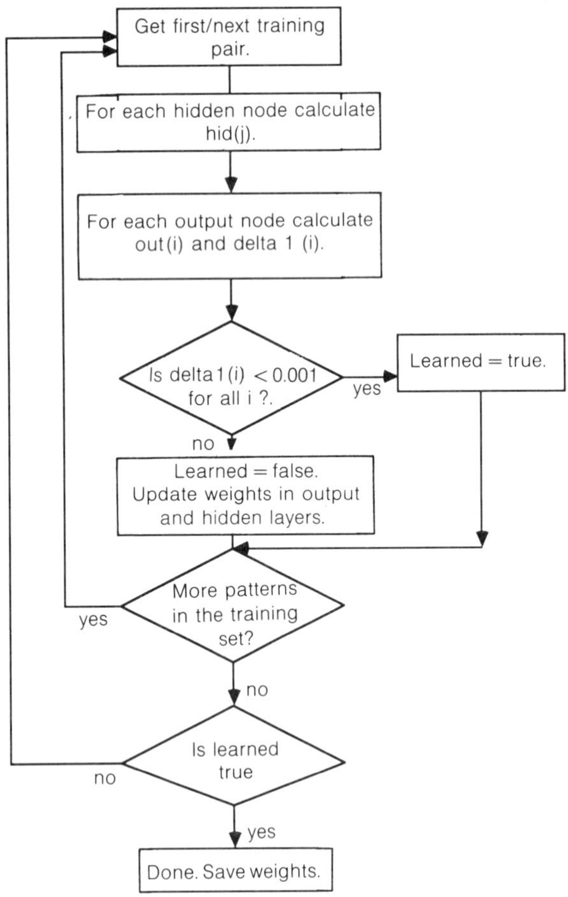

Fig. 16.3 Flowchart of training program.

Then the error associated every hidden layer node j is calculated as

$$\text{delta2}(j) = \text{hid}(j) \times (1 - \text{hid}(j)) \times \text{sum}$$

where

$$\text{sum} = \sum_{\substack{\text{all output}\\ \text{nodes } i}} \text{delta1}(i) \times W2(j, i)$$

and the weight matrix $W1$ between input layer and hidden layer is changed using (16.1) with delta2 instead of delta1. This procedure is repeated until errors at all nodes, with all the patterns, are less than a Z value of 0.001. A flowchart of the program is shown in Fig. 16.3. The simulation program was written in the C programming language and run on a Vax 8550 computer.

Four versions of the program were implemented; each simulations gave different training times as shown in Table 16.4.

Note the difference in the training time between different output representations; the effect of input representation is not as pronounced. Representation Out2 took longer to train due to the nature of the patterns – only one node is active at a time with all others required to be suppressed (made equal to 0). On the other hand, the patterns of the Out1 representation are not completely arbitrary; they were derived by trying to compress the 54-bit input pattern into a 9-bit pattern. Each subpattern used for training, namely LL, HH and N, which are 6 bits long, is compressed into one bit as follows:

subpattern LL: 100000 is compressed to 0
subpattern HH: 101111 is compressed to 1
subpattern N: 101010 is compressed to either 0 or 1, arbitrarily.

These learning times confirm our claim that pairs of training set patterns that are 'close' to each other take less time for the neural network to learn. For the training set pairs of Out1, the output pattern is a compressed version of the input pattern, whereas the output patterns of Out2 do not bear any relationship with the input patterns.

16.6.2 Operational phase

Actual diagnostics operation takes place in this phase. The values of the weights are fixed, as per the output of the training phase. The simulation of

Table 16.4 Training time of neural networks

Representations	*Training time*
In 1–Out 1	3 min 25 sec
In 2–Out 1	2 min 43 sec
In 1–Out 2	5 min 08 sec
In 2–Out 2	6 min 04 sec

the operational phase varies depending on the representations used. The most important modification is the ouput decoding technique. For Out1, a string comparison technique is used; for Out2, a search for the highest activation value is needed.

The operational phase is similar to the training phase, but without error calculations or weight modifications. In other words, the input pattern is presented, processed by the hidden layer and the output layer, until the output pattern appears and is decoded to determine the diagnostic message.

16.6.3 The testing procedure

After the neural networks are trained, they can be used in real-time operational mode. In this mode, a neural network inputs parameter readings/values and outputs a message about the failure status, (or it confirms Normal operational status). In addition, when the network does not identify the failure scenario, a message is displayed. The simulation of this diagnosis process is implemented in the C language and run on the Vax 8550 computer.

During real-time operations, readings of the parameters are continuous. Therefore, it is necessary to convert each to the corresponding binary representation, before input to the neural network.

For the purpose of testing, however, a test case has been developed where parameter readings are in binary form, assuming that conversion has already been done. Since there are nine parameters, each of which can be at any of seven levels, there are 7^9 possible state descriptions, i.e. 40 353 607 cases could theoretically occur, but most of these are unlikely, and represent unrealistic scenarios. The only recognizable state descriptions are those that fit one of the scenarios described in Table 16.3. To get an estimate of the number of most probable state descriptions, note that, on the average, in identifying a scenario, five of the nine parameters dominate. Further, any of these parameters can be in any of four levels (rather than seven) if it is to fit into a defined description of a scenario. (For example, if a parameter should be 'lower' as specified in the scenario description table, Table 16.3, then it is more likely to be in levels L, LL, LLL, or N rather than any of the high H levels, for identification of the scenario). Therefore the approximate number of these state descriptions is 12×5^4 or 7500. This number is still too large, for the analysis of all possible cases. Therefore, we included 100 state descriptions in the test case, representing a variety of hypothetical and probable situations in real time operation. Table 16.5 outlines groups of scenarios considered in this test case. We feel that the test case contains sufficient state descriptions to provide insights into the performance and limitations of diagnostic neural networks.

16.6.4 Results

Since each test case was derived from the pattern of a known scenario, each should be classified as belonging to that particular scenario. The objective of

Table 16.5 Constituents of the test case

Group	*Group characteristics*	*No. of cases*
1	State descriptions used for training	12 all failures plus normal op.
2	All relevant oob parameters are at their most level (LLL or HHH)	11 all failures
3	All relevant oob parameters are at their least severe level (L or H)	11 all failures
4	The first parameter is distorted and is two levels off value used for training	11 all failures
5	The first and second parameters are distorted and fall two levels off value used for training	11 all failures
6	The first parameter is off by three levels from value used for training	11 all failures
7	The first and the second parameter are off by three levels	11 all failures
8	One parameter is off by four levels (e.g. from LL to HH). The parameter chosen is the first that can be offset by four levels and still fall into one of the seven predefined levels	11 all failures
9	Two parameters are off by four levels. (same observation as group 8)	11 all failures

our simulation was to determine whether the neural network was able to accomplish this or not. The test cases were presented to the neural network under different input and output representation schemes (In1-Out1, In1-Out2, In2-Out1, In2-Out2). In each case, it could be that the input data are:

1. classified consistent with the expectations;
2. unidentified; or
3. wrongly classified.

We derive a qualitative measure of the performance of the neural networks as follows. A neural network's performance is qualified as being:

1. excellent if 100% of the cases are correctly identified;
2. good if 60–90% of the cases are correctly identified, the rest being unidentified or having at most 1% wrongly classified; and
3. poor if it does not satisfy either of the above criteria.

Table 16.6 summarizes the output performance of the simulated neural networks for all nine groups of test data. Since two input representation schemes (In1 and In2) and two output representation schemes (Out1 and Out2) were defined, four different neural diagnostic systems are simulated: Out1-In1,

Table 16.6 Output of simulated neural networks

Group	*Expected response*	*Out1-In1*	*Out1-In2*	*Out2-In1*	*Out2-In2*
1	N	N	N	N	N
	1	1	1	1	1
	2	2	2	2	2
	3	3	3	3	3
	4	4	4	4	4
	5	5	5	5	5
	6	6	6	6	6
	7	7	7	7	7
	8	8	8	8	8
	9	9	9	9	9
	10	10	10	10	10
	11	11	11	11	11
2					
	1	1	1	1/2	1/10
	2	2	2	2	2
	3	3	3	3	3
	4	4	4	4	4
	5	5	5	5/N	5
	6	6	6	6	6
	7	7	7	7	7
	8	8	8	8	8
	9	9	9	9/11	9
	10	10	10	10	10
	11	11	11	11	11
3	1	U	1	N/1	1/3
	2	1	2	2/N	2/1
	3	N	U	N/3	3/N
	4	U	4	N/4	4/N
	5	U	5	N/5	5/N
	6	U	6	6/N	6/10
	7	U	7	7/N	7/N
	8	8	8	N/5	8/N
	9	U	U	N/9	9/N
	10	U	10	N/10	10/N
	11	U	11	11/N	11/N

Group	*Expected response*	*Out1-In1*	*Out1-In2*	*Out2-In1*	*Out2-In2*
4	1	1	1	1/5	1/5
	2	1	1	2	2
	3	U	U	3/N	3/N
	4	4	4	4	4/N
	5	5	5	5	5/1
	6	6	6	6/1	6
	7	7	7	7/8	7
	8	8	8	8	8/1
	9	9	9	9	9
	10	10	10	10	10/6
	11	11	11	11	11
5	1	U	1	5/1	5/1
	2	U	U	2/5	2
	3	N	U	N/3	N/3
	4	N	N	N/4	N/4
	5	5	1	1/5	1/5
	6	6	6	6/1	6
	7	7	U	7/3	7
	8	8	1	8/1	8/1
	9	9	9	9	9
	10	10	U	10/1	10/3
	11	11	U	11/1	11/9
6	1	1	1	1/5	1/8
	2	1	1	2	2/8
	3	3	3	3/N	3/N
	4	4	4	4	4/N
	5	5	5	5	5/1
	6	6	6	6/1	6
	7	7	7	7/8	7
	8	8	8	8	8/1
	9	9	9	9	9
	10	10	10	10	10/7
	11	11	11	11	11/8

Group	*Expected response*	*Out1-In1*	*Out1-In2*	*Out2-In1*	*Out2-In2*
7					
	1	5	U	5/1	1/5
	2	5	U	2	2/8
	3	N	U	3/N	N/3
	4	N	U	N/4	N/4
	5	U	1	1/5	1/5
	6	6	6	6/1	6
	7	7	U	7/3	7
	8	8	1	8/1	8/1
	9	9	9	9	9
	10	U	U	10/1	10/3
	11	U	U	11	11/9
8	1	1	1	1/5	1/8
	2	2	2	2	2/8
	3	3	3	3/6	3/N
	4	4	4	4/11	N/4
	5	5	U	5	5/6
	6	6	U	10/6	10/6
	7	7	7	7/10	9/7
	8	8	U	10/8	8/11
	9	U	U	9/11	9/11
	10	U	U	10	10/8
	11	11	U	9/11	11/9
9	1	5	5	5/11	1/8
	2	U	U	2	2/8
	3	N	U	4/3	N/3
	4	N	U	2/7	N/3
	5	5	U	5	4/5
	6	U	U	10/1	5/10
	7	U	N	N/7	N/10
	8	U	U	10	10/8
	9	U	U	10/7	10/9
	10	U	U	10/9	N/10
	11	U	U	9/11	9/10

Out1-In2, Out2-In1, and Out2-In2 and are shown in Table 16.6 over four columns. U stands for unidentified. When two scenarios are given, such as 1/5, the two scenarios 1 and 5 are suspected and the first (scenario 1) has a higher confidence level (see Fig. 16.4 for a sample output of neural network using output representation Out2).

Out1-In1 network

This network's performance can be classified into three categories:

1. excellent performance (100% correct response) for groups 1 and 2;
2. poor performance for group 3 (least extreme values of parameters) where it only identified one scenario, and for group 7 (two parameters off by three levels) where it only identified four scenarios, and for group 9 (two parameters off by four levels) where it did not identify any scenario; and
3. good performance for the rest of the groups: from 10 out of 11 to 7 out of 11 diagnoses being as expected.

In addition, the output reveals that:

1. the scenarios in those groups that did not get proper identification are mainly scenarios 1, 2, and 3;
2. the network's overall performance worsens as more parameters are distorted. However, diagnostics is not much affected when only one parameter is distorted, even when the distortion is by four levels.

Out1-In2 network

With reference to column 3 of Table 16.6, we see that this network's performance is much better than the previous one for group 3 scenarios; least extreme values of parameters: 9/11 success. However, it performed worse for groups 5, 7, and 9, which have two parameters distorted. Otherwise, performance is comparable to that of the Out1-In1 network. In addition the same remarks apply: scenarios 1, 2, and 3 are the main scenarios that the network confuses, and performance degrades more when more parameters are distorted

Table 16.7 Output summary of simulated neural networks

	In 1	*In 2*
total	100	100
hits	67	63
misses	11	9
Unident.	22	28
% hit	67%	63%
% miss	11%	9%

than when distortion of one parameter is more severe. For overall comparison of the Out1-In1 and the Out1-In2 networks, see Table 16.7.

Out2-In1 network

Table 16.6 summarizes the output of this network. Since this network uses output representation scheme 2, the output message is formed by searching the highest activation level among the output nodes. This activation level is equal to the score of the failure scenario represented by that node. Since the network is binary, the score ranges between 0 and 1. When the highest score is higher than 0.9, the corresponding scenario number is given and the diagnosis is confident enough. Otherwise the two highest scores are identified and the two corresponding scenario numbers are suggested. The one listed first is the scenario with the higher score, which also represents the confidence level. A sample of the actual detailed output is shown in Fig. 16.4. In this figure, the binary input pattern is echoed, followed by the diagnostic message.

Referring back to Table 16.6, we see that this network had high confidence diagnostics for test groups 1, 2, 4, and 6. These are the scenarios corresponding to training data, extreme values (i.e. LLL and HHH), one parameter off by two levels, and one parameter off by three levels, respectively. These are also the groups for which the previous two networks performed well.

The most important aspect of this network's performance is that 94% of the time, when the correct scenario is not identified, it is given as one of the two suspected scenarios with lower confidence levels. As long as the correct scenario is included in the diagnosis message, the diagnosis can be considered successful. Using this criterion, the network had a success rate of 94%. There were six misses, five from group 9, which contains the most distortion.

```
10101010000001010101010101010101010000001010101010101010
No scenario selected. Not confident enough.
Suspect scenario 3 scored 0.863340.
suspect NORMAL OPERATION scored 0.228807.

100000101010101010101000000100000010101010000001000000100000
FAILURE: scenario 5 most likely occurred.

100000101010101011111011111100000101010101111100000101010
No scenario selected. Not confident enough.
Suspect scenario 6 scored 0.769618.
Suspect scenario 1 scored 0.127447.

1000001010101010111110111111010101010101011111010101010111 1
No scenario selected. Not confident enough.
Suspect scenario 7 scored 0.867172.
Suspect scenario 8 scored 0.074162.
```

Fig. 16.4 Sample output of an Out2 network.

Out2-In2 network

The last column of Table 16.6 shows the output of the Out2-In2 network. The performance of this network is comparable to that of the Out2-In1 network, except that:

- for group 3 test cases, the network gives the correct scenario as the first (most confident) guess, whereas in the Out2-In1 network output, it was a second guess; and
- the overall success ratio is higher: 96%. The only four scenarios that did not get identified belong to group 9, which contains the most distortion.

16.6.5 Analysis of results

The significant differences in output of the four neural networks shows that knowledge representation, both in input and output data, plays a major role in the diagnostic process.

Input representation

The input representation reflects the relationship between the different parameter levels, and, therefore the relationship between the scenarios. With In2, the overall success rate is slightly better, but the rate of unidentified scenarios is higher. A further analysis of In2's output reveals that this network's performance is considerably better for scenarios where the input patterns are not severely distorted, and worse for scenarios with relatively larger distortions. (These scenarios with large distortions caused the higher unidentified rate.) From this, we can conclude that an In2 network is more rigorous in its classification. It makes more distinctions between patterns and classifies scenarios correctly only when they fit 'close enough' to the class of failures. In other words, the 'close enough' criteria are enforced more strictly by an In2 network. This is due to the fact that representation scheme In2 reflects a stricter relationship between the parameter levels (than In1): out-of-bounds (oob) levels are represented with a large distance from the representation of the normal level – they differ by two bits instead of one. As a consequence, the H-type oob representations (H, HH, and HHH) are all the more different from the L-type oob representations (L, LL, and LLL). Due to this, under representation In2, the following applies.

1. Least extreme oob parameters, levels L and H, will be classified closer to levels LL and HH, respectively, rather than the N level, since: distance (L, LL) = 1 and distance (L, N) = 2. Similarly, distance (H, HH) = 1 and distance (H, N) = 2. This accounts for the improved identification in test data where parameters are in levels L and H (group 3).
2. When input data contain a parameter reading that is disturbed two levels off the level used for training, for example from level HH to level N or from level N to level LL, the distance involved in the change is 3. This distance

is more than it would be under representation In1 (a distance of 2). In general, all disturbances involve more difference bits under In2 than under In1, except for disturbances inside the same oob type. This explains the low rate of recognition for more disturbed data under representation In2 (groups 5, 8).

The choice of input representation determines the limits on system performance, namely, the maximum amount of distortion the neural network can take and still recognize the scenario. The amount of distortion is measured as the number of bits by which an input pattern differs from the corresponding pattern that was used to train the network. We call a neural network more *tolerant* when it can take larger amounts of distortion. We can thus say that a neural network using In1 is more tolerant than one using In2. Tolerance may or may not be a desirable feature depending on the application and the desired response of the user. A tolerant network may cause some important features of a pattern to be overlooked and considered as noise.

We feel that for this particular application, a less tolerant network is desirable, since it calls attention to distortions which can signal the occurrence of a new type of failure hitherto unknown.

In conclusion, before designing a neural network for diagnostics, experts in the domain should set up criteria for allowable distortions and tolerances.

Output representation

The overwhelming advantage of using Out2 is the high success rate, since any diagnosis that includes the correct scenario, either as the first suspected or as the second suspected scenario, is considered a success. Furthermore, the ability to suggest a second suspected failure scenario and provide confidence levels is very desirable and helpful to the user. On the other hand, the effect of using the two different input representations In1 and In2, when Out2 is used, is not as pronounced as when Out1 is used, except for better performance for group 3 test cases (least extreme). Here, we consider only situations where the correct scenario is given as the first suspected scenario (higher confidence level). The fact that the effect of the two different input representations did not show when Out2 is used can be explained by recalling that Out2 looks at the real valued score of each output node and is therefore more accurate in presenting the output. In Out1, a conversion from real to binary must take place. During this conversion, details can be lost, resulting in the difference in performance between the two input representations.

Overall analysis

Two interesting phenomena were noticed during the simulations of the four neural networks. First, notice that the scenarios that are consistently difficult to identify, sometimes being confused with each other, are scenarios 1, 2, and 3. This is particularly true for networks that use output representation Out1.

To understand this phenomenon, let's look back at Table 16.3. The scenario definitions for scenarios 1, 2, and 3 are

scenario	*P*1	*P*2	*P*3	*P*4	*P*5	*P*6	*P*7	*P*8	*P*9
*S*1	L	L		L	L		L	L	
*S*2	L	L		L	L	H	L	L	
*S*3	L	L				L			

In the neural network simulations, scenario 2 is sometimes diagnosed as scenario 1 (for example, look at groups 1, 2, and 4 of the test case). From Table 16.3, we see that scenarios 1 and 2 have almost identical descriptions: the only difference is in parameter *P*6. Therefore, the status of parameter 6 is critical in distinguishing between scenarios 1 and 2. However, our input knowledge representation scheme does not reflect the fact that any particular parameter is more critical than others. When an input that is supposed to represent scenario 2 is presented to the network, the state of parameter 6 is considered a distortion and therefore the input pattern was found to fit scenario 1's description. This is particularly true for networks that use output representation Out1. With this output method, there is conversion from real to binary numbers at the output nodes during output decoding, and some details are lost during the conversion. On the other hand, Out2 reports the exact real valued score of each output node and allows one to see exactly how close the input pattern is to each of the scenarios 1 and 2.

The second phenomenon that we have noticed is the effect of additional parameters being distorted. We noticed that performance of the four diagnostic systems when one parameter is off by three levels is better than when two parameters are off by two levels, and performance when one parameter is off by four levels is better than when two parameters are off by two or three levels. Generalizing these facts, we can expect performance to degrade with more parameters being distorted. However, when only one parameter is distorted, we do not expect performance to go worse, regardless of how severe the distortion is. One might suspect that this is due to increased number of difference bits involved, but this is not necessarily true. For example, there are less difference bits when two parameters are off by two levels (4 bits) than when one parameter is off by four levels (6 bits for In2, 4 for In1). This behavior can be attributed to the manner in which information is stored in a neural network. Information is distributed throughout the network in the form of the connection weight matrices. When a localized disturbance occurs in the input such as one parameter reading being distorted, it is not likely to disrupt the neural network as a whole. When more parameters are distorted, the resulting disturbance is distributed over the network structure, and is therefore more likely to cause degradation in performance.

16.6.6 Processing speed analysis

Table 16.8 shows the average response times for a diagnostic operation of each of the four networks. These processing times satisfy real-time operation

Table 16.8 Processing times of one diagnostic operation

Neural Network	*Processing time*
Out1-In1	29.3 msec
Out1-In2	29.7 msec
Out2-In1	33.3 msec
Out2-In2	42.8 msec

requirements, unlike previous approaches to this problem, where response time was about 10 seconds. This improvement in speed over previous conventional AI approaches is attributed to the following:

1. No search technique is involved in solving this problem; only a small number of calculations are done; and
2. Unlike conventional AI techniques, these calculations are numeric rather than symbolic. Conventional computers are much faster with numerical calculations since they were designed for that purpose, and since symbolic computations are converted to a sequence of numerical computations before being done on a conventional computer.

This processing time is expected to be orders of magnitude better when the neural network is implemented in hardware.

Looking at Table 16.8, we can see differences in the relative processing times of the four neural network architectures. When In2 is used, processing time is slightly longer than when In1 is used because there are a few more nodes in the input layer. This difference will disappear when the neural network is implemented in hardware, since all the nodes in one layer will be operating in parallel, and, consequently, the processing time will depend not on the number of nodes but on the number of layers, which is fixed for this model. Of course, when there are more input nodes, the nodes in the hidden layer will have more weight-input products to add, but this effect is expected to be very small.

We also notice the larger processing time involved with using Out2. As mentioned earlier, this is due to the output decoding technique. In this technique, all output nodes must be searched to identify the maximum score. In addition, the larger difference in processing time between Out2-In1 and Out2-In2 is due to the use of more hidden nodes for In2.

16.7 SUMMARY AND CONCLUSIONS

In this chapter, a neural network model and representation schemes were suggested for continuous process diagnostics. The knowledge representation issue is fundamental, since the expert knowledge about the system has to be encoded in binary patterns suitable for processing by a neural network.

Different input and output representation (In1, In2, Out1, Out2) were derived and simulated for the TMI-2 nuclear reactor. Performance testing and results of our diagnostic neural networks were discussed. The neural network performed well under a variety of input situations. The closest match was found and reported even under noisy conditions. However, when the noise exceeded the tolerance of the network, inaccurate responses were recorded. Significant differences in performance were found when different input and output representations were used. Performance is best when In2 and Out2 are used because:

1. Out2 gives a better interpretation of the output by looking at the real valued score of each output node, and provides confidence levels and second guesses for a best match; and
2. In2 provides a better input representation scheme by making the out-of-bounds levels more distinct from each other, making it harder for a neural network to tolerate large disturbances in the input.

This improved performance of the Out2-In2 network is obtained at the expense of slightly increased processing time and hardware.

REFERENCES

Ballard, D.H. (1986) Cortical connections and parallel processing: Structure and function. *Behavioural and Brain Sciences*, **9**, 67–120.

Ballard, D.H. (1987) Interpolation coding: A representation for numbers in neural models. *Biological Cybernetics*, **57**, 389–402.

Bourlard, H. and Wellekens, C. (1987) Multilayer perceptron and speech perception, in *1st International Conference on Neural Networks*, June 1987, IEEE, pp. iv–407.

Burr, D.J. (1986) A neural network digit recognizer, in *Proceedings of the International Conference on Systems, Man, and Cybernetics*, IEEE.

Dietz, W.E., Kiech, E.L. and Ali, M. (1989) Jet and rocket engine fault diagnosis in real time. *Journal of Neural Network Computing*, Summer, 5–18.

Elman, J.L. and Zipser, D. (1987) Learning the hidden structure of speech, ICS Report 8701, Institute of Cognitive Science, University of California, San Diego, La Jolla, CA, February.

Feldman, J. (1985) Connectionist models and parallelism in high level vision. *Computer Vision, Graphics, and Image Processing*, **31**, 178–200.

Lippmann, R.P. and Gold, B. (1987) Neural classifiers useful for speech recognition, in *1st International Conference on Neural Networks*, June 1987, IEEE, pp. iv–417.

Milne, R. (1987) Strategies for diagnosis, *IEEE Transactions on Systems, Man, and Cybernetics*, **SMC-17**(3), May/June.

Nann, S.R. (1988) A decision support system for control and automation of dynamical processes, unpublished M.S. thesis in Mechanical Engineering, Penn State, December.

Rumelhart, D.E., McClelland, J.L. and the PDP Research Group (1986) *Parallel Distributed Processing, Explorations in the Microstructure of Cognition, Volume 1: Foundations*, Cambridge: MIT Press.

Yoon, Y.O., Brobst, R.W., Bergstresser, P.R. and Peterson, L.L. (1989) A desktop neural network for dermatology diagnosis. *Journal of Neural Network Computing*, Summer, 43–50.

Index